Optimisation, Optimal Control and Nonlinear Dynamics in Electrical Power, Energy Storage and Renewable Energy Systems

Optimisation, Optimal Control and Nonlinear Dynamics in Electrical Power, Energy Storage and Renewable Energy Systems

Editors

Victor M. Becerra
Ahmed Rachid

MDPI • Basel • Beijing • Wuhan • Barcelona • Belgrade • Manchester • Tokyo • Cluj • Tianjin

Editors

Victor M. Becerra
University of Portsmouth
UK

Ahmed Rachid
University of Picardie Jules
Verne
France

Editorial Office
MDPI
St. Alban-Anlage 66
4052 Basel, Switzerland

This is a reprint of articles from the Topic published online in the open access journals *Energies* (ISSN 1996-1073), *Electronics* (ISSN 2079-9292), *Batteries* (ISSN 2313-0105), and *Electricity* (ISSN 2673-4826) (available at: https://www.mdpi.com/topics/optimisation).

For citation purposes, cite each article independently as indicated on the article page online and as indicated below:

LastName, A.A.; LastName, B.B.; LastName, C.C. Article Title. *Journal Name* **Year**, *Volume Number*, Page Range.

ISBN 978-3-0365-5585-0 (Hbk)
ISBN 978-3-0365-5586-7 (PDF)

Cover image courtesy of Victor M. Becerra

Contents

About the Editors

Victor M. Becerra

Professor Victor M. Becerra is currently a Professor of Power Systems Engineering at the School of Energy and Electronic Engineering, University of Portsmouth, UK, where he has been since 2015. He obtained his PhD in Control Engineering from the City University of London in 1994. During 1989–1991, he carried out power systems analysis and control systems design at C.V.G. Edelca, Caracas, Venezuela. Between 1994 and 1999, he was a post-doctoral Research Fellow at the Control Engineering Research Centre at the City University of London. During the period between 2000 and 2015, he was an academic at the School of Systems Engineering, University of Reading, UK, where he became a Professor of Automatic Control in 2012. His current research interests consider a range of issues related to power systems and smart grids, as well as the methods and applications of automatic control. During his career, he has received funding for research and innovation from the EPSRC, Innovate UK, the Royal Academy of Engineering, the European Space Agency, and the EU. He has published over 150 refereed papers, one research monograph and two edited volumes. Professor Becerra is a Fellow of the IET and a Senior Member of the IEEE.

Ahmed Rachid

Ahmed Rachid received an engineering degree in Cybernetics in 1983, an MBA in 1985, a PhD in electrical engineering in 1986 and a Habilitation degree in 1991 from the University of Nancy (France). He has been teaching since 1984 in different universities and engineering schools mainly in the fields of control engineering and signal processing. Since 1992, he has been a professor at the University of Picardie Jules Verne (France), where he created the institution's first laboratory in Automatic Systems. His current research includes processes and systems modeling, simulation, control, diagnosis and observation with applications in blast furnaces, vehicle dynamics, diesel engines, induction electrical drives, stepper motors, greenhouses, magnetic levitation, pantograph-catenary, co-generation, electric vehicles, solar power systems, etc. He has supervised 24 PhD theses and has co-authored 2 patents, 2 books and over 150 scientific papers. He has organized and chaired several international conferences and has coordinated various European research projects and industrial R&D contracts.

energies

MDPI

Article

Hardware Implementation of Novel Shade Dispersion PV Reconfiguration Technique to Enhance Maximum Power under Partial Shading Conditions

Vinaya Chandrakant Chavan [1], Suresh Mikkili [1,*] and Tomonobu Senjyu [2]

1 Department of Electrical and Electronics Engineering, National Institute of Technology Goa, Ponda 403401, India; chavanvinaya19@gmail.com
2 Department of Electrical and Electronics Engineering, University of the Ryukyus, Nishihara 903-0213, Japan; b985542@tec.u-ryukyu.ac.jp
* Correspondence: mikkili.suresh@nitgoa.ac.in

Abstract: Partial shade condition is a significant factor contributing to the PV panel performance in mismatch losses and power generation. The technique suggested in this study allows the physical rearrangement of the PV panel to distribute the shade on the entire PV array. MPPT, selecting suitable inverter topology, or PV panel reconfiguration enhances the performance of the PV panel. This study proposes a new shade dispersing method, novel shade dispersion (NSD). It compares the performance of the NSD method with conventional configurations (CCs). This research article models and simulates 6 × 6 PV array configurations such as Series-Parallel (SP), Total-Cross-Tide (TCT), Bridge-Linked (BL), Honey-Comb (HC), and the newly proposed NSD method under non-shading and nine different partial shading cases. The performance indices used for comparative analysis are global maximum power points, efficiency, power enhancement, open circuit voltage, short circuit current, and number of crests. The Soltech 1 STH-215-P PV module was selected in the MATLAB/Simulink environment to simulate PV array arrangements. Hardware experiments validate the performance of the NSD method.

Keywords: partial shading; conventional configuration; reconfiguration techniques

Citation: Chavan, V.C.; Mikkili, S.; Senjyu, T. Hardware Implementation of Novel Shade Dispersion PV Reconfiguration Technique to Enhance Maximum Power under Partial Shading Conditions. *Energies* **2022**, *15*, 3515. https://doi.org/10.3390/en15103515

Academic Editor: Alberto Reatti

Received: 7 April 2022
Accepted: 9 May 2022
Published: 11 May 2022

1. Introduction

Renewable energy has become one of the most critical factors to preserve the climate and the earth's assets for the coming generation. Power has a crucial part in the financial growth and the welfare of a nation. From India's perspective, solar energy increases power generation and generates energy reliably considering ecological, communal, and financially beneficial properties. India has significant potential, availability, and other features of solar power. Therefore, the government also emphasizes the promotion of photovoltaic generation in the Indian power sector [1]. Series and parallel connected modules form an array to obtain power per requirement. PV material, temperature, irradiance, dust, array structure, maximum power point tracking (MPPT) method, converter topology, etc., are the different parameters responsible for the performance of PV array [2,3]. This uncertainty affects the estimation of a PV cell's series and shunt resistances (namely, Rs and Rsh). Ref. [4] provided the uncertainty analysis of the measured currents, voltages, temperatures, and irradiances. At a constant temperature, the output current increases with an increase in irradiance with negligible effect on the voltage. Voltage is substantially reduced at constant irradiance, and the current slightly increases with increased temperature up to the rated temperature [5]. Any object that obstructs the sun rays' path to the PV panel or uneven falling sunlight on the earth creates a partial shading condition (PSC) [6]. A bypass diode is connected antiparallel to the PV module to avoid the hot spot issues created by PSC. Under PSC, the difference between the minimum irradiance received by non-overlapped and overlapped solar cells determines the overcurrent [7]. A diode adds antiparallel either to a PV module or a set of PV series modules [8]. It inserts the number of peaks in the PV curve, among which the rise consisting of the highest value is known as the global maximum power point (GMPP). Traditional MPPT struggles to obtain true MPP in rapidly changing PSCs. In this area, extensive research considers improved conventional methods, new metaheuristic methods, etc., which increases the complexity and cost of the MPPT system [9]. Ref. [10] performed the experimental validation of TCT with the standard deviation of PSC and found a better configuration. Results demonstrated that 15% and 25% reduced irradiance. The power yield degraded to 4% and 7.5%. Refs. [11,12] explored the performance of series (S), parallel (P), SP, TCT, BL, and HC PV topologies and concluded, without considering bypass diode BL, and with bypass diode TCT generates maximum power. Reconfiguration is an effective way to improve power generation [13]. This was classified into two broad categories: electrical and physical reconfiguration. Both methods have their boons and downsides [14]. Ref. [15] proposed the maximum and minimum algorithm dynamic reconfiguration method, which uses a double pole double throw switch to change the electrical connections. Ref. [16] provided the dynamic array reconfiguration for water pumping applications. The authors of [17] investigated efficiency and robustness of the dynamic array reconfiguration for hybrid PV system application. Refs. [15–17] efficiently used the switch matrix to reconfigure the PV modules in the array for different applications. Ref. [18] proposed an electrical repositioning on a 4 $\times$ 5 arrangement with 12 switches to make adaptive cross ties to enhance the power generation. For case I the proposed method improved GMPP, fill factor (FF), and power enhancement (PE) compared with SP, and minimized power losses (PL). It was found best at 291.3 W, 0.586, 2.71%, 55.6 W, and 83.97%, for TCT in MATLAB/Simulink and experimentation with the same performance index under all PSCs. Complexity, sensor requirement, switches, and their controlling structure, sensor speed dependency, cost, etc., are the significant limitations of dynamic reconfiguration. Therefore, many researchers go for static array reconfiguration. The authors of [19] proposed and demonstrated a Sudoku-based physical reconfiguration system on a 9 $\times$ 9 array. It improves the power generation from 4 to 26% for different PSCs. The authors of [20] proposed an improved Sudoku method on a 9 $\times$ 9 array and compared it to the CC. Ref. [21] offered a magic square method and compared it with TCT, Sudoku, and the optimal Sudoku method on an 81 modules array structure. Ref. [22] proposed the zigzag method and compared it with the TCT and Sudoku method on a 6 $\times$ 6 PV array. Ref. [23] related competence square and the dominance square method with TCT on a 9 $\times$ 9 PV array. Ref. [24] proposed a clue-based skyscraper puzzle method and compared it with TCT, Sudoku method, and the puzzle-based competence square method. This study reviews many static reconfiguration (SR) methods and analyzes their performance. SR is a robust, simple, less costly, and effective way to deal with partial shading issues and improve the performance of the PV array. Still, these methods have some limitations.

The novelty of work is:

Electrical networks persist similarly to TCT and the physical position of units in the PV array changed.

- Sudoku is limited to a 9 $\times$ 9 array and has different logic patterns. Calculating a higher-order magic matrix is very difficult in the magic square method. Square puzzle patterns may require more computational burden, complex wiring, and losses; puzzle-based methods such as the skyscraper have clue dependency, and clue choosing is tedious. This study proposes a new method of static reconfiguration to address these issues;
- The proposed novel shade dispersion (NSD) method is a two-step method. In the first step, a magical submatrix is formed, and in the second step, logical shifting is performed to distribute the shade and increase the power generation of the PV array. This method considers magic sub-matrix shifting to reposition the panels in the same column, maintaining the least possible wiring;

- The proposed scheme was analyzed under ten shading cases. The results validated with SP, TCT, BL, and HC compared the global maximum power point (GMPP), mislead power (MP), output power (%O/P), efficiency, fill factor, and % power enhancement for NSD under-considered shading pattern. Figure 1 gives the workflow of this paper.

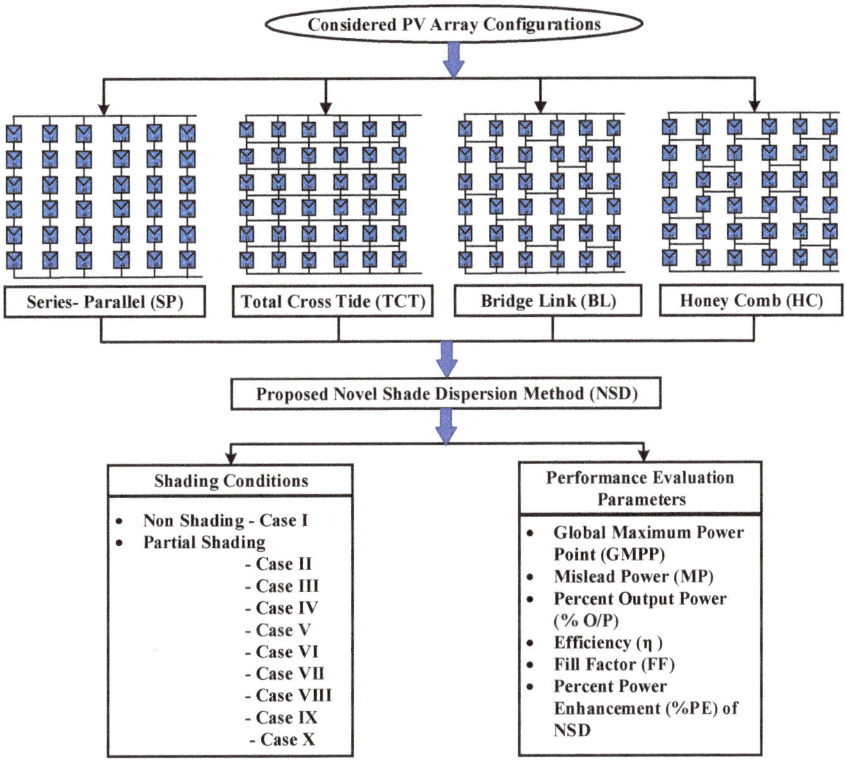

Figure 1. Representation of PV array connection structures under partial shading conditions with performance evaluation parameters.

2. Modeling and Simulation of NSD Method

This section explains the modeling and simulation of conventional PV array configurations and the proposed novel shade dispersion (NSD) method in MATLAB/Simulink. The panel electrical parameters has mentioned in Appendix A. The flowchart for panel repositioning is provided in Figure 2. The present work consists of 36 modules placed in six rows and six column patterns (shown in Figure 3a) for conventional configurations and the NSD method (Figure 3b). The TCT connection-based MATLAB simulation is shown in Figure 4. Mathematical modeling for output voltage, output current, and output power of all the traditional PV array structures are provided in Table 1. Here, the module, row, and array output voltages are V_j, V_R, and V_{PV}, respectively. The module, string, and array output currents are I_j, I_S, and I_{PV}. P_{PV} is the PV generation of the array. A novel shade dispersion method is proposed to reconfigure the panel of a variety to improve the performance through shade dispersion. In the proposed technique, electrical connections remain the same as in TCT [20]; only the physical location of the modules changes. It uses voltage, current, and power equations for TCT, as mentioned in Table 1. The basic principle used for this method is row current equalization. It proposes a magic submatrix arrangement with the magic number 21 and its logical shifting to reposition the panels within a column, as shown in Figure 3b.

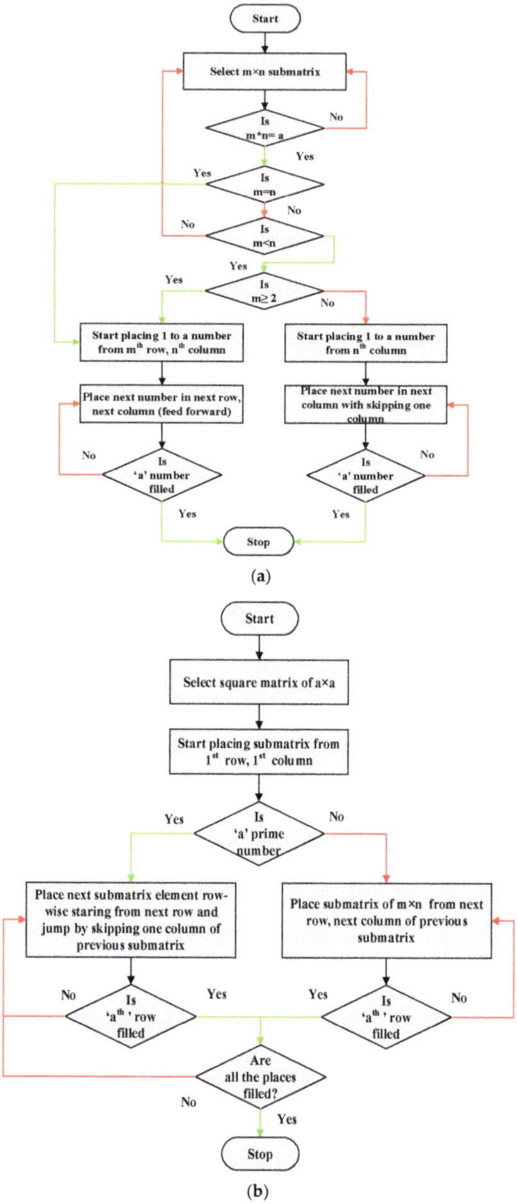

Figure 2. Flowchart for panel repositioning of NSD method (**a**) Step 1: submatrix formation; (**b**) Step 2: main matrix formation.

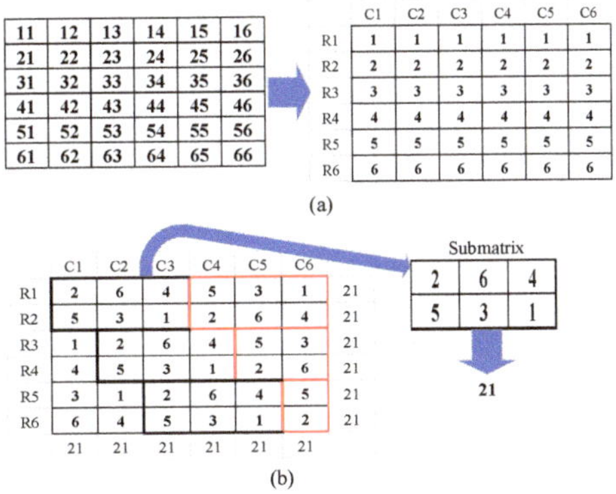

Figure 3. Panel positions in PV array (**a**) conventional methods, (**b**) NSD method.

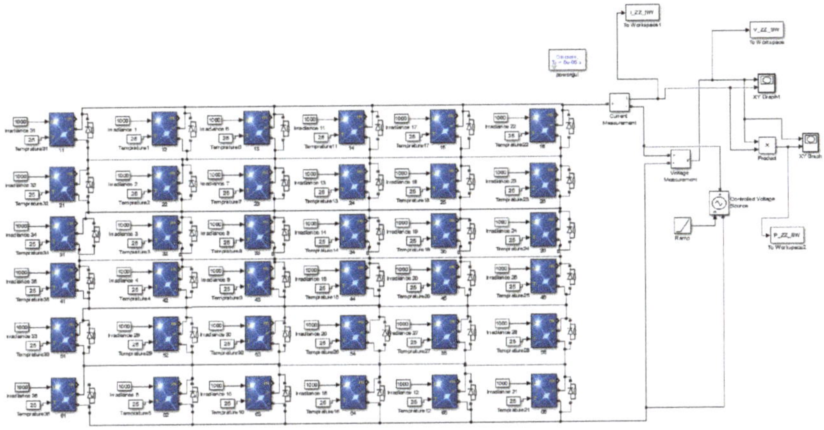

Figure 4. TCT-based MATLAB simulation for the NSD method.

Table 1. The output voltage, current, and power for conventional PV array configurations.

Connection	Output Voltage	Output Current	Output Power
SP	$V_{PV} = \sum_{j=1}^{6} V_j = 6V_j$	$I_{PV} = \sum_{j=1}^{6} I_{Sj} = 6I_S$	$P_{PV} = 36V_jI_S$
TCT	$V_{PV} = \sum_{j=1}^{6} V_j = 6V_j$	$I_{PV} = I_1 + I_7 + I_{13} + I_{17} + I_{23} + I_{29} = 6I_S$	$P_{PV} = 36V_jI_S$
BL	$V_{PV} = \sum_{j=1}^{6} V_j = 6V_j$	$I_{PV} = I_{S1} + I_{S2} + I_{S3} + I_{S4} + I_{S5} + I_{S6} = 6I_S$	$P_{PV} = 36V_jI_S$
HC	$V_{PV} = \sum_{j=1}^{6} V_j = 6V_j$	$I_{PV} = I_{S1} + I_{S2} + I_{S3} + I_{S4} + I_{S5} + I_{S6} = 6I_S$	$P_{PV} = 36V_jI_S$

The steps involved to build the proposed 6 × 6 NSD matrix are:
Step I: Formation of the submatrix

1. The submatrix is proposed so that rows and column multiplication of the submatrix is 6;
2. Therefore, numbers 1 to 6 are placed in the 2 × 3 submatrix to summation equal to the magic number 21;

Step II: Formation of the primary matrix

3. The submatrix placing is started in the 1st row, 1st column position. The next submatrix is placed at the next (3rd) row, next (2nd) column position. This process continues until the last (6th) row is filled by the considered submatrix;
4. Submatrix rows are interchanged, and that submatrix starts to fill in vacant places of the primary matrix, starting from the next column of previously filled areas. Here, the interchanged submatrix is filled from 1st row, 4th column. The next submatrix is placed in the 3rd row and 4th column. The last column of the submatrix is placed in the same row and vacant column serially;
5. It is performed so that the summation of row numbers, as well as column numbers, is equal to the magic number 21, as shown in Figure 3b;
6. The process continues until vacant places in the primary matrix are filled, and the fifth step is achieved.

3. Partial Shadings Conditions and Shade Dispersion Effect

For CCs the shade does not scatter, as shown in Figure 5. Through NSD, the partial shading condition spread on a complete array, as shown in Figure 6. Under case I (i.e., non-shading condition), the performance of the NSD method is the same as the CC (i.e., without reconfiguration), as shown in Figure 7a. The performance indicators are the global maximum power point (GMPP), mislead nine PSCs as power (MP), output power (%O/P), efficiency, fill factor, and % power enhancement of NSD. The %PE was calculated for NSD and compared with all the CCs. For case I, all 36 modules of the PV array have 1000 W irradiance; the remaining PSCs have four groups of irradiance.

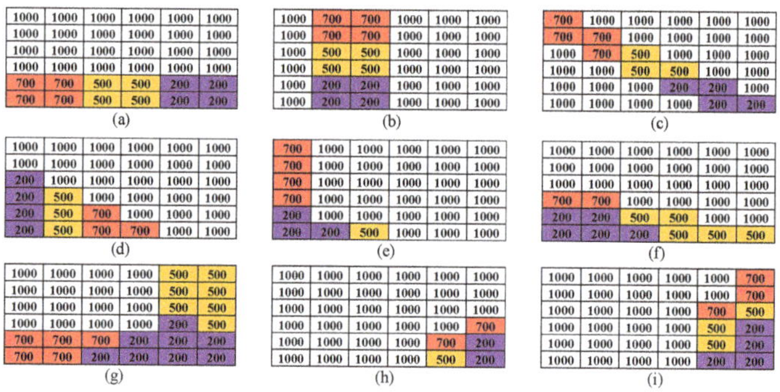

Figure 5. Partial shading conditions (**a**) case II, (**b**) case III, (**c**)case IV, (**d**) case V, (**e**) case VI, (**f**) case VII, (**g**) case VIII, (**h**) case IX, (**i**) case X.

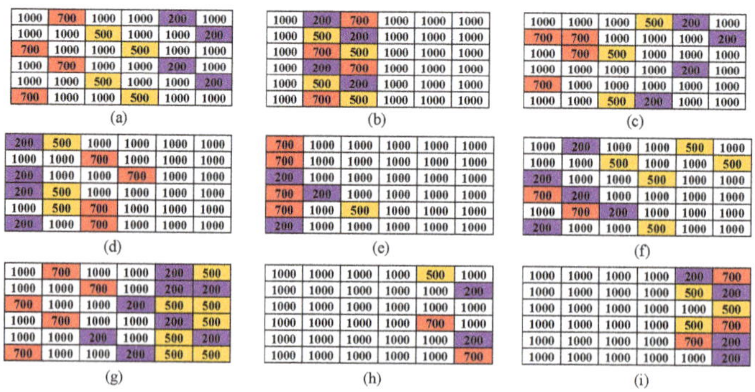

Figure 6. Shade spreading due to NSD method (**a**) case II, (**b**) case III, (**c**)case IV, (**d**) case V, (**e**) case VI, (**f**) case VII, (**g**)case VIII, (**h**) case IX, (**i**) case X.

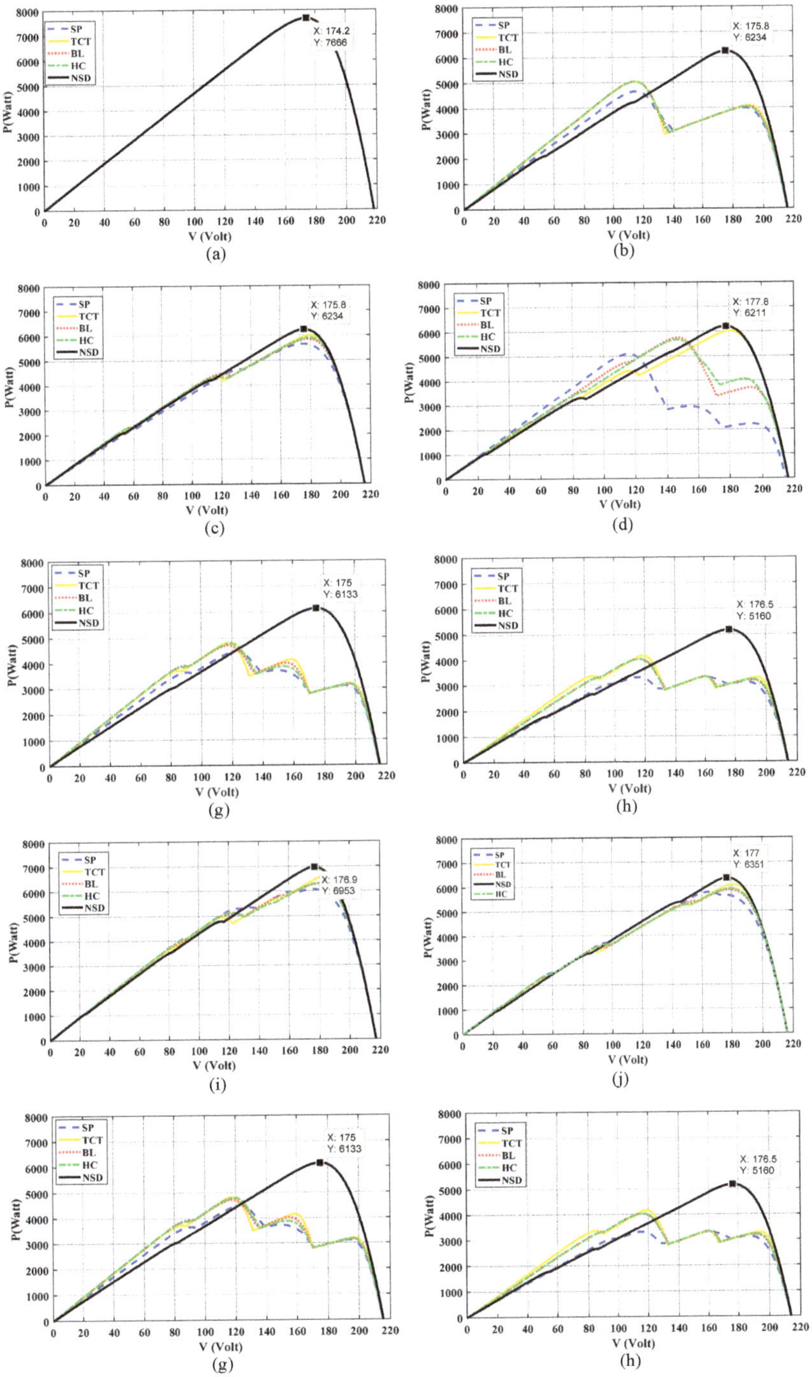

Figure 7. PV curves under various irradiance levels (**a**) case I, (**b**) case II, (**c**)case III, (**d**) case IV, (**e**) case V, (**f**) case VI, (**g**)case VII, (**h**) case VIII, (**i**) case IX, (**j**) case X.

- Case II to case IV have group 1 of 24 panels with 1000 W irradiance; groups 2, 3, and 4 have 200 W, 500 W, and 700 W irradiance, respectively, with four panels in each group
- In case V, 10 modules have PSCs, group 1 with 1000 W has 26 panels. Group 2 with 200 W, has four modules, and groups 3 and 4 have 500 W, 700 W irradiance respectively, with three panels;
- Case VI has groups 1, 2, 3, and 4 with 1000 W, 200 W, 500 W, and 700 W irradiance on 28, 3, 1, and 4 modules, respectively;
- In case VII, 24 panels came in group 1. Groups 2 and 3, have 200 W and 700 W irradiance, respectively. Group 4 has two panels that have 700 W irradiance;
- In case VIII, groups 1, 2, 3, and 4 have five panels in each group of 1000 W, 200 W, 500 W, and 700 W irradiance with 16, 8, 7, and five panels, respectively;
- In case IX, only five modules are partially shaded, and 31 modules have 1000 W irradiance in group 1. It has two 200 W modules in group 2, one 500 W module in group 3 and two 700 W irradiance panels in group 4;
- In case X, with 1000 W, group 1 has 25 panels. Group 2, with 200 W has four panels, and groups 3 and 4 have 500 W and 700 W irradiance, respectively, with three panels.

4. Result and Discussion

The performance evaluation of PV array CCs and proposed reconfiguration was performed under various shading situations on MATLAB/Simulink. The assessment has two segments, (1) power-voltage profile and (2) computational parameters.

4.1. Power-Voltage Profiles under Various PSC

The sketches in Figure 7 represent the variation in PV curves for different conventional configurations and the proposed NSD method of reconfiguration under various irradiance levels. Under case I of the non-shading condition, all the panels have the same standard irradiance of 1000 W. Figure 7a highlights that all the configuration and NSD methods have the same ideal performance under case I with only one peak on the PV curve.

Under case II, conventional configurations have two peaks on the PV curve, while the NSD method has only one extreme. TCT, BL, and HC have identical extreme power, and it is more than SP, while for NSD, the maximum capacity is much more enhanced, as revealed in Figure 7b. Figure 7c indicates that SP has the lowest power output for case III. It offers continuous improvement for BL, HC, TCT, and NSD. NSD offers 6234 W as the maximum power output. In Figure 7d, SP has three peaks, and other methods have two heights. BL and HC show improved performance over SP but lesser than TCT and NSD. NSD provides the highest output of 6234 W, as shown. Figure 7e portrays that BL and HC have six peaks, SP and TCT have five peaks, while NSD is reduced to four peaks and the highest power generation equal to 6303 W for case V. In Figure 7f case VI, HC and SP have the same maximum power generation as BL and TCT with three peaks; in contrast, the NSD has the highest power generation of 6646 W with three small deep peaks. Figure 7g showcases the case VII PV curves for SP, TCT, BL, HC, and NSD. NSD has a single top with 6133 W as the highest power generation. In Figure 7h SP and TCT have six and four peaks. BL, HC, and NSD structures have three peaks with the highest power, 5160 W, for the NSD method. Figure 7i portrays case IX. Under this shade, TCT has maximum power extraction of 6563 W in conventional configuration while NSD has improved power extraction with 6953 W.

4.2. Computational Parameters

The performances of all CC and NSD were investigated in the following six computational parameters, as observed in Appendix B, which highlights the efficacy of NSD over CC. Tables 2–6 give the performance of SP, TCT, BL, HC and NSD method under ten PSCs respectively.

Table 2. SP performance under PSC.

Shade Cases	V_{OC} (V)	I_{SC} (A)	No. of Peaks	P_M (W)	V_M (V)	I_M (A)
Case I	217.80	47.16	1	7665.80	174.00	44.00
Case II	217.80	43.49	2	4654.30	115.15	40.42
Case III	217.80	38.64	2	5655.20	174.98	32.31
Case IV	217.80	47.15	4	4711.90	115.76	44.00
Case V	217.80	45.23	5	5005.40	160.34	31.21
Case VI	216.31	44.81	3	5818.10	153.06	38.01
Case VII	214.91	43.49	4	4496.00	123.36	36.45
Case VIII	213.54	37.49	6	3309.10	115.42	28.67
Case IX	216.81	47.16	4	6068.50	175.65	34.55
Case X	215.89	44.80	5	5768.50	164.53	35.06

Table 3. TCT performance under PSC.

Shade Cases	V_{OC} (V)	I_{SC} (A)	No. of Peaks	P_M (W)	V_M (V)	I_M (A)
Case I	217.70	47.16	1	7665.80	174.00	44.00
Case II	215.41	47.16	2	5039.80	114.48	44.02
Case III	215.76	42.42	3	6001.20	179.43	33.45
Case IV	215.90	44.74	4	6028.20	180.12	33.47
Case V	215.30	47.14	5	4627.30	189.10	27.74
Case VI	216.48	44.80	3	5833.80	147.02	39.68
Case VII	215.09	47.15	4	4807.50	116.94	41.11
Case VIII	214.02	42.44	4	4172.10	118.73	35.14
Case IX	216.95	47.15	3	6563.30	181.91	36.08
Case X	216.01	44.78	4	6093.50	180.53	33.75

Table 4. BL performance under PSC.

Shade Cases	V_{OC} (V)	I_{SC} (A)	No. of Peaks	P_M (W)	V_M (V)	I_M (A)
Case I	217.00	47.16	1	7665.80	174.00	44.05
Case II	215.36	47.16	2	5039.80	114.48	44.02
Case III	215.73	42.43	3	5849.50	178.21	32.82
Case IV	216.03	44.79	3	5749.20	146.56	41.22
Case V	216.30	47.15	6	5036.20	184.84	27.24
Case VI	216.40	44.80	3	5706.00	148.28	38.48
Case VII	215.01	47.15	4	4702.90	117.56	40.00
Case VIII	213.57	39.29	3	4033.20	115.99	34.77
Case IX	216.90	47.15	4	6300.60	179.86	35.03
Case X	215.97	44.79	3	5862.10	178.79	32.79

Table 5. HC performance under PSC.

Shade Cases	V_{OC} (V)	I_{SC} (A)	No. of Peaks	P_M (W)	V_M (V)	I_M (A)
Case I	217.78	47.16	1	7665.80	174.00	44.00
Case II	215.33	47.16	2	5039.80	114.48	44.00
Case III	215.75	42.43	4	5913.70	179.34	32.97
Case IV	215.63	44.79	2	5674.10	146.28	38.78
Case V	216.40	47.15	6	5102.40	185.66	27.48
Case VI	216.38	44.81	3	5873.30	151.80	38.69
Case VII	214.98	47.16	4	4797.10	119.08	40.29
Case VIII	213.59	39.30	3	4042.00	116.37	34.74
Case IX	216.90	47.16	2	6320.40	179.78	35.16
Case X	216.04	47.14	5	5915.20	179.48	32.96

Table 6. NSD performance under PSC.

Shade Cases	V_{OC} (V)	I_{SC} (A)	No. of Peaks	P_M (W)	V_M (V)	I_M (A)
Case I	217.79	47.17	1	7665.80	174.00	44.06
Case II	215.79	40.86	1	6234.00	175.95	35.43
Case III	215.79	40.86	1	6234.00	175.95	35.43
Case IV	215.94	44.75	3	6211.00	177.56	34.98
Case V	216.04	44.75	4	6303.20	177.23	35.56
Case VI	216.54	44.78	3	6645.70	177.48	37.44
Case VII	215.54	39.25	1	6132.90	174.90	35.07
Case VIII	213.98	34.57	3	5160.20	176.34	29.26
Case IX	216.97	47.10	2	6952.80	176.76	39.34
Case X	216.04	43.18	2	6350.60	177.09	35.86

4.2.1. Global Maximum Power Point (GMPP)

The PV curve has multiple peaks due to PSC, as shown in Figure 8. GMPP is the maximum power of all the crests. Figure 8a indicates that GMPP for all CCs and NSD have concentrated 7665 W. For case II, TCT, BL, and HC provide equal power at 5039 W in CC, and NSD improves it up to 6234 W. Case III also NSD has GMPP at 6234 W. Under case IV in CC, TCT provides maximum output at 6028 W while NSD has 6211 W. Under case V, NSD has 6303 W as a GMPP; this is 1201 W more than HC, which offers the highest power of all CCs. Under case VI to case X in CC, TCT offers the uppermost GMPP while NSD has GMPP at 6646 W, 6133 W, 5160 W, 6953 W, and 6350 W for respective PSC. It is 812 W, 1325 W, 988 W, 389 W, and 257 W more than TCT under the same PSCs.

4.2.2. Mislead Power (MP)

Figure 8b shows that the proposed NSD method provides the least misleading power as the red NSD trace is at the innermost on the radar pattern. Case I does not have any MP. From case II to case X it is 1431 W, 1431.8 W, 1454 W, 1363 W, 1020 W, 1533 W, 2506 W, 713 W, and 1315 W continuously. NSD reduces the MP by about 11 to 46% under these shading cases.

4.2.3. Percent Output (%O/P)

Figure 8c Highlights that all considered structure provides 100% output for case I. NSD provides 81% O/P, the maximum from cases II, III, and IV. In CC, HC provides the highest %O/P at 66.5% under case V, while NSD improves it to 82.69%. In CC under case VI to case X, TCT offers the uppermost %O/P, and NSD improved it by 10.5%, 17.3%, 12.9%, 5%, and 3.4% compared with TCT for the respective shading cases.

4.2.4. Efficiency (η)$^-$

Figure 8d shows the NSD structure is more efficient than TCT, which has the same electrical connections; under-considered PSCs. The radar pattern representation in Figure 8d clearly shows that NSD offers efficiency near 21%. It is approximately equal to the PV array efficiency for case I.

4.2.5. Fill Factor (FF)$^-$

Figure 8e highlights that the proposed method offers 2–5% improvement in FF for cases III, IV, IX, and X; 15.19% more FF for case V, 8% improvement for case VI, and 20–25% enhancement for case II, VII, and VIII, compared with maximum FFs in CC.

4.2.6. Percent Power Enhancement (%PE)

The %PE ensures the power increment by the proposed reconfiguration. Figure 8f shows that NSD improves power extraction compared with CC. Compared with SP, TCT, BL, and HC, NSD has a 25.9, 15.79, 17.44, and 16.45% power boost.

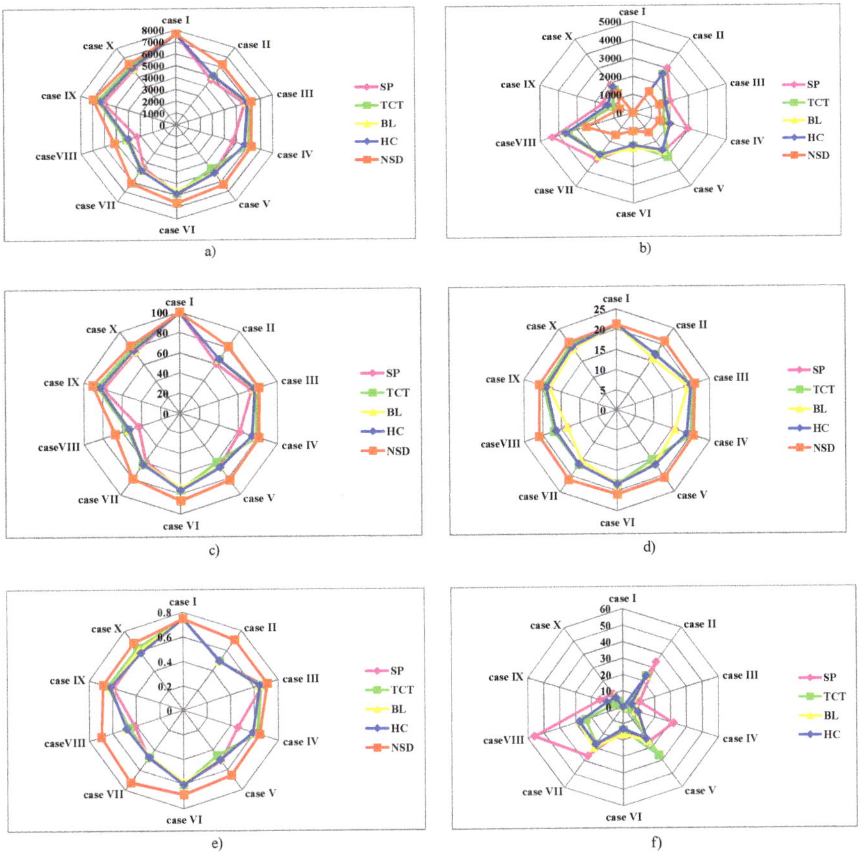

Figure 8. Performance evaluation parameter (**a**) GMPP (W), (**b**) MP(W), (**c**)%O/P, (**d**) η, (**e**) FF, (**f**) %PE.

5. Experimental Implementation and Result Analysis

5.1. Experimental Setup

Towards real-time implementation of the proposed NSD method, a prototype laboratory experiment with 16 PV modules and a 4 × 4 array structure was conducted on the PV chroma 62050 H-600 S simulator (Taiwan). The chroma 63203 E DC (Taiwan) electronic load was used to develop an experimental setup, as shown in Figure 9. The electrical parameters for each 5 Watt panel are, V_{OC} = 21.6 V, V_M = 17.5 V, I_{SC} = 0.3 A, I_M = 0.29 V. Under non-shading conditions, this PV array provides 80 W power generation.

As shown in Figures 10 and 11, two types of irradiance circumstances were examined in the experiment. The shading patterns considered for experimentation were horizontal (case I) and triangular corner (case II). These are similar to case II and case V in Section 2 but with different irradiance levels and array sizes. The shade concentrates before reconfiguring this array structure, as shown in Figure 10, for all the conventional configurations. After applying the proposed NSD reconfiguration, the shadow was distributed on the array, as shown in Figure 11.

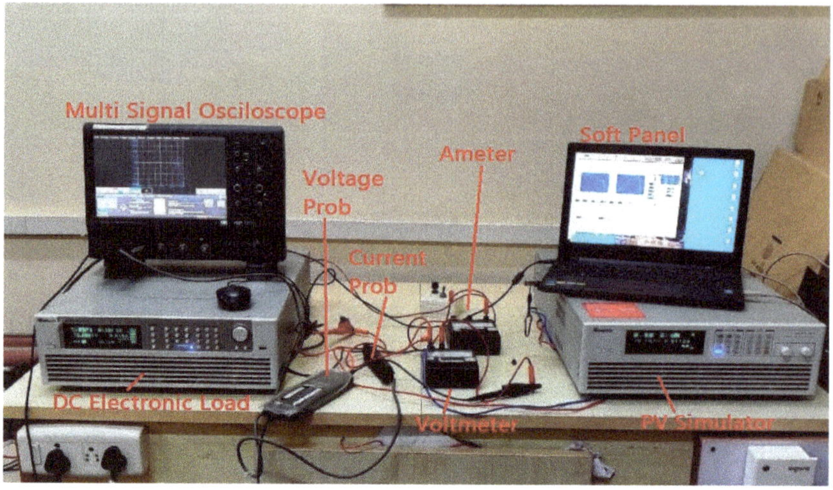

Figure 9. Experimental setup.

1000	1000	1000	1000
1000	1000	1000	1000
200	300	400	500
1000	1000	1000	1000

a)

1000	1000	1000	1000
700	1000	1000	1000
500	700	1000	1000
300	500	700	1000

b)

Figure 10. PSC before reconfiguration (**a**) case I, (**b**) case II.

200	1000	1000	1000
1000	300	1000	1000
1000	1000	1000	500
1000	1000	400	1000

a)

500	1000	700	1000
1000	700	1000	1000
700	500	1000	1000
300	1000	1000	1000

b)

Figure 11. PSC after reconfiguration (**a**) case I, (**b**) case II.

5.2. Experimental Result Analysis

To perform the analysis of conventional configurations, SP, BL, HC, TCT, and the proposed NSD reconfiguration, the PV curve nature was obtained by changing the load resistance through DC Electronic load from 4000 Ω to 1 Ω. The curve nature was observed on an HDO4034A (USA) 350 MHz high-definition oscilloscope.

5.2.1. PSC Case I

Figure 12 depicts the PV curves for each conventional configuration, NSD reconfiguration scheme, and the global MPP (the dot on the PV curve) under PSC case I. The PV curve shows GMPP for SP, BL, HC, TCT, and NSD at 59.22 W, 49.79 W, 49.79 W, 59.23 W, and 66.57 W, respectively. Moreover, SP and TCT have two peaks on the PV curve, BL and HC have three heights, and NSD reconfiguration reduces the number of peaks.

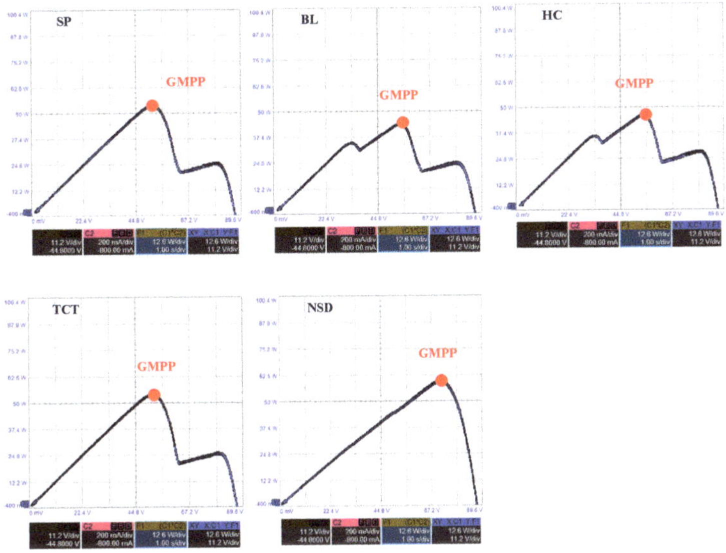

Figure 12. PV curves under PSC case I.

5.2.2. PSC Case II

Under this shading condition, Figure 13 shows that all conventional configurations have three peak PV curves. In contrast, the proposed NSD reconfiguration has one minor local peak and GMPP at a higher level with an improved curve nature. SP, BL, HC, TCT, and NSD provide GMPP at 52.92 W, 54.63 W, 53.75 W, 56.04 W, and 70.03 W, respectively. Figure 13 highlights that there is much improvement in the PV curve nature under this shading condition and GMPP location.

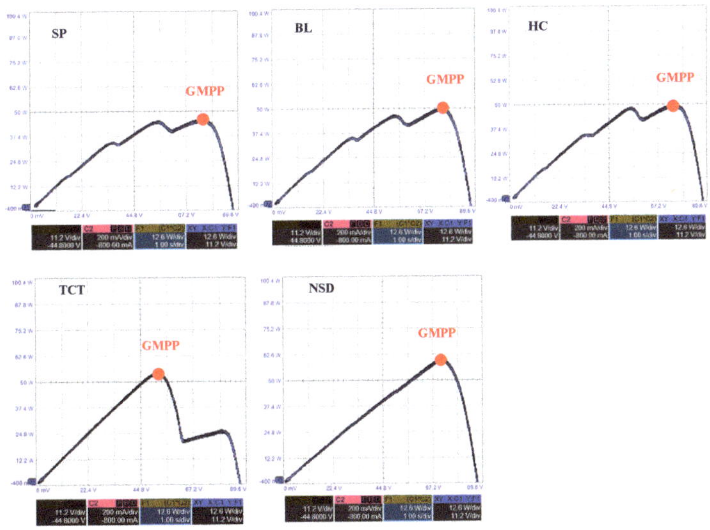

Figure 13. PV curves under PSC case II.

6. Conclusions

From the discussion mentioned above, this study concludes that:

- In partial shading cases, for the least irradiating case VIII and most irradiating case IX, NSD has maximum power extraction of 5160 W and 6952.8 W, respectively; the highest efficiency of 20.97% and 20.87%, respectively; FF is 0.6976 and 0.6803, respectively; and the lowest ML of 2505.6 W and 713 W among all configurations, respectively;
- Under PSC for the considered PV array, TCT has the highest GMPP, %O/P, efficiency, FF, %PE, and least %MP at 5683 W, 74%, 18.54%, 0.5784, and 28.73%. Compared with TCT, NSD has 705.77 W, 9.2%, and 0.1139 enhancement in GMPP, %O/P, and FF, and a 2.41% reduction in MP;
- The results obtained from the first simulation show that the NSD method achieved the most significant output power and overall performance compared with all conventional configurations under nine partial shading cases;
- Moreover, experimental research was carried out. The suggested NSD PV array reconfiguration design beats the other conventional PV array configurations (SP, BL, HC, and TCT) in power increase and several peaks. The experimental setup validated the influence of panel shifting on various PV array topologies.

Author Contributions: Conceptualization—V.C.C. and S.M.; Methodology, Software, Validation—V.C.C., S.M. and T.S.; Investigation, Resources, Writing—Original Draft Preparation, Writing—Review and Editing, Visualization; Supervision; Project Administration, V.C.C., S.M. and T.S. All authors have read and agreed to the published version of the manuscript.

Funding: The authors gratefully acknowledge the support offered by the Science and Engineering Research Board (SERB), Department of Science and Technology, Government of India, under the Grant number: EEQ/2021/000294 for this research work.

Institutional Review Board Statement: Not applicable.

Informed Consent Statement: Not applicable.

Data Availability Statement: Not applicable.

Conflicts of Interest: The authors declare no conflict of interest.

Appendix A

Table A1. PV Module electrical characteristics.

Symbol	Electrical Characteristics	Ratings
P_{max}	Maximum power	231.15 W
V_{oc}	Open circuit voltage	36.3 V
I_{sc}	Short circuit current	7.84 A
I_{max}	Maximum power point current	7.35 A
V_{max}	Maximum power point voltage	29 V
-	Temperature coefficient of Voc	−0.3609%/deg. °C
-	Temperature coefficient of Isc	0.102%/deg. °C

Appendix B

Definitions:

- **Global Maximum Power Point (GMPP):** GMPP is the single maximum point on the PV curve.

$$\text{GMPP} = \max(\text{multiple power Peaks})$$

- **Mislead Power (MP)**: MP denotes the difference between GMPP in the non-shading (NS) condition and GMPP in the shaded (PSC). It creates the hot spot effect and, due to that, power loss. It is provided by:

$$MP = GMPP_{NS} - GMPP_{PSC}$$

- **Percent Output (%O/P):** The %O/P is calculated as the maximum power ratio under PSC to extreme power under non-shading (NS) conditions. It provides how much energy is generated under PSC compared with power at standard irradiance. It reaches the %O/P of the PV array under PSC and NS conditions. It provides how much percentage of PSC affects the PV array output.

$$\%O/P = \frac{P_{max(PSC)}}{P_{max(NS)}}$$

- **Efficiency (η):** It is provided by the ratio output power (power generation) P_{Out} to input power (solar insolation) P_{in} On the PV panel. It is provided by:

$$\eta = \frac{P_{Out}}{P_{in}}$$

- **Fill Factor (FF):** The maximum power generated at PSC to the total rated capacity of a solar PV array is FF. It is also known as the usefulness of PV array. It is always 0 to 1. It is provided by:

$$FF = \frac{V_M I_M}{V_{OC} I_{SC}}$$

- **Percent Power Enhancement (%PE)**: %PE is provided by the difference between the GMPP for the proposed NSD method and the GMPP for conventional configuration (CC) under partial shading conditions. It provides how much power is enhanced by NSD compared with CC under various irradiance conditions.

$$\%\ PE = \frac{GMPP_{NSD} - GMPP_{CC}}{GMPP_{CC}}$$

References

1. Sahoo, S.K. Renewable and sustainable energy reviews solar photovoltaic energy progress in India: A review. *Renew. Sustain. Energy Rev.* **2016**, *59*, 927–939. [CrossRef]
2. Fouad, M.M.; Shihata, L.A.; Morgan, E.S.I. An integrated review of factors influencing the performance of photovoltaic panels. *Renew. Sustain. Energy Rev.* **2017**, *80*, 1499–1511. [CrossRef]
3. Venkateswari, R.; Sreejith, S. Factors influencing the efficiency of photovoltaic system. *Renew. Sustain. Energy Rev.* **2019**, *101*, 376–394. [CrossRef]
4. Attivissimo, F.; Di Nisio, A.; Savino, M.; Spadavecchia, M. Uncertainty analysis in photovoltaic cell parameter estimation. *IEEE Trans. Instrum. Meas.* **2012**, *61*, 1334–1342. [CrossRef]
5. Hysa, A. Modeling and Simulation of the Photovoltaic Cells for Different Values of Physical and Environmental Parameters. *Emerg. Sci. J.* **2019**, *3*, 395–406. [CrossRef]
6. Praveen Kumar, B.; Prince Winston, D.; Cynthia Christabel, S.; Venkatanarayanan, S. Implementation of a switched PV technique for rooftop 2 kW solar PV to enhance power during unavoidable partial shading conditions. *J. Power Electron.* **2017**, *17*, 1600–1610. [CrossRef]
7. Ziar, H.; Nouri, M.; Asaei, B.; Farhangi, S. Analysis of overcurrent occurrence in photovoltaic modules with overlapped by-pass diodes at partial shading. *IEEE J. Photovolt.* **2014**, *4*, 713–721. [CrossRef]
8. Mäki, A.; Valkealahti, S. Power losses in long string and parallel-connected short strings of series-connected silicon-based photovoltaic modules due to partial shading conditions. *IEEE Trans. Energy Convers.* **2012**, *27*, 173–183. [CrossRef]
9. Hussan, R.; Sarwar, A. Maximum power point tracking techniques under partial shading condition-A review. In Proceedings of the 2018 2nd IEEE International Conference on Power Electronics, Intelligent Control and Energy Systems, New Delhi, India, 22–24 October 2018; pp. 293–298. [CrossRef]
10. Bana, S.; Saini, R.P. Experimental investigation on power output of different photovoltaic array configurations under uniform and partial shading scenarios. *Energy* **2017**, *127*, 438–453. [CrossRef]
11. Pendem, S.R.; Mikkili, S. Modelling and performance assessment of PV array topologies under partial shading conditions to mitigate the mismatching power losses. *Sol. Energy* **2018**, *160*, 303–321. [CrossRef]

12. Bingöl, O.; Özkaya, B. Analysis and comparison of di ff erent PV array con fi gurations under partial shading conditions. *Sol Energy* **2018**, *160*, 336–343. [CrossRef]
13. Raju, V.B.; Chengaiah, C. Power Enhancement of Solar PV Arrays under Partial Shading Conditions with Reconfiguration Methods. In Proceedings of the 2019 Innovations in Power and Advanced Computing Technologies (i-PACT), Vellore, India 22–23 March 2019; pp. 1–7. [CrossRef]
14. Ajmal, A.M.; Sudhakar Babu, T.; Ramachandaramurthy, V.K.; Yousri, D.; Ekanayake, J.B. Static and dynamic reconfiguration approaches for mitigation of partial shading influence in photovoltaic arrays. *Sustain. Energy Technol. Assess.* **2020**, *40*, 100738 [CrossRef]
15. Manjunath, M.; Reddy, B.V.; Lehman, B. Performance improvement of dynamic PV array under partial shade conditions using M2 algorithm. *IET Renew. Power Gener.* **2019**, *13*, 1239–1249. [CrossRef]
16. Varma Gadiraju, H.K.; Reddy Barry, V.; Maraka, I. Dynamic Photovoltaic Array Reconfiguration for Standalone PV Water Pumping System. In Proceedings of the 2019 8th International Conference on Power Systems: Transition towards Sustainable Smart and Flexible Grids (ICPS), Jaipur, India, 20–22 December 2019; IEEE: Piscataway, NJ, USA, 2019; pp. 29–32. [CrossRef]
17. Bularka, S.; Gontean, A. Dynamic PV array reconfiguration under suboptimal conditions in hybrid solar energy harvesting systems. In Proceedings of the 2017 IEEE 23rd International Symposium for Design and Technology in Electronic Packaging Constanta, Romania, 26–29 October 2017; pp. 419–422. [CrossRef]
18. Pachauri, R.K.; Alhelou, H.H.; Bai, J.; Golshan, M.E.H. Adaptive Switch Matrix for PV Module Connections to Avoid Permanent Cross-Tied Link in PV Array System under Non-Uniform Irradiations. *IEEE Access* **2021**, *9*, 45978–45992. [CrossRef]
19. Rani, B.I.; Ilango, G.S.; Nagamani, C. Enhanced power generation from PV array under partial shading conditions by shade dispersion using Su Do Ku configuration. *IEEE Trans. Sustain. Energy* **2013**, *4*, 594–601. [CrossRef]
20. Sai Krishna, G.; Moger, T. Improved SuDoKu reconfiguration technique for total-cross-tied PV array to enhance maximum power under partial shading conditions. *Renew. Sustain. Energy Rev.* **2019**, *109*, 333–348. [CrossRef]
21. Sreekantha Reddy, S.; Yammani, C. A novel Magic-Square puzzle based one-time PV reconfiguration technique to mitigate mismatch power loss under various partial shading conditions. *Optik* **2020**, *222*, 165289. [CrossRef]
22. Vijayalekshmy, S.; Bindu, G.R.; Rama Iyer, S. Performance comparison of Zig-Zag and Su do Ku schemes in a partially shaded photo voltaic array under static shadow conditions. In Proceedings of the 2017 Innovations in Power and Advanced Computing Technologies (i-PACT), Vellore, India, 21–22 April 2017; pp. 1–6. [CrossRef]
23. Dhanalakshmi, B.; Rajasekar, N. A novel Competence Square based PV array reconfiguration technique for solar PV maximum power extraction. *Energy Convers. Manag.* **2018**, *174*, 897–912. [CrossRef]
24. Nihanth, M.S.S.; Ram, J.P.; Pillai, D.S.; Ghias, A.M.Y.M.; Garg, A.; Rajasekar, N. Enhanced power production in PV arrays using a new skyscraper puzzle based one-time reconfiguration procedure under partial shade conditions (PSCs). *Sol. Energy* **2019**, *194*, 209–224. [CrossRef]

energies

Article

Multiobjective Load Dispatch for Coal-Fired Power Plants under Renewable-Energy Accommodation Based on a Nondominated-Sorting Grey Wolf Optimizer Algorithm

Yue Cao [1,*], Tao Li [1], Tianyu He [1], Yuwei Wei [2], Ming Li [3] and Fengqi Si [1]

1 Key Laboratory of Energy Thermal Conversion and Control of Ministry of Education, School of Energy and Environment, Southeast University, Nanjing 210096, China; 220180461@seu.edu.cn (T.L.); hetianyu@seu.edu.cn (T.H.); fqsi@seu.edu.cn (F.S.)
2 Guangxi Special Equipment Inspection and Research Institute, Nanning 530219, China; gxtjybgs@gxtjy.org
3 Changyuan Hanchuan Power Generation Company Limited, China Energy Investment Corporation, Xiaogan 431600, China; ming.li.jd@chnenergy.com.cn
* Correspondence: ycao@seu.edu.cn; Tel.: +86-158-2182-1826

Abstract: Coal-fired power plants are widely used to achieve a power balance in grids with renewable energy, which leads to new requirements for speediness in load dispatch. This paper presents a nondominated-sorting grey wolf optimizer algorithm (NSGWO) for the multiobjective load dispatch of coal-fired power plants that employed efficient nondominated sorting, a reference-point selection strategy, and a simulated binary crossover operator. The optimization results of the benchmark functions indicated that the NSGWO algorithm had a better accuracy and a better distribution than the traditional multiobjective grey wolf optimizer algorithm. Regarding the load dispatch of economy, environmental protection, and speediness strategies, the NSGWO had the best performance of all the simulated algorithms. The optimal-compromise solutions of the economy and speediness strategies of the NSGWO algorithm had a good distribution, which elucidated that this novel algorithm was favorable to allowing coal-fired power plants to accommodate renewable energy.

Keywords: load dispatch; multiobjective optimization; grey wolf optimizer; nondominated sorting; coal-fired power plant

Citation: Cao, Y.; Li, T.; He, T.; Wei, Y.; Li, M.; Si, F. Multiobjective Load Dispatch for Coal-Fired Power Plants under Renewable-Energy Accommodation Based on a Nondominated-Sorting Grey Wolf Optimizer Algorithm. *Energies* **2022**, *15*, 2915. https://doi.org/10.3390/en15082915

Academic Editor: Elyas Rakhshani

Received: 21 March 2022
Accepted: 13 April 2022
Published: 15 April 2022

1. Introduction

Modern coal-fired power plants are clean and highly efficient, and their emissions of NOx and coal consumption are controlled at remarkably low levels [1]. Regarding the national energy distribution in China, coal-fired power plants still need to output power for basic loads in the grid. A coal-fired power plant may have different units (such as water-cooling and air-cooling units [2]) and units of various scales. Differences in unit characteristics may result in variable performances; therefore, it is necessary to optimize the operation strategy of coal-fired power plants with multiple units on the basis of both the economy and environmental-protection objectives. Furthermore, regarding the long-term objective of carbon neutrality [3], coal-fired power plants will play a new role in the power grid. Considering the development of renewable energy including solar power and wind power, coal-fired power plants, which have a strong capability to accommodate renewable energy, are widely used to achieve a power balance in power grids. In terms of the uncertainty of renewable energy [4], it is necessary to operate coal-fired power plants under conditions of rapid load change. Therefore, the optimal operation of coal-fired power plants should be further investigated while considering the factors of the economy, environmental protection, and speediness.

1.1. Load-Dispatch Strategy

Units of various types and scales in a coal-fired power plant usually receive the load command from the grid, which limits the optimal operation of the coal-fired power plant Regarding the load-dispatch strategy, the coal-fired power plant receives the total load command from the grid, and then dispatches it to different units [5]. Based on previously mentioned factors, namely the economy, environmental protection, and speediness, the load dispatch of a coal-fired power plant is a multiobjective optimization of load command between different units.

Several researchers have employed optimal load-dispatch strategies for power generation systems. Dai et al. [6] proposed a multiobjective economic load-dispatch method for large coal-fired power plants using a hierarchical clustering and retrieval strategy based on the fuzzy c-means clustering algorithm. They concluded that the data-mining-based method was able to achieve plant-level optimal load dispatching while meeting the actual requirements of the grid. Ishraque et al. [7] introduced various optimization strategies of load dispatch for microgrid power systems with renewable energy sources. They considered that the proposed load-following method was the best dispatch strategy to achieve lower operating costs and lower pollutant emissions. Cui et al. [8] considered that the heat load and power load could be optimized simultaneously in microgrids. Regarding the differences in real-time response of heat load and power load, they performed short-term and long-term simulations and elucidated that the total cost could be reduced, as could the peak load of the power grid. Li et al. [9] also focused on investigating multiobjective load dispatch between the heat load and power load, and concluded that the proposed method could balance the production cost and pollutant-emission objectives under the fluctuation of the thermoelectric load demand. Xu et al. [10] introduced the load-dispatch strategy to save coal for coal-fired power plants using a support-vector machine. Their results indicated that the proposed load-dispatch strategy performed better than the normal separate load demand on a single unit in a coal-fired power plant. In addition, novel load-dispatch strategies have been studied for similar power-generation systems. Liao et al. [11] investigated the short-term load-dispatching method for a hydropower plant with multiple turbines, and simulation results indicated that it could achieve optimal operation while simultaneously meeting the practical operation requirements of the hydraulic and electric constraints. Jamal et al. [12] introduced a bioinspired computational heuristic algorithm for dispatch loads for a power-generation system that included coal-fired thermal-generating units and wind-power units. They considered that the proposed algorithm had advantages in accuracy, convergence, and robustness in solving load-dispatch problems. Furthermore, Bie et al. [13] conducted a load-dispatch investigation based on the uncertainty of renewable energy, and the results showed that the real-time load-dispatch model could calculate the load command while considering the primary and secondary regulation of the grid.

In general, much research has been published on load-dispatch optimization, but few studies considered the present application of a coal-fired power plant with renewable-energy accommodation, which is significant in terms of the multiobjective load dispatch of modern coal-fired power plants.

1.2. Multiobjective Optimization Algorithms

Among the load-dispatch strategies, researchers consider the optimization algorithm to be the key technology needed to achieve the objectives of energy saving and reducing pollutant emissions. In addition to the traditional optimization methods (including dynamic programming, Lagrange relaxation, and the exhaustive method), intelligent optimization algorithms have already been introduced in recent investigations.

Intelligent optimization algorithms usually employ an evolutionary mechanism or heuristic algorithm to enhance the convergence speed and robustness ability of the optimal solution procedure. Ghasemi et al. [14] proposed a differential evolution algorithm (DE) based on the mathematical model of sociopolitical evolution. They performed comparative simulations to verify the effectiveness of the algorithm, and concluded that this algorithm was a reliable tool for solving the load-dispatch problem. Neto et al. [15] proposed a method that combined a differential-evolution algorithm and the greedy randomized adaptive search procedure algorithm. The simulation results showed that the algorithm could solve the nonsmooth load-dispatch problem by considering the effect of valve-point loading. Secui et al. [16] proposed a new heuristic algorithm called modified symbiosis (MSOS) that could solve the large-scale-unit load-optimization problem with a valve-point effect. They concluded that the proposed algorithm had a good performance based on the simulation results of five power-generation systems of different constraints and dimensions.

Furthermore, other heuristic algorithms, including the bat algorithm and the population extremum algorithm, have been introduced to solve the multiobjective optimization of load dispatch. Kavousi-Fard et al. [17] proposed an improved bat algorithm to solve the nonconvex load-optimization problem, which was then examined on IEEE test systems to show its high abilities and satisfying performance. Chen et al. [18] used the feasible-solution method to deal with constraints, and proposed a constrained multiobjective population extremal optimization algorithm called CMOPEO-EED. They considered the proposed algorithm to be a viable option to solve the load-dispatch problem with renewable power generation.

Particle-swarm optimization (PSO) is another commonly used algorithm in multiobjective optimization. Hosseinnezhad et al. [19] used an improved PSO algorithm (θ-PSO) to solve the economic load-optimization problem that could easily deal with various constraints. The findings of this study elucidated that the advantage of computational efficiency of the θ-PSO algorithm could be a promising alternative approach to solving the load-dispatch problem in practical power systems. Regarding the combined advantages of the PSO algorithm and other intelligent algorithms, researchers have investigated hybrid algorithms based on PSO. Narimani et al. [20] proposed a hybrid algorithm based on the PSO algorithm and leapfrog algorithm for multiregional economic and environmental-protection load distribution. They verified the effectiveness of the hybrid algorithm using different test systems to balance the multiobjectives of generation costs and environmental issues. Zou et al. [21] proposed a new global particle-swarm-optimization algorithm, which adopted an updated method that relied on the global optimal particle to guide all particle-search activities and a slightly perturbed randomization method for particle flight trajectories with uniform distribution. Mandal et al. [22] introduced variable acceleration coefficients into the PSO algorithm and then proposed a self-organized hierarchical particle-swarm-optimization algorithm. They found that the results of the proposed method were superior in terms of fuel costs, emission output, and losses.

In general, intelligent optimization algorithms have advantages in solving the multi-objective optimization problems of load dispatch at power plants, though their application under the conditions of renewable energy accommodation should be further investigated.

1.3. Grey Wolf Optimizer

Former studies show that various intelligent algorithms have been employed to solve the multioptimization issues in power-generation systems and similar applications. However, considering the rapid load-change conditions present under renewable-energy accommodation, the load dispatch of current coal-fired power plants requires rapid and accurate convergence.

The grey wolf optimizer (GWO) is a swarm-intelligence optimization algorithm proposed by Mirjalili et al. [23]. Compared with other natural heuristic optimization algorithms such as particle-swarm optimization and differential evolution, GWO has attracted attention due to its fast convergence speed, strong robustness, and high solution accuracy with a simple structure. Zhang et al. [24] used a GWO to plan the path of an unmanned combat aerial vehicle, and they concluded that it was more competent than other state-of-the-art evolutionary algorithms in terms of quality, speed, and stability of solutions. Although the grey wolf algorithm has superior performance on most data sets, it may not have the best performance on all. In view of the defects of a GWO, many scholars have made relevant improvements to the performance of the GWO algorithm. Madhiarasan et al. [25] optimized the initial population of a GWO to improve the diversity of the population. Saremi et al. [26] improved the search mechanism of a GWO. The results indicated that the proposed GWO algorithm was able to significantly improve the performance of multiobjective optimization. Long et al. [27] proposed an updating method for the key parameter of a GWO in the application of constrained optimization problems. They considered that the simulations for both benchmark functions and engineering applications could demonstrate the good performance of the proposed GWO algorithm. Moreover, hybrid optimization algorithms combining GWO with other intelligent optimization algorithms have already been introduced by researchers [28–30]. However, few scholars have conducted in-depth research on the evolution mechanism of the GWO for the present application of load dispatch optimization. Song et al. [31] introduced a GWO to solve the economic emission problem of two different power systems, and used a price penalty factor to combine the goals of economic dispatch and emission dispatch into a single function. Wong et al. [32] solved the economic scheduling problem under practical constraints with a GWO and tested them on two test systems with practical constraints. Moradi et al. [33] used a GWO to solve nonlinear and nonconvex economic scheduling problems while considering the valve-point effect and transfer loss, and compared the obtained results with some existing heuristic methods. Jangir et al. [34] investigated the economic constrained-emission dispatch problem based on a nondominated-sorting GWO, and they considered that it could be used in a power grid with wind power. However, they did not consider the speediness requirement of units in the coal-fired power plant, which indicated that the solving procedure of nondominated-sorting GWO should be further optimized.

The above shows that limited work has been conducted on the multiobjective optimization algorithm while considering the rapid and accurate convergence requirements under the unit rapid load-change condition. The more constraints that are considered, the more complex the load-dispatch optimization will be. Therefore, further research on the GWO needs to be carried out according to the specific problems of the multiobjective optimization of the load dispatch of coal-fired power plants.

1.4. Present Investigation

On the basis of the previous literature review, although much research has been published on load-dispatch optimization, few studies considered the present application of a coal-fired power plant with a renewable-energy accommodation. Moreover, limited work has been conducted on the multiobjective optimization algorithm that considered the rapid and accurate convergence requirements under the unit's rapid load-change condition. Therefore, an investigation into the multiobjective optimization of the load dispatch of a coal-fired power plant was conducted using a novel grey wolf algorithm.

In the present paper, the proposed nondominated-sorting grey wolf algorithm showed that it may help solve various optimization problems in similar applications. The organization of this study was arranged as follows. Section 2 shows the methods for the multiobjective optimization problem of load dispatch. Section 3 details the solving procedure of the novel grey wolf algorithm. Section 4 presents the simulation results and further discussion. Section 5 presents the conclusions of this study.

2. Methods

A theoretical model for the multiobjective optimization of the load dispatch of a coal-fired power plant, as well as optimization constraints and performance evaluation indices, were proposed in order to achieve the requirements related to the economy, environmental protection, and speediness.

2.1. Multiobjective Optimization Problem of a Coal-Fired Power Plant

2.1.1. Problem Description

The load-dispatch problem of a coal-fired power plant can generally be described as per Equation (1):

$$\begin{cases} \min \mathrm{F}(x) = \min(\mathrm{f}_1(x), \mathrm{f}_2(x), \ldots, \mathrm{f}_m(x))^T \\ st.\begin{cases} \mathrm{g}_i(x) \leq 0, i = 1, 2, \ldots p \\ \mathrm{h}_j(x) = 0, j = 1, 2, \ldots, q \end{cases} \end{cases} \tag{1}$$

where F(x) is the objective function of the economy performance, and the sub-functions represent the economy characteristics of units in the coal-fired power plant. Further, $\mathrm{g}_i(x)$ and $\mathrm{h}_j(x)$ represent inequality constraints and equality constraints, respectively. Specifically, pollutant emission is an inequality constraint when it is not selected as an optimization objective.

For a coal-fired power plant, the relationship between the pollutant emission g_i and the unit load P_i can be expressed as the quadratic function relationship shown below

$$\mathrm{g}_i(P_i) = \alpha_i P_i^2 + \beta_i P_i + \gamma_i \tag{2}$$

where α_i, β_i, and γ_i are the emission-characteristic coefficients of the unit.

Regarding the requirements of the relevant national policies of environmental protection, it is necessary to employ the environmental protection index as an optimization object instead of an inequality constraint. Thus, the objective function of load dispatch when considering environmental protection is defined as shown in Equation (3):

$$\min \mathrm{G} = \min \sum_{i=1}^{N} \mathrm{g}_i(P_i) \tag{3}$$

where G represents the pollutant discharge of the whole plant.

The power grid has specific requirements for the load response rate of coal-fired power plants, specially under the condition of renewable-energy accommodation. Therefore, the speediness of load variation should be considered in the load dispatch of coal-fired power plants. The minimum time consumption of load dispatch under ideal conditions T_{ideal} is shown in Equation (4):

$$T_{ideal} = \left| P_D - \sum_{i=1}^{N} P_{i,now} \right| / \sum_{i=1}^{N} v_i \tag{4}$$

where P_D is the total load command of all units, MW; and v_i represents the load changing speed of unit i, MW/min.

The sum of squares of the differences between the load-dispatch times of each unit and the minimum time under ideal conditions T(P) can be calculated as follows:

$$\mathrm{T}(P) = \sum_{i=1}^{N} ((P_i - P_{i,now})/v_i - T_{ideal})^2 \tag{5}$$

Therefore, combining former indicators with the economic indicators, the multiobjective optimization of load dispatch for a coal-fired power plant can be described as shown in Equation (6):

$$\begin{gathered} \min(F, G, T), \min(F, G), \min(F, T) \\ st.\begin{cases} \sum_{i=1}^{N} P_i - P_D \\ P_i^{\min} \leq P_i \leq P_i^{\max} \\ \left| P_i - P_i^{now} \right| / t_i \leq v_i \end{cases} \end{gathered} \tag{6}$$

2.1.2. Constraint-Violation Degree

The constraint-violation degree is introduced to show whether an individual x violates the i-th inequality constraint or the j-th equality constraint, which is expressed in Equations (7) and (8):

$$Con_i(x) = \max\{g_i(x), 0\}, 1 \leq i \leq p \tag{7}$$

$$Con_j(x) = \max\{h_j(x), 0\}, 1 \le j \le q \quad (8)$$

Thus, the total constraint-violation degree of individual x is shown in Equation (9):

$$v(x) = \sum_{i=1}^{p} Con_i(x) + \sum_{j=1}^{q} Con_j(x) \quad (9)$$

In the multiobjective optimization of the load dispatch for a coal-fired power plant, the equation constraint describes that the sum of the loads of each unit must be equal to the total load command, while the inequality constraint shows the load-regulation range.

2.1.3. Feasible and Infeasible Solutions

When an individual x satisfies both the inequality constraints and the equality constraints in the load-dispatch model, this individual is called a feasible solution; otherwise, it is an infeasible solution. The region composed of all feasible solutions is called the feasible region Ω, and the set composed of infeasible solutions is called the infeasible region.

2.1.4. Dominance

For the two specific solutions x_1 and x_2, the economy, speediness, and environmental-protection indicators corresponding to solution x_1 are better than those corresponding to solution x_2. Thus, solution x_1 is called a nondominated solution, while solution x_2 is called a dominant solution, which can be written as $x_1 \succ x_2$.

All individuals in the solution set that cannot be dominated by other individuals are called nondominated solutions, and this set of nondominated solutions is called the Pareto front (PF) of optimization solutions.

2.2. *Constraints for Multiobjective Optimization*

Constraint processing should be conducted to balance the relationship between the objective function and the constraints. In this paper, the penalty-function method was used to deal with constraints. By adding the penalty function $F(x, M)$ into the objective function, the constrained optimization problem was transformed into an unconstrained optimization problem. Parameter M in the penalty function represents a large positive number, called the penalty factor, which acts as a penalty for the solution that violates the constraints. Using the external penalty function, the multiobjective function with the penalty function is expressed as follows:

$$\min F(x, M) = \min(f_1(x) + M_1 v(x), f_2(x) + M_2 v(x), \ldots, f_m(x) + M_m v(x)) \quad (10)$$

2.3. *Evaluation Index of Multiobjective Optimization*

For the multiobjective optimization of a load dispatch, evaluation indices were employed to compare the simulation results, which may have shown the convergence performance and comprehensive performance of the optimization algorithm.

2.3.1. Convergence Evaluation Index

The generational distance (GD) was introduced to evaluate the convergence performance, which represented the degree of the solution set P to the real Pareto frontier P^*, as shown in Equation (11):

$$GD(P, P^*) = \frac{1}{|P|}\sqrt{\sum_{i=1}^{|P|} d_i^2} \quad (11)$$

where $|P|$ is the number of solutions in P, and d_i represents the Euclidean distance from the i-th solution in P to the nearest solution in P^*. Therefore, a small GD value meant the convergence performance was better.

2.3.2. Comprehensive Evaluation Index

The inverse generational distance (*IGD*) was used to represent the degree of the real Pareto frontier P^* to the solution set P [35]. It evaluated the convergence and diversity of the algorithm simultaneously, which was defined as:

$$IGD(P,P^*) = \frac{1}{|P^*|}\sqrt{\sum_{i=1}^{|P^*|} d_i'^2} \quad (12)$$

where $d_i'^2$ represents the nearest Euclidean distance from the i-th individual in P^* to the solution in P. Here, a small *IGD* showed that the optimization algorithm was better in both the convergence performance and the diversity of solutions.

2.3.3. Optimal-Compromise Solution

For the multiobjective optimization of a load dispatch, the fuzzy set theory is suggested to find an optimal-compromise solution from the Pareto fronts. The fuzzy membership degree for the load-dispatch optimization can be expressed as:

$$\varphi_{pi} = \begin{cases} 1 & f_{pi} \le f_i^{\min} \\ (f_i^{\max} - f_{pi})/(f_i^{\max} - f_i^{\min}) & f_i^{\min} < f_{pi} < f_i^{\max} \\ 0 & f_{pi} \ge f_i^{\max} \end{cases} \quad (13)$$

where φ_{pi} represents the satisfaction of the i-th objective of the p-th optimal solution, $i \in \{1, 2, \ldots, m\}$; f_{pi} represents the value of the i-th objective of the p-th optimal solution; and $f_i^{\max}$ and $f_i^{\min}$ represent the maximum and minimum values of the i-th objective in all optimal solutions, respectively. When φ_{pi} was equal to 1, it meant that the individual was completely satisfied with the i-th target value.

The standardized satisfaction calculation method of each solution p in the nondominated solution set is shown in Equation (14), where the solution corresponding to the largest φ_{pi} was the optimal-compromise solution:

$$\varphi_p = \sum_{k=1}^{m} \varphi_{pk} / \sum_{p=1}^{N} \sum_{k=1}^{m} \varphi_{pk} \quad (14)$$

Regarding the previous objectives and constraints of the multiobjective optimization of the load dispatch for a coal-fired power plant, the sequence diagram is shown in Figure 1.

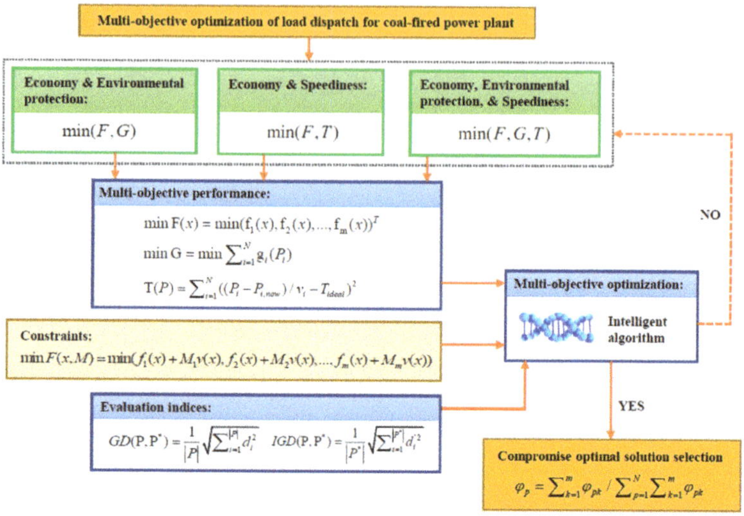

Figure 1. Sequence diagram of multiobjective optimization of load dispatch of coal-fired power plant.

3. Solving Procedure of the Optimization Algorithm

The multiobjective grey wolf algorithm was employed to find the load-dispatch optimization solution for a coal-fired power plant. However, the issues of local optimum and insufficient stability may limit the performance of this optimization algorithm. In addition, the influence of the alpha wolf may result in a poor distribution of the obtained solutions., and the traditional nondominated sorting may consume more time in the iteration process. Considering the multiobjective optimization of a load dispatch with multiple complex constrains, a nondominated-sorting grey wolf optimizer (NSGWO) algorithm was proposed that employed efficient nondominated sorting (ENS), a reference-point selection strategy, and a simulated binary crossover operator.

3.1. Fast Nondominated Sorting Based on ENS

Fast nondominated sorting is a cyclic stratification process. If this process is conducted in sequence, it will require too much time to converge. In fact, the frontier number of individuals is determined by the dominated individual, which could infer the following equation:

$$front(p) = 1 + \max_{q \in Q} front(q) \tag{15}$$

where $front(p)$ represents the number of the frontier where individual p is located, and Q is the set composed of all individuals that dominate individual p.

The first step of the ENS was to arrange all individuals in the population according to their first-dimension target values from small to large. If the first-dimension target values were the same, we compared their second-dimension index values, and so on. In the sorted population, it was impossible for the individual in the latter order to dominate the individual in front of it. This meant that the individual that dominated p must have been assigned to the frontier when examining p. The number of the frontier where p is located was calculated using Equation (15).

On this basis, starting from the first frontier F_1, it was judged in turn whether each frontier contained an individual that dominated p. The first frontier that does not contain the dominant p was the frontier where p resides. If there were individuals dominating p up to the last frontier F_k, then p was assigned to a new frontier F_{k+1}. If all individuals were mutually nondominant, then a nondominant comparison was required between them. In this case, since there was only one frontier, the complexity was $T_{worst} = m \cdot N(N -$

$1)/2 = O(mN^2)$. The algorithm could achieve the best performance if the population satisfied the following conditions: there were $\sqrt{N}$ frontiers in total, and each frontier contained $\sqrt{N}$ individuals. Moreover, an individual was dominated by all individuals whose frontier number was smaller than its under this condition. For a specific individual p, if it belonged to frontier F_i, it was necessary to find an individual dominating it in each frontier $F_i(j < i)$. This indicated that at least $i-1$ nondominated comparisons were required. For all $\sqrt{N}$ individuals in F_i, a total of $\sqrt{N}(i+1)$ comparisons were required. In addition, all individuals of F_i needed to be compared pairwise, which meant that a total of $\sqrt{N}(\sqrt{N}-1)/2$ comparisons were required. Considering all frontiers, the best period of complexity is shown in Equation (16):

$$T_{best} = m \cdot \sum_{i=1}^{\sqrt{N}} \left[\sqrt{N}(i-1) + \sqrt{N}(\sqrt{N}-1)/2\right] = m \cdot \left[N(\sqrt{N}-1)/2 + N(\sqrt{N}-1)/2\right] = O(MN^{1.5}) \quad (16)$$

3.2. Selection Strategy Based on Reference Point

To select N individuals from the set of offspring and parents $R_t = P_t \cup Q_t$, it should firstly divide R_t into multiple nondominated layers $(F_1, F_2, \ldots, F_k)$ by nondominated sorting. Then, a new population S_t is established from the start of F_1 until its size is equal to or exceeds N for the first time. In order to maintain the diversity of solutions, a selection strategy based on reference point was used to select $n_L = N - \sum_{i=1}^{L-1} N_i$ individuals, where N_i is the number of individuals in the i-th frontier, and n_L is the number of individuals that needed to be selected from the last frontier satisfying the former requirements. Moreover, the systematic approach was introduced to generate a set of uniform weight vectors $\{\lambda_1, \lambda_2, \ldots, \lambda_R\}$ so that the reference points could be determined. In order to compare the target values of each dimension, it was necessary to adaptively normalize the target value of each dimension. The ideal point $\overline{z} = (z_1^{\min}, z_2^{\min}, \ldots, z_m^{\min})$ of defining population S_t was composed of the minimum value $z_1^{\min}, i = 1, 2, \ldots, m$ of each dimension of S_t. Based on the ideal point, the objective function could be transformed to $f_i'(x) = f_i(x) - z_i^{\min}$. Then, a hyperplane could be built by extra points obtained from the objective function; this process can be expressed as:

$$ASF(x, w) = \max_{i=1}^{m} f_i'(x)/w_i, x \in S_t \quad (17)$$

$$Z^{i,\max} = s : \operatorname{argmin}_{s \in S_t} ASF(s, w), w = (\tau, \ldots, \tau), \tau = 10^{-6} \quad (18)$$

For the i-th target, an extra target vector $z^{i,\max}$ was generated. Thus, m extra vectors were generated by m targets which further constituted an m-dimensional linear hyperplane, and then the intercept $a_i, i = 1, 2, \ldots, m$ could be calculated. The objective function can be normalized as shown in Equation (19):

$$f_i^n(x) = (f_i'(x) - z_i^{\min})/(a_i - z_i^{\min}) \quad (19)$$

3.3. Simulated Binary Crossover Operator

Under the influence of the top three levels of wolves, a GWO easily falls into a local optimum. Therefore, a simulated binary crossover (SBX) operator was introduced in this paper. Assuming that the two parent individuals are $x^1(x_1^1, x_2^1, \ldots x_n^1)$ and $x^2(x_1^2, x_2^2, \ldots x_n^2)$, the SBX operator could generate the offspring individuals $c^1(c_1^1, c_2^1, \ldots c_n^1)$ and $c^2(c_1^2, c_2^2, \ldots c_n^2)$ using the following equations:

$$\begin{cases} c_i^1 = 0.5 \times [(1+\beta) \cdot x_i^1 + (1-\beta) \cdot x_i^2] \\ c_i^2 = 0.5 \times [(1-\beta) \cdot x_i^1 + (1+\beta) \cdot x_i^2] \end{cases} \quad (20)$$

where β is randomly determined by the distribution factor η according to Equation (21):

$$\beta = \begin{cases} (rand \times 2)^{1/(1+\eta)}, rand \leq 0.5 \\ (1/(2 - rand \times 2))^{1/(1+\eta)}, rand > 0.5 \end{cases} \quad (21)$$

With the increase in iterations, the probability of generating individuals according to the grey wolf hunting mechanism increased, thereby enhancing the development ability and convergence of the algorithm. In this paper, the variable a was controlled by a nonlinear decreasing strategy to balance the exploration and development capabilities of the algorithm, as shown in Equation (22):

$$a = 2(1 - (t/iter)^2) \tag{22}$$

where *iter* is the maximum number of iterations.

3.4. Solving Procedure of NSGWO

The solving procedure of the proposed nondominated-sorting grey wolf optimizer algorithm is presented in Figure 2. There were five specific steps of this algorithm, which are detailed as follows.

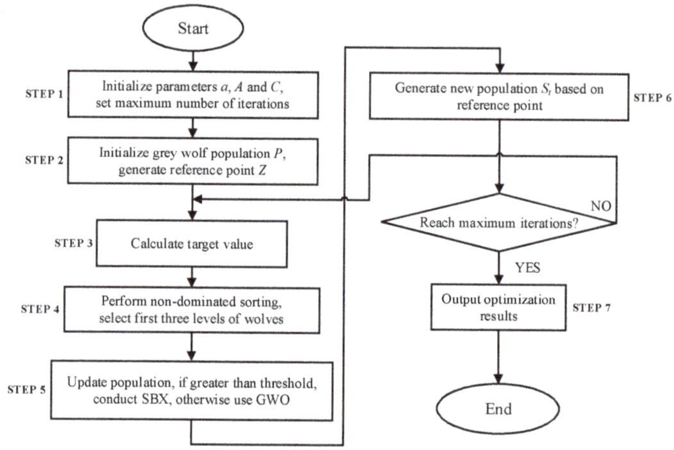

Figure 2. Solving procedure of nondominated-sorting grey wolf optimizer algorithm.

Step 1: Set the number N of individuals in the grey wolf population and the maximum number of iterations *iter*. Initialize parameters a, A, and C.

Step 2: Initialize the grey wolf population P, and generate a reference point Z.

Step 3: Calculate the target value of each dimension of individuals in the population.

Step 4: Perform the nondominated sorting based on ENS, and select the first three levels of wolves α, β, δ.

Step 5: Update the population P_t and generate the offspring Q_t. When the random number is greater than the set probability threshold, binary crossover mutation is conducted. Otherwise, update the population based on the evolution mechanism of GWO.

Step 6: Select the first N individuals based on the reference point in the merged population $R_t = P_t \cup Q_t$, and then generate a new population S_t.

Step 7: If the maximum number of iterations is not reached, repeat steps (2–4). Otherwise, output the optimization results.

4. Results and Discussion

4.1. NSGWO Algorithm Performance Test

In order to test and evaluate the performance of the NSGWO algorithm, three benchmark functions [36] were employed in this paper, the detailed structures of which are shown in Table 1.

Table 1. Benchmark functions for multiobjective optimization test of NSGWO algorithm.

Function Name	Objective Function	Variable Range
ZDT1	$\begin{cases} f_1(x) = x_1 \\ f_2(x) = g(x)h(f_1(x), g(x)) \\ g(x) = 1 + \frac{9}{29}\sum_{i=2}^{30} x_i \\ h(f_1(x), g(x)) = 1 - \sqrt{f_1(x)/g(x)} \end{cases}$	$0 \le x_i \le 1, 1 \le i \le 30$
UF2	$\begin{cases} f_1 = x_1 + \frac{2}{\lvert J_1 \rvert} \sum\limits_{j \in J_1} y_j^2 \\ f_2 = 1 - \sqrt{x} + \frac{2}{\lvert J_2 \rvert} \sum\limits_{j \in J_2} y_j^2 \\ J_1 = \{j \mid j is\ even\ number, 2 \le j \le n\}, J_2 = \{j \mid j is\ odd\ number, 2 \le j \le n\} \\ y_j = \begin{cases} x_j - [0.3x_1^2 \cos(24\pi x_1 + 4j\pi/n) + 0.6x_1] \cos(6\pi x_1 + j\pi/n), j \in J_1 \\ x_j - [0.3x_1^2 \cos(24\pi x_1 + 4j\pi/n) + 0.6x_1] \cos(6\pi x_1 + j\pi/n), j \in J_2 \end{cases} \end{cases}$	$0 \le x_i \le 1, 1 \le i \le 30$
UF4	$\begin{cases} f_1 = x_1 + \frac{2}{\lvert J_1 \rvert} \sum\limits_{j \in J_1} \mathrm{h}(y_j) \\ f_2 = 1 - x_2 + \frac{2}{\lvert J_2 \rvert} \sum\limits_{j \in J_2} \mathrm{h}(y_j) \\ y_j = x_j - \sin(6\pi x_1 + j\pi/n), j = 2, 3, \ldots, n \\ h(t) = \lvert t \rvert / (1 + e^{2\lvert t \rvert}) \end{cases}$	$0 \le x_i \le 1, 1 \le i \le 30$

Then, test simulations were conducted for the multiobjective optimization algorithm; the results for the *GD* and *IGD* values of the former benchmark functions are presented in Tables 2 and 3, respectively. Here, the results of the multiobjective grey wolf optimizer (MOGWO) and the multiobjective particle-swarm optimizer (MOPSO) are also shown for performance comparisons. For the unimodal function ZDT1, the *IGD* value of the NSGWO optimization result was 0.00690, which was an order of magnitude smaller than the results of other optimization algorithms, indicating that the improved algorithm had a strong ability to jump out of the local optimum. The UF series functions were all multimodal functions, which provided many local optimal solutions, and therefore the requirements for the algorithm were relatively high. Regardless of the *IGD* value or the *GD* value, the results of the improved algorithm were better than those of the other algorithms in the UF4 function. However, the *IGD* of the improved algorithm was smaller, as was the difference between the maximum and minimum values of the *GD* value, indicating that the improved algorithm had a better stability. The above test results showed that the improved algorithm NSGWO had good stability and the ability to jump out of the local optimum.

Table 2. *GD* values of optimization results for benchmark functions.

Function	Algorithm	Mean Value	Worst Value	Optimal Value
ZDT1	NSGWO	0.00490	0.01250	0.00090
	MOGWO	0.00754	0.06144	0.00119
	MOPSO	0.06150	0.36040	0.00370
UF2	NSGWO	0.00500	0.00600	0.00420
	MOGWO	0.05949	0.11078	0.02178
	MOPSO	0.10420	0.15110	0.06570
UF4	NSGWO	0.04160	0.04260	0.04000
	MOGWO	0.05756	0.06230	0.05198
	MOPSO	0.07910	0.08800	0.07030

Table 3. *IGD* values of optimization results for benchmark functions.

Function	Algorithm	Mean Value	Worst Value	Optimal Value
ZDT1	NSGWO	0.00690	0.01340	0.00390
	MOGWO	0.20824	0.69615	0.01391
	MOPSO	0.06310	0.34060	0.00820
UF2	NSGWO	0.01600	0.01980	0.01250
	MOGWO	0.06961	0.08307	0.05542
	MOPSO	0.09580	0.11190	0.08190
UF4	NSGWO	0.03430	0.03530	0.03220
	MOGWO	0.06200	0.06890	0.05861
	MOPSO	0.07733	0.08944	0.06682

Figure 3 shows the Pareto fronts of optimization results of the benchmark function in which the horizontal and vertical coordinates represent a certain target value of the benchmark function. For the ZDT1 function, the solutions of MOGWO and NSGWO were close to the real Pareto fronts. For the UF2 function, most of the MOGWO solutions were far from the real Pareto fronts, while most of the NSGWO solutions were concentrated on the real Pareto fronts. Moreover, the results indicated that the UF4 function was relatively complex. Although the optimization results of the NSGWO algorithm in Figure 3c were not close to the real Pareto fronts, they were still better than those of MOGWO. In general, the NSGWO had better accuracy and a better distribution of the obtained solutions.

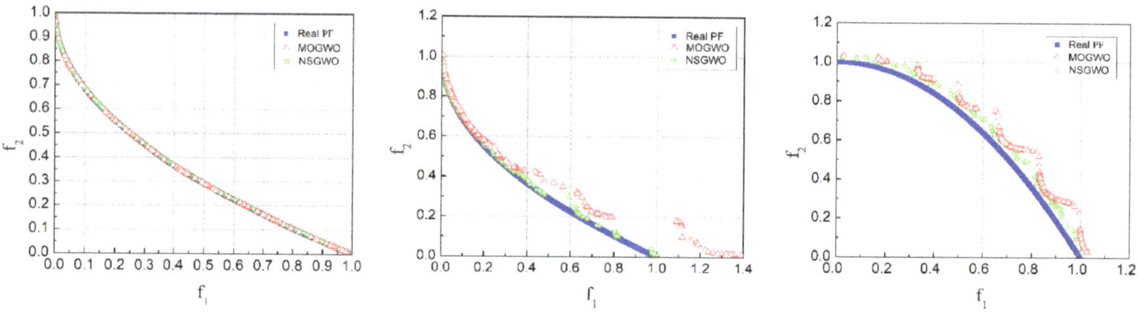

(a) (b) (c)

Figure 3. Pareto fronts of multiobjective optimization algorithms for benchmark functions. (**a**) ZDT1; (**b**) UF2; (**c**) UF4.

4.2. Multiobjective Optimization of Load Dispatch Based on NSGWO

The NSGWO algorithm was employed in the multiobjective optimization of the load dispatch for coal-fired power plants while considering the economy, environmental protection, and speediness. To better show the performance of the NSGWO algorithm in load dispatch, four intelligent algorithms, including the real-number coding genetic algorithm (RCGA), particle-swarm optimizer (PSO), modified nondominated-sorting genetic algorithm (NSGA-II), and MOGWO, were employed in the following simulations. Their configurations in terms of the population size and maximum number of iterations are summarized in Table 4, and were selected through analyses of presimulations.

Table 4. Configurations of NSGWO and other intelligent algorithms used in simulations of multiobjective optimization.

	Simulation I [1]			Simulation II [1]			Simulation III [1]	
	NSGWO	**RCGA**	**PSO**	**NSGWO**	**NSGA-II**	**MOGWO**	**NSOGWO**	**MOGWO**
Population size	200	200	200	100	100	100	100	100
Number of iterations	500	500	500	300	300	300	1000	1000

[1] Simulations I–III represent the optimization of economy and environmental protection, the optimization of economy and speediness for various power-scale units, and the optimization of economy and speediness for similar-scale units, respectively.

4.2.1. Optimization of Economy and Environmental Protection

The NSGWO algorithm was employed for the multiobjective optimization of the load dispatch for coal-fired power plants while considering economy and environmental protection. The penalty function was used in the following simulations to deal with the equality constraints in the load-dispatch model. Firstly, a coal-fired power plant with six units was investigated; its coal consumption and pollutant-emission characteristics are listed in Table 5 [37]. The total command of these six units was 283.4 MW, and their load-changing speed was limited to 5% of the current output power. To better show the performance of the NSGWO algorithm in the load dispatch, the RCGA and PSO were introduced for a comparison with the optimization results. Then, the simulations were conducted; the results are shown in Tables 6 and 7.

Table 5. Characteristics of dispatched units based on economy and environmental protection.

Unit	Coal Consumption Characteristics	Pollutant Emission Characteristics	P_i^{min} /MW [1]	P_i^{max} /MW [1]
1	$f_1(P_1) = 0.010P_1^2 + 2.0P_1 + 10 + \lvert 1.5\sin[6.28(P_1^{min} - P_1)] \rvert$	$g_1(P_1) = 6.49 \times 10^{-4} P_1 - 0.056P_1 + 4.091$	5	50
2	$f_2(P_2) = 0.012P_2^2 + 1.5P_2 + 10 + \lvert 1.0\sin[8.98(P_2^{min} - P_2)] \rvert$	$g_2(P_2) = 5.64 \times 10^{-4} P_2 - 0.061P_2 + 2.543$	5	60
3	$f_3(P_3) = 0.004P_3^2 + 1.8P_3 + 20 + \lvert 1.0\sin[14.78(P_3^{min} - P_3)] \rvert$	$g_3(P_3) = 4.59 \times 10^{-4} P_3 - 0.051P_3 + 4.258$	5	100
4	$f_4(P_4) = 0.006P_4^2 + 1.0P_4 + 10 + \lvert 0.5\sin[20.94(P_4^{min} - P_4)] \rvert$	$g_4(P_4) = 6.38 \times 10^{-4} P_4 - 0.030P_1 + 5.326$	5	120
5	$f_5(P_5) = 0.004P_5^2 + 1.8P_5 + 20 + \lvert 0.5\sin[25.13(P_5^{min} - P_5)] \rvert$	$g_5(P_5) = 4.59 \times 10^{-4} P_5 - 0.051P_5 + 4.258$	5	100
6	$f_6(P_6) = 0.010P_6^2 + 1.5P_6 + 10 + \lvert 0.5\sin[18.48(P_6^{min} - P_6)] \rvert$	$g_6(P_6) = 5.15 \times 10^{-4} P_1 - 0.056P_1 + 6.131$	5	60

[1] Variables P_i^{min} and P_i^{max} represent lower limit and upper limit of output power.

Table 6. Optimization results of economic load dispatch for coal-fired power plant without valve-point effect of steam turbines.

Algorithm	P_1 /MW	P_2 /MW	P_3 /MW	P_4 /MW	P_5 /MW	P_6 /MW	Coal Consumption Cost/(USD h^{-1})	Pollution Emission /(th $^{-1}$)
NSGWO	15.7	33.0	58.4	94.4	46.7	35.2	94.5382	0.2321
RCGA	11.5	30.6	59.9	98.2	51.3	35.5	95.6516	0.2199
PSO	12.8	27.0	55.5	100.5	45.4	44.5	95.6517	0.2207

The results showed that the performance of the NSGWO algorithm was better than that of the other algorithms in reducing power-plant coal consumption and power-plant emissions, which indicated the advantages of the NSGWO algorithm in the optimal dispatch of multiobjective economic and environmental-protection loads. Tables 6 and 7 also show that the valve-point effect of steam turbines had a great influence on the optimal dispatch results. The optimization results of Unit 1 and Unit 6 in the two models were quite different. When the valve-point effect was considered, the optimal dispatch results of these two units were 21.4 MW and 41.0 MW, respectively. However, the optimal dispatch results changed to 15.7 MW and 35.2 MW without considering the valve-point effect of the steam turbines.

Table 7. Optimization results of load dispatch under economy and environmental-protection strategies for coal-fired power plant considering valve-point effect of steam turbines.

Item	Economy Strategy			Environmental-Protection Strategy		
	NSGWO	RCGA	PSO	NSGWO	RCGA	PSO
Cost/(USD h $^{-1}$)	95.2184	96.1129	99.0861	100.7577	101.5641	101.9237
Emission/(th $^{-1}$)	0.229	0.214	0.214	0.193	0.194	0.195
P_1/MW	21.4	11.5	9.9	40.5	41.1	37.1
P_2/MW	30.9	30.6	36.3	40.7	46.3	46.7
P_3/MW	50.5	59.9	48.4	50.2	54.4	56.4
P_4/MW	93.0	98.2	87.4	37.7	39	36.5
P_5/MW	46.7	51.3	66.4	58.8	54.4	52.2
P_6/MW	41.0	35.5	39.0	55.5	51.5	57.8

In order to further illustrate the advantages of NSGWO, the Pareto fronts were determined for the NSGWO and MOGWO algorithms without considering the valve-point effect, as shown in Figure 4. The solutions of NSGWO were more widely distributed, and the solutions of MOGWO were basically dominated by the solutions of NSGWO. The comparison results show that the NSGWO algorithm had a certain application value in the multiobjective optimization of the load dispatch for coal-fired power plants.

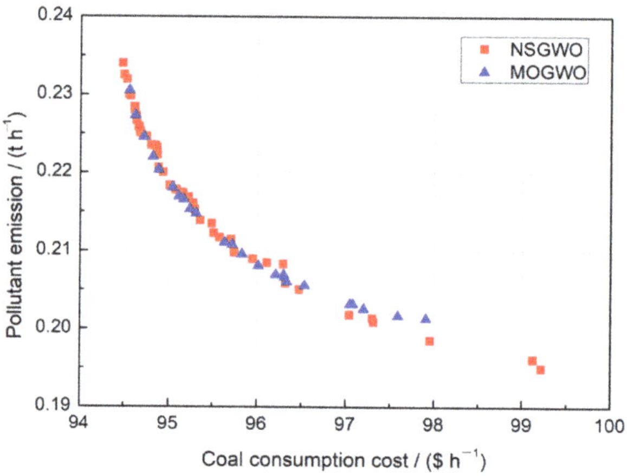

Figure 4. Pareto fronts of coal-consumption cost and pollutant emissions of load-dispatch optimization under NSGWO and MOGWO algorithms.

4.2.2. Optimization of Economy and Speediness for Various Power-Scale Units

Considering the requirements of rapid load change of coal-fired power plants under a renewable-energy accommodation, simulations were conducted to show the performance of the NSGWO algorithm in the multiobjective optimization of economy and speediness. Here, 10 units of various power scales are selected to be dispatched in the following simulations, the characteristics of which are shown in Table 8 [38]. The total dispatch command of the 10-unit power plant was 1500 MW, and the penalty-factor values of the two-dimensional objective of economy and speediness in the NSGWO algorithm were 0.5 and 2.1, respectively. Table 9 presents the optimization results of the load dispatch under the economy and speediness strategies. To better explain the advantages of nondominated sorting, the optimization of the NSGA-II was also performed, as shown in Table 9.

Table 8. Characteristics of 10 dispatched units with various power scales based on economy and speediness.

Unit	Coal-Consumption Characteristics	P_{now}/MW	v_i/(MW min $^{-1}$)	P_i^{min}/MW	P_i^{max}/MW
1	$f_1(P_1) = 0.00510P_1^2 + 2.2034P_1 + 15$	15	7	15	60
2	$f_2(P_2) = 0.00396P_2^2 + 1.9101P_2 + 25$	20	8	20	80
3	$f_3(P_3) = 0.00393P_3^2 + 1.8518P_3 + 40$	30	10	30	100
4	$f_4(P_4) = 0.00382P_4^2 + 1.6966P_4 + 32$	35.8	15	25	120
5	$f_5(P_5) = 0.00212P_5^2 + 1.8015P_5 + 29$	50.3	8	50	150
6	$f_6(P_6) = 0.00261P_6^2 + 1.5354P_6 + 72$	75.4	6	75	280
7	$f_7(P_7) = 0.00289P_7^2 + 1.2643P_7 + 49$	124	11	120	320
8	$f_8(P_8) = 0.00148P_8^2 + 1.2130P_8 + 82$	251.85	10	125	445
9	$f_9(P_9) = 0.00127P_9^2 + 1.1954P_9 + 105$	298.75	12	250	520
10	$f_{10}(P_{10}) = 0.00135P_{10}^2 + 1.1285P_{10} + 100$	298.9	9	250	550

Table 9. Optimization results of load dispatch under economy and speediness strategies for 10 units of various power scales.

Item	Optimal-Compromise Solution			Economy Strategy			Speediness Strategy		
	NSGWO	NSGA-II	MOGWO	NSGWO	NSGA-II	MOGWO	NSGWO	NSGA-II	MOGWO
f_h/(th $^{-1}$) [1]	3118.20	3120.77	3124.15	3115.02	3115.22	3123.25	3130.11	3131.29	3126.27
T/min [1]	4.91	4.98	4.49	7.48	7.12	5.13	3.52	3.59	3.77
P_1/MW	15.64	20.96	24.48	15.14	15.00	23.50	34.21	35.94	25.66
P_2/MW	37.58	37.88	45.04	26.00	27.33	45.24	44.566	45.50	46.30
P_3/MW	45.16	48.03	48.21	36.20	34.21	45.48	59.17	61.00	52.29
P_4/MW	65.82	69.80	76.64	58.34	55.86	76.33	80.19	70.07	77.91
P_5/MW	79.48	76.81	76.43	76.74	82.85	75.24	77.36	78.82	78.27
P_6/MW	101.45	94.49	97.10	109.45	109.63	95.54	94.91	96.26	97.04
P_7/MW	159.34	156.95	152.18	146.84	149.06	152.97	157.04	155.41	156.39
P_8/MW	298.22	294.80	296.78	302.62	306.39	301.44	284.90	286.40	289.48
P_9/MW	354.25	358.46	345.97	364.46	356.67	339.19	339.36	341.79	344.03
P_{10}/MW	343.06	341.81	337.18	364.15	363.01	345.06	328.23	328.81	332.64

[1] Variables f_h and T represent coal consumption per hour and adjustment time, respectively.

The results indicated that the optimal solutions of the NSGWO algorithm had the minimum adjustment times. For the minimum coal consumption of the economy strategy, the optimal solutions of the NSGWO algorithm were better than those of the other algorithms. Moreover, we found that the optimal-compromise solution of the NSGWO algorithm was able to dominate the solutions of the NSGA-II algorithm.

Figure 5 shows the Pareto fronts of the NSGWO algorithm and the MOGWO algorithm under the combined dispatch of economy and speediness. Here, a speediness index was employed to better present the optimal solutions that was defined as the square of the adjustment time. It can be seen that the optimization results of the NSGWO algorithm were more widely distributed than those of the MOGWO algorithm. The solution that dominated the optimization results of the MOGWO algorithm was found in the Pareto fronts of the NSGWO algorithm. In general, this indicated that the performance of the NSGWO algorithm in the load-dispatch optimization of economy and speediness was the best among the simulated algorithms.

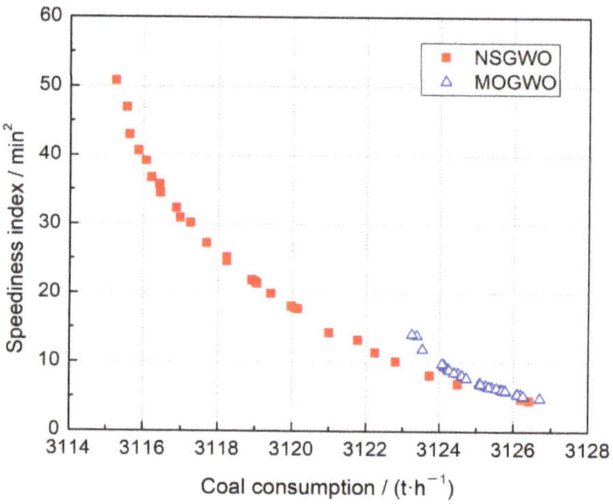

Figure 5. Pareto fronts of coal consumption and speediness index of load-dispatch optimization for 10 units with various power scales under NSGWO and MOGWO algorithms.

4.2.3. Optimization of Economy and Speediness for Similar-Scale Units

Moreover, a real coal-fired power plant with eight 600 MW units and two 660 MW units was investigated under a renewable-energy accommodation, which may further explain the performance of the NSGWO algorithm in the load dispatch. These 10 units had similar power scales, so a higher optimization accuracy was needed in the algorithm. The coal-consumption characteristics, as well as other initial conditions and constraints, are shown in Table 10, in which the unit for coal consumption is grams of coal per kWh. In addition, the penalty-factor values of the two-dimensional targets of economy and speediness in the algorithms were 0.5 and 1.1, respectively, and the total power command of this 10-unit power plant was 4000 MW.

Table 10. Characteristics of 10 dispatched units based on economy and speediness with 10 similar-scale units.

Unit	Coal-Consumption Characteristics	P_{now}/MW	v_i/(MW min $^{-1}$)	P_i^{min}/MW	P_i^{max}/MW
1	$f_1(P_1) = 0.000128P_1^2 - 0.10413P_1 + 361.508$	409.164	10	240	600
2	$f_2(P_2) = 0.000118P_2^2 - 0.10897P_2 + 360.777$	343.598	10	240	600
3	$f_3(P_3) = 0.000104P_3^2 - 0.11813P_3 + 348.179$	321.472	12	240	600
4	$f_4(P_4) = 0.000115P_4^2 - 0.11699P_4 + 350.690$	368.000	10	240	600
5	$f_5(P_5) = 0.000190P_5^2 - 0.12183P_5 + 352.533$	359.787	10	240	600
6	$f_6(P_6) = 0.000154P_6^2 - 0.11906P_6 + 350.421$	420.531	8	240	600
7	$f_7(P_7) = 0.000150P_7^2 - 0.11031P_7 + 352.498$	362.807	8	240	600
8	$f_8(P_8) = 0.000144P_8^2 - 0.10936P_8 + 346.564$	333.882	10	240	600
9	$f_9(P_9) = 0.000272P_9^2 - 0.12038P_9 + 342.560$	400.494	5	264	660
10	$f_{10}(P_{10}) = 0.000292P_{10}^2 - 0.12938P_{10} + 340.501$	427.870	5	264	660

Table 11 shows the optimization results for the real coal-fired power plant with 10 similar-scale units. It indicates that the minimum coal consumption of NSGWO was 328 g/(kWh), which was better than that of MOGWO. The minimum adjustment time of NSGWO was 1.36 min, which was also shorter than that of MOGWO. Furthermore, the solutions of the MOGWO algorithm for these two minima were dominated by the NSGWO algorithm, which meant that the distribution of the solutions of the NSGWO algorithm was better than that of the MOGWO algorithm.

Table 11. Optimization results of load dispatch under economy and speediness strategies for a real coal-fired power plant with 10 similar-scale units.

Item	Optimal-Compromise Solution		Economy Strategy		Speediness Strategy	
	NSGWO	MOGWO	NSGWO	MOGWO	NSGWO	MOGWO
$f_w/(\text{gkWh}^{-1})$ [1]	329.70	330.34	328.02	328.19	331.30	331.71
T/min [1]	11.16	8.17	22.85	23.21	5.32	6.68
P_1/MW	378.15	405.97	302.75	323.48	434.55	385.14
P_2/MW	394.44	420.79	353.30	415.80	379.64	376.09
P_3/MW	455.35	419.57	595.65	600.00	385.35	401.64
P_4/MW	462.83	414.01	527.16	543.27	413.77	412.89
P_5/MW	353.74	351.06	382.36	309.99	359.38	372.00
P_6/MW	450.43	438.11	447.16	427.13	455.26	465.65
P_7/MW	368.90	371.26	376.00	350.80	383.84	379.13
P_8/MW	385.39	408.06	383.56	347.93	371.17	384.94
P_9/MW	366.62	376.87	310.88	336.77	392.15	393.99
P_{10}/MW	383.84	394.51	321.08	344.28	425.31	421.73

[1] Variables f_w and T represent coal consumption per kWh and adjustment time, respectively.

Figure 6 illustrates the Pareto fronts of coal consumption and adjustment time of the former simulations for NSGWO and MOGWO. It shows that the optimization results of NSGWO were basically at the lower left of MOGWO, indicating that the optimization effect of NSGWO was better. Most of the solutions of the MOGWO algorithm are concentrated in the upper left, while the solutions of the NSGWO algorithm are more uniformly distributed. This indicates that the solutions of the NSGWO algorithm had a better distribution in the combined load dispatch while considering the requirements of both economy and speediness for coal-fired power plants.

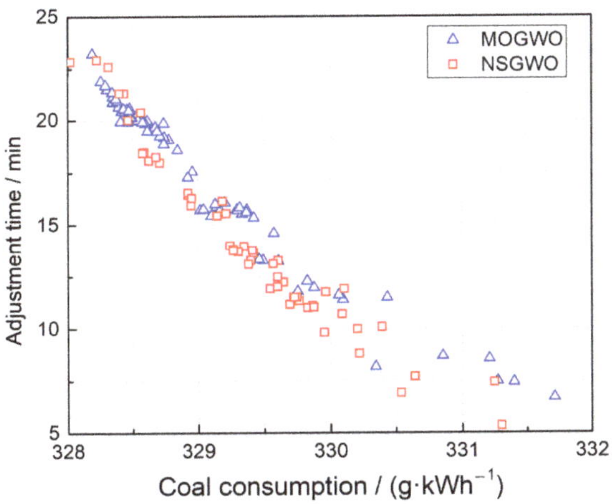

Figure 6. Pareto fronts of coal consumption and adjustment time of load-dispatch optimization for coal-fired power plant with 10 similar-scale units under NSGWO and MOGWO algorithms.

The results of this study showed that the proposed nondominated-sorting GWO algorithm achieved high accuracy and had a good distribution of solutions, which may extend its applications in multiobjective optimization. We caution that the presented analyses for the NSGWO and other intelligent algorithms were somewhat simplified, and further investigations are needed. The optimization procedure proposed in this study may help researchers conduct other investigations into the optimization of multiobjective problems and the operation of power-generation systems.

5. Conclusions

Multiobjective optimizations of load dispatch under a renewable energy accommodation were conducted based on the NSGWO algorithm. On the basis of the performed analyses, the conclusions of this paper can be summarized as follows:

(1) The proposed NSGWO algorithm employed efficient nondominated sorting, a reference point selection strategy, and a simulated binary crossover operator in order to avoid falling into the local optimum, enhance the solution diversity, and shorten the convergence time. The optimization results of benchmark functions indicated that the NSGWO algorithm had a better accuracy and a better distribution than the MOGWO algorithm.

(2) Regarding the load dispatch of economy and environmental-protection strategies, the performance of the NSGWO algorithm was better than those of the other simulated algorithms in reducing power-plant coal consumption and emissions. In addition, the widely distributed solutions of the NSGWO algorithm could dominate those of the MOGWO algorithm, which elucidated the application value of the NSGWO algorithm in the multiobjective load dispatch of coal-fired power plants.

(3) The NSGWO algorithm could achieve lower coal consumption and a shorter adjustment time than the MOGWO algorithm could for coal-fired power plants with various power-scale units and similar-scale units. The optimal-compromise solutions of the NSGWO algorithm had a better distribution in the combined load dispatch of economy and speediness strategies; thus, it is favorable for coal-fired power plants to accommodate renewable energy.

Author Contributions: Conceptualization, Y.C. and F.S.; methodology, Y.C. and T.L.; software, T.L. and T.H.; validation, T.H.; formal analysis, Y.W. and M.L.; investigation, Y.C. and T.L.; writing—original draft preparation, Y.C.; writing—review and editing, T.H. and F.S. All authors have read and agreed to the published version of the manuscript.

Funding: This research was funded by the Natural Science Foundation of Jiangsu Province (grant number BK20210240).

Institutional Review Board Statement: Not applicable.

Informed Consent Statement: Not applicable.

Data Availability Statement: The data presented in this study are available upon request from the corresponding author.

Conflicts of Interest: The authors declare no conflict of interest.

References

1. Yao, X.; Zhang, M.; Kong, H.; Lyu, J.; Yang, H. Investigation and control technology on excessive ammonia-slipping in coal-fired plants. *Energies* **2020**, *13*, 4249. [CrossRef]
2. Zhu, M.; Xiao, W.; Shen, J.; Lee, K. Dynamic modeling, validation and analysis of direct air-cooling condenser with integration to the coal-fired power plant for flexible operation. *Energy Conv. Manag.* **2021**, *245*, 114601. [CrossRef]
3. Chen, X.; Lin, B. Towards carbon neutrality by implementing carbon emissions trading scheme: Policy evaluation in China. *Energy Policy* **2021**, *157*, 112510. [CrossRef]
4. Li, Y.; Wang, C.; Li, G.; Chen, C. Optimal scheduling of integrated demand response-enabled integrated energy systems with uncertain renewable generations: A Stackelberg game approach. *Energy Conv. Manag.* **2021**, *235*, 113996. [CrossRef]
5. Hou, H.; Xue, M.; Xu, Y.; Tang, J.; Zhu, G.; Liu, P.; Xu, T. Multiobjective joint economic dispatching of a microgrid with multiple distributed generation. *Energies* **2018**, *11*, 3264. [CrossRef]
6. Dai, B.; Wang, F.; Chang, Y. Multi-objective economic load dispatch method based on data mining technology for large coal-fired power plants. *Control Eng. Pract.* **2022**, *121*, 105018. [CrossRef]
7. Ishraque, M.F.; Shezan, S.A.; Ali, M.M.; Rashid, M.M. Optimization of load dispatch strategies for an islanded microgrid connected with renewable energy sources. *Appl. Energy* **2021**, *292*, 116879. [CrossRef]
8. Cui, H.; Xia, W.; Yang, S.; Wang, X. Real-time emergency demand response strategy for optimal load dispatch of heat and power micro-grids. *Int. J. Electr. Power Energy Syst.* **2020**, *121*, 106127. [CrossRef]
9. Li, L.; He, D.; Jin, J.; Yu, B.; Gao, X. Multi-objective load dispatch control of biomass heat and power cogeneration based on economic model predictive control. *Energies* **2021**, *14*, 762. [CrossRef]

0. Xu, J.; Gu, Y.; Chen, D.; Li, Q. Data mining based plant-level load dispatching strategy for the coal-fired power plant coal-saving: A case study. *Appl. Therm. Eng.* **2017**, *119*, 553–559. [CrossRef]
1. Liao, S.; Zhao, H.; Li, G.; Liu, B. Short-term load dispatching method for a diversion hydropower plant with multiple turbines in one tunnel using a two-stage model. *Energies* **2019**, *12*, 1476. [CrossRef]
2. Jamal, R.; Men, B.; Khan, N.; Raja, M. Hybrid bio-inspired computational heuristic paradigm for integrated load dispatch problems involving stochastic wind. *Energies* **2019**, *12*, 2568. [CrossRef]
3. Bie, P.; Zhang, B.; Lu, J.; Li, H.; Wang, Y.; Luan, L.; Chen, G. Chance-constrained real-time dispatch with renewable uncertainty based on dynamic load flow. *Energies* **2017**, *10*, 2111. [CrossRef]
4. Ghasemi, M.; Taghizadeh, M.; Ghavidel, S.; Abbasian, A. Colonial competitive differential evolution: An experimental study for optimal economic load dispatch. *Appl. Soft. Comput.* **2016**, *40*, 342–363. [CrossRef]
5. Neto, J.X.V.; Reynoso-Meza, G.; Ruppel, T.H.; Mariani, V.C.; Coelho, L.S. Solving non-smooth economic dispatch by a new combination of continuous GRASP algorithm and differential evolution. *Int. J. Electr. Power Energy Syst.* **2017**, *84*, 13–24. [CrossRef]
6. Secui, D.C. A modified Symbiotic Organisms Search algorithm for large scale economic dispatch problem with valve-point effects. *Energy* **2016**, *113*, 366–384. [CrossRef]
7. Kavousi-Fard, A.; Khosravi, A. An intelligent θ-modified bat algorithm to solve the non-convex economic dispatch problem considering practical constraints. *Int. J. Electr. Power Energy Syst.* **2016**, *82*, 189–196. [CrossRef]
8. Chen, M.; Zeng, G.; Lu, K. Constrained multi-objective population extremal optimization based economic-emission dispatch incorporating renewable energy resources. *Renew. Energy* **2019**, *143*, 277–294. [CrossRef]
9. Hosseinnezhad, V.; Babaei, E. Economic load dispatch using θ-PSO. *Int. J. Electr. Power Energy Syst.* **2013**, *49*, 160–169. [CrossRef]
20. Narimani, H.; Razavi, S.; Azizivahed, A.; Naderi, E.; Fathi, M.; Ataei, M.H.; Narimani, M.R. A multi-objective framework for multi-area economic emission dispatch. *Energy* **2018**, *154*, 126–142. [CrossRef]
21. Zou, D.; Li, S.; Li, Z.; Kong, X. A new global particle swarm optimization for the economic emission dispatch with or without transmission losses. *Energy Conv. Manag.* **2017**, *139*, 45–70. [CrossRef]
22. Mandal, K.K.; Mandal, S.; Bhattacharya, B.; Chakraborty, N. Non-convex emission constrained economic dispatch using a new self-adaptive particle swarm optimization technique. *Appl. Soft. Comput.* **2015**, *28*, 188–195. [CrossRef]
23. Mirjalili, S.; Mirjalili, S.M.; Lewis, A. Grey Wolf Optimizer. *Adv. Eng. Softw.* **2014**, *69*, 46–61. [CrossRef]
24. Zhang, S.; Zhou, Y.; Li, Z.; Pan, W. Grey wolf optimizer for unmanned combat aerial vehicle path planning. *Adv. Eng. Softw.* **2016**, *99*, 121–136. [CrossRef]
25. Madhiarasan, M.; Deepa, S.N. Long-term wind speed forecasting using spiking neural network optimized by improved modified grey wolf optimization algorithm. *Int. J. Adv. Res.* **2016**, *4*, 356–368. [CrossRef]
26. Saremi, S.; Mirjalili, S.Z.; Mirjalili, S.M. Evolutionary population dynamics and grey wolf optimizer. *Neural Comput. Appl.* **2015**, *26*, 1257–1263. [CrossRef]
27. Long, W.; Liang, X.; Cai, S.; Jiao, J.; Zhang, W. A modified augmented Lagrangian with improved grey wolf optimization to constrained optimization problems. *Neural Comput. Appl.* **2016**, *28*, 421–438. [CrossRef]
28. Zhang, S.; Luo, Q.; Zhou, Y. Hybrid grey wolf optimizer using elite opposition-based learning strategy and simplex method. *Int. J. Comput. Intell. Appl.* **2017**, *16*, 1750012. [CrossRef]
29. Singh, N.; Hachimi, H. A new hybrid whale optimizer algorithm with mean strategy of grey wolf optimizer for global optimization. *Math. Comput. Appl.* **2018**, *23*, 14. [CrossRef]
30. Zhang, X.; Kang, Q.; Cheng, J.; Wang, X. A novel hybrid algorithm based on biogeography-based optimization and grey wolf optimizer. *Appl. Soft. Comput.* **2018**, *67*, 197–214. [CrossRef]
31. Song, H.M.; Sulaiman, M.H.; Mohamed, M.R. An application of grey wolf optimizer for solving combined economic emission dispatch problems. *Int. Rev. Mod. Simul.* **2014**, *7*, 838–844.
32. Wong, L.I.; Sulaiman, M.H.; Mohamed, M.R. Solving economic dispatch problems with practical constraints utilizing grey wolf optimizer. *Appl. Mech. Mater.* **2015**, *785*, 511–515. [CrossRef]
33. Moradi, M.; Badri, A.; Ghandehari, R. Non-convex constrained economic dispatch with valve point loading effect using a grey wolf optimizer algorithm. In Proceedings of the 2016 6th Conference on Thermal Power Plants (CTPP), Tehran, Iran, 19–20 January 2016; pp. 96–104.
34. Jangir, P.; Jangir, N. A new non-dominated sorting grey wolf optimizer (NS-GWO) algorithm: Development and application to solve engineering designs and economic constrained emission dispatch problem with integration of wind power. *Eng. Appl. Artif. Intell.* **2018**, *72*, 449–467. [CrossRef]
35. Bosman, P.; Thierens, D. The balance between proximity and diversity in multiobjective evolutionary algorithms. *IEEE Trans. Evol. Comput.* **2003**, *7*, 174–188. [CrossRef]
36. Das, I. Normal-boundary intersection: A new method for generating the Pareto surface in nonlinear multicriteria optimization problems. *Soc. Ind. Appl. Math.* **1998**, *8*, 631–657. [CrossRef]
37. Hemamalini, S.; Simon, S.P. Emission constrained economic dispatch with valve-point effect using particle swarm optimization. In Proceedings of the TENCON 2008-2008 IEEE Region 10 Conference, Hyderabad, India, 19–21 November 2008; pp. 1500–1505.
38. Wang, Z.; Liu, J.; Tan, W.; Yang, G. Multi-objective optimal load distribution based on speediness and economy in power plants. *Proc. Chin. Soc. Electr. Eng.* **2006**, *26*, 86–92.

energies

MDPI

Article

Reserve-Constrained Unit Commitment Considering Adjustable-Speed Pumped-Storage Hydropower and Its Economic Effect in Korean Power System

Woo-Jung Kim [1], Yu-Seok Lee [1], Yeong-Han Chun [1,*] and Hae-Seong Jeong [2]

[1] Department of Electronic and Electrical Engineering, Hongik University, Seoul 04066, Korea; dnwnd9581@mail.hongik.ac.kr (W.-J.K.); lys9392601@mail.hongik.ac.kr (Y.-S.L.)

[2] Master's Space, Gwangmyeong-si 14348, Korea; econohs@masterspace.co.kr

* Correspondence: yhchun@hongik.ac.kr

Abstract: The Korean government has declared the goal of net-zero-carbon emissions with a focus on renewable energy expansion. However, a high proportion of baseload generators and an increasing proportion of variable renewable energy (VRE) may cause problems in the power system operation owing to the low cycling capability of baseload generators and variability of VRE. To maintain system reliability, the government is planning to construct pumped-storage hydropower (PSH) plants, which can provide flexibility to the system. This study evaluated the operating cost savings obtained by different types of PSH: the adjustable-speed PSH (AS-PSH) and fixed-speed PSH (FS-PSH), based on the duck-curve phenomenon and the increase in spinning reserve requirement. In this study, the reserve-constrained unit commitment was formulated using a mixed-integer-programming considering the operational characteristics of AS-PSH and conventional generators. To consider the duck-shaped net-load environment, the projected VRE output data were calculated through physical models of wind turbines and photovoltaic modules. The operating costs for the non-PSH, FS-PSH, and AS-PSH construction scenarios were KRW 43,129.38, 40,038.44, and 34,030.46, respectively. The main factor that derived this difference was determined to be the primary reserve of AS-PSH's pumping mode.

Keywords: adjustable-speed pumped-storage hydropower; fixed-speed pumped-storage hydropower; variable renewable energy; duck-curve; reserve-constrained unit commitment; spinning reserve requirement; nuclear power generators

Citation: Kim, W.-J.; Lee, Y.-S.; Chun, Y.-H.; Jeong, H.-S. Reserve-Constrained Unit Commitment Considering Adjustable-Speed Pumped-Storage Hydropower and Its Economic Effect in Korean Power System. *Energies* **2022**, *15*, 2386. https://doi.org/10.3390/en15072386

Academic Editor: Massimiliano Renzi

Received: 3 January 2022
Accepted: 18 March 2022
Published: 24 March 2022

1. Introduction

1.1. Research Motivation

The Korean government has declared the goal of achieving net-zero carbon emissions by 2050. Therefore, Korea is planning to significantly increase the proportion of renewable energy resources. In the generation expansion plan, which determines the Korean investment in power generation facilities, 2030 was set as the target year to ensure that 20% of the annual power is generated by renewable energy resources [1]. Among the renewable energy resources, 75% is expected to be generated from wind and solar energy, which are variable renewable energy (VRE) resources. However, VRE resources may cause difficulties in balancing the supply and demand within a power system owing to their output variability and uncertainty [2]. Specifically, solar power generation makes the system net-load have the characteristics of a duck-curve, which is a concept first introduced by the California Independent System Operator [3]. It has a duck-shaped net-load pattern, which is carried by conventional generators. The VRE's long-term and short-term variability require increased cycling capabilities of the generator and operating reserves, respectively. Therefore, it is difficult for the system operator to maintain system reliability while maintaining the operating proportion of baseload generators during the daytime.

In 2020, the first output reduction of nuclear power generators was conducted by the system operator to maintain reliability in the Korean power system. Shinkori nuclear power generators (#3 and #4) have a total capacity of 2800 MW. The power system operator was scheduled to reduce the total output power to 600 MW for 8 h, and the scheduled output power was maintained for 13 h. The system operator predicted that the system demand was lower than usual owing to long special holidays and analyzed that it would be difficult to meet the system frequency maintenance range if a nuclear power generator with the maximum output (1400 MW) is tripped. There are similar issues with implementing net-zero carbon emissions. Although the coal-fired and nuclear power generators are expected to gradually phase out, the proportion of these baseload generators is still expected to be high in 2030. It is especially difficult to expect frequent output adjustments or startup/shutdown operations with nuclear power generators. In addition, existing nuclear power generators in Korea do not provide frequency response capabilities. Accordingly, situations can occur where several nuclear power generators would need to be operated at a minimum generation level or even excluded from scheduling. In those cases, the operating cost of the power system can increase significantly, and the existing baseload generators may even become stranded assets.

Grid-scale energy storage systems, such as battery energy storage systems (BESS) and pumped-storage hydropower (PSH), can provide solutions to operating costs by mitigating the effect of the duck-curve phenomenon of the net-load and increasing the utilization of nuclear power generators. Although BESSs are capable of providing load shifting and fast frequency responses, PSH is technically proven and more desirable in that it has a large energy capacity and the capability of providing operating reserves. The Korean government is planning to build PSH plants [1]. However, there are three types of PSH plants: fixed-speed (FS), adjustable-speed (AS), and ternary (T) types. As a quick-start resource, a conventional FS-PSH can be rapidly deployed to the system in case of a generator's forced outage or insufficient reserve [4]. AS-PSHs and T-PSHs can provide the same operational advantages as FS-PSHs. In addition, AS-PSH can control its rotor speed using power electronic devices, which enables a rapid power adjustment in the generating and pumping modes and is therefore expected to further improve the flexibility of the power system. Given that the pump and turbine of T-PSH can rotate simultaneously on the same shaft in the same direction, it has the advantage of adjusting the pumping power in the hydraulic short-circuit mode and requires less time to transition between the generating and pumping modes [5]. Therefore, we needed to decide which type will be of benefit to our system when they are constructed, which was the motivation of our research. In this paper, we focused on the economic effect obtained by AS-PSH using the unit commitment (UC) optimization approach in a future Korean power system. In the case of T-PSH, there are three modes: generating mode, pumping mode, and short-circuit mode, which each have different operational characteristics. Therefore, future research is needed to model the operational characteristics of T-PSH in UC formulation.

1.2. Related Research

Typically, PSH charges at night when the system demand is low and discharges during peak-demand hours. However, owing to the large-scale integration of VREs, the economic benefits that can be obtained from the PSH have been studied from new perspectives [6–12]. In [6], the impact of reducing the operating cost achieved by the input of FS-PSH in a power system with high wind energy was analyzed, and FS-PSH was found to reduce the scheduling cost by reducing the startup and shutdown costs of thermal power generators. In [7], the effect of reducing operating costs through FS-PSH in a wind energy environment was analyzed through a genetic algorithm. The paper derived the reduction in energy purchase cost and environmental cost of the power system. In [8], it was found that as the initial upper reservoir volume of PSH increases, the overall operating cost consisting of grid power purchase cost, demand response program purchase cost, and solar power sale income decreases. In [9], the effect of reduction in the wind power curtailment and the total operating cost consisting of energy cost, ancillary service cost, and startup cost were derived using FS-PSH as a reserve resource. In [10], the energy cost in a standalone hybrid wind/solar/biomass system was lower in the scenario when PSH was the input than when battery storage was the input. In [11], the authors evaluated the potential contribution of PSH to the Greek power system's future scenario. Through UC simulation, it was deduced that the PSH contributes to energy balance by absorbing the over-generation of wind and solar energy. Ref. [12] proposes a stochastic scheduling approach considering price-based demand response and PSH in a microgrid. In [12], the operating cost was reduced by PSH's discharging at the peak time and charging at the off-peak time. However, given the proposed methods in [6–12], it is difficult to verify the economic contribution obtained from the flexibility of AS-PSH.

The reduction in operating costs due to the flexibility of AS-PSH was studied in [13–15]. In [13], the dynamic dispatch optimization problem was solved with the reserve and network constrained in a wind energy environment, and an economic analysis was performed on FS-PSH, T-PSH, and AS-PSH. However, in [13], it was difficult to verify probable changes in operating conditions within the duck-curve phenomenon, given that the influence on solar energy was not considered. In [14], the economic impact of AS-PSH was analyzed under the conditions of minimizing fuel, cycling, and startup cost. It was modeled as having an operating range in the pumping mode and was derived to reduce the operating cost. In [15], the cost benefits obtained with AS-PSH from the viewpoint of daily operation were compared with those of FS-PSH. However, the study focused on the AS-PSH's flexible operating range in pumping. Therefore, it is still necessary to focus on the reserve capability of AS-PSH that can respond to VRE's short-term variations.

In addition to PSH-related studies, UC optimization research is ongoing from various perspectives. Authors in [16–18] developed methods to efficiently find the UC solution at a high time resolution to cope with the variability and intermittent nature of renewable energy. Ref. [16] proposes a thermal unit commitment algorithm through dynamic programming that combines a priority list method to perform UC and ED at a time resolution of less than 60 min. In [17], the authors pointed out that the branch and cut strategy may exceed the allowable value in the sub-hourly UC problem due to large numbers of virtual transactions. In [17], the authors reduced the calculation time by including the concept of ordinal-optimization in surrogate absolute-value Lagrangian relaxation [19]. Ref. [18] proposes an approach to reduce calculation time by implementing soft constraints of reserve and transmission capacity within the surrogate absolute-value Lagrangian relaxation framework. References [20,21] analyzed the impact of greenhouse gas emissions on the UC problem. In [20], the authors formulated the environmental UC problem considering greenhouse gas emission constraints and carbon taxes in the existing UC approach. In [21], the authors adopted the piecewise linearized greenhouse gas function of the generator's loading level to the UC based on the MIP method. The greenhouse gas function was applied in the form of cost by applying a weighting factor.

1.3. Contribution

Previous studies on PSH verified its effects on the operating cost savings of power systems with integrated VRE and PSH. However, the majority of these studies focused on the traditional FS-PSH's contribution to power systems in terms of energy saving. The other studies [13–15] on AS-PSH did not consider its capability to supply primary and secondary reserves, particularly in the pumping mode. In a system with a considerable amount of solar energy resources, system operators need large operating reserves, and most generation sources need to be off-lined owing to an over-generation; on the other hand, most PSHs are required to operate in pumping mode. Therefore, to analyze the economic contribution of AS-PSH in a power system with a high proportion of renewable energy resources, it is necessary to model the AS-PSH to provide various reserves, including the primary reserve. Based on the above discussion, the purpose of this study is to verify the economic contribution of AS-PSH's reserve capability compared with that of FS-PSH in the Korean power system.

To achieve the objective of this study, a reserve-constrained UC (RCUC) was formulated using mixed-integer-programming (MIP) to consider the cycling characteristics of conventional thermal generators in the duck-curve environment and reflect the Korean operating reserve services. The primary, secondary, and regulation reserves defined by the Korean power market were modeled in the RCUC formulation. This formulation can be widely used in other power systems with a similar scheme of system frequency restoration by primary and secondary reserves. The time-series data of the VRE were calculated through physical models, including the output model of wind turbines and photovoltaic (PV) modules. The short-term variation of the VRE was extracted from the time-series data. The increment in the spinning reserves was considered under the n-sigma criterion, assuming that the distribution of the VRE's short-term output variation follows a Gaussian distribution. With the operation range set in the generating and pumping modes, the AS-PSH was modeled to provide primary, secondary, and regulation reserves. Simulations were performed on the 2030 generation mix in Korea. Scenarios were set depending on the types of new PSH plants, namely, non-PSH, FS-PSH, and AS-PSH plants. The simulation was implemented using the optimization solver Xpress-MP.

The main contributions in this paper are listed below:

- An MIP-based RCUC problem was formulated. FS-PSH was modeled to provide the primary and secondary/regulation reserves in the generation mode, while AS-PHS was modeled to provide the primary and the secondary/regulation reserve both in the generating and pumping modes.
- A comparative analysis was performed on the Korean power system considering the generation mix in 2030 using the developed RCUC. The Korean power system is expected to consist of more than 200 large generators, including eighteen nuclear power generators with a capacity of 20.4 GW and a peak demand of 110 GW in 2030.
- The comparative study showed that the AS-PSH was superior to the FS-PSH in terms of operating costs, which are mainly dependent on the procurement of primary and secondary reserves. It is partly because the system is isolated and partly because the share of non-flexible nuclear power generators is high. We verified that the most critical factor is the capability of supplying the primary reserve.

2. Variable Renewable Energy in Future Korean Power System

Before analyzing the influence of the new PSH on the future power system in Korea, a methodology to generate the time-series data of the 2030 VRE is introduced. Based on the data, the increment in the spinning reserve due to the short-term variation of the VRE was estimated.

2.1. *Modeling of Variable Renewable Energy's Output Power*

The 2030 target capacity of the VRE is about 54.2 GW, and wind and solar energy account for 17.7 GW and 36.5 GW, respectively [1]. Based on the capacity of the VRE to be installed, the 2030 output power can be calculated using historic regional meteorological data. From the Korea Meteorological Administration (KMA), the regional wind speed, solar irradiance, and temperature data measured every minute were used [22].

2.1.1. Wind Energy Output Power Model

To calculate the total output power of wind energy in 2030, it is necessary to determine the location and capacity of each wind farm. To solve the problem, it was assumed that the total capacity of wind energy in which the location is not determined would be distributed to areas with high energy potentials. Therefore, the capacity was distributed proportionally to the locations with high average annual wind speed. Finally, a total of 94 locations of the 2030 wind farms and their capacity were determined.

In a wind farm, each wind turbine is installed at a distance. Therefore, input data conversion was performed without applying the same wind speed to all wind turbines in the same wind farm. Each wind farm consists of N turbine units, and the wind speed after $(n-1)$ minutes is applied to calculate the output power of the nth wind turbine at time t.

$$P_{WF}(t) = \sum_{n=1}^{N} P_{WT,n}(t) = \sum_{n=1}^{N} P_{WT,n}(V_h(t-n+1)) \tag{1}$$

To calculate the output power of the wind turbine, Equations (2) and (3) were used [23]. For the convenience of calculation, each wind turbine group was modeled with one unit. The data in Table 1 were used for the specifications of onshore and offshore wind turbines [24].

$$P_{WT,n}(t) = \begin{cases} 0 & for\ V_h(t-n+1) < V_{in}\ and\ V_h(t-n+1) > V_{out}, \\ P_{curvefit}(V_h(t-n+1))\cdot\frac{C_{WF}}{P_r\cdot N} & for\ V_{in} \le V_h(t-n+1) \le V_r, \\ \frac{C_{WF}}{N} & for\ V_r < V_h(t-n+1) \le V_{out}. \end{cases} \tag{2}$$

$$P_{curvefit}(V_h(t)) = a_1 V_h(t)^6 + a_2 V_h(t)^5 + a_3 V_h(t)^4 + a_4 V_h(t)^3 + a_5 V_h(t)^2 + a_6 V_h(t) + a_7 \tag{3}$$

Table 1. Parameters of wind turbines (onshore and offshore).

	Onshore	Offshore
Rated power	3 MW	8 MW
Cut-in wind speed	3 m/s	3.5 m/s
Rated wind speed	10 m/s	10 m/s
Cut-out wind speed	20 m/s	25 m/s
Tower hub height	120 m	130 m

Given that wind speed increases with an increase in altitude, the wind speed used as input data for Equations (2) and (3) should be measured at the tower height. However, the wind speed data of the KMA are measured at an altitude of 10 m. Therefore, using Equation (4) based on the wind power law, we estimated wind speed data at the tower height [25].

$$V_h = V_m \left(\frac{Z_h}{Z_m} \right)^{\alpha} \tag{4}$$

2.1.2. Solar Energy Output Power Model

By 2030, 36.5 GW of PV plants is expected to be installed. As in the case of wind power, it is necessary to determine the capacity and the location of each PV plant. However, most PV plants are expected to be installed at a small scale and dispersed over the nation. Therefore, it is necessary to estimate insolation data for areas where KMA does not observe the insolation. To address this problem, we expanded the number of observation stations using a two-dimensional interpolation method [26]. The capacity of each PV plant was assigned in proportion to historical accumulated solar insolation.

The solar insolation data were converted to the irradiance data. The PV output power, which is determined by the irradiance and temperature on the PV module surface, can be expressed as Equation (5) [27].

$$P_{PV,M} = P^{*}_{PV,M}\cdot\frac{G}{G^{*}}[1-\gamma\cdot(T_C-25)] \tag{5}$$

Given that each PV plant consists of PV arrays of several PV modules, the output power of each PV plant can be calculated using Equation (6).

$$P_{PV} = P_{PV,M}\cdot C_{PV}/P^{*}_{PV,M} \tag{6}$$

Given that the temperature on the PV module surface used in Equation (5) is difficult to measure, it can be approximated using Equation (7) based on the irradiance and the air temperature [28].

$$T_C = T_A + \frac{T_{NOCT} - 20}{800} G \tag{7}$$

2.2. Characteristics of Variable Renewable Energy in Korea

In this subsection, the output power characteristic of the 2030 VRE is described. Figure 1 depicts the calculated time-series output power of the VRE and its duration curve for the year 2030.

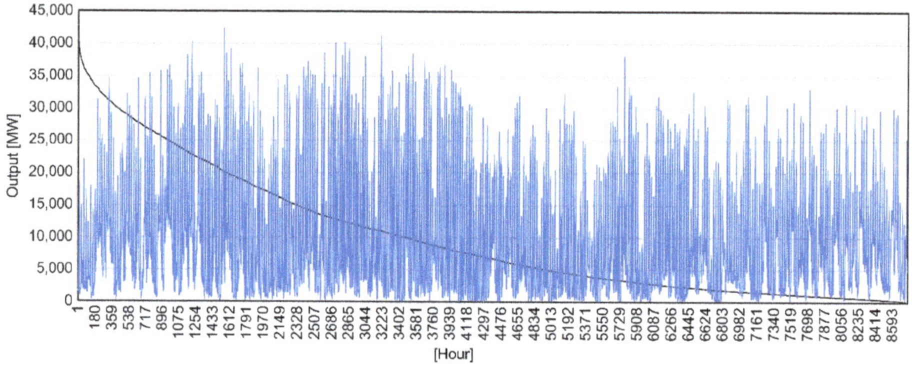

Figure 1. Output and duration curve of VRE for 2030.

Figure 2 presents the average daily wind and solar power generation of each month. The average daily wind power generation is highest in winter and lowest in summer. The average daily solar power generation is highest in spring and lowest in winter. Based on these results, the average daily VRE power generation is highest in spring and lowest in autumn.

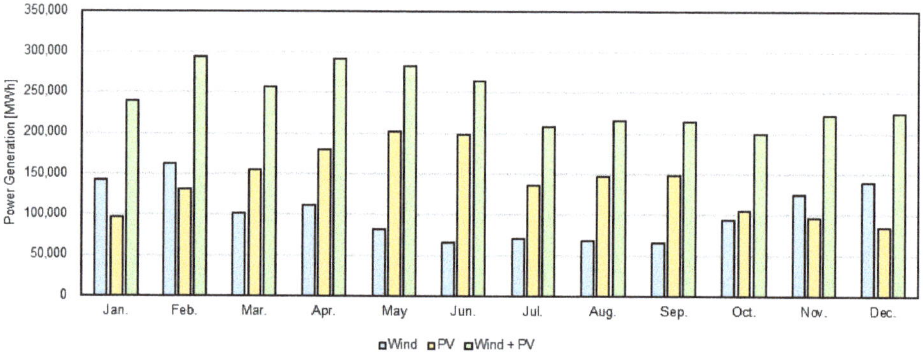

Figure 2. Average daily VRE power generation of each month for 2030.

Figure 3 illustrates the hourly average output power for wind and solar energy throughout the year 2030. Wind energy demonstrates the highest output power between 14:00 and 15:00, and the highest PV output power and both wind and solar output power are between 12:00 and 13:00.

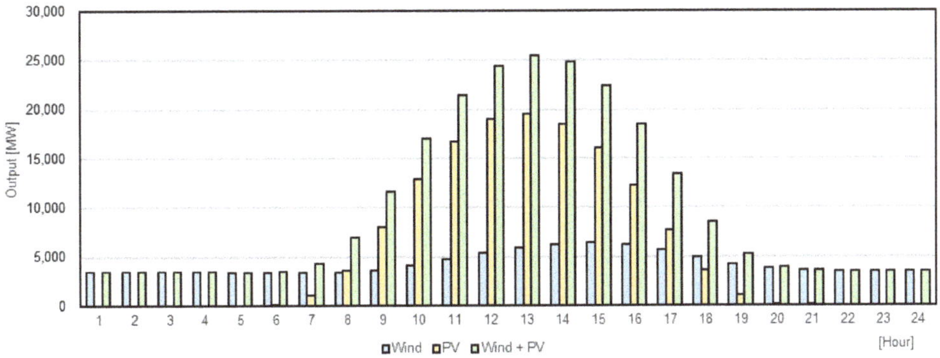

Figure 3. Hourly average output (MW) of wind and PV power projected for 2030.

2.3. Influence of Variable Renewable Energy on Reserve Requirements of the Korean Power System

Traditionally, most power system operators have used the N−1 contingency method, which secures a reserve equal to one or more largest units [29]. The system operator of the Korean power system uses the traditional reserve procurement approach to cover the largest generator outage (1400 MW). Table 2 outlines the operating reserve requirements for the Korean power market [30].

Table 2. Operating reserve requirements for the Korean power system.

Operating Reserves		Requirement	State	Activated by	Secured for
Regulation reserve		700 MW	Spinning	AGC signal	Short-term load variation
Frequency restoration reserves	Primary reserve	1000 MW	Spinning	Governing system	Largest unit loss
	Secondary reserve	1400 MW	Spinning	AGC signal	Largest unit loss
	Tertiary reserve	1400 MW	Standstill	Manual	Reserve restoration
Quick-start reserve		2000 MW	Standstill	Manual	Load forecast error, etc.

The primary reserve prevents the system frequency from dropping immediately after an instantaneous event, such as a generator's forced outage by the synchronized generator's governor response, and the secondary reserve is used to restore the system frequency to the nominal frequency (60 Hz). The required primary reserve in Table 2 was calculated based on the value at the time of a quasi-steady state of the system frequency after the outage of the largest unit (1400 MW) in Korea. In this process, an unused capacity of 1400 MW is required as the secondary reserve to restore the system frequency to the nominal frequency (60 Hz). Immediately after restoration to the nominal frequency by the secondary reserve, the primary reserve is automatically resecured. The regulation reserve is estimated as 700 MW to respond to non-instantaneous events, such as short-term load fluctuations. The tertiary reserve is secured by resources in the standstill state to cover the case of reserve shortages, and the quick-start reserve responds to load forecasting errors.

The unit capacity of VRE resources is lower than that of conventional generators, and the short-term variation of the VRE can influence the system frequency. Therefore, the traditional approach of securing a spinning reserve equal to one or more of the largest units is not effective, and it is necessary to adopt an approach that can cover the short-term variation of the VRE.

Based on the operating reserve concept presented in [2], we divided the VRE output fluctuation into event and non-event cases. The reserve for an event is assumed necessary for instantaneous events, such as generator outages and short-term output variations of VRE. These events can be covered by the governor response of the generator. The non-event reserve can be provided by the generator ramping capability according to AGC signals or manual operation. In this study, a 1 min variation in the VRE was assigned to the primary reserve and a 5 min variation to the secondary and regulation reserve. The 1 min and 5 min variations were calculated using Equations (8) and (9), respectively, based on the VRE output data calculated at 1 min intervals, as presented in the previous subsection.

$$P(t) - P(t-1) = P_{1\text{min}} \quad (8)$$

$$P(t) - P(t-5) = P_{5\text{min}} \quad (9)$$

As outlined in Table 3, the n-sigma criterion was used to calculate the increment in each spinning reserve requirement due to the output fluctuation of the VRE. The standard deviation σ refers to the time series variability of the VRE. Assuming that the output variability of the VRE follows a normal distribution, an additional reserve to cover ±σ of the variability can respond to 68% of the statistical data, a reserve of ±2σ to 95%, and that of ±3σ to 99.7% [31]. Table 3 presents the specifications of the method for allocating additional spinning reserve requirements under the assumption that the current spinning reserve requirement is maintained.

Table 3. Allocation method of the increment amount of reserve requirement.

Reserves	Activated by	Current Standard	Increments of Reserve
Primary reserve	Governing system	Largest unit loss	nσ of one-min. variation of wind and solar output
Secondary reserve	AGC signal	Largest unit loss	nσ of five-min. variation of wind and solar output
Regulation reserve		3σ of five-min. variation of load	

The simple arithmetic addition of the estimated reserve requirement for VRE to the existing reserve requirement for load variability can result in an overestimation, which can be costly. This problem can be solved by using geometric additions [32]. Figure 4 illustrates the method of calculating the primary reserve requirement secured through the governor response, and Figure 5 presents the calculation method for the requirement of the secondary reserve and regulation reserve activated by AGC signals.

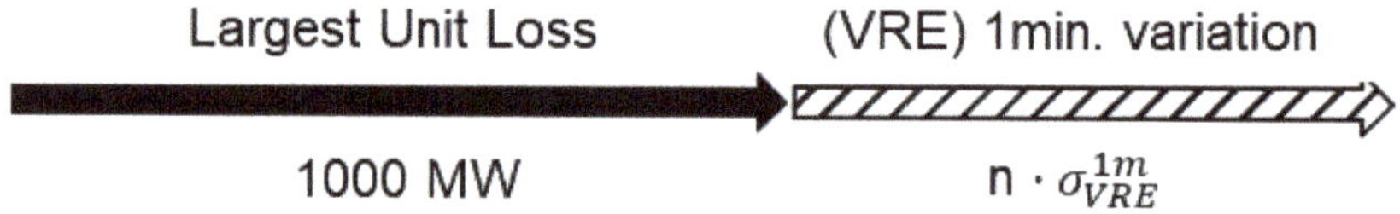

Figure 4. Primary reserve requirement for governor response.

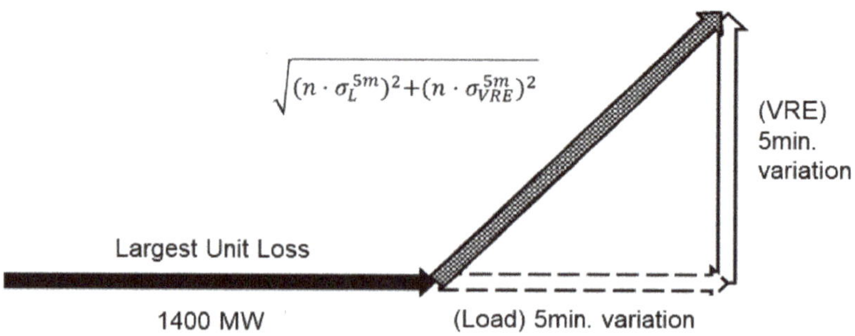

Figure 5. Secondary and regulation reserve requirement for AGC operation.

3. RCUC Considering Adjustable-Speed Pumped-Storage Hydropower

This section introduces the RCUC formulation. Section 3.1 presents the features of AS-PSH. Section 3.2 presents the objective function for the optimization, which is to minimize the operating costs of the power system. Section 3.3 presents the constraints considering AS-PSH's operational characteristics.

3.1. Flexibility of Adjustable-Speed Pumped-Storage Hydropower

AS-PSH is a resource that can provide a higher quality of flexibility to power systems than FS-PSH by adjusting the rotor speed and electrical output/input of the generator/motor. With respect to power system operations, AS-PSH has the following advantages when compared with FS-PSH.

(1) AS-PSH can be operated over a wider operational range in the generating mode, and the input power can be controlled in the pumping mode.
(2) AS-PSH can provide primary, secondary, and regulation reserves in the generating and pumping modes.

The critical factors for the advantages of AS-PSH in power system operation are its operation range and spinning reserve capacity. Therefore, it is necessary to determine the operation range and spinning reserve capacity of AS-PSH. In this study, the operation range and spinning reserve capacity were determined based on an investigation of the technical reports and AS-PSH plants in practice.

The operation range is defined by the difference between the minimum and maximum powers. The operation ranges of FS-PSH and AS-PSH were reviewed in reports [5] and [33], and the comparison is shown in Table 4. However, the operation ranges of practical PSH plants vary. For example, in the Kazunogawa PSH plant [34], Unit #4 is an AS-type, and Units #1 and #2 are FS-type. In the generating mode, the operation range of Unit #1 is wider than that of Units #1 and #2 by 32.5%. In the pumping mode, the operation range of Unit #1 is approximately 32%. In this study, the operation range of future AS-PSH plants was set as 30–100% in the generating mode and 60–100% in the pumping mode.

Table 4. Example of pumped-storage hydropower (PSH) operation range.

Operation Range		FS-PSH	AS-PSH
Generating mode	Argonne Report [5]	16–100%	16~100%
	JICA Report [33]	30–100%	30~100%
Pumping mode	Argonne Report	60–100%	60~100%
	JICA Report	70–100%	70~100%

The primary reserve keeps the system frequency within operation range directly after a disturbance. Therefore, the capacity with which AS-PSH can instantaneously control the power within a few seconds was assumed to be the primary reserve. The range of the primary reserve was set based on the example of the AS-PSH in practical operation. Unit #4 of the Okawachi PSH in Japan is capable of a step response output change of 32 MW (10% of the maximum rated power) within 0.2 s in the generating mode, and 80 MW (20% of the maximum rated power) within 0.2 s in the pumping mode [35]. Accordingly, we set the range of the primary reserve to be secured by AS-PSH to 10% of the maximum rated power in the generating mode and 20% in the pumping mode. The range of secondary/regulation reserves can be set based on the capacity to control the power for several minutes. Given that AS-PSH has a high ramping capability, the maximum capacity of the secondary/regulation reserves was set as the difference between the minimum and maximum power. The operation and reserve range of each PSH type are shown in Figures 6 and 7 and Table 5. The reserves were divided into upward and downward directions.

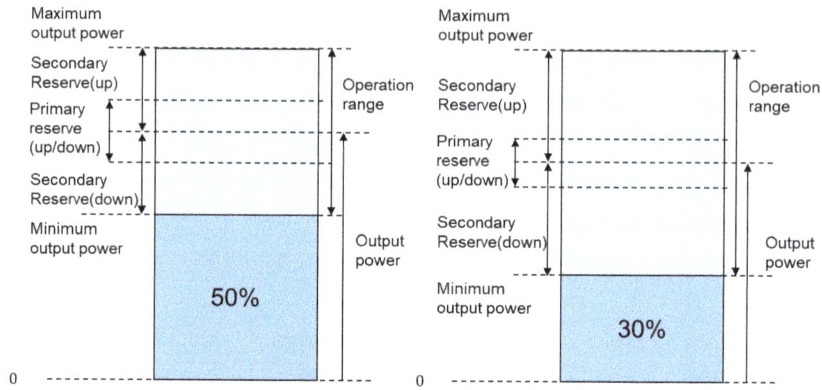

Figure 6. Operation and reserve ranges of PSH (generating mode).

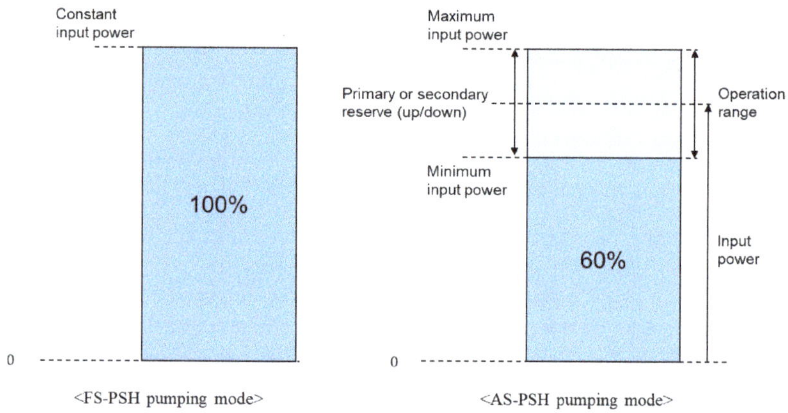

Figure 7. Operation and reserve ranges of PSH (pumping mode).

Table 5. Operation and reserve range of FS-PSH and AS-PSH.

		Operation Range	Max. Primary Reserve (Up/Downward)	Max. Secondary/Regulation Reserve (Up/Downward)
FS-PSH	Generating mode	50~100%	10%	50%
	Pumping mode	Constant at 100%	-	-
AS-PSH	Generating mode	30~100%	10%	70%
	Pumping mode	60~100%	20%	40%

3.2. Objective Function to Minimize the Operating Cost of the Power System

To predict the output pattern of the conventional generators of the future power system, it was assumed that the generators produced an economic output under the spinning reserve constraints and generator cycling characteristics. Accordingly, we formulated an objective function to minimize the operating cost of the power system. The system operating cost was divided into the fuel and startup costs of the generator. To implement the constraints of the minimum output, minimum downtime, and minimum uptime of generators, the objective function containing integer variables can be defined as follows:

$$Min: \sum_{i \in GU} \sum_{t=1}^{T} \left\{ FC_i\left(p_{i,t}^{g}\right) + STC_i \cdot u_{i,t} \right\} \tag{10}$$

In general, the power generation cost function is expressed as a quadratic function. However, in this study, we modeled the cost function of the quadratic function as a piecewise linearization function [36,37].

3.3. Constraints Considering AS-PSH

AS-PSH increases the total system load during its pumping mode and can provide primary, secondary, and regulation reserves in the pumping mode, which should be considered in the constraints. This subsection presents the constraints of the RCUC containing the modeling of AS-PSH.

3.3.1. Integer Variable Constraints

Constraints (11) and (12) represent the startup and shutdown states of a generator and prevent the simultaneous occurrence of the startup and shutdown states. Constraint (13) prevents the concurrent occurrence of the PSH generating and pumping modes.

$$u_{i,t} - d_{i,t} = i_{i,t} - i_{i,t-1} \tag{11}$$

$$u_{i,t} + d_{i,t} \leq 1 \tag{12}$$

$$i_{i,t} + l_{i,t} \leq 1 \tag{13}$$

3.3.2. Load Balance Constraints

At each hour, the total generated power and the difference between the load and VRE power were set as equal. In addition, considering the pumping input power of the PSH plants, the load balance constraint can be expressed by Constraint (14).

$$\mathrm{P_t^{Load}} - \mathrm{P_t^{VRE}} + \sum_{i \in PP} p_{i,t}^{p} = \sum_{i \in GU} p_{i,t}^{g} \tag{14}$$

3.3.3. Power Limit Constraints

The output power of generators and the input power of AS-PSH plants in the pumping mode can range from the minimum to maximum power, as expressed by Constraints (15) and (16).

$$P_i^{min} \times \mathrm{i_{i,t}} \leq p_{i,t}^{g} \leq P_i^{max} \times \mathrm{i_{i,t}} \tag{15}$$

$$PP_i^{min} \times \mathrm{l_{i,t}} \leq p_{i,t}^{p} \leq PP_i^{max} \times \mathrm{l_{i,t}} \tag{16}$$

3.3.4. Minimum Uptime/Downtime Constraints

During the daytime, when the power generated by VRE resources increases, a valley is created in the net-load curve, and generator cycling operations can be required. In the case of a coal-fired power generator in an offline state during the period wherein the net-load is lowered, the offline state should be maintained for a minimum of 8–12 h. Therefore, it is difficult to provide ramping capabilities during the period wherein the net-load increases due to the decrease in the VRE's output. The minimum downtime and minimum up-time constraints were thus considered through Constraints (17) and (18).

$$\sum_{t=k}^{k+MU_i-1} i_{i,t} \geq MU_i \times u_{i,k} \tag{17}$$

$$\sum_{t=k}^{k+MD_i-1} (1 - i_{i,t}) \geq MD_i \times d_{i,k}) \tag{18}$$

3.3.5. Generator Ramp-rate Constraints

The difference in output power between hour t and $t-1$ of a generator is limited by the ramp rate and expressed by Constraints (19) and (20).

$$\mathrm{p}_{\mathrm{i,t}}^{\mathrm{g}} - \mathrm{p}_{\mathrm{i,t-1}}^{\mathrm{g}} \leq 60 \times RU_i \tag{19}$$

$$\mathrm{p}_{\mathrm{i,t-1}}^{\mathrm{g}} - \mathrm{p}_{\mathrm{i,t}}^{\mathrm{g}} \leq 60 \times RD_i \tag{20}$$

3.3.6. Spinning Reserve Provision Constraints

The spinning reserve requirements should be secured under the constraints of the generator operation range. The target value of a synchronized generator for governor response is determined by the system frequency deviation and its droop characteristic [38]. To maintain the system frequency, the governor response that each unit can provide should be within the operation range of the system frequency. Therefore, the maximum capacity of the primary reserve of each unit can be expressed by Constraint (21) and the primary reserve of a generator at hour t by Constraints (22) and (23).

$$GR_i^{CG,up,max} = GR_i^{CG,down,max} = {P_i^{max} \cdot SF_l} \Big/ {R_i \cdot SF_s} \tag{21}$$

$$0 \leq gr_{i,t}^{CG,up} \leq GR_i^{CG,up,max} \times i_{i,t} \tag{22}$$

$$0 \leq gr_{i,t}^{CG,down} \leq GR_i^{CG,down,max} \times i_{i,t} \tag{23}$$

As discussed in the previous subsection, the maximum primary reserve provided in the generating and pumping modes of the AS-PSH is expressed by Constraints (24) and (25). The primary reserve provided by an AS-PSH at hour t is expressed by Constraints (26), (27), (28), and (29).

$$PR_i^{g,up,max} = PR_i^{g,down,max} = P_i^{max} \times 10\% \tag{24}$$

$$PR_i^{p,up,max} = PR_i^{p,down,max} = PP_i^{max} \times 20\% \tag{25}$$

$$0 \leq pr_{i,t}^{g,up} \leq PR_i^{g,up,max} \times i_{i,t} \tag{26}$$

$$0 \leq pr_{i,t}^{g,down} \leq PR_i^{g,down,max} \times i_{i,t} \tag{27}$$

$$0 \leq pr_{i,t}^{p,up} \leq PR_i^{p,up,max} \times l_{i,t} \tag{28}$$

$$0 \leq pr_{i,t}^{p,down} \leq PR_i^{p,down,max} \times l_{i,t} \tag{29}$$

The secondary/regulation reserves are provided by the ACG signals according to the ramp-rate of each generator. Therefore, a ramp-rate should be reflected in the ACG capacity available to each generator. The UC problem in this study considers the constraints of secondary and regulation reserves at 5 min intervals. The secondary and regulation reserves available to each generator can be calculated using Constraints (30) and (31), and the AGC capacity secured by a generator at hour t is expressed by Constraints (32) and (33).

$$AGC_i^{CG,up,max} = 5 \times RR_i^{up} \tag{30}$$

$$AGC_i^{CG,down,max} = 5 \times RR_i^{down} \tag{31}$$

$$0 \leq agc_{i,t}^{CG,up} \leq AGC_i^{CG,up,max} \times i_{i,t} \tag{32}$$

$$0 \leq agc_{i,t}^{CG,down} \leq AGC_i^{CG,down,max} \times i_{i,t} \tag{33}$$

The secondary/regulation reserve capacities of the AS-PSH are defined by Constraints (34) and (35). The reserves of the AS-PSH in the generating and pumping modes at hour t are expressed by Constraints (36)–(39).

$$AGC_i^{g,up,max} = AGC_i^{g,down,max} = P_i^{max} - P_i^{min} \tag{34}$$

$$AGC_i^{p,up,max} = AGC_i^{p,down,max} = PP_i^{max} - PP_i^{min} \tag{35}$$

$$0 \leq agc_{i,t}^{g,up} \leq AGC_i^{g,up,max} \times i_{i,t} \tag{36}$$

$$0 \leq agc_{i,t}^{g,down} \leq AGC_i^{g,down,max} \times i_{i,t} \tag{37}$$

$$0 \leq agc_{i,t}^{p,up} \leq AGC_i^{p,up,max} \times l_{i,t} \tag{38}$$

$$0 \leq agc_{i,t}^{p,down} \leq AGC_i^{p,down,max} \times l_{i,t} \tag{39}$$

3.3.7. Power Limit Constraints Considering Spinning Reserve

The output range of a conventional generator should consider its spinning reserve at hour t, as expressed by Constraints (40) and (41).

$$p_{i,t}^{g} + agc_{i,t}^{CG,up} + gr_{i,t}^{CG,up} \leq P_i^{max} \tag{40}$$

$$p_{i,t}^{g} - agc_{i,t}^{CG,down} - gr_{i,t}^{CG,down} \geq P_i^{min} \tag{41}$$

The output and input power range of AS-PSH can be further expressed by Constraints (42), (43), (44), and (45).

$$p_{i,t}^{g} + agc_{i,t}^{g,up} + pr_{i,t}^{g,up} \leq P_i^{max} \times i_{i,t} \tag{42}$$

$$p_{i,t}^{g} - agc_{i,t}^{g,down} - pr_{i,t}^{g,down} \geq P_i^{min} \times i_{i,t} \tag{43}$$

$$p_{i,t}^{p} - agc_{i,t}^{p,up} - pr_{i,t}^{p,down} \leq PP_i^{max} \times l_{i,t} \tag{44}$$

$$p_{i,t}^{p} + agc_{i,t}^{p,down} + pr_{i,t}^{p,down} \geq PP_i^{min} \times l_{i,t} \tag{45}$$

3.3.8. Spinning Reserve Requirement Constraints

The spinning reserve procured at each hour should satisfy Constraints (46) and (47), which specify the requirements of the primary reserve and secondary/regulation reserves, respectively. In this study, the up-spinning and down-spinning reserve requirements were assumed to be identical. The primary reserve requirement in the Korean power market is secured for the largest unit loss as the change in the generators' output power in a quasi-steady state after a generator outage. The Korean power system has an average governor response of 1000 MW at the quasi-steady state for the largest unit outage [39]. The requirement of a 1400 MW secondary reserve is the capacity required to remove the steady-state error of the system frequency after the largest unit loss. In addition, the standard deviations of VRE are set to vary according to its output power, preventing the overestimation of spinning reserves during the night when the output power of the VRE is low. Based on the reserve requirement determination method in [40], we derived the standard-deviation function of VRE's output.

$$SR_t^{PR,up} = SR_t^{PR,down} = PR^{LU} + n \cdot \sigma_{VRE}^{1min}\left(P_t^{VRE}\right) \tag{46}$$

$$SR_t^{AGC,up} = SR_t^{AGC,down} = SR^{LU} + \sqrt{\left(n \cdot \sigma_L^{5min.}\right)^2 + \left(n \cdot \sigma_{VRE}^{5min}\left(P_t^{VRE}\right)\right)^2} \tag{47}$$

The spinning reserve secured by AS-PSHs and conventional generators should satisfy the reserve requirements at each hour, as expressed by Constraints (48), (49), (50), and (51).

$$\sum_{i\in AP^c} gr_{i,t}^{CG,up} + \sum_{i\in AP} (pr_{i,t}^{g,up} + pr_i^{p,up}) \geq SR_t^{PR,up} \tag{48}$$

$$\sum_{i\in AP^c} gr_{i,t}^{CG,down} + \sum_{i\in AP} \left(pr_{i,t}^{g,down} + pr_i^{p,down}\right) \geq SR_t^{PR,down} \tag{49}$$

$$\sum_{i\in AP^c} agc_{i,t}^{CG,up} + \sum_{i\in AP} (agc_{i,t}^{g,up} + agc_{i,t}^{p,up}) \geq SR_t^{AGC,up} \tag{50}$$

$$\sum_{i\in AP^c} agc_{i,t}^{CG,down} + \sum_{i\in AP} (agc_{i,t}^{g,down} + agc_{i,t}^{p,down}) \geq SR_t^{AGC,down} \tag{51}$$

3.3.9. Upper Reservoir of PSH Constraint

The PSH has an energy constraint based on its upper reservoir volume. Constraints (52) and (53) consider the efficiency when the PSH charges and discharges. The upper reservoir volume at hour t was measured based on electrical energy.

$$v_{i,t} = v_{i,t-1} - p_{i,t-1}^{g} + p_{i,t-1}^{p} \times EFF_i \tag{52}$$

$$0 \leq v_{i,t} \leq V_i^{max} \tag{53}$$

4. Simulation

This section presents the scenarios, simulation results, and discussion.

4.1. Scenarios

The construction of three new PSH plants by 2030, with reference to [1], was assumed; and the following scenarios were set. Scenario A denotes no PSH plant construction, Scenario B denotes FS-PSH plant construction, and Scenarios C and D denote AS-PSH plant construction. To verify the contribution of the AS-PSH operating in the pumping mode to primary reserve provision, Scenario C constrained the AS-PSH to provide only secondary/regulation reserves in pumping mode, and the AS-PSH in Scenario D was modeled to provide all the reserves. Table 6 outlines each scenario. Both FS-PSH and AS-PSH were assumed to have the same capacity for the generating and pumping modes.

Table 6. Scenario description of RCUC simulation.

Scenario	Existing PSH Plants	New PSH Plants in 2030		Total Capacity of PSH Plants, MW
	FS-Type, MW	FS-Type, MW	AS-Type, MW	
A	4700	0	0	4700
B	4700	600 × 3 = 1800	0	6500
C	4700	0	600 × 3 = 1800	6500
D	4700	0	600 × 3 = 1800	6500

Given the low frequency of the high output power of VRE, in addition to the low investment efficiency of transmission and distribution facilities required, we considered curtailing approximately 5% of the target power generation of VRE. If the VRE output is limited to a maximum of 26 GW in 2030, 5% of the total generation can be curtailed, and the VRE output curve in Figure 1 can be depicted as shown in Figure 8.

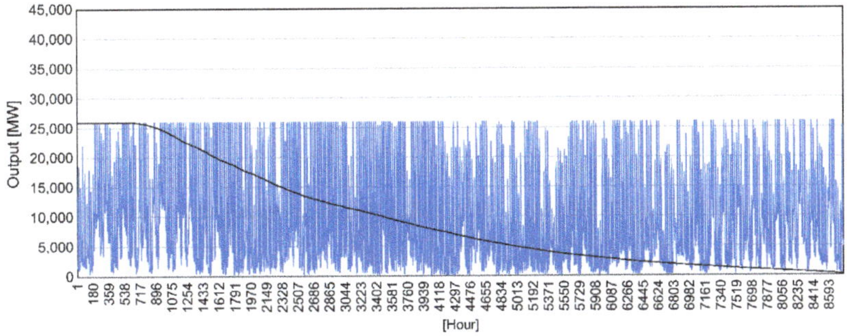

Figure 8. Output and duration curve of 2030 VRE with an output limit of 26 GW.

The calculated output of VRE in Korea differs by month, and the generation of VRE is highest in spring. Generally, the load curve on weekends tends to be lower than that on weekdays. The effects of VRE were assumed to be significant on weekends. Therefore, UC simulation was performed on a weekend daily load curve with high VRE generation in spring 2030. To verify the VRE effect in a typical load curve pattern, special days, such as traditional holidays, were excluded.

The daily load, net-load, and output power of the VRE considering the output limit are shown in Figure 9. The requirements of the spinning reserves with respect to Constraints (46) and (47), which adopt the 2σ criterion of the short-term variation of the VRE, are plotted in Figure 10. The current standard adopting the 3σ criterion for the short-term variation of the load was maintained.

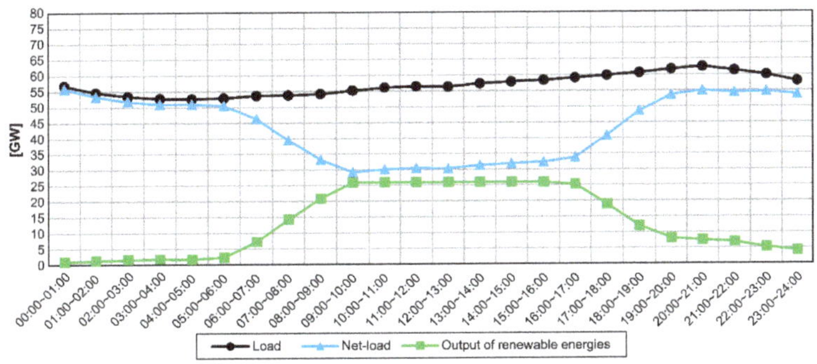

Figure 9. Curves of load, net-load, and VRE output power on spring Sunday, 2030.

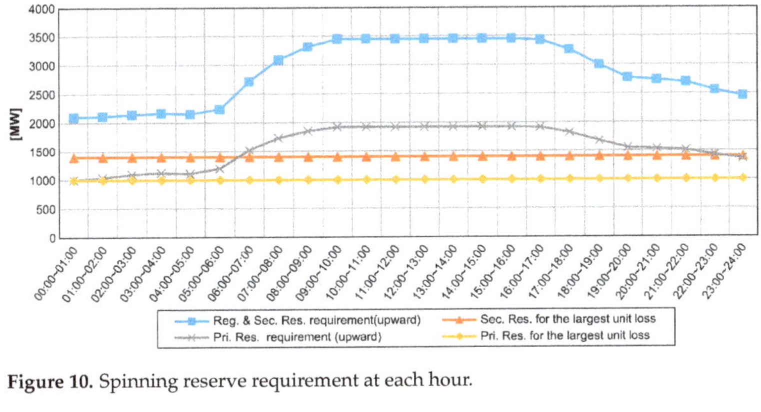

Figure 10. Spinning reserve requirement at each hour.

4.2. Results

In the UC simulation, the following two assumptions were made:

(1) All of the available nuclear power generators are maintained in an online state for 24 h.
(2) Nuclear power generators do not provide spinning reserves.

Assumption 1 implies that there is no startup or shutdown operation of nuclear power generators for 24 h. Nuclear power generators are set to have a significantly limited ramp-rate. Assumption 2 implies that the UC results do not count as a spinning reserve even if nuclear power generators have unused capacity. The total available capacity of nuclear power generators was assumed to be 16.4 GW out of 20.4 GW in 2030, with reference to their historical planned outage rates. Starting with the case where all available capacity of nuclear power generators is operable, the feasible solution was found by excluding the online state nuclear power generators. Table 7 presents the feasible solutions to the optimization problem according to the capacity of the nuclear power generators in the online state for each scenario.

Table 7. Feasibility of each scenario.

Scenario	Total Online Capacity of Nuclear Power Generators, GW											
	16.4	**15.4**	**14.4**	**13.4**	**12.4**	**11.4**	**10.4**	**9.4**	**8.4**	**7**	**5.6**	**4.2**
A	Infeasible											Feasible
B	Infeasible									Feasible		
C	Infeasible									Feasible		
D	Infeasible				Feasible							

In all scenarios, the UC problem was found to be infeasible when nuclear power generators were employed at the maximum capacity of 16.4 GW. In Scenarios A and B, the maximum available capacities of the nuclear power generators were derived as 4.2 GW and 7 GW, respectively. Similarly, the maximum available capacities of the nuclear power generators in Scenarios B and D were 7 GW and 12.4 GW, respectively. Figures 11–14 present the diagram of the UC solution for each scenario when the feasible solution is operated as per the maximum capacity of the nuclear power plant, as derived in Table 7.

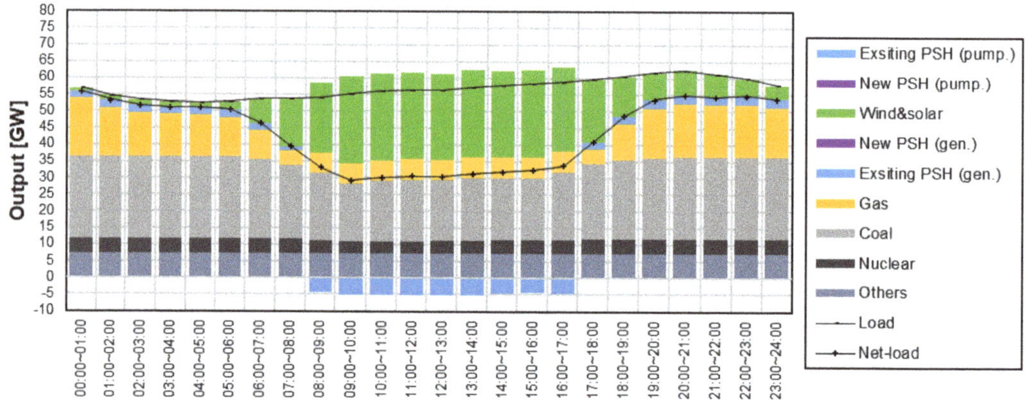

Figure 11. The UC solution diagram of Scenario A.

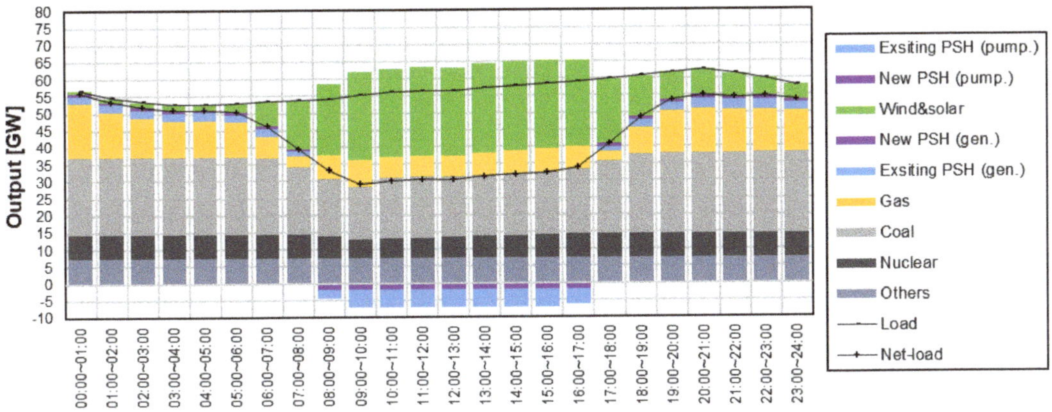

Figure 12. The UC solution diagram of Scenario B.

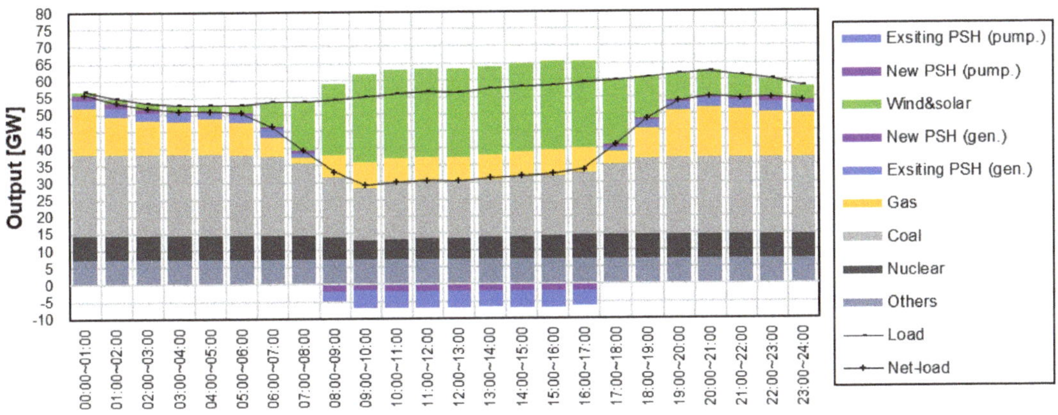

Figure 13. The UC solution diagram of Scenario C.

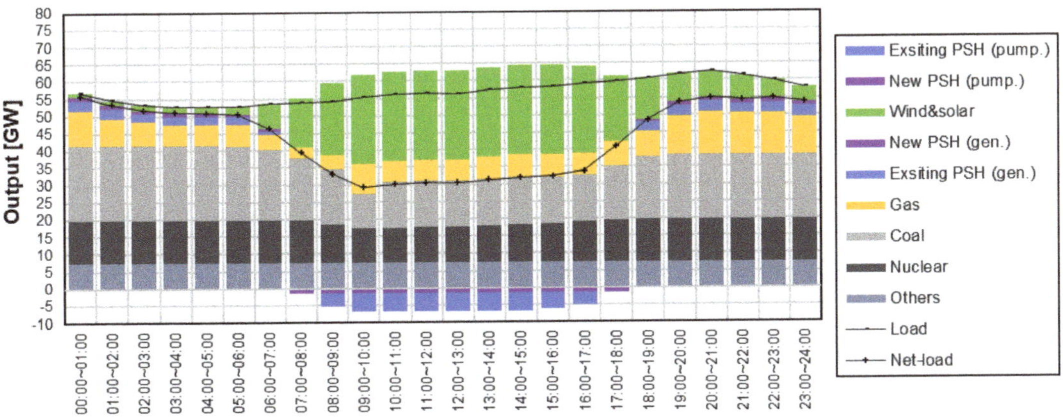

Figure 14. The UC solution diagram of Scenario D.

Table 8 presents the operating cost results of each scenario. Given the characteristics of the Korean power market, which is based on fuel cost, the costs of VRE and hydro energy were not considered. In Figures 11–14, the resources named 'others' consists of new energy resources, such as fuel cells and by-product gases. Using the expected power generation of the new energy resources in 2030, the output power was assumed to be uniform throughout the day [1]. Given that the 24 h power generation pattern of the energy sources was applied in all scenarios, their costs were neglected.

Table 8. Cost result of each scenario.

Fuel-Type	Scenario A		Scenario B		Scenario C		Scenario D	
	Cost, KRW	**Share, %**	**Cost, KRW**	**Share, %**	**Cost, KRW**	**Share, %**	**Cost, KRW**	**Share, %**
Nuclear	493.24	6.8	822.84	11.4	825.00	11.4	1,427.16	19.9
Coal	25,078.89	38.0	23,594.92	35.5	23,453.91	35.4	19,604.10	29.4
Gas	17,557.25	17.3	15,620.68	15.3	15,076.00	15.4	12,999.19	12.9
Total costs	43,129.38	62.1	40,038.44	62.2	39,354.92	62.2	34,030.46	62.2

The total operating costs decreased in the order of Scenarios A, B, C, and D. The proportion of nuclear power generation was 6.8%, 11.4%, 11.4%, and 19.9% in Scenarios A, B, C, and D, respectively, and that of thermal power generation was calculated as 55.3%, 50.8%, 50.8%, and 42.3%, respectively. Based on the operating cost result of Scenario A, the operating cost savings for scenarios B, C, and D were KRW 3090.94, 3774.46, and 9098.92, respectively.

4.3. Discussion

The results shown in Table 8 and Figures 11–14 were found to be strongly related to the generator output limit in Constraints (15), (40), and (41); the generator's spinning reserve provision in Constraints (21)–(23) and (30)–(33); and the spinning reserve requirement in Constraints (46)–(51). If nuclear power generators with a total capacity of 16.4 GW are operating, the thermal power generators carry a given net-load, excluding the output power of the nuclear power generators. However, in this case, the spinning reserve requirement is not satisfied. In contrast, if the nuclear power generator is operating at 16.4 GW and the thermal power generators procure the requirement of spinning reserves, the supply and demand balance of Constraint (14) is violated.

Therefore, the AS-PSH operating in pumping mode during low net-load and providing primary reserves was the main factor for the result of Scenario D. The primary reserve from a conventional generator is dominated by its droop and not its ramp-rate characteristics. As mentioned above, the primary reserve that can be secured for each generator is considerably limited when compared with the secondary/regulation reserve. For example, under Constraint (21), if the system-frequency maintenance range is ± 0.2 Hz, the primary reserve that can be secured by a 500 MW generator with droop 6% is merely 25 MW (5% of the maximum rated power of the generator). In this case, with a 25 MW increase in the primary reserve, 500 MW of thermal power generation must be committed sequentially.

An increase in the primary reserve requirements requires an increased number of thermal power generators in the online state, but the duck-curve phenomenon limits the number of online generators. The FS-PSH operated in pumping mode, contributing only to mitigating the duck-curve phenomenon, while the AS-PSH operated in pumping mode and contributed to both primary reserve provision and mitigating the duck-curve phenomenon. As shown in Figure 15, the results of the AS-PSH securing all the unused capacity as the primary reserve in the low net-load period were derived. Therefore, the maximum available capacity of nuclear power generators, which are economic resources, was the highest in Scenario D.

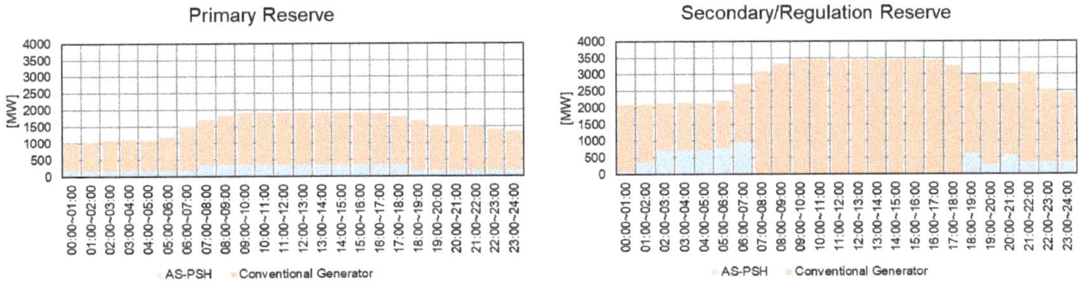

Figure 15. Reserves of AS-PSHs and conventional generators in Scenario D.

5. Conclusions

In this study, we performed RCUC simulations for the Korean power system projected for 2030 and analyzed the economic effect obtained by the flexibility of AS-PSH. Prior literature focused on the traditional advantages of PSHs, such as energy storage and quick-start capabilities, and provided insufficient analysis on the economic effect from the primary reserve capability of AS-PSH. In this study, we focused on the reserve capability of AS-PSH, considering that AS-PSH in pumping mode can provide primary, secondary, and regulation reserves. To analyze the economic effect obtained by AS-PSH in environments with a high share of VRE, we considered the short-term variations of VRE in the spinning reserve requirement defined by the Korean power market. We then divided the flexibility provided by AS-PSH into primary, secondary, and regulation reserves and modeled them in the RCUC formulation to allow for a range of operations in the pumping mode. In the simulation, it was postulated that the spinning reserve is procured only through generators, including PSHs. We simulated four scenarios: Scenario A denoted no PSH plant construction, Scenario B denoted FS-PSH plant construction. Scenarios C and D denoted AS-PSH plant construction. Scenario D showed the best cost result.

The key findings of this study are as follows.

- In the duck-curve environment with an increased spinning reserve requirement, it was difficult to operate nuclear power generators at their maximum capacities. In the results of Scenario A, the maximum available capacity of the nuclear power generators was 4.2 GW out of 16.4 GW.
- New PSH plants contributed to mitigating the duck-curve phenomenon and increased the maximum available capacity of the nuclear power generators, resulting in operating cost savings.
- The costs in Scenarios A, B, C, and D were KRW 43,129.38, 40,038.44, 39,354.92, and 34,030.46. The results were mainly derived from the provision of the primary reserve by AS-PSH's pumping mode.

This study is limited in that the patterns of load curves vary depending on weekdays or weekends and are influenced by the seasons. The characteristics of VRE are also influenced by the seasons. Therefore, it is necessary to evaluate the economic effect over a longer timeframe in a follow-up study. In a long-term evaluation, a more simplified model for long-term economic analysis should be developed because the computational burden of the MIP model increases significantly as the analysis period increases. However, in a simplified model, the results of a short-term analysis model may not be guaranteed. Therefore, in a future long-term economic evaluation model, it will be necessary to develop a novel algorithm capable of considering the operational characteristics derived from the short-term economic evaluation model.

Author Contributions: Conceptualization, W.-J.K.; data curation, W.-J.K.; formal analysis, W.-J.K.; funding acquisition, Y.-H.C.; investigation, W.-J.K. and Y.-S.L.; methodology, W.-J.K. and Y.-S.L.; software, W.-J.K. and H.-S.J.; supervision, Y.-H.C.; validation, W.-J.K. and Y.-S.L.; writing—original draft, W.-J.K., Y.-S.L., and H.-S.J.; writing—review and editing, W.-J.K. and Y.-H.C. All authors have read and agreed to the published version of the manuscript.

Funding: This work was supported by the Korea Institute of Energy Technology Evaluation and Planning (KETEP) and the Ministry of Trade, Industry and Energy (MOTIE) of the Republic of Korea (No. 2019371010006B).

Conflicts of Interest: The authors declare no conflict of interest.

Nomenclature

Sets and indices

GU	Set of all generating units
PP	Set of PSH units, PP ⊂ GU
AP	Set of AS-PSH units, AP ⊂ PP ⊂ GU
T	Operation period, index by t
i	Index for unit, $i \in GU$, PP, or AP
t	Index for time interval, $t = 1, \ldots, T$

Parameters and functions

$FC_i(\cdot)$	Fuel cost function of unit i at hour t $
STC_i	Startup cost of Unit i, $
$\mathrm{P}_t^{\mathrm{Load}}$	System load at Time t, MW
$\mathrm{P}_t^{\mathrm{VRE}}$	Output of renewable energy sources at Time t, MW
P_i^{min}	Minimum output power of Unit i, MW
P_i^{max}	Maximum output power of Unit i, MW
PP_i^{min}	Minimum pumping input power of PSH plant i, MW
PP_i^{max}	Maximum pumping input power of PSH plant i, MW
MU_i	Minimum up-time of Unit i, hour
MD_i	Minimum down-time of Unit i, hour
RU_i	Up-ramping limit of Unit i, MW/minute
RD_i	Down-ramping limit of Unit i, MW/minute
R_i	Droop of Unit i, p.u.
SF_i	Operational limit deviation of system frequency, Hz
SF_s	Standard system frequency, Hz
PR^{LU}	Requirement of primary reserve respond to largest unit loss, MW
SR^{LU}	Requirement of secondary reserve respond to largest unit loss, MW
σ_{VRE}^{1min}	Standard deviation of VRE's 1 min variation
σ_{VRE}^{5min}	Standard deviation of VRE's 5 min variation
$\sigma_L^{5min.}$	Standard deviation of load's 5 min variation
EFF_i	Pumping efficiency of PSH plant i
V_i^{max}	Maximum volume of upper reservoir of PSH plant i, MWh
$AGC_i^{CG,up,max}$	Maximum up-secondary and regulation reserve of unit i, MW
$AGC_i^{g,up,max}$	Maximum up-secondary and regulation reserve of AS-PSH i (generating), MW
$AGC_i^{p,up,max}$	Maximum up-secondary and regulation reserve of AS-PSH i (pumping), MW
$GR_i^{CG,up,max}$	Maximum up-primary reserve of unit i, MW
$PR_i^{g,up,max}$	Maximum up-primary reserve of AS-PSH (generating) i, MW
$PR_i^{p,up,max}$	Maximum up-primary reserve of AS-PSH (pumping) i, MW
$AGC_i^{CG,down,max}$	Maximum down-secondary and regulation reserve of unit i, MW
$AGC_i^{g,down,max}$	Maximum down-secondary and regulation reserve of AS-PSH i (generating), MW
$AGC_i^{p,down,max}$	Maximum down-secondary and regulation reserve of AS-PSH i (pumping), MW
$GR_i^{CG,down,max}$	Maximum down-primary reserve of unit i, MW
$PR_i^{g,down,max}$	Maximum down-primary reserve of AS-PSH (generating) i, MW
$PR_i^{p,down,max}$	Maximum down-primary reserve of AS-PSH (pumping) i, MW
$SR_t^{PR,up}$	Up-primary reserve requirement at hour t, MW
$SR_t^{PR,down}$	Down-primary reserve requirement at hour t, MW
$SR_t^{AGC,up}$	Up-secondary and regulation reserve requirement at hour t, MW
$SR_t^{AGC,down}$	Down- secondary and regulation reserve requirement at hour t, MW

Variables

$u_{i,t}$	Integer variable for startup, 1 when off to on at hour t
$d_{i,t}$	Integer variable for shutdown, 1 when on to off at hour t
$\mathrm{i}_{i,t}$	Integer variable for on/off, on = 1/off = 0 at hour t
$\mathrm{l}_{i,t}$	Integer variable for on/off (pumping mode of PSH), on = 1/off = 0 at hour t
$p_{i,t}^{g}$	Output of unit i at hour t, MW
$p_{i,t}^{p}$	Pumping output of PSH plant i at hour t, MW
$agc_{i,t}^{CG,up}$	Up-secondary and regulation reserve of unit i at hour t, MW
$agc_{i,t}^{g,up}$	Up-secondary and regulation reserve of AS-PSH (generating) i at hour t, MW
$agc_{i,t}^{p,up}$	Up-secondary and regulation reserve of AS-PSH (pumping) i at hour t, MW
$gr_{i,t}^{CG,up}$	Up-primary reserve of unit i at hour t, MW.
$pr_{i,t}^{g,up}$	Up-primary reserve of AS PSH (generating) i at hour t, MW
$pr_{i}^{p,up}$	Up-primary reserve of AS PSH (pumping) i at hour t, MW
$agc_{i,t}^{CG,down}$	Down-secondary and regulation reserve of unit i at hour t, MW
$agc_{i,t}^{g,down}$	Down-secondary and regulation reserve of AS-PSH (generating) i at hour t, MW
$agc_{i,t}^{p,down}$	Down-secondary and regulation reserve of AS-PSH (pumping) i at hour t, MW
$pr_{i,t}^{g,down}$	Down-primary reserve of AS-PSH (generating) i at hour t, MW
$pr_{i}^{p,down}$	Down-primary reserve of AS-PSH (pumping) i at hour t, MW
$gr_{i,t}^{CG,down}$	Down-primary reserve of unit i at hour t, MW
$v_{i,t}$	Volume of PSH plant i at hour t

Solar and wind power model parameters

n	Index for a wind turbine in a wind farm, $n = 1, 2, 3, \ldots, N$
$P_{WT,n}$	Output power of nth wind turbine in a wind farm, MW
$P_{curvefit}$	Polynomial function of output power for ($V_{in} \leq V \leq V_r$), MW
P_r	Rated output power of wind turbine, MW
V_{in}	Cut-in wind speed, m/s
V_r	Rated wind speed, m/s
V_{out}	Cut-out wind speed, m/s
P_{WF}	Output power of wind farm, MW
C_{WF}	Capacity of wind farm, MW
V_{m}	Wind speed at height Z_{m}, m/s
V_{h}	Wind speed at height Z_{h}, m/s
Z_{m}	Height 1 (lower height, 10m), m
Z_{h}	Height 2 (upper height, Tower hub height), m
α	Power-law exponent
$P_{PV,M}$	Output power of PV module, W
$P_{PV,M}^{*}$	Rated output power of PV module in STC, W
G	Actual solar irradiance, W/m^2
G^{*}	Solar irradiance in STC, 1000 W/m^2
γ	Module maximum power temperature coefficient, 1/°C
T_C	PV module temperature, °C
P_{PV}	Output power of PV power plant, MW
C_{PV}	Capacity of PV power plant, MW
T_A	Air temperature, °C
T_{NOCT}	Nominal operating cell temperature, °C

References

1. Ministry of Trade, Industry and Energy of South Korea. *The 9th Basic Plan for Long-Term Electricity Supply and Demand*; Ministry of Trade Industry and Energy Republic of Korea: Sejong, Korea, 2020.
2. Ela, E.; Milligan, M.; Kirby, B. *Operating Reserves and Variable Generation*; Technical Report; National Renewable Energy Laboratory: Golden, CO, USA, 2011. [CrossRef]
3. California, ISO. Fast Fact, What the Duck Curve Tells us about Managing a Green Grid. 2016. Available online: https://www.caiso.com/Documents/FlexibleResourcesHelpRenewables_FastFacts.pdf (accessed on 29 October 2020).

4. Kim, W.J.; Lee, Y.S.; Chun, Y.H. Assessment of Primary Control Reserve Requirement from Generator Considering a Battery Energy Storage System. *J. Electr. Eng. Technol.* **2020**, *15*, 315–324. [CrossRef]
5. Koritarov, V.; Veselka, T.D.; Gasper, J.; Bethke, B.M.; Botterud, A.; Wang, J.; Mahalik, M.; Zhou, Z.; Milostan, C.; Feltes, J. Modeling and Analysis of Value of Advanced Pumped Storage Hydropower in the United States. *Science* **2014**, *246*, 25–31. [CrossRef]
6. Pérez-Díaz, J.I.; Jiménez, J. Contribution of a Pumped-Storage Hydropower Plant to Reduce the Scheduling Costs of an Isolated Power System with High Wind Power Penetration. *Energy* **2016**, *109*, 92–104. [CrossRef]
7. Al-Masri, H.M.K.; Al-Quraan, A.; AbuElrub, A.; Ehsani, M. Optimal Coordination of Wind Power and Pumped Hydro Energy Storage. *Energies* **2019**, *12*, 4387. [CrossRef]
8. Bakhshaei, P.; Alireza, A.; Reza, A. Operation optimization of a grid-connected photovoltaic/pumped hydro storage considering demand response program by an improved crow search algorithm. *J. Energy Storage* **2021**, *44*, 103326. [CrossRef]
9. Rahmati, I.; Foroud, A.A. Pumped-storage units to address spinning reserve concerns in the grids with high wind penetration. *J. Energy Storage* **2020**, *31*, 101612. [CrossRef]
10. Alturki, F.A.; Awwad, E.M. Sizing and Cost Minimization of Standalone Hybrid WT/PV/Biomass/Pump-Hydro Storage-Based Energy Systems. *Energies* **2021**, *14*, 489. [CrossRef]
11. Dianellou, A.; Christakopoulos, T.; Caralis, G.; Kotroni, V.; Lagouvardos, K.; Zervos, A. Is the Large-Scale Development of Wind-PV with Hydro-Pumped Storage Economically Feasible in Greece? *Appl. Sci.* **2021**, *11*, 2368. [CrossRef]
12. Ma, Y.; Li, C.; Zhou, J.; Zhang, Y. Comprehensive stochastic optimal scheduling in residential micro energy grid considering pumped-storage unit and demand response. *J. Energy Storage* **2020**, *32*, 101968. [CrossRef]
13. Nag, S.; Lee, K.Y. Network and Reserve Constrained Economic Analysis of Conventional, Adjustable-Speed and Ternary Pumped-Storage Hydropower. *Energies* **2020**, *13*, 4140. [CrossRef]
14. Aburub, H.; Basnet, S.; Ward, T.J. On the use of adjustable-speed pumped hydro storage operation in the U.S. electricity market. *J. Energy Storage* **2019**, *23*, 495–503. [CrossRef]
15. Feng, C.; Zheng, Y.; Li, C.; Mai, Z.; Wu, W.; Chen, H. Cost advantage of adjustable-speed pumped storage unit for daily operation in distributed hybrid system. *Renew. Energy* **2021**, *176*, 1–10. [CrossRef]
16. Kazemi, M.; Siano, P.; Sarno, D.; Goudarzi, A. Evaluating the impact of sub-hourly unit commitment method on spinning reserve in presence of intermittent generators. *Energy* **2016**, *113*, 338–354. [CrossRef]
17. Wu, J.; Luh, P.; Chen, Y.; Bragin, M.; Yan, B. A novel optimization approach for sub-hourly unit commitment with large numbers of units and virtual transactions. *IEEE Trans. Power Syst.* **2021**. [CrossRef]
18. Raghunathan, N.; Bragin, M.A.; Yan, B.; Luh, P.B.; Moslehi, K.; Feng, X.; Yu, Y.; Yu, C.-N.; Tsai, C.-C. Exploiting Soft Constraints within Decomposition and Coordination Methods for Sub-hourly Unit Commitment. *Int. J. Electr. Power Energy Syst.* **2022**, *139*, 108023. [CrossRef]
19. Bragin, M.A.; Luh, P.B.; Yan, B.; Sun, X. A scalable solution methodology for mixed-integer linear programming problems arising in automation. *IEEE Trans. Autom. Sci. Eng.* **2018**, *16*, 531–541. [CrossRef]
20. Yurdakul, O.; Sivrikaya, F.; Albayrak, S. Quantification of the Impact of GHG Emissions on Unit Commitment in Microgrids. In Proceedings of the IEEE/PES Transmission and Distribution Conference and Exposition: Latin America, Montevideo, Uruguay, 28 September 2020. [CrossRef]
21. Nicolosi, F.F.; Alberizzi, J.C.; Caligiuri, C.; Renzi, M. Unit commitment optimization of a micro-grid with a MILP algorithm: Role of the emissions, bio-fuels and power generation technology. *Energy Rep.* **2021**, *7*, 8639–8651. [CrossRef]
22. Korea Meteorological Administration. Available online: https://data.kma.go.kr/cmmn/main.do (accessed on 15 January 2020).
23. Marčiukaitis, M.; Žutautaitė, I.; Martišauskas, L.; Jokšas, B.; Gecevičius, G.; Sfetsos, A. Non-Linear Regression Model for Wind Turbine Power Curve. *Renew. Energy* **2017**, *113*, 732–741. [CrossRef]
24. Doosan Heavy Industries & Construction. Available online: https://www.doosanheavy.com/download/pdf/products/energy/DHI_Wind_Power_Brochure_Eng.pdf (accessed on 21 July 2021).
25. Irwin, J.S. A Theoretical Variation of the Wind Profile Power-Law Exponent as a Function of Surface Roughness and stability. *Atmos. Environ.* **1979**, *13*, 191–194. [CrossRef]
26. Cherenack, P. Conditions for Cubic Spline Interpolation on Triangular Elements. *Comp. Math. Appl.* **1984**, *10*, 235–244. [CrossRef]
27. Osterwald, C.R. Translation of Device Performance Measurements to Reference Conditions. *Solar Cells* **1986**, *18*, 269–279. [CrossRef]
28. Stultz, J.W. Thermal and Other Tests of Photovoltaic Modules Performed in Natural Sunlight. *J. Energy* **1979**, *3*, 363–372. [CrossRef]
29. Billinton, R.; Allan, R.N. Operating reserve. In *Reliability Evaluation of Power Systems*, 2nd ed.; Springer: Boston, MA, 1996. [CrossRef]
30. Korea Power Exchange 2021. The Rules and Regulation of Power Market Operation. Available online: https://www.kpx.or.kr/www/selectBbsNttList.do?bbsNo=114&key=29 (accessed on 19 November 2021).
31. Holttinen, H. Impact of Hourly Wind Power Variations on the System Operation in the Nordic Countries. *Wind. Energy* **2005**, *8*, 197–218. [CrossRef]
32. Holttinen, H.; Milligan, M.; Ela, E.; Menemenlis, N.; Dobschinski, J.; Rawn, B.; Bessa, R.J.; Flynn, D.; Gomez-Lazaro, E.; Detlefsen, N.K. Methodologies to Determine Operating Reserves Due to Increased Wind Power. *IEEE Trans. Sustain. Energy* **2012**, *3*, 713–723. [CrossRef]

33. Tokyo Electric Power Company. *Final Report on Feasibility Study on Adjustable Speed Pumped Storage Generation Technology*; Japan International Cooperation Agency: Tokyo, Japan, 2012. Available online: https://openjicareport.jica.go.jp/643/643/643_000_12044822.html. (accessed on 6 January 2020).
34. Takashi, F.; Hirotaka, T.; Hajime, K. Commencement of Commercial Operation of Kazunogawa Hydroelectric Power Station Unit 4 of Tokyo Electric Power Co., Inc. Employing World's Largest Capacity Adjustable-Speed System. *Toshiba Rev.* **2015**, *70*, 7–10. Available online: https://www.global.toshiba/jp/technology/corporate/review/abstract/2015-01.html. (accessed on 24 August 2021).
35. Kawabara, T.; Shibuya, A.; Furata, H.; Kita, E.; Mitsuhasi, K. Design and dynamic response characteristics of 400 MW adjustable speed pump storage built for Ohkawachi Power Station. *IEEE Trans. Energy Convers.* **1998**, *11*, 376–384. [CrossRef]
36. Carrión, M.; Arroyo, J.M. A Computationally Efficient Mixed-Integer Linear Formulation for the Thermal Unit Commitment Problem. *IEEE Trans. Power Syst.* **2006**, *21*, 1371–1378. [CrossRef]
37. Ostrowski, J.; Anjos, M.F.; Vannelli, A. Tight Mixed Integer Linear Programming Formulations for the Unit Commitment Problem. *IEEE Trans. Power Syst.* **2012**, *27*, 39–46. [CrossRef]
38. Kundur, P. Control of active power and reactive power. In *Power System Stability and Control*; McGraw-Hill: New York, NY, USA, 1994.
39. Park, M.-S.; Kim, W.-J.; Chun, Y.-H. Calculation of Primary Reserve Requirements Based on Measurements of Frequency Trajectories of Korean Power System. *Trans. Korean Inst. Electr. Eng.* **2019**, *68*, 399–404. [CrossRef]
40. Corbus, D.; King, J.; Mousseau, T. *Eastern Wind Integration and Transmission Study*; US Department of Energy, National Renewable Energy Laboratory: Golden, CO, USA, 2010. Available online: https://www.energy.gov/eere/wind/downloads/eastern-wind-integration-and-transmission-study-ewits-revised. (accessed on 21 January 2022).

electronics

Article

Techno-Economic Evaluation of Hybrid Energy Systems Using Artificial Ecosystem-Based Optimization with Demand Side Management

Hammed Olabisi Omotoso *, Abdullah M. Al-Shaalan, Hassan M. H. Farh and Abdullrahman A. Al-Shamma'a *

Electrical Engineering Department, College of Engineering, King Saud University, Riyadh 11421, Saudi Arabia; shaalan@ksu.edu.sa (A.M.A.-S.); hfarh1@ksu.edu.sa (H.M.H.F.)
* Correspondence: homotoso@ksu.edu.sa (H.O.O.); ashammaa@ksu.edu.sa (A.A.A.-S.)

Abstract: Electrification of remote rural areas by adopting renewable energy technologies through the advancement of smart micro-grids is indispensable for the achievement of continuous development goals. Satisfying the electricity demand of consumers while adhering to reliability constraints with docile computation analysis is challenging for the optimal sizing of a Hybrid Energy System (HES). This study proposes the new application of an Artificial Ecosystem-based Optimization (AEO) algorithm for the optimal sizing of a HES while satisfying Loss of Power Supply Probability (LPSP) and Renewable Energy Fraction (REF) reliability indices. Furthermore, reduction of surplus energy is achieved by adopting Demand Side Management (DSM), which increases the utilization of renewable energy. By adopting DSM, 28.38%, 43.05%, and 65.37% were achieved for the Cost of Energy (*COE*) saving at 40%, 60%, and 80% REF, respectively. The simulation and optimization results demonstrate the most cost-competitive system configuration that is viable for remote-area utilization. The proposed AEO algorithm is further compared to Harris Hawk Optimization (HHO) and the Future Search Algorithm (FSA) for validation purpose. The obtained results demonstrate the efficacy of AEO to achieve the optimal sizing of HES with the lowest *COE*, the highest consistent level, and minimal standard deviation compared with HHO and FSA. The proposed model was developed and simulated using the MATLAB/code environment.

Keywords: Demand Side Management; Artificial Ecosystem-based Optimization; optimal sizing; Hybrid Energy System; Future Search Algorithm; Harris Hawk optimization; Renewable Energy Fraction; surplus energy; Loss of Power Supply Probability

Citation: Omotoso, H.O.; Al-Shaalan, A.M.; Farh, H.M.H.; Al-Shamma'a, A.A. Techno-Economic Evaluation of Hybrid Energy Systems Using Artificial Ecosystem-Based Optimization with Demand Side Management. *Electronics* **2022**, *11*, 204. https://doi.org/10.3390/electronics11020204

Academic Editor: Carlos Andrés García-Vázquez

Received: 9 November 2021
Accepted: 4 January 2022
Published: 10 January 2022

1. Introduction

The reduction of conventional energy sources coupled with increasing global warming have accelerated the growth of renewable energy sources (RES) such as solar and wind [1]. RES can be exploited for both grid connection and for off-grid, especially in rural areas with restricted grid connection [2–4]. This will reduce fossil fuel dependency, harmful emissions, and consumption costs. In spite of their benefits, RES performance is often limited due to the intermittent and unpredictable nature of their output power [5]. These factors affects both energy production and the operational costs of the system. Connecting two or more RES is essential to sustain energy for remote areas by preserving the quality and reliability of power [6]. Likewise, incorporating diesel generators with RES can ensure service quality and reliability and results in less battery maintenance [7,8]. Therefore, it is essential to optimize Hybrid Energy System (HES) size to minimize installation and maintenance cost.

Several studies have adopted analytical, probabilistic, and heuristic techniques in achieving HES optimization. The probabilistic technique is modelled on the random prospect of a particular system [9,10]. However, this method is not robust for obtaining optimal result with dynamic change in RES. The analytical technique is only capable of handling simple and precise methods [11,12]. The potent results of the heuristic methods [13–20] make them dependable for complex optimal sizing of HES with sufficient computational time. In [13], the Harmony Search (HS) algorithm was utilized for sizing PV/diesel energy sources. The obtained result in [13] demonstrated the robustness of HS in reducing pollution and system cost. Masoud et al. [14] demonstrated the effectiveness of Dynamic Multi-Objectives Particle Swarm Optimization (DMOPSO) for sizing the HES. By utilizing the DMOPSO algorithm, the results revealed a decrease in PV panel cost and NPC. In [15], a genetic algorithm combined with particle swarm optimization was used for the optimal design of a hybrid wind–PV–battery system. Loss of Load Expected (LOLE) and Loss of Energy Expected (LOEE) were subjected to constraints to minimize operation cost using multi-objective particle swarm optimization [16]. Eftichios et al. [17] presented a genetic algorithm using the system cost function subject to Loss of Load Hours (LOLH) constraints for HES optimization. Minimization of *COE* is presented in [18] to optimize the size of the HES subject to Renewable Energy Fraction (REF) and LPSP constraints. In [17], the levelized cost of energy was obtained through optimal sizing of PV, wind, battery, and diesel generator. Wang et al. [20] proposed a non-dominated sorting algorithm II, incorporated with re-ranking based genetic operators to determine system reliability, greenhouse gas emission, and lifetime cost for optimal sizing of the HES. Crow Search (CS) and Particle Swarm Optimization (PSO) were introduced in [21,22] for optimal sizing and allocation of renewable distributed generations. Morteza et al. presented a solar-hydrogen model for hybrid energy sizing using Demand Side Management (DSM) [23]. The results revealed the decrease in electricity cost; however, the model did not utilize system constraints. Hassan et al. [24] presented a strategy to allocate renewable distributed generations using CS with PSO.

The majority of the optimal sizing of HES in the literature did not apply DSM concepts. DSM is the future of power systems, incorporating cutting-edge technologies and distributions system for supplying electricity in efficient and smart ways [25]. DSM buttresses smart grid in several ways, such as control, electricity cost reduction, monitoring, and management of energy resources [26]. DSM can help to reduce load peak, which increases the efficiency of grid operation, promotes a decrease in the greenhouse gas effect, and leads to electricity bill reduction. Effective DSM can avoid the upgrading of electrical infrastructures such as transmission lines and distribution networks. Some of the generic method of DSM [27] are valley filling, load conservation, peak clipping, load growth or load establishment, and load shifting. Peak clipping is the reduction of peak loads without shifting it to off-peak hours. Valley filling involves encouraging consumption during the off-peak period. Load conservation involves encouraging consumers to make use of efficient appliances to reduce energy wastage. Load growth is adopted in a situation of surplus energy that might be the result of the integration of renewable energy. Load shifting is most commonly utilized and involves the shifting of load from peak periods to off-peak periods or periods where the energy cost is cheap [27]. Cost saving is the major purpose of the load-shifting process [28]. This method will be utilized in this study for the optimal sizing of the HES. Figure 1a–d shows a pictorial representation of various DSM strategies.

Thus, the foremost objective of this study is to extend the work presented in [29] by incorporating DSM for the optimal sizing of HES for remote communities of Al Sulaymaniyah village, Saudi Arabia, using a new optimization algorithm. The evaluation of three algorithms: Artificial Ecosystem-based Optimization (AEO) [30], Future Search Algorithm (FSA) [31], Harris Hawk Optimization (HHO) [32], were evaluated for optimal sizing. These algorithms were analyzed for optimizing Cost of Energy (*COE*) at zero Loss of Power Supply Probability (LPSP). Additionally, the Renewable Energy factor is proposed to ensure a hybrid energy utilization among the energy sources: diesel, battery storage, PV, and wind. The results demonstrated the efficacy of AEO in obtaining the lowest *COE* with and without DSM.

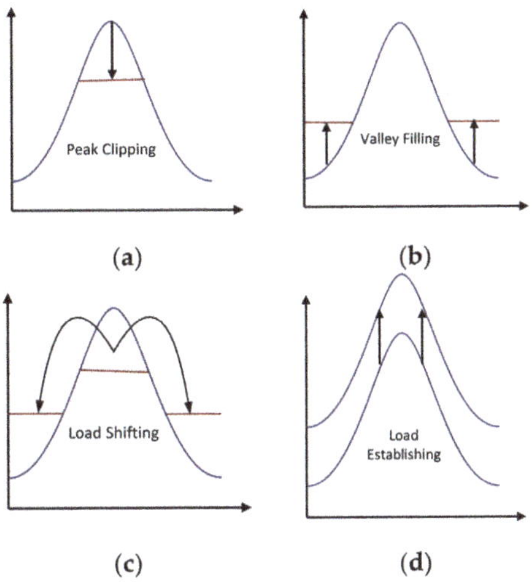

Figure 1. (**a**) Peak Clipping; (**b**) Valley Filling; (**c**) Load Shifting; (**d**) Load establishment.

The novelty and contribution of this paper can be summarized as follows:

- A new application of Artificial Ecosystem-based Optimization is utilized for the first time to achieve optimal sizing of HES by minimizing the Cost of Energy (*COE*);
- Renewable Energy Fraction (REF) and Loss of Power Supply Probability (LPSP) are utilized to achieve stand-alone HES consisting of PV/WTGs/battery/diesel for Al Sulaymaniyah village, Saudi Arabia;
- A load-shifting strategy based on the available renewable is employed for the DSM to achieve a minimal cost of energy;
- AEO is compared to FSA and HHO with DSM and without DSM in achieving the *COE* and to verify its efficacy;
- Different values of REF at 40%, 60%, and 80% are utilized as constraints to determine the *COE* with DSM and without DSM;
- The results demonstrated the effectiveness of AEO to achieve the lowest *COE*, both with DSM and without DSM.

The rest of the paper is arranged as follows: Section 2 describes the Al Sulaymaniyah site and load; Section 3 proposes the HES configuration; Section 4 discusses the AEO algorithm; Section 5 describes the power management approach; Section 6 introduces the developed DSM; Section 7 describes the reliability indices; Section 8 illustrates the objective function; Section 9 discusses the results; and Section 10 presents the conclusion.

2. Al Sulaymaniyah Site Description and Meteorological Data

The geographical location of Al Sulaymaniyah is shown in Figure 2 [29]. The village is located in the northern part of Saudi Arabia at an elevation of 1820 ft above sea level. The location has enormous availability of wind and solar energy. Al Sulaymaniyah village comprises mosques, a water pump, a small hospital, and recreation facilities. The village is supplied through diesel generators by the Saudi Electricity Company (SEC). The climatic conditions at the village are continental. The village's temperature ranges between 0 °C and 45 °C throughout the year. The summer period is usually from May to September, while the winter season is from November to February. The day period varies throughout the year. The lengthiest day occurs in June, with 14 h of sunlight, and the shortest day occurs in December, with 10 h of sunlight.

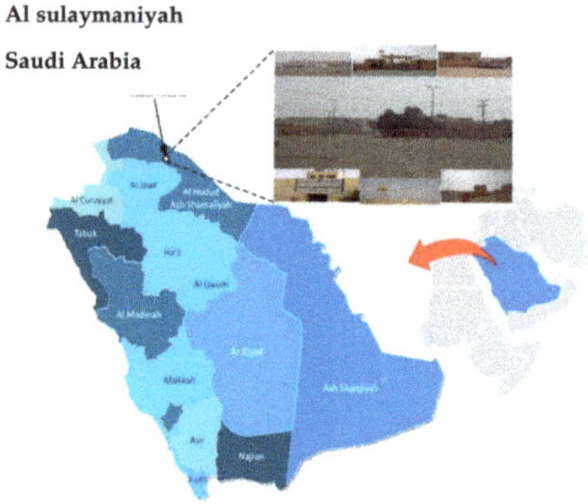

Figure 2. Al Sulaymaniyah village, Saudi Arabia.

The measured average hourly energy consumption at Al Sulaymaniyah village comprises 8760 data points. Currently, the village is powered by three 456 kW diesel generators connected in parallel to a common bus. The generated power is connected to a 1250 kVA step-up transformer. The full description of Al Sulaymaniyah meteorological data and load pattern are described in [29].

3. Configuration of the HES

The proposed structure for the Hybrid Energy System is presented in Figure 3 [33]. This comprises battery storage, the inverter for converting DC energy sources to AC, diesel generators, load, PV, and wind energy sources. In this study, PV and wind are considered as the two major sources of energy. Either the diesel generator or storage energy system compensates the disparity between PV and wind. Al Sulaymaniyah's low PV and wind coefficient of correlation implies that PV and wind might not be able to adequately supply the energy required without the utilization of energy storage and diesel sources. PV and storage system are connected to the DC side of the bus bar, while the diesel generator and wind turbines are linked to the AC side of the bus bar. The diesel generator serves as an alternative source of energy whenever the PV, wind, and storage system are unable to meet the load demand.

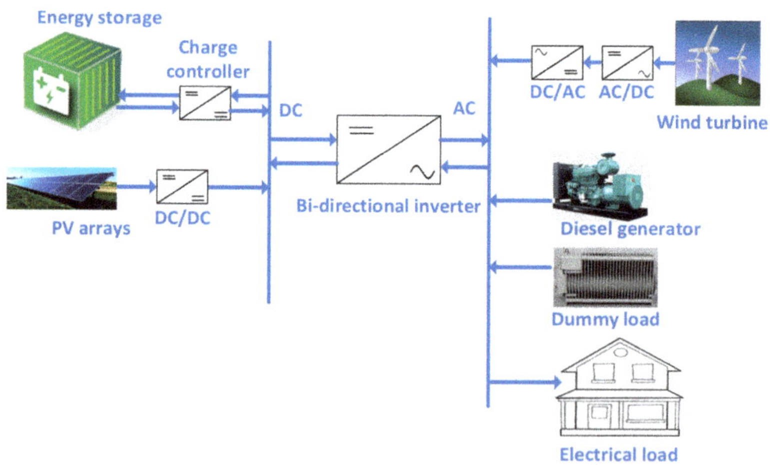

Figure 3. Overview of the proposed Hybrid Energy System.

The modelling of the PV, wind, storage system, and diesel generator is introduced as follows.

3.1. PV Modelling

Solar irradiance, PV array manufacturer's data, and temperature are major factors that influence the output power generated from PV systems, as shown in (1) [34].

$$P_{PV}= P_r f pv\left[\frac{\overline{G_T}}{\overline{G_{T,STC}}}\right](1+\alpha p(T_c - T_{c,STC})) \tag{1}$$

where P_r, $T_{c,STC}$, T_c, $\overline{G_T}$, $\overline{G_{T,STC}}$, αp, and fpv are the PV rated power, PV temperatures under STC and normal conditions, global irradiance under normal and Standard Test Conditions (STCs), power temperature coefficient, and de-rating factor, respectively. The temperature of the PV steady-state operation is described in (2).

$$T_c = \frac{T_a + (NOCT - T_{a,\,NOCT})(1 - 1.11_{\eta\mathrm{MPP}}(1 - \alpha p T_{c,STC}))\left(\frac{\overline{G_T}}{\overline{G_{T,NOCT}}}\right)}{1 + 1.11(\alpha\rho\eta_{\mathrm{MPP},STC})\ (NOCT - T_{a,NOCT})\left(\frac{\overline{G_T}}{\overline{G_{T,NOCT}}}\right)} \tag{2}$$

where $NOCT$ and T_a are operating cell nominal and ambient temperature, respectively. $G_{T,NOCT}$ and $T_{a,NOCT}$ are, respectively, the solar irradiance and ambient temperature with respect to $NOCT$; η_{MPP} and $\eta_{\mathrm{MPP},STC}$ are the Maximum Power Point (MPP) efficiency of the PV module and efficiency under STCs, respectively.

3.2. Wind Turbine Generator (WTG) Modelling

The approximated curve of power output of a wind turbine system can be expressed as follows [35]:

$$P_{WT}(u) = P_r \times \begin{cases} 0 & u \le u_c \text{ or } u > u_f \\ \frac{u^2 - u_c^2}{u_r^2 - u_c^2} & u_c < u \le u_r \\ 1 & u_r \le u \le u_f \end{cases} \tag{3}$$

where u, u_c, P_r, u_r, and u_f are wind speed, cut-in/starting speed of wind turbine, rated power of wind turbine, rated speed, and cut-out/furling speed, respectively. It is evident from Equation (3) that wind turbine output power depends on five parameters, which are the four wind turbine parameters and the site wind speed (u). The four wind turbine parameters are u_c, P_r, u_r, and u_f. Based on a quadratic model, a typical wind turbine output power characteristic curve is shown in Figure 4 [29].

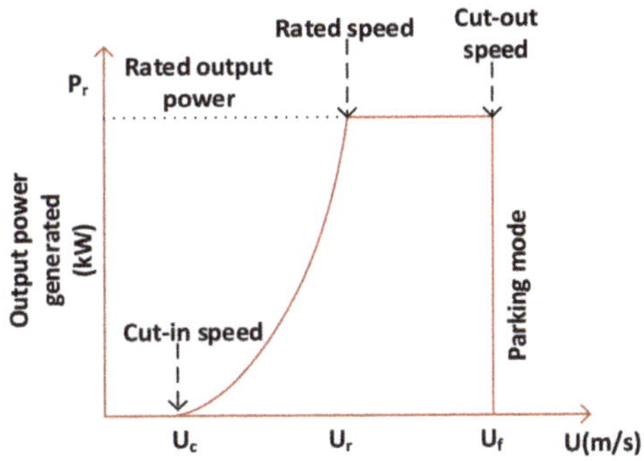

Figure 4. Wind turbine output power characteristics curve.

3.3. Storage System Modelling

One of the critical aspects of RES is the storage system because of the stochastic nature of RES output power. The depth of discharge and battery bank nominal capacity are important factors to be considered in storage system. The load demand determines the energy flow through the storage system. The State of Charge (*SOC*) of the battery bank is represented as (4) [29].

$$SOC(t) = SOC(t-1)(1-\sigma) + \left(P_{GA}(t) - \frac{P_L(t)}{\eta_{inv}}\right)\eta_{battery} \quad (4)$$

where $P_L(t)$, η_{inv}, $\eta_{battery}$, $P_{GA}(t)$, and σ are total load demand, net inverter efficiency, round trip efficiency of battery within [0.5, 0.95], total output power, and self-discharging rate, respectively. The charging and discharging operation of the storage system is described by the positive and negative second term of Equation (4). The constraints for the storage system State of Charge (*SOC*) are described in (5).

$$SOC(t) = \begin{cases} SOC_{min} & SOC(t) < SOC_{min} \\ SOC(t) & SOC_{\min} < SOC(t) < SOC_{max} \\ SOC_{max} & SOC(t) > SOC_{max} \end{cases} \quad (5)$$

The *SOC* cannot exceed the maximum limit of SOC_{max}, and likewise the minimum *SOC* must be equal to or exceed the minimum allowable limit, SOC_{min}.

3.4. Diesel Generator

The diesel generator is utilized whenever the storage and RES is insufficient to sustain load demand. The annual fuel consumption and the cost is represented as (6) [35].

$$C_{Dsl} = C_F \sum_{t=1}^{8760} A \times P_{Dsl}(t) + B \times P_R \quad (6)$$

where $C_{Dsl,}$, P_R, C_F, and $P_{Dsl}(t)$ are the annual fuel consumption (L/h), nominal power, cost of fuel per liter, and generated power (kW) of diesel, respectively. A, B are the fuel constants (L/kW). The two main import factor that influences the fuel consumption are both the output power and nominal power. The output power of the diesel generator is recommended by the manufacturers between a preset minimum level and nominal power.

4. Artificial Ecosystem-Based Optimization (AEO)

AEO is a type of population-based optimization algorithm. It imitates the production, decomposition, and consumption behavior of living organisms. This concept depicts energy flow in the ecosystem of the Earth. In the Earth's ecosystem, producers use water, sunlight, and carbon dioxide to make food energy. Similarly, the AEO production stage is used to improve the balance between exploration and exploitation. The second stage, consumers, just like animals, cannot produce their own food. They acquire their energy and nutrients from fellow consumers or producers. This stage improves the exploration of the algorithm. Decomposition is the last stage. Decomposers feed on both consumers and producers. The exploitation of the algorithm is enriched at this stage [30].

4.1. Producer

In the search space at this stage, a random individual (x_{randi}) and a best individual (x_q) are randomly produced. The decomposer (best individual) and the search space's upper and lower boundaries update the producer (worst individual). This update will facilitate other individuals to hunt for separate regions. The mathematical representation for this is illustrated in (7) [30].

$$x_1(ti+1) = (1-a)x_q(ti) + ax_{randi}(ti) \quad (7)$$

where

$$a = \left(1 - \frac{ti}{It}\right) * r_1 \quad (8)$$

$$x_{randi} = r * (Up - Lw) + Lw \quad (9)$$

$$x_2(ti+1) = x_2(ti) + K[x_2(ti) - x_1(ti)] \quad (10)$$

where r_1 and r are random numbers within 0 and 1, It is maximum iterations, a is linear weighting coefficient, q is the population number, and Lw and Up are the lower and upper boundaries, respectively. The flight levy, K, is for enhancement of the exploration level. This is represented in (11) [36]:

$$K = 0.5 * v_1 / |v_2| \quad (11)$$

where $v_2 = N(0,1)$ and $v_1 =\sim N(0,1)$. $N(0,1)$ denotes normal distribution.

4.2. Consumption

This stage enhances the exploration by letting the algorithms upgrade the solution of individuals. The consumers can be classified as herbivore, carnivore, or omnivore. The herbivore feeds on both the consumers and the producers. The second feed on the consumers with an advanced energy level. The last feeds on the producers and/or consumers with an advanced energy level. The mathematical equation for herbivore as a consumer is represented as follows [36]:

$$x_i(ti+1) = x_i(ti) + K[x_i(ti) - x_i(ti)],\ i \in [3, \ldots, n] \quad (12)$$

The mathematical equation for consumer as a carnivore is represented as follows:

$$\begin{cases} x_i(ti+1) = x_i(ti) + K\left[x_i(ti) - x_j(ti)\right], \\ i \in [3, \ldots, n],\ j = randi([2\ i-1]) \end{cases} \quad (13)$$

The mathematical equation for an omnivore consumer is denoted as:

$$\begin{cases} x_i(ti+1) = x_i(ti) + K\left[r_2(x_i(ti) - x_1(ti)) + (1-r_2)(x_i(ti) - x_j(ti))\right], \\ i \in [3, \ldots, n],\ j = randi([2\ i-1]) \end{cases} \quad (14)$$

where r_2 is within the range of 0 and 1.

4.3. Decomposers

The decomposition stage is very important because it feeds the producers and encompasses the food chain. In Earth's ecosystem, when a consumer passes away, the decomposers feed on its leftovers. The mathematical Equation (15) for the decomposer model introduces new factors. The factors are weight coefficients (w_e and h_e) and decomposition factor (D_e). The decomposition supports the exploitation of the algorithm by updating individual locations based on the best solution, as described in (16) [30].

$$x_i(ti+1) = x_n(ti) + D_e(w_e x_n(ti) - h_e x_i(tt)),\ i = 1, \ldots, n \quad (15)$$

The newly introduced factors are defined by Equation (16), shown below.

$$\begin{cases} D_e = 3u, & u \sim N(0,1) \\ w_e = r_3 * randi([1,2]) - 1 \\ h_e = 2 * r_3 - 1 \end{cases} \quad (16)$$

where r_3 is within the range of 0 and 1.

4.4. Termination

During this stage, x_q is updated after the fitness is obtained for each individual. Subsequently, the termination condition is evaluated; if it is met, the execution is terminated and x_q is returned, or else the first stage is repeated. Figures 5 and 6 show the typical flow of energy in Earth's ecosystem and the AEO ecosystem [30], and Algorithm 1 summarized the pseudocode AEO algorithm.

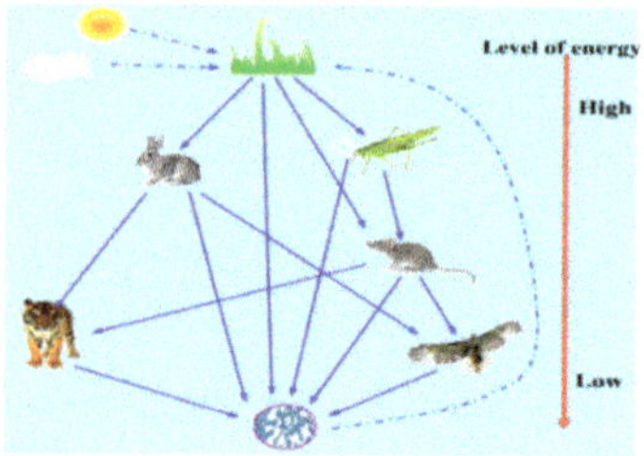

Figure 5. Energy flow in an ecosystem.

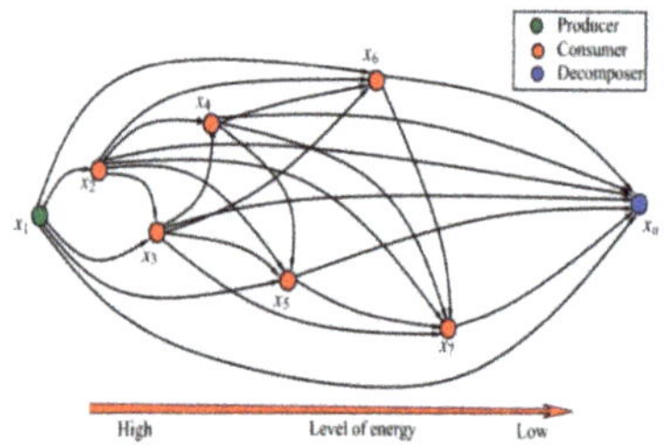

Figure 6. AEO ecosystem.

Algorithm 1: Pseudocode of Artificial Ecosystem-based Optimization for optimal the sizing of HES

---

1. Initialization: Random initialization of AEO ecosystem, x_1, and evaluation of fitness, f_{fi}; x_q = best solution established so far.
2. **While** the halt condition is not obtained, perform:
  First Stage: Production
    Individual x_1, update its position with (7).
  Second Stage: Consumption
    Individual x_1 (i = 2, ... ,n)
  Herbivorous act occurs
    If $rand < \frac{1}{3}$, the individual update is carried out using (12)
  Omnivorous act occurs
    Else if 1/3 and *rand* < 2/3 the update to individual is carried out using (14)
  Carnivorous act occurs
    Else the individual update is carried out with (13)
    End if
  End if
  Third stage: Decomposition
   Individual update is carried out with (15)
   Individual fitness is calculated
   Best position found so far is updated, x_q
   End while
  Fourth Stage: Termination
    Return x_q

---

5. Power Management Approach

The unstable nature of renewable resources creates a multifaceted power management scheme for HES. Due to the fact that the quantity of energy generated from resources is paltry, the capability of the generator cannot be instantaneously amplified to balance the increase in load. Moreover, occasionally, a dump load is required to dispel the surplus when the quantity of electricity generated is more than the demand. Moreover, this will avoid the overcharging of the batteries. To design an effective system, it is essential to adopt a power management methodology. Table 1 shows the parameters used for the simulation [35]. The following cases are considered for the power management methodology.

Case 1: The required energy is supplied by renewable generation and the surplus is utilized for charging the storage facility.

Case 2: As in Case 1, if the excess energy is greater than the storage capacity and load, the surplus is dissipated via a dump load.

Case 3: The storage facility is utilized as a priority when the renewable energy is insufficient to meet the load rather than making use of diesel.

Case 4: The generator is used to charge the battery and to supply the load when the renewable energy is insufficient to sustain the load and the storage is depleted.

Table 1. Parameters of the HES system.

	Parameters	Values	Unit
Battery	Depth of Discharge	60	%
	Round-trip efficiency	80	%
	O&M Cost	5	USD/kWh/year
	Capital cost	200	USD/kWh
	Life span	5	years
Photovoltaic Module	O&M Cost	15	USD/kW/year
	Capital cost	1000	USD/kW
	Efficiency	16	%
	Life span	20	years
WTG	Hub-height	60	M
	Cut-in/cut-off/ rated Speed	3/25/9.5	m/s
	O&M Cost	30.33	USD/KW/year
	Capital cost	1300	USD/kW
	Life span	20	years
DC/AC Converter	Life span	10	years
	Capital cost	133	USD/kW
Diesel Generator	Life span	15,000	hours
	Capital cost	300	USD/kW
	O&M Cost	0.012	USD/kWh
Project Factors	Interest rate	3	%
	Life span	20	years
	Inflation rate	2	%

6. Demand Side Management (DSM)

Demand Side Management modifies the consumption patterns of consumer electricity to yield the desired alterations in the load structure of power distributions. The resulted changes are expected to reflect the planned objectives. This study presents a load-shifting technique to maximize the utilization of renewable energy resources and reduce the energy supply from the diesel. The objective load curve is aimed at reducing the load during periods of insufficient renewable energy and likewise increasing the load during periods of surplus renewable energy. The DSM strategy cannot be scheduled beyond a day, and the shifted load is limited to 20% of the total load. If the renewable energy is less than the load and within 24 h, as represented in (17), the load shift is described as (18). If (18) is greater than 20% of the load, the load shift is donated as (19). More so, by considering the future surplus available renewable energy, the load shift is limited to (20) if the available renewable energy is less than the expected shifted load in (19). The load at time, t, and future time, $t+i$, are represented as (21) and (22), respectively. The methodology also applies when there is surplus renewable energy.

$$P(t) + P(w) < \left(\frac{P1(t)}{univ}\right) \& \, t < 24 \text{ h} \tag{17}$$

$$lt = (\left(\frac{P1(t)}{univ}\right) - (Pw(t) + Pp(t)) \tag{18}$$

$$lt = 0.2 * P1(t) \tag{19}$$

$$lt = P(ren) \tag{20}$$

$$P1(t) = P1(t) - lt \tag{21}$$

$$P1(t) = P1(t) - lt \tag{22}$$

where $P(t)$, $P(w)$, $P1(t)$, $univ, P(ren)$, and lt represents PV power, wind power, load, inverter efficiency, surplus renewable energy, and load shift at time, t, respectively.

7. Reliability Indices

7.1. Loss of Power Supply Probability

The probability of power supply failure can be measured using one of the statistical reliability indices, Loss of Power Supply Probability (LPSP). Loss of power can be as a result of technical failure or low energy supply from the generating source that is not able to meet the demand [30]; 0% and 1% LPSP demonstrate, respectively, that the load will not be supplied and the load will be supplied. This can be calculated as described in (23) [35].

$$LPSP = \frac{\sum(P_{load} - P_{pv} - P_{wind} + P_{soc_{min}} + P_{diesel})}{\sum P_{load}} \tag{23}$$

where P_{load}, $P_{pv}, P_{wind}, P_{soc_{min}}$, $and\ P_{diesel}$, are load demand, PV power, wind power, minimum battery power, and diesel power, respectively.

7.2. Renewable Energy Fraction (REF)

This is the ratio of diesel energy generated compared to energy generated from both the wind and the PV system, as shown below. The fraction is an indication of how much renewable energy is served. A fraction of 100% represents an ideal system, with only renewable energy resources. An equal sum of PV and wind power to diesel power represents a fraction of 0%. A fraction between 0% and 100% represent a hybrid energy source. *REF* is represented as (24) [29].

$$REF = \left(1 - \frac{\sum P_{diesel}}{\sum P_{wind} + P_{pv}}\right) * 100 \tag{24}$$

8. Objective Function

Appropriate sizing of the energy system with high quality units can lead to resourceful energy management and low-cost electricity for the consumers. The aim of the optimization is to obtain a minimal Cost of Energy (*COE*). *COE* is a widely used indicator for HES financial profitability [31]. This is defined as the ration of cost to unit price of electricity, as described in (25), subject to constraints (26) [29,35].

$$COE\left(\frac{\$}{kWh}\right) = \frac{Annualized\ cost(\$)}{annual\ energy\ delivered\ by\ the\ system\ (kWh)} = \frac{Total\ Net\ Present\ Cost\ (\$)}{P_{load}(kW)\left(\frac{8760h}{year}\right)} * CRF \quad (25)$$

$$\text{Constraints}\begin{cases} LPSP \leq LPSP_{desired} \\ REF \leq REF_{desired} \\ 0 \leq P_{Wind} \leq P_{wind,max} \\ 0 \leq P_{pv} \leq P_{pv,max} \\ 0 \leq P_{bat} \leq P_{bat,max} \\ 0 \leq P_{DSL} \leq P_{DSL,max} \end{cases} \quad (26)$$

The total NPC comprises installed capital costs. The installed capital costs includes the replacement cost, operation cost, maintenance cost, and present cost. P_{load} is the power consumed per hour. The current worth of HES components for a specific duration with the interest rate is calculated with the Capital Recovery Factor (*CRF*), which is shown below in (27) [35].

$$CRF = \frac{i * (1+i)^n}{(1+i)^n - 1} \quad (27)$$

where n is the system life period, which is represented by PV life duration because of its lengthier life, and i represents interest rate. P_{wind}, P_{pv}, P_{DSL}, and P_{bat} represent the rated power of wind, solar, diesel, and battery (in kWh), respectively, with their maximum denoted as $P_{wind,max}$, $P_{pv,max}$, $P_{DSL,max}$, and $P_{bat,max}$. $REF_{desired}$ and $LPSP_{desired}$ denote the defined *REF* and LPSP, respectively.

9. Simulation Results and Discussions

The proposed AEO, in addition to HHO and FSA, were employed to determine the optimal sizing of the stand-alone Hybrid Energy System to obtain the minimal Cost of Energy (*COE*). The hourly average load, wind speed, and solar irradiance of Al Sulaymaniyah village, Saudi Arabia, were selected for the study of optimal energy sizing in this study. The variables for the optimization are the capacity of the PV energy source, batteries, diesel generators, and wind turbine energy. The performance of the algorithms was evaluated under four performance indicators, which are Standard Deviation (STD), mean, best solution, and worst solution. The metaheuristics algorithm involves a stochastic model in their computational analysis. As a result, each run of the algorithm will not yield the same result. This creates an enormous challenge to determine the quality of the solution. Hence, it is crucial to investigate the consistency of the solution. By running the algorithms for 30 runs, a unique distinct pattern is obtained for the algorithms, which implies that the parameters always unite at a definite location. To obtain a rational comparison, the algorithms were evaluated under the same population sizes, iterations, boundary conditions, and number of runs. The simulation was developed using the MATLAB environment. Moreover, different levels of REF at 40%, 60%, and 80% at an LPSP of 0% were employed to achieve the *COE* and optimal sizing.

9.1. Optimal Sizing of the HES without DSM

One substantial standard for evaluating the performance of any optimization method is the assessment of the obtained fitness results. Tables 2–4 show the Standard Deviation (STD), mean, best solution, and worst solution obtained for REF at 40%, 60%, and 80%, respectively. These results were evaluated during 30 runs, 300 iterations, and a population size of 50 without DSM at 0% LPSP. Figures 7–9 show the fitness curve of the three algorithms without DSM at 40%, 60%, and 80% REF, respectively. From the results, it is obvious that AEO obtained the minimum *COE* for all three stages. The algorithm further maintains a low STD value in obtaining the *COE*, as shown in Tables 2–4.

Table 2. Optimal, Minimum, Maximum, and STD for AEO, FSA, and HHO at 40% REF and LPSP of 0%.

		COE (USD/kWh)	P_{batt} (kW)	PV (kW)	Pw (kW)	P_{diesel} (kW)	Surplus (%)
AEO	Optimal	1.383630	62.46	167.98	102.14	168.56	35.05
	Mean	1.393066	59.96	154.99	108.26	168.50	35.82
	Max	1.467197	61.53	81.25	142.95	171.01	40.19
	STD	0.016362	10.03	25.08	10.71	1.29	1.11
FSA	Optimal	1.463650	0.00	166.14	111.88	164.88	36.17
	Mean	1.500501	34.71	161.68	109.69	167.88	37.12
	Max	1.993242	37.52	156.82	109.91	166.11	40.00
	STD	0.23622	28.89	67.40	19.13	8.76	4.07
HHO	Optimal	1.392138	60.77	147.83	110.09	168.35	35.32
	Mean	1.500154	57.73	122.63	123.51	173.22	38.55
	Max	1.799334	47.34	7.50	190.97	189.51	51.14
	STD	0.106058	26.01	65.86	29.47	6.33	4.05

Table 3. Optimal, Minimum, Maximum, and STD for AEO, FSA, and HHO at 60% REF and LPSP of 0%.

		COE (USD/kWh)	P_{batt} (kW)	PV (kW)	Pw (kW)	P_{diesel} (kW)	Surplus (%)
AEO	Optimal	1.133689	69.55	190.20	129.39	162.26	39.27
	Mean	1.143942	65.13	180.08	135.52	162.00	40.39
	Max	1.206799	50.35	102.23	177.15	162.26	46.53
	STD	0.016446	8.80	28.97	14.17	1.00	1.83
FSA	Optimal	1.134227	66.36	185.97	131.59	161.82	40.39
	Mean	1.154109	27.32	198.22	133.23	159.85	41.24
	Max	1.177233	0.00	250.00	118.59	159.41	39.19
	STD	0.015713	31.73	19.06	7.21	1.66	2.11
HHO	Optimal	1.138241	56.38	211.12	123.02	161.48	39.65
	Mean	1.206398	62.38	157.37	146.20	165.00	42.25
	Max	1.596142	8.81	39.77	190.56	175.87	50.89
	STD	0.095449	29.07	58.09	26.11	5.46	3.37

Table 4. Optimal, Minimum, Maximum, and STD for AEO, FSA, and HHO at 80% REF and LPSP of 0%.

		COE (USD/kWh)	P_{batt} (kW)	PV (kW)	Pw (kW)	P_{diesel} (kW)	Surplus (%)
AEO	Optimal	0.832235	71.22	244.09	189.75	150.20	49.71
	Mean	0.837487	66.17	224.90	200.15	149.78	51.3
	Max	0.853710	63.39	178.53	225.41	149.48	53.77
	STD	0.004773	10.14	18.05	9.25	1.23	2.53

Table 4. *Cont.*

		COE (USD/kWh)	P_{batt} (kW)	*PV* (kW)	*Pw* (kW)	P_{diesel} (kW)	*Surplus* (%)
FSA	Optimal	1.224227	75.22	200.09	180.75	123.20	48.71
	Mean	1.247038	60.17	221.90	210.15	159.78	50.3
	Max	1.336750	60.39	188.53	221.42	159.28	53.77
	STD	0.234455	22.41	67.05	29.25	8.23	3.13
HHO	Optimal	0.835613	74.05	239.33	190.54	150.56	49.61
	Mean	0.922779	73.36	203.67	190.57	155.05	47.8
	Max	1.702104	78.25	11.09	199.96	189.03	44.47
	STD	0.156130	20.19	53.99	11.74	7.85	2.33

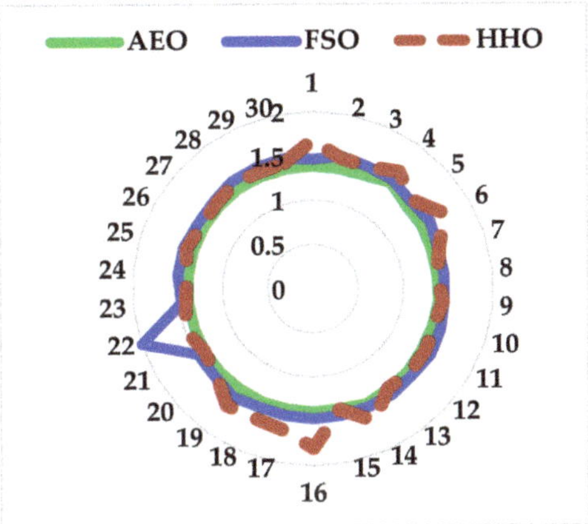

Figure 7. Fitness for cost of energy without DSM at 40% REF and 0% LPSP.

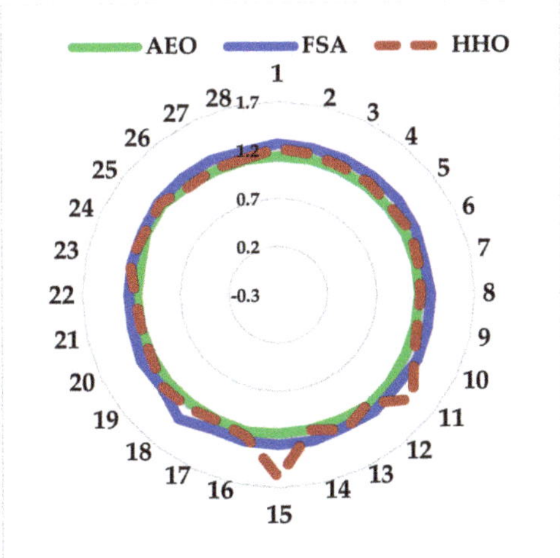

Figure 8. Fitness for cost of energy without DSM at 60% REF and 0% LPSP.

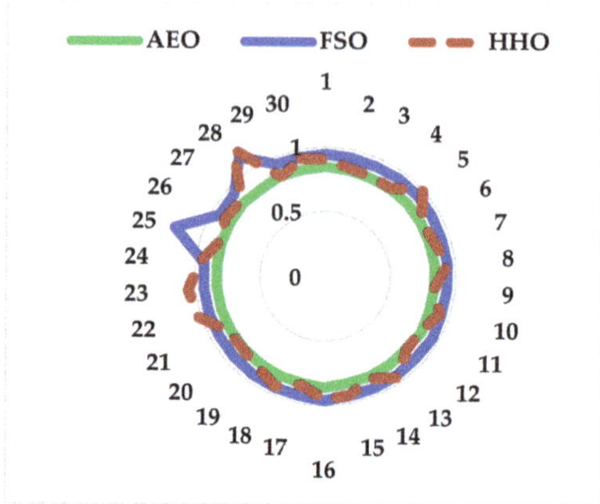

Figure 9. Fitness for cost of energy without DSM at 80% REF and 0% LPSP.

Figure 10 shows the percentage energy utilization for each of the algorithms without DSM. The run with the lowest *COE* for each algorithm was selected for the energy utilization plot. AEO utilizes the right combination of PV and wind energy sources to obtain the minimal cost of energy. Figure 10 shows that FSA utilizes the highest percentage for diesel energy. This influences its *COE* to be highest among the algorithms.

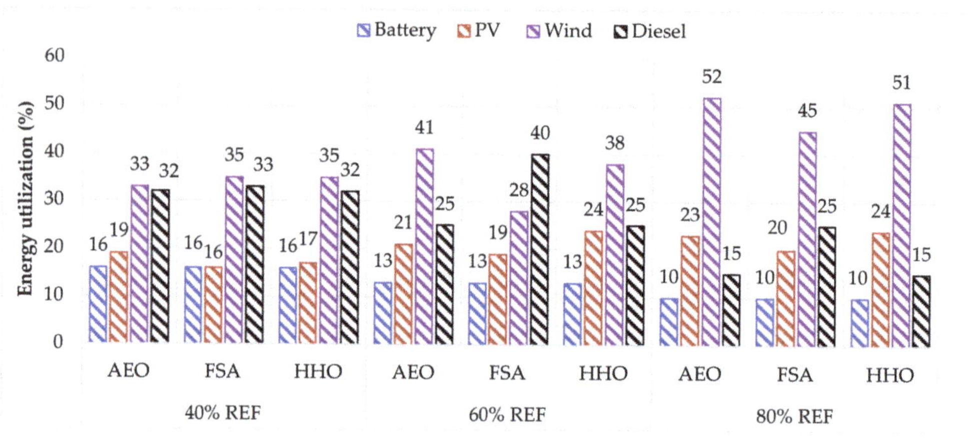

Figure 10. Energy utilization percent for AEO, FSA, and HHO without DSM and at 0% LPSP.

9.2. Optimal Sizing of the HES with DSM

The DSM is performed to optimize the usage of the available renewable energy resources. The loads are shifted in two categories: if there is insufficient renewable energy to meet the load or when there is surplus renewable energy. In the case of the surplus energy, future loads that are less than the available renewable energy are shifted to present time. All these load shifting are limited to a 24 h time frame. Tables 5–7 show the Standard Deviation (STD), mean, best solution, and worst solution obtained for 30 runs under 300 iterations and a population size of 50 for the three stages of Renewable Energy Fraction (REF). Figures 11–13 show the fitness curve (*COE*) for the 30 runs of each algorithm at 40%, 60%, and 80% REF, respectively. It is obvious from Tables 5–7 and Figures 11–13 that AEO achieved the minimal cost of energy at very low STD for all the 30 runs at 0% LPSP.

This further demonstrates the ability of AEO to maintain the lowest *COE* with the highest consistent level and minimal STD.

Table 5. Optimal, Minimum, Maximum, and STD for AEO, FSA, and HHO with DSM at 40% REF and 0% LPSP.

		COE (USD/kWh)	P_{batt} (kW)	*PV* (kW)	*Pw* (kW)	P_{diesel} (kW)	*Surplus* (%)
AEO	Optimal	1.077723	22.41	228.15	62.25	143.73	29.24
	Mean	1.119163	41.46	164.24	86.23	149.94	29.93
	Max	1.267866	58.48	50.08	139.18	159.44	37.23
	STD	0.047366	12.68	50.36	21.67	3.68	2.59
FSA	Optimal	1.087369	26.59	220.29	64.32	144.40	28.94
	Mean	1.195678	10.33	215.97	68.28	143.48	30.38
	Max	1.445218	57.70	129.56	98.27	153.31	29.84
	STD	0.092633	16.50	77.73	31.46	12.79	5.96
HHO	Optimal	1.087369	16.40	202.62	71.30	144.81	29.43
	Mean	1.199288	53.89	123.47	103.10	156.13	32.11
	Max	1.445218	41.38	11.77	154.75	168.20	41.77
	STD	0.091652	24.77	67.14	29.99	7.48	4.23

Table 6. Optimal, Minimum, Maximum, and STD for AEO, FSA, and HHO with DSM at 60% REF and 0% LPSP.

		COE (USD/kWh)	P_{batt} (kW)	*PV* (kW)	*Pw* (kW)	P_{diesel} (kW)	*Surplus* (%)
AEO	Optimal	0.792494	29.27	247.04	76.52	127.85	29.27
	Mean	0.811080	37.90	208.59	90.92	132.09	37.90
	Max	0.922186	64.84	91.04	146.83	144.38	64.84
	STD	0.027468	15.01	35.78	15.93	3.78	3.01
FSA	Optimal	0.816321	59.17	217.05	79.51	117.45	30.17
	Mean	0.962236	37.90	208.59	90.92	132.09	36.91
	Max	1.152819	61.85	91.12	136.13	124.18	66.81
	STD	0.092889	15.01	35.78	15.93	3.78	6.02
HHO	Optimal	0.814320	59.70	128.63	127.71	143.48	30.45
	Mean	0.910234	59.70	128.63	127.71	143.48	35.15
	Max	1.115856	101.25	19.23	191.81	160.80	35.17
	STD	0.068200	24.95	52.29	27.79	6.51	5.12

Table 7. Optimal, Minimum, Maximum, and STD for AEO, FSA, and HHO with DSM at 80% REF and 0% LPSP.

		COE (USD/kWh)	*Pbatt* (kW)	*PV* (kW)	*Pw* (kW)	P_{diesel} (kW)	*Surplus* (%)
AEO	Optimal	0.503245	37.28	242.48	123.52	106.58	36.50
	Mean	0.511628	36.54	215.42	136.24	108.43	37.82
	Max	0.528746	33.89	171.15	159.52	110.70	41.07
	STD	0.007268	17.74	22.49	10.57	2.36	1.42

Table 7. *Cont.*

		COE (USD/kWh)	*Pbatt* (kW)	*PV* (kW)	*Pw* (kW)	P_{diesel} (kW)	*Surplus* (%)
FSA	Optimal	0.644275	74.29	217.11	199.87	125.87	50.0
	Mean	0.644275	77.73	172.01	192.91	147.81	46.33
	Max	0.781750	90.51	21.71	209.00	138.48	42.66
	STD	0.319604	19.94	70.91	9.09	12.90	4.3
HHO	Optimal	0.627125	54.28	247.16	196.86	135.86	49.83
	Mean	0.787858	78.79	176.08	194.92	147.84	45.34
	Max	1.353082	89.50	20.70	200.00	178.48	41.65
	STD	0.217370	18.93	74.90	8.06	11.89	3.3

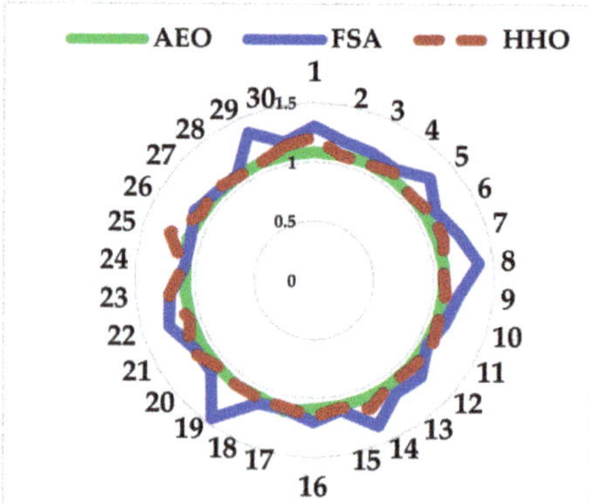

Figure 11. Fitness for cost of energy with DSM at 40% REF and 0% LPSP.

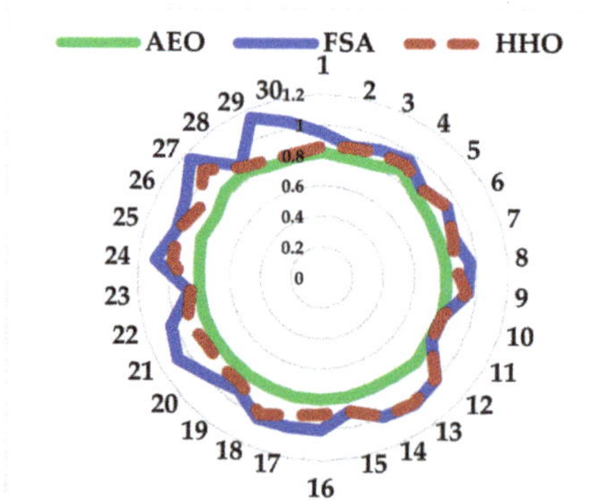

Figure 12. Fitness for cost of energy with DSM at 60% REF and 0% LPSP.

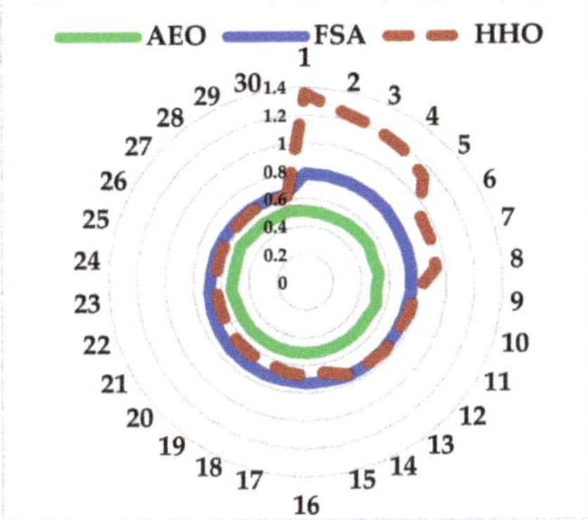

Figure 13. AEO, FSA, and HHO Cost of Energy with DSM at 80% REF and 0% LPSP.

Figure 14 shows the percentage energy utilization for each of the algorithms with DSM. The run with the lowest *COE* for each algorithm was selected for the energy utilization plot. From Tables 5–7 and Figures 11–13, AEO utilizes the right combination of PV and wind energy sources to obtain the minimal cost of energy compared to the other algorithms. In the Figure, FSA utilizes the highest percentage for diesel energy. This influences its *COE* to be highest among the algorithms.

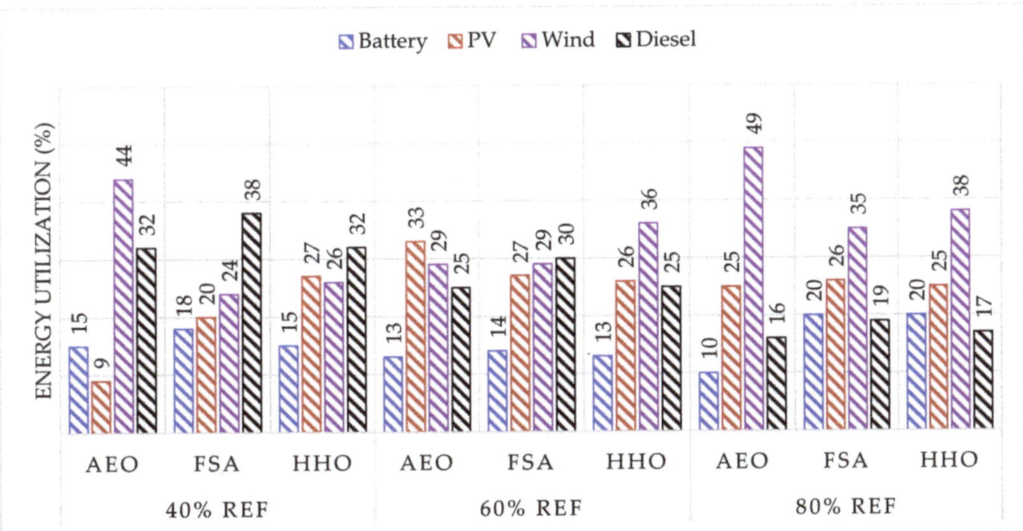

Figure 14. Energy utilization percent for AEO, FSA, and HHO with DSM and at 0% LPSP.

9.3. COE with DSM and without DSM

The DSM in this study is aimed at reducing the surplus renewable energy. Comparing the several results obtained, it can be observed that reducing the surplus energy decreases the *COE*. Because AEO produces the optimal results among the algorithms, Table 8 shows the percentage energy cost savings and *COE* at different REF when DSM is introduced as obtained from AEO. It can be observed that, with the introduction of DSM, greater energy saving is achieved. Figures 15–17 show a typical one-day load profile for the AEO algorithm compared with and without Demand Side Management. The load profile shows how the surplus energy is utilized when there is low renewable energy and surplus renewable energy. In the figures, during the early hours of the day with low renewable energy supply, the loads are shifted to a later period in the day when there is a surplus renewable energy supply.

Table 8. Percentage energy cost saving and *COE* at different REF.

REF (%)	*COE* with DSM (USD/kWh)	Surplus Energy with DSM (%)	*COE* without DSM (USD/kWh)	Surplus Energy without DSM (%)	Percentage Energy Cost Saving (%)
40	1.077723	29.24	1.38363	35.05	28.38
60	0.792494	29.27	1.133689	39.27	43.05
80	0.503245	36.5	0.832235	49.71	65.37

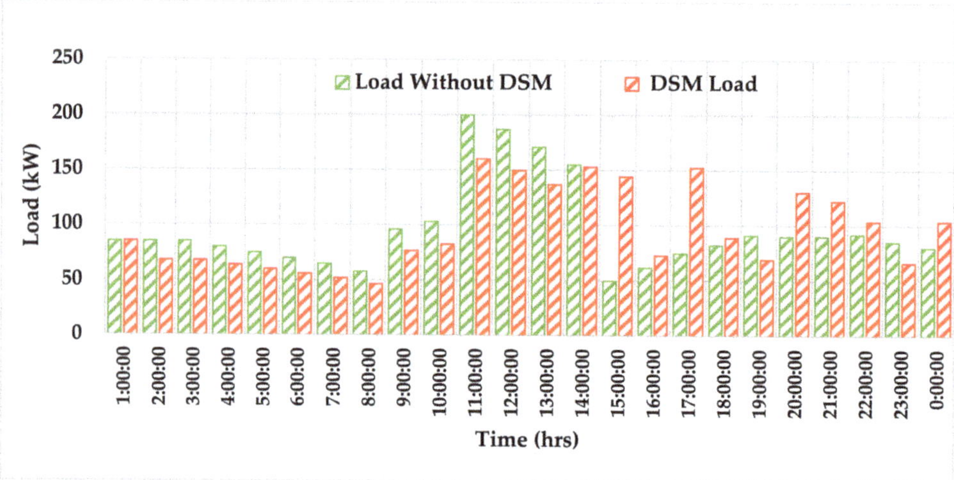

Figure 15. Daily load profile of AEO at 40% REF.

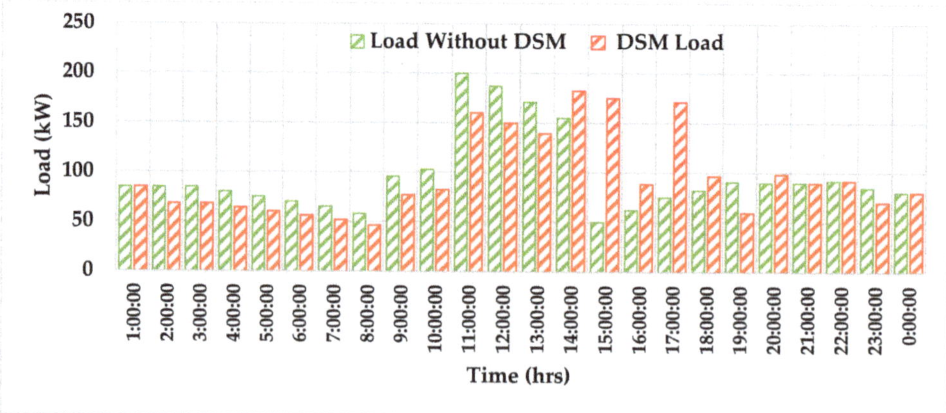

Figure 16. Daily load profile of AEO at 60% REF.

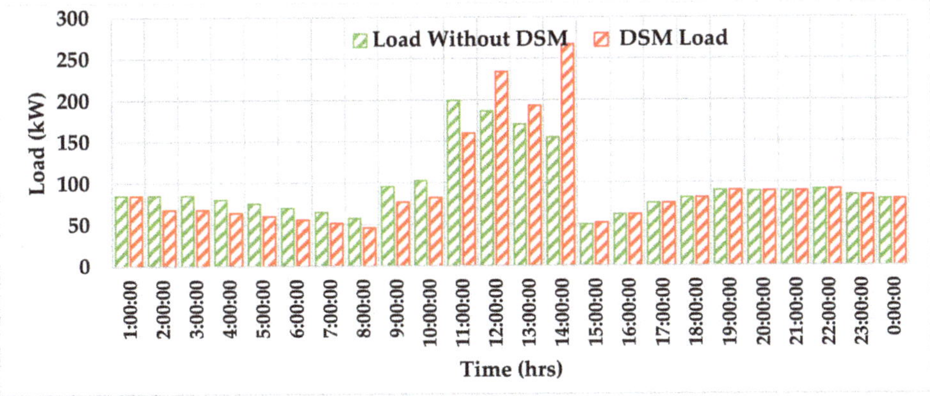

Figure 17. Daily load profile of AEO at 80% REF.

10. Conclusions

This paper presents a new application of AEO in the optimal sizing of a stand-alone Hybrid Energy System (HES). Al Sulaymaniyah, a rural area in Saudi Arabia, was selected for the optimal sizing of the HES. Cost of Energy (*COE*) was considered as the objective function of this model using both REF and LPSP as the reliability indices. Because of the abundance of renewable energy at the location, the optimal sizing selected both wind and solar energy as the major sources of energy. Furthermore, the optimal sizing was evaluated with and without Demand Side Management (DSM) under different Renewable Energy Fractions. It is obvious for both scenarios that AEO outperformed both HHO and FSA. By performing the DSM, more renewable energy was utilized while reducing the surplus unutilized energy. The DSM strategy achieved 28.38%, 43.05%, and 65.37% in *COE* saving at 40%, 60%, and 80% REF, respectively. The results stressed the importance of utilizing DSM for optimal sizing by further reducing the CEO. While the AEO algorithm introduced in this study has been shown to be efficient, it may not be efficient in other optimization problems. The efficacy of the algorithm can also be extended to the optimal load flow and optimal sitting of distributed generation.

Author Contributions: Conceptualization, H.O.O., A.M.A.-S. and A.A.A.-S.; methodology, H.O.O., A.A.A.-S. and H.M.H.F.; software, H.O.O.; validation, H.O.O., H.M.H.F. and A.A.A.-S.; formal analysis, H.O.O.; investigation, H.O.O.; resources, H.O.O. and A.A.A.-S.; data curation, H.O.O.; writing—original draft preparation, H.O.O.; writing—review and editing, H.O.O., A.M.A.-S., H.M.H.F. and A.A.A.-S.; visualization, H.O.O., A.M.A.-S., H.M.H.F. and A.A.A.-S.; supervision, H.O.O., A.M.A.-S., H.M.H.F. and A.A.A.-S.; project administration, H.O.O. and A.M.A.-S.; funding acquisition, A.M.A.-S. All authors have read and agreed to the published version of the manuscript.

Funding: This research received no external funding.

Acknowledgments: The authors would like to acknowledge the Researchers Supporting Project number (RSP-2021/337), King Saud University, Riyadh, Saudi Arabia.

Conflicts of Interest: The authors declare no conflict of interest.

Abbreviations

The following are abbreviations used.

AEO	Artificial Ecosystem-based Optimization
COE	Cost of Energy
CRF	Capital Recovery Factor
DMOPSO	Dynamic Multi-Objectives Particle Swarm Optimization
DSM	Demand Side Management
FSA	Future Search Optimization
HES	Hybrid Energy System
HHO	Harris Hawk Optimization
HS	Harmony Search
LOEE	Loss of Energy Expected
LOLE	Loss of Load Expected
LOLH	Loss of Load Hours
LPSP	Loss of Power Supply Probability
MPP	Maximum Power Point
PFT	Power Failure Time
REF	Renewable Energy Fraction
RES	Renewable Energy Resources
SOC	State of Charge
STCs	Standard Test Conditions
STD	Standard Deviation
WTG	Wind Turbine Generator
a	Linear weighting coefficient
A, B	Fuel constants
σ	Total output power
αp	Power temperature coefficient
C_{Dsl}	Annual fuel consumption
C_F	Cost of fuel per liter
D_e	Decomposer model
$E_l(t)$	Total load demand
h_e	Weight coefficients
$\overline{G_T}$	Global irradiance under normal conditions
$\overline{G_{T,STC}}$	Global irradiance under STC
$G_{T,NOCT}$	Solar irradiance with respect to NOCT
fpv	De-rating factor
η_{MPP}	Maximum Power Point (MPP) efficiency of PV module
$\eta_{\mathrm{MPP},STC}$	Efficiency under STCs
η_{inv}	Net inverter efficiency
$\eta_{battery}$	Round trip efficiency of battery
NOCT	Operating cell nominal temperature
Lw	Lower boundary
lt	Load shift
P_r	PV rated power
$P_{Dsl}(t)$	Generated power of diesel
P_R	Nominal power
$P(t)$	PV power
$P(w)$	Wind power
$P1(t)$	Load
P_{wind}	Wind rated power
P_{pv}	Rated solar power
P_{bat}	Rated battery power
$P(ren)$	Surplus renewable energy
q	Population number
r, r_1	Random number
$T_{c,STC}$	PV temperatures under STC
T_c	PV temperatures under normal conditions

T_a	Ambient Temperature
$T_{a,NOCT}$	Ambient temperature with respect to *NOCT*
w_e	Weight coefficients
u_c	Cut-in speed
u_r	Rated speed
x_{randi}	Random individual
U_p	Upper boundary
x_q	Best individual
u_f	Cut-out frequency
$univ$	Inverter efficiency

References

1. Fekete, H.; Kuramochi, T.; Roelfsema, M.; Elzen, M.; den Forsell, N.; Höhne, N.; Luna, L.; Hans, F.; Sterl, S.; Olivier, J.; et al. A review of successful climate change mitigation policies in major emitting economies and the potential of global replication. *Renew. Sustain. Energy Rev.* **2021**, *137*, 110602. [CrossRef]
2. Clausen, L.T.; Rudolph, D. Renewable energy for sustainable rural development: Synergies and mismatches. *Energy Policy* **2020**, *138*, 111289. [CrossRef]
3. Alturki, F.A.; Omotoso, H.O.; Al-Shamma'a, A.A.; Farh, H.M.H.; Alsharabi, K. Novel Manta Rays Foraging Optimization Algorithm Based Optimal Control for Grid-Connected PV Energy System. *IEEE Access* **2020**, *8*, 187276–187290. [CrossRef]
4. Al-Shamma'a, A.A.; Omotoso, H.O.; Noman, A.M.; Alkuhayli, A.A. Grey Wolf Optimizer Based Optimal Control for Grid-Connected PV System. In Proceedings of the IECON 2020 the 46th Annual Conference of the IEEE Industrial Electronics Society, Singapore, 19–21 October 2020; pp. 2863–2867.
5. Ramesh, M.; Saini, R.P. Dispatch strategies based performance analysis of a hybrid renewable energy system for a remote rural area in India. *J. Clean. Prod.* **2020**, *259*, 120697. [CrossRef]
6. Das, B.K.; Hassan, R.; Tushar, M.S.H.K.; Zaman, F.; Hasan, M.; Das, P. Techno-economic and environmental assessment of a hybrid renewable energy system using multi-objective genetic algorithm: A case study for remote Island in Bangladesh. *Energy Convers. Manag.* **2021**, *230*, 113823. [CrossRef]
7. Das, M.; Singh, M.A.K.; Biswas, A. Techno-economic optimization of an off-grid hybrid renewable energy system using metaheuristic optimization approaches–Case of a radio transmitter station in India. *Energy Convers. Manag.* **2019**, *185*, 339–352. [CrossRef]
8. Ramesh, M.; Saini, R. Demand Side Management based techno-economic performance analysis for a stand-alone hybrid renewable energy system of India. *Energy Sources Part A Recover. Util. Environ. Eff.* **2021**, 1–29. [CrossRef]
9. Lujano-Rojas, J.M.; Dufo-López, R.; Bernal-Agustín, J.L. Probabilistic modelling and analysis of stand-alone hybrid power systems. *Energy* **2013**, *63*, 19–27. [CrossRef]
10. Alsayed, M.; Cacciato, M.; Scarcella, G.; Scelba, G. Design of hybrid power generation systems based on multi criteria decision analysis. *Sol. Energy* **2014**, *105*, 548–560. [CrossRef]
11. Hung, D.Q.; Mithulananthan, N.; Bansal, R.C. Analytical strategies for renewable distributed generation integration considering energy loss minimization. *Appl. Energy* **2013**, *105*, 75–85. [CrossRef]
12. Luna-Rubio, R.; Trejo-Perea, M.; Vargas-Vázquez, D.; Ríos-Moreno, G.J. Optimal sizing of renewable hybrids energy systems: A review of methodologies. *Sol. Energy* **2012**, *86*, 1077–1088. [CrossRef]
13. Askarzadeh, A. Distribution generation by photovoltaic and diesel generator systems: Energy management and size optimization by a new approach for a stand-alone application. *Energy* **2017**, *122*, 542–551. [CrossRef]
14. Sharafi, M.; ElMekkawy, T.Y. A dynamic MOPSO algorithm for multiobjective optimal design of hybrid renewable energy systems. *Int. J. Energy Res.* **2014**, *38*, 1949–1963. [CrossRef]
15. Ghorbani, N.; Kasaeian, A.; Toopshekan, A.; Bahrami, L.; Maghami, A. Optimizing a hybrid wind-PV-battery system using GA-PSO and MOPSO for reducing cost and increasing reliability. *Energy* **2018**, *154*, 581–591. [CrossRef]
16. Baghaee, H.R.; Mirsalim, M.; Gharehpetian, G.B.; Talebi, H.A. Reliability/cost-based multi-objective Pareto optimal design of stand-alone wind/PV/FC generation microgrid system. *Energy* **2016**, *115*, 1022–1041. [CrossRef]
17. Koutroulis, E.; Kolokotsa, D.; Potirakis, A.; Kalaitzakis, K. Methodology for optimal sizing of stand-alone photovoltaic/wind-generator systems using genetic algorithms. *Sol. Energy* **2006**, *80*, 1072–1088. [CrossRef]
18. Al-Shamma'a, A.A.; Addoweesh, K.E. Techno-economic optimization of hybrid power system using genetic algorithm. *Int. J. Energy Res.* **2014**, *38*, 1608–1623. [CrossRef]
19. Kolhe, M.L.; Ranaweera, K.M.I.U.; Gunawardana, A.G.B.S. Techno-economic sizing of off-grid hybrid renewable energy system for rural electrification in Sri Lanka. *Sustain. Energy Technol. Assess.* **2015**, *11*, 53–64. [CrossRef]
20. Wang, R.; Li, G.; Ming, M.; Wu, G.; Wang, L. An efficient multi-objective model and algorithm for sizing a stand-alone hybrid renewable energy system. *Energy* **2017**, *141*, 2288–2299. [CrossRef]
21. Farh, H.M.H.; Al-Shaalan, A.M.; Eltamaly, A.M.; Al-Shamma'a, A.A. A novel severity performance index for optimal allocation and sizing of photovoltaic distributed generations. *Energy Rep.* **2020**, *6*, 2180–2190. [CrossRef]

22. Farh, H.M.H.; Al-Shaalan, A.M.; Eltamaly, A.M.; Al-Shamma'A, A.A. A Novel Crow Search Algorithm Auto-Drive PSO for Optimal Allocation and Sizing of Renewable Distributed Generation. *IEEE Access* **2020**, *8*, 27807–27820. [CrossRef]
23. Kaluthanthrige, R.; Rajapakse, A.D.; Lamothe, C.; Mosallat, F. Optimal Sizing and Performance Evaluation of a Hybrid Renewable Energy System for an Off-Grid Power System in Northern Canada. *Technol. Econ. Smart Grids Sustain. Energy* **2019**, *4*, 4. [CrossRef]
24. Farh, H.M.H.; Eltamaly, A.M.; Al-Shaalan, A.M.; Al-Shamma'a, A.A. A novel sizing inherits allocation strategy of renewable distributed generations using crow search combined with particle swarm optimization algorithm. *IET Renew. Power Gener.* **2021**, *15*, 1436–1450. [CrossRef]
25. Moslehi, K.; Kumar, R. A Reliability Perspective of the Smart Grid. *IEEE Trans. Smart Grid* **2010**, *1*, 57–64. [CrossRef]
26. Logenthiran, T.; Srinivasan, D.; Khambadkone, A.M.; Aung, H.N. Multiagent System for Real-Time Operation of a Microgrid in Real-Time Digital Simulator. *IEEE Trans. Smart Grid* **2012**, *3*, 925–933. [CrossRef]
27. Logenthiran, T.; Srinivasan, D.; Shun, T.Z. Demand Side Management in Smart Grid Using Heuristic Optimization. *IEEE Trans. Smart Grid* **2012**, *3*, 1244–1252. [CrossRef]
28. Kumar, S.; Kaur, T.; Upadhyay, S.; Sharma, V.; Vatsal, D. Optimal Sizing of Stand Alone Hybrid Renewable Energy System with Load Shifting. *Energy Sources Part A Recover. Util. Environ. Eff.* **2020**, 1–20. Available online: https://www.tandfonline.com/doi/full/10.1080/15567036.2020.1831107 (accessed on 8 November 2021). [CrossRef]
29. Alturki, A.F.; Farh, M.H.H.; Al-Shamma'a, A.A.; Al Sharabi, K. Techno-Economic Optimization of Small-Scale Hybrid Energy Systems Using Manta Ray Foraging Optimizer. *Electronics* **2020**, *9*, 2405. [CrossRef]
30. Zhao, W.; Wang, L.; Zhang, Z.; Al Sharabi, K. Artificial ecosystem-based optimization: A novel nature-inspired meta-heuristic algorithm. *Neural Comput. Appl.* **2020**, *32*, 9383–9425. [CrossRef]
31. Elsisi, M.; Wang, L.; Zhang, Z.; Al Sharabi, K. Future search algorithm for optimization. *Evol. Intell.* **2019**, *12*, 21–31. [CrossRef]
32. Heidari, A.A.; Mirjalili, S.; Faris, H.; Aljarah, I.; Mafarja, M.; Chen, H. Harris hawks optimization: Algorithm and applications. *Futur. Gener. Comput. Syst.* **2019**, *97*, 849–872. [CrossRef]
33. Belfkira, R.; Zhang, L.; Barakat, G. Optimal sizing study of hybrid wind/PV/diesel power generation unit. *Sol. Energy* **2011**, *85*, 100–110. [CrossRef]
34. Akhtari, M.R.; Baneshi, M. Techno-economic assessment and optimization of a hybrid renewable co-supply of electricity, heat and hydrogen system to enhance performance by recovering excess electricity for a large energy consumer. *Energy Convers. Manag.* **2019**, *188*, 131–141. [CrossRef]
35. Borhanazad, H.; Mekhilef, S.; Gounder Ganapathy, V.; Modiri-Delshad, M.; Mirtaheri, A. Optimization of micro-grid system using MOPSO. *Renew. Energy* **2014**, *71*, 295–306. [CrossRef]
36. Eid, A.; Kamel, S.; Korashy, A.; Khurshaid, T. An Enhanced Artificial Ecosystem-Based Optimization for Optimal Allocation of Multiple Distributed Generations. *IEEE Access* **2020**, *8*, 178493–178513. [CrossRef]

energies

MDPI

Article

Optimal Power Dispatch in Energy Systems Considering Grid Constraints

Alejandro Rubio *, Frank Schuldt, Peter Klement and Karsten von Maydell

DLR Institute of Networked Energy Systems, Carl-von-Ossietzky-Str. 15, 26129 Oldenburg, Germany; frank.schuldt@dlr.de (F.S.); peter.klement@dlr.de (P.K.); karsten.maydell@dlr.de (K.v.M.)

* Correspondence: alejandro.rubio@dlr.de; Tel.: +49-441-99906-481

Abstract: As a consequence of the increasing share of renewable energies and sector coupling technologies, new approaches are needed for the study, planning, and control of modern energy systems. Such new structures may add extra stress to the electric grid, as is the case with heat pumps and electrical vehicles. Therefore, the optimal performance of the system must be estimated considering the constraints imposed by the different sectors. In this research, an energy system dispatch optimization model is employed. It includes an iterative approach for generating grid constraints, which is decoupled from the linear unit commitment problem. The dispatch of all energy carriers in the system is optimized while considering the physical electrical grid limits. From the considered scenarios, it was found that in a typical German neighborhood with 150 households, a PV penetration of ~5 kW_p per household can lead to curtailment of ~60 MWh per year due to line loading. Furthermore, the proposed method eliminates grid violations due to the addition of new sectors and reduces the energy curtailment up to 45%. With the optimization of the heat pump operation, an increase of 7% of the self-consumption was achieved with similar results for the combination of battery systems and electrical vehicles. In conclusion, a safe and optimal operation of a complex energy system is fulfilled. Efficient control strategies and more accurate plant sizing could be derived from this work.

Keywords: sector coupling; optimal power flow; energy system optimization; grid flexibilization; oemof-Solph; PowerFactory

Citation: Rubio, A.; Schuldt, F.; Klement, P.; von Maydell, K. Optimal Power Dispatch in Energy Systems Considering Grid Constraints. *Energies* **2022**, *15*, 192. https://doi.org/10.3390/en15010192

Academic Editor: Tomonobu Senjyu

Received: 2 December 2021
Accepted: 23 December 2021
Published: 28 December 2021

1. Introduction

1.1. Background

Nowadays, there are several initiatives and international efforts to reduce the CO_2 emissions in all energy sectors as part of the Paris agreement [1]. In Germany, the so-called "Energiewende" establishes the goals for the energy transformation towards a zero-emission national energy system [2]. A fundamental step to achieve such ambitious goals is the electrification of the residential heat and transport sectors that accounted in 2016 for ~10% and ~18.2% of the total emissions in Germany, respectively [3]. By 2050, an increase of 50% of the district heating in Europe is expected, with approximately 30% of that demand being covered by heat pumps [4]. However, combined heat and power is expected to serve as a bridge technology coupling electricity and heat sectors [5]. Additionally, a fleet of around six million electrical vehicles is planned by the German government by 2030 [6]. This makes the analysis of modern and future energy systems more challenging, due to the added complexity of the new technologies and energy sectors. Along with a significant deployment of renewable sources in the electricity grid, the integration of these sector coupling technologies may add an extra burden to the existing distribution grids [7,8]. Therefore, new concepts and techniques are needed to properly study and optimize the grid structure for adequate operation of such new energy systems [9]. In this research, an approach that enables the ease evaluation and optimization of energy systems considering the power grid limits is developed. This methodology was based on the energy system for the "Energetisches Nachbarschaftsquartier Fliegerhorst Oldenburg", which will be a living laboratory in the city of Oldenburg, Germany [10].

1.2. Optimization of Energy Systems

Typically, the aggregation approach has been of common use in the literature when it comes to the analysis and optimization of energy systems [11]. Then, the optimization problem is independent of its actual physical characteristics. Mathematically, the energy system optimization problem can be expressed as:

$$\begin{aligned} min: &\sum_{t\in T}\Big(\sum_{(s,e)\in E}\sum_{i\in I_1} c^i_{(s,e)}(t)\cdot p^i_{s,e}(t)\cdot\tau\Big) \\ &+\sum_{(s,e)\in E}\sum_{i\in I_2} c^i_{(s,e)}\cdot p^i_{(s,e)} \\ &+\sum_{t\in T}\Big(\sum_{n\in N}\sum_{i\in I_3} c^i_n(t)\cdot p^i_n(t)\cdot\tau\Big) \\ &+\sum_{n\in N}\sum_{i\in I_4} c^i_n\cdot p^i_n \end{aligned} \tag{1}$$

$$s.t. \sum_{n\in N} a^j_n(t)\cdot p_n(t) \geq \sum_{m\in M} d^j_m(t), \quad \forall j\in J, \forall t\in T \tag{2}$$

$$0 \leq p^i_{(s,e)}(t) \leq \bar{p}^i_{(s,e)}(t), \quad \forall(s,e)\in E, \forall t\in T \tag{3}$$

$$0 \leq p^i_n(t) \leq \bar{p}^i_n(t), \quad \forall n\in N, \forall t\in T \tag{4}$$

The Equation (1) minimizes the cost in a period T at a resolution τ. The set E represents the energy flow between nodes, whereas the set N represents the flows between nodes and sources. The sets I_1 to I_4 establish the alternative to have multiple flows with different associated costs. Fixed and time-varying costs are included in c. The variable p denotes the power that flows between nodes and sources. The set J constitutes the types of sectors to consider (e.g., heat and electricity), whereas the set M represents the demand associated with the j_{th} sector. The inequality constraint (2) ensures that the sum of the M demands for each j_{th} sector is fulfilled by the sum of the sources multiplied by the coupling factor a corresponding to the n_{th} source and the j_{th} sector. The constraints (3) and (4) represent the boundary conditions for flows and sources.

Many approaches are found in the literature dealing with the optimization of energy systems considering the actual electrical grid topology. For power systems, the implementation of linearized power flows approaches to relax the optimization problem is of common use in the researched literature [12,13]. Whereas the typical DC power flow neglects reactive power, some studies have enhanced the method by adding the consideration of reactive power and with the implementation of data-driven approaches to determine voltage magnitudes and angles [14]. Novoa et al. [15] have applied a decoupled linearized power flow [16] in combination with a mixed-integer linear problem to find the optimal allocation of PV and batteries within an energy system. However, these methods are less accurate than the AC power flow and more complex than the DC power flow. The combination of a commercial tool to solve power flow problems with a linear unit commitment is presented by Nolden et al [17]. In this approach, only an electrical system without storage at a single time step is considered. Similarly, Fortenbacher et al. [11] propose a distributed model predictive control within sub-grids to solve a multi-period dispatch optimization problem.

When an entire energy system with multiple energy carriers is optimized, generally the physical structures are disregarded and the systems are simplified. Some solutions to work around the over-simplification of the models have been developed. Lohmeier et al have proposed specialized tools for each energy sector that can be implemented together in a co-simulation [18]. As the physical constraints of the energy systems are considered, the complexity in solving such problems also increases. For instance, non-convex systems result when hydraulic equations are taken into consideration in an energy hub with multiple energy carriers [19]. The time-step dependency from energy storage may add extra complexity to the already mentioned systems. The utilization of the Newton–Rapson method is inefficient in time-domain equations. Therefore, Levron et al. [20] have used in combination with a power flow solver, dynamic programming to tackle the time-dependent functions. As the share of renewable energies increases at the medium and low voltage of the grid, some power quality issues such as line over-loading and over-voltages may appear [7,9]. For this reason, the potential of CHP and P2G options has been considered to provide grid flexibilization [21,22].

The main contribution of this research is the development of a decoupled approach for the dispatch optimization of a sector coupled energy system. The method takes into consideration the electric grid limits to optimally integrate different energy sectors into an energy system. In order to assess the method, some scenarios comprising an energy system with high share of renewables, storage systems, and sector coupling technologies were considered. As a main result, it was found that high shares of PV without storage lead to a higher energy curtailment. Additionally, the proposed method was capable of reducing energy curtailment and increase self-consumption of the energy system by adding flexibility through the implementation of heat pumps and electric vehicles.

2. Methodology

In this section, the implemented methodology to consider grid limitations in dispatch optimization problems in energy systems is presented. The idea behind this work is to reduce the complexity of the optimization problem when considering grid constraints. The unit commitment problem and the power flow solutions are decoupled, the solution to the unit commitment is realized considering a mixed-integer linear problem (MILP) using *oemof-Solph* [23], whereas an AC optimal power flow (OPF) calculation is carried out in *PowerFactory* [24] to verify grid quality standards compliance. From this OPF, grid congestion and voltage violations are avoided; the measurements to achieve this are re-implemented in the linear problem as constraints. Figure 1 shows this iterative process to obtain the solution of the optimization problem in the energy system. Even though this approach could be applied to any energy system, regardless of its dimension, this work is primarily intended for its implementation on mid and low voltage systems as described in Section 1.

2.1. Overview: Iterative Process

Since sector coupling and energy storage technologies will be present in the energy systems of the future, an iterative approach is needed when decoupling the unit dispatch optimization and the grid constraint generation. This is due to the fact that the constraint generation from the OPF may influence the dispatch of the sector coupling technologies and storage such as heat pumps, heat water tank, and batteries.

This section presents an overview of the main steps depicted in Figure 1.

1. Energy system structure: Here, the energy sectors to be considered are defined as well as the relevant technologies and their models. Natural renewable resources and demands time series must also be included. Time steps have to be big enough to make valid the steady-state assumption of the different energy sectors [18]. In this work, only the electric and heat sector are considered and a time step of an hour during a year is evaluated. Although several technologies, markets, and demands could be present in a distributed energy system [25], only the ones presented in Table 1 were considered in this study. Furthermore, the energy system is seen as a system aggregator from the grid perspective. This means that the system can buy and sell electricity in the market.

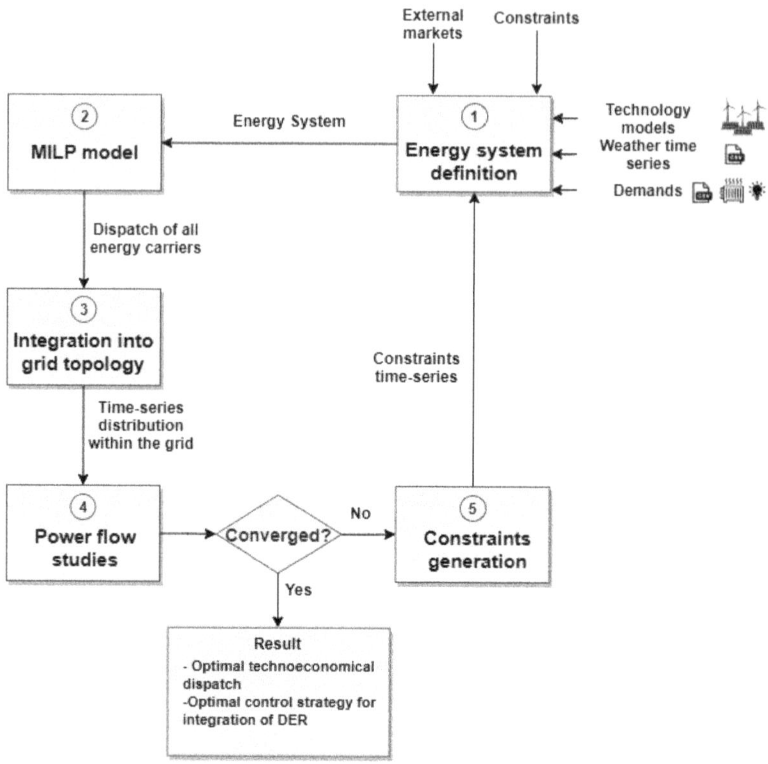

Figure 1. Steps for constraints generation in the optimization process.

Table 1. Summary of the considered energy system components.

Source	Market	Storage	Coupling	Demand
Photovoltaic (PV)	Natural Gas	Battery	Heat Pump	Electricity
Cogeneration (CHP)	Electricity	Hot Water Storage	CHP	Heat
Gas Boiler			Electro-Vehicles	

2. Linear unit commit: Using a holistic approach, an abstraction of the energy system structure is created in *oemof-Solph*. This abstraction contains all possible energy flows between sources and sinks, and between energy sectors through the coupling technologies as depicted in Figure 2.

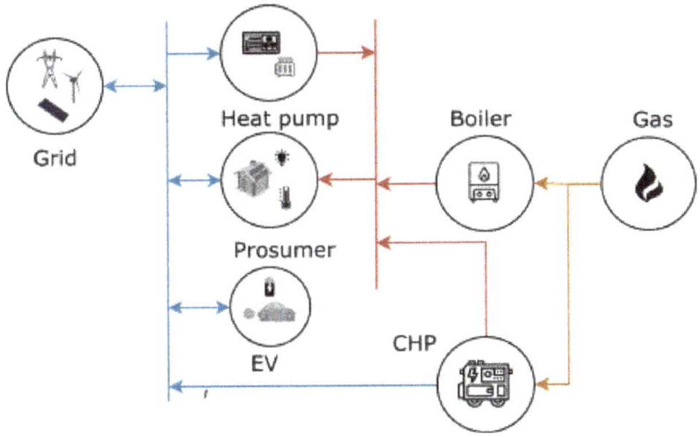

Figure 2. Simplified example of an energy system structure representation in *oemof-Solph* and the energy flow between sectors. The blue lines represent the electrical part of the system, whereas the red and orange represent the heat and gas sectors, respectively.

Costs and additional constraints can be associated with each of the energy flows [23]. Here, the system will only consider the economical constraints of the energy flows. Therefore, the objective function is the cost minimization. From Equation (1), the following objective equation is derived for the considered energy system:

$$min: \sum_{t \in T} \Big(p_{elec\_ext}(t) \cdot c_{day\_ahead}(t) \tag{5}$$

$$+ p_{in\_chp}(t) \cdot c_{gas} + p_{in\_boiler}(t) \cdot c_{gas} \Big) \cdot \tau \tag{6}$$

where p_{elec_ext} represents the imported or exported electric power from the external grid, c_{day_ahead} the day ahead price for a given time, p_{in_chp} and p_{in_boiler} the gas import for the CHP and boiler, respectively, and c_{gas} represents the cost of gas. The power given or provided to the external grid at a given time is given by $p_{ext} = p_{elec\_loads} - p_{elec\_gen}$. The electric generation p_{elec_gen} and electric loads p_{elec_loads} can be defined as follows:

$$p_{elec\_gen}(t) = p_{pv}(t) + p_{elec\_chp}(t) \tag{7}$$

$$p_{elec\_loads}(t) = p_{elec\_demand}(t) + p_{battery}(t) + p_{ev}(t) + p_{elec\_heat\_pump}(t) \tag{8}$$

In Equations (7) and (8), the parameters p_{pv} and p_{elec_chp} refer to the PV generation and the CHP electric power generation within the system, whereas the parameters p_{elec_demand}, $p_{battery}$, p_{ev}, and $p_{elec\_heat\_pump}$ refer to the electric demand, the battery power consumption, electrical vehicle charging, and the heat pump demand, respectively. Every storage unit (including EVs) has been considered as a load. That means that a negative power represents a power injection to the system. The same convention has been used for the external grid power flow. As constraints for the optimization problem, the thermal and electric demands must be supplied at any time. Equations (9) and (10) depict such constraints where p_{th_demand}, p_{elec_demand}, and p_{pv} are fixed time series.

$$p_{th\_demand}(t) = p_{th\_heat\_pump}(t) + p_{th\_boiler}(t) + p_{th\_chp}(t) - p_{th\_storage}(t) \tag{9}$$

$$p_{elec\_demand}(t) = p_{elec\_ext} + p_{pv}(t) + p_{elec\_chp}(t) - p_{battery}(t) - p_{ev}(t) - p_{elec\_heat\_pump}(t) \tag{10}$$

As a result, the optimized power dispatch for each non-fixed source to supply the demand at each time step is obtained. However, up to this point, just the total installed capacities and demands have been considered as per the inequality (4). The actual topology of the electrical and thermal grid has been also disregarded as a common practice in the optimization of energy systems [19].

3. Grid topology consideration: Before integrating these flows into the electrical grid in PowerFactory, the flow distribution must be considered. This applies only to the electric sources, sinks, and sector coupling technologies. The distribution of heat technologies is disregarded since only the electrical grid topology is being considered in this study.
   To represent a typical low voltage grid, one of the so-called *"Merit Order Netz-Ausbau 2030"* (MONA) reference grids is used. The ONT_8003 grid model is used as reference [26] and depicted in Figure 3. This topology represents a typical low voltage grid in Germany.

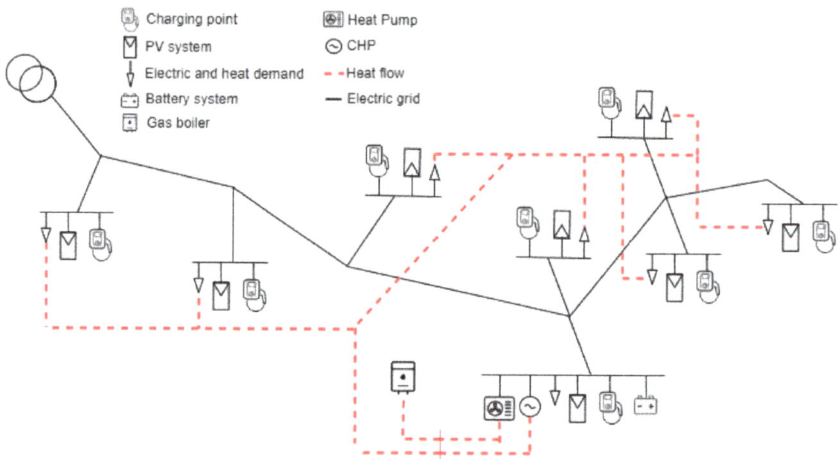

Figure 3. Representation of the ONT_8003 MONA grid with the addition of the technologies mentioned in Table 1.

   The energy flows for each node per technology are derived from the optimization results and the capacity installed per building.
4. Power flow calculation: Once the flow distributions at each node have been obtained, these energy flows must be added as time characteristics to the corresponding elements in PowerFactory.
   To determine if the optimized dispatch complies with the grid standards, a quasi-dynamic power flow (QDPF) study is performed for a year with 1 h time steps. From the results of the QDPF, lines exceeding 100% loading and voltage variation outside of the range of $\pm$10% of the nominal voltage are considered as grid violations [8].
   The time steps containing such violations will be re-optimized by PowerFactory to avoid line congestion and bus voltage violations. For the system to converge, a dispatchable source is considered as slack, so it has enough power to cover the demand in case that the renewable sources are curtailed due to system violations.
5. Constraints generation: Similar to the constraint generation for the linear optimization [27], the *oemof-Solph* model will be limited by constraints generated from the OPF as denoted in Equation (4). Power time-series are the link between the two tools. Therefore, PowerFactory and *oemof-Solph* will exchange information about the active power flows, being the reactive power flow after the last OPF calculation considered as optimal.

To build the OPF problem, the following objective function and constraints are considered:

$$min: \sum_{t \in T_{opf}} \Big(p^{opf}_{elec\_ext}(t) \cdot c_{day\_ahead}(t) + p^{opf}_{in\_chp}(t) \cdot c_{gas} \Big) \cdot \tau \tag{11}$$

$$s.t. S^l_{line}(t) \leq S^l_{nom}(t), \forall l \in L \tag{12}$$

$$S_{line} = \sqrt{P^2_{line} + Q^2_{line}} \tag{13}$$

$$0.9 \cdot V_{nom} \leq |V^b_{bus}(t)| \leq 1.1 \cdot V_{nom}, \forall b \in B \tag{14}$$

$$\frac{-e_{battery}(t-\tau)}{\tau} \leq p^{opf}_{battery}(t) \leq \frac{k_{battery} - e_{battery}(t-\tau)}{\tau} \tag{15}$$

$$0 \leq p^{opf}_{pv}(t) \leq p_{pv}(t) \tag{16}$$

$$S^n \leq S^n_{nom}, \forall n \in N \tag{17}$$

In Equation (11), the term for the thermal boiler is not considered. This is due to the fact that only electric units are taken into consideration in the OPF. The set T_{opf} is a sub set of T comprising only the periods of time that led to grid violations from step 4. The variables $p^{opf}_{elec\_ext}$ and $p^{opf}_{elec\_chp}$ denote the optimized power flows to avoid such violations. The constraints (12) and (13) ensure that the nominal capacity S_{nom} of any line of the system is not violated by the actual power flow S_{line}. The set of all lines in the energy system is denoted by L. The actual power flow in the line is given by the square root of the sum of squares of the real power P_{line} and the reactive power Q_{line}. To ensure voltage compliance, the constraint (14) is included. This keeps the voltage magnitude $|V_{bus}|$ of all buses in the system, denoted by the set B, within $\pm 10\%$ of the nominal voltage V_{nom}. The constraint (15) limits the power that can be drawn by the batteries. The difference between the battery system capacity $k_{battery}$ and the battery system energy content $e_{battery}$ at the previous time-step $t - \tau$ provides the available energy to be drawn for charging the batteries. The energy content is provided from the linear optimization performed in step 2. The difference is divided by τ in order to get the charging power limit. A lower energy bound is given to allow the discharge of the batteries. This lower bound is equal to the negative of the energy content. In the OPF additional flexibility is provided with the PV system. The optimizer can reduce the power output p^{opf}_{pv} as per constraint (16). Furthermore, to keep the thermal limits of all technologies within acceptable ranges, constraint (17) is included. Where the apparent power S^n of each technology of the set N must not exceed the nominal apparent power S^n_{nom}. Similarly to the linear optimization, constraint (10) must be complied by the OPF, therefore:

$$p_{elec\_demand}(t) = p^{opf}_{elec\_ext}(t) + p^{opf}_{pv}(t) + p^{opf}_{elec\_chp}(t) - p^{opf}_{battery}(t) - p^{opf}_{ev}(t) - p^{opf}_{elec\_heat\_pump}(t) \tag{18}$$

In Equation (18), the superscript opf denotes the power flow obtained from the OPF from the respective source or sink. Considering that $COP_{heat\_pump} = \frac{p_{th\_heat\_pump}}{p_{elec\_heat\_pump}}$ and $p_{th\_sources} = p_{th\_heat\_pump} + p_{th\_boiler} + p_{th\_chp}$, Equations (18) and (9) can be combined to yield:

$$\frac{p_{th\_demand} - p_{th\_sources} - p_{th\_heat\_pump} - p_{th\_storage}}{COP_{heat\_pump}} = p^{opf}_{elec\_ext} + p^{opf}_{elec\_gen} - p^{opf}_{elec\_loads} + p^{opf}_{elec\_heat\_pump} \tag{19}$$

Equation (19) shows how the changes in the OPF have to be compensated by the heat elements in the energy system. However, as previously indicated, the OPF only takes into consideration the electric components and the sector coupling technologies. Therefore, the changes in the OPF have to be reflected in the linear optimization. Equation (20) is employed to determine the deviation between the OPF and the initial power flow.

$$\Delta p_n = |p_n^{opf} - p_n|, \forall n \in N \quad (20)$$

The power flow per n_{th} technology resulting from the OPF is represented by p_n^{opf}, whereas the initial power from the linear optimization is represented by p_n. Whenever $\Delta p_n \neq 0$, the value of the OPF is passed as an upper bound in (4) for the linear optimization in the next iteration.

2.2. Evaluation Scenarios

In order to assess the method described in Section 2, three main scenarios are evaluated:

1. High generation scenario: The size of the PV installation is fixed to 1500 kW_p and no storage or flexible loads are considered. The influence of line loading and voltage levels in the optimization is evaluated.
2. Heat pump and heat storage scenario: The influence of heat pump and heat storage is analyzed. Here, the size of the PV installation is reduced to 700 kW. Heat pumps and heat storage are added with 600 kW_{th} and 150 m^3 of capacity, respectively. The potential for flexibilization services from the heat sector is evaluated in this scenario through the implementation of the proposed method.
3. Electromobility scenario: A fleet of 62 EVs and 500 kWh of battery storage are added to the *Heat pump and heat storage scenario*. These are connected at eight different points within the network. It is assumed that the EVs are only connected from 6 p.m. to 7 a.m. of the next day [28]. Additionally, it is assumed that the daily required demand of the EVs is around 10 kWh, which is approximately the double required per EV per day [6]. Therefore, the state of charge is not the constraint for charging at the end of the charging period in the optimization, but to ensure enough daily coverage in a daily basis.

3. Results

In this section, the main results obtained from the implementation of the electric grid constraints into the optimal operation of an energy system are presented.

3.1. High Generation Scenario

3.1.1. Line Loading Constraint

As described in Section 2, in this scenario, a significant capacity of PV is considered to be installed in the system. In Figure 4a, the loading on the main feeder during one year is shown. It is observed that the line can be overloaded up to 150% of its capacity. Figure 4b shows the new line loading after the OPF has generated the corresponding grid constraints at each time step. With the consideration of the violated hours shown in Figure 4a as constraints, the optimization reduces the load in the line at around 100% of its nominal capacity.

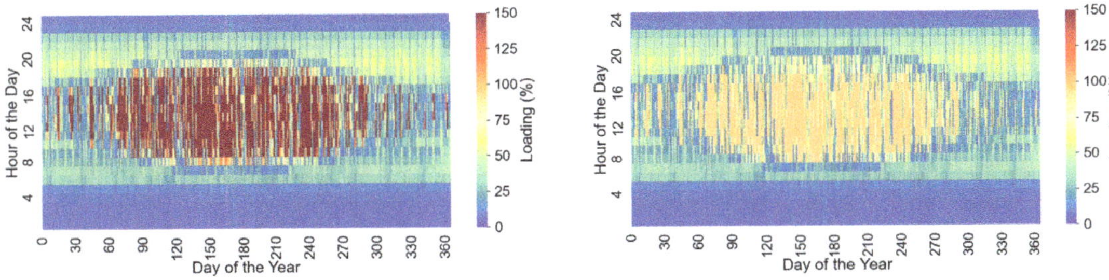

(a) Without grid constraints. (b) With grid constraints.

Figure 4. Main feeder loading with and without the consideration of the grid constraints into the optimization.

3.1.2. Voltage Constraint

Similarly, Figure 5a,b depict the voltage behavior on the PV bus before and after the grid constraints generation.

Following the same pattern than the over-loading caused by the high in-feed depicted in Figure 4a, voltages above of the nominal voltage occur along with the main feeder over-loading.

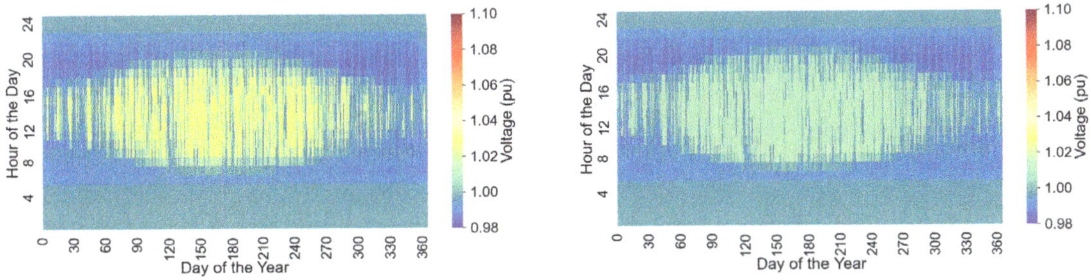

(a) Without grid constraints. (b) With grid constraints.

Figure 5. PV bus voltage at node number 4 with and without the consideration of the grid constraints into the optimization.

Figure 6a,b show the power dispatched by the PV plant before and after the consideration of the voltage and loading constraints of the grid. A noticeable curtailment is needed to maintain the grid quality parameters of loading and voltage within the admissible ranges. The absence of means to store or shift loads during such midday peaks leads to the unavoidable PV curtailment.

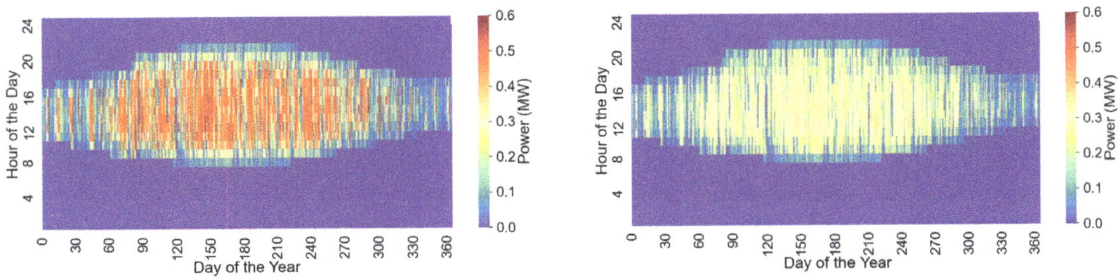

(a) Without grid constraints. (b) With grid constraints.

Figure 6. PV power dispatch optimization with and without the consideration of the grid constraints.

3.2. Heat Pump and Heat Storage

In this section, the proposed methodology is tested by adding some flexibilization technologies such as heat storage and heat pumps.

Even though the installed PV capacity has been decreased to 700 kW_p for this scenario, the mismatch between the electric demand and the generation will still create energy flows back to the grid. Such energy flows are big enough to overload the main feeder. As the heat pumps and heat storage are considered in the energy system, the optimizer has the possibility to transfer the energy surplus from the PV system into another energy carrier. Figure 7 shows how the line loading constraint due to the high PV feed-in affects the power dispatch of the heat pumps and PV. The flexibilization of the heat pump activation avoids that the locally generated energy leaves the system. This diminishes the energy injected into the grid, avoiding in this manner the line overloading and diminishing the PV curtailment.

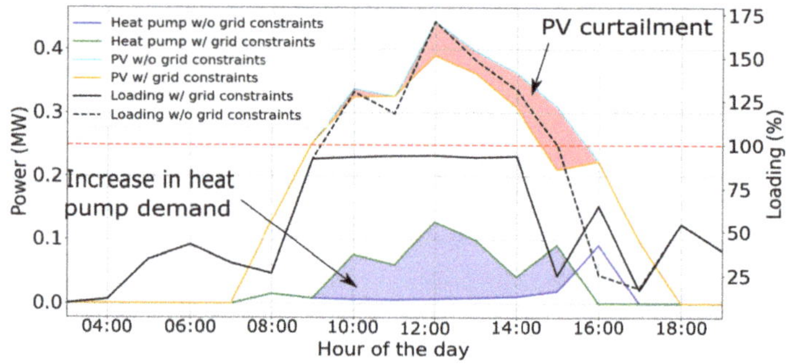

Figure 7. Flexibilization provided by heat pumps and heat storage. The blue area represents the energy otherwise curtailed without the grid constraints implementation in the dispatch optimization.

Figure 8 shows the variation in energy content in the water storage tank. A noticeable difference is found especially in the months of summer. This is in accordance with the fact that higher values of sun irradiance and less thermal demand are present during such months. Therefore, the heat pumps are continuously being used to tackle PV generation surplus when grid constraint violations occur.

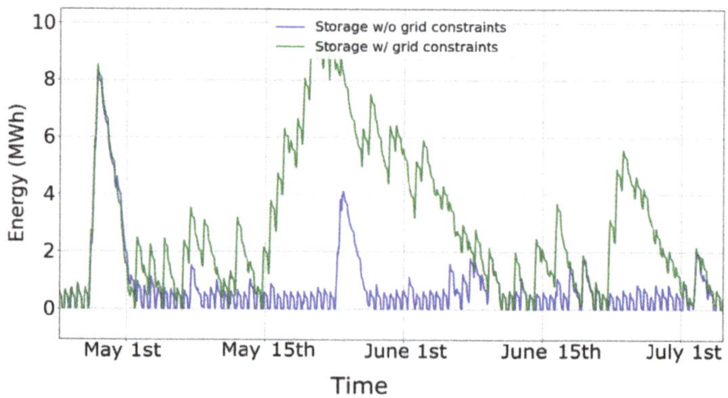

Figure 8. Change in energy content of the water heat storage.

3.3. Electromobility Scenario

As the fleet of electric cars is added to the energy system, the yearly demand is increased by around 300 MWh. This increment in demand causes line overloads throughout the network. In Figure 9, it can be seen how the main feeder is over-loaded while supplying energy to the EVs during the charging phase. It can be noted that the EV batteries support the grid after they have been connected at 6 pm, which is typically a time of the day with high demand. After the implementation of the constraints into the proposed method, the optimizer re-schedules the EV's batteries charging and grid feed-in to avoid the overload of the lines.

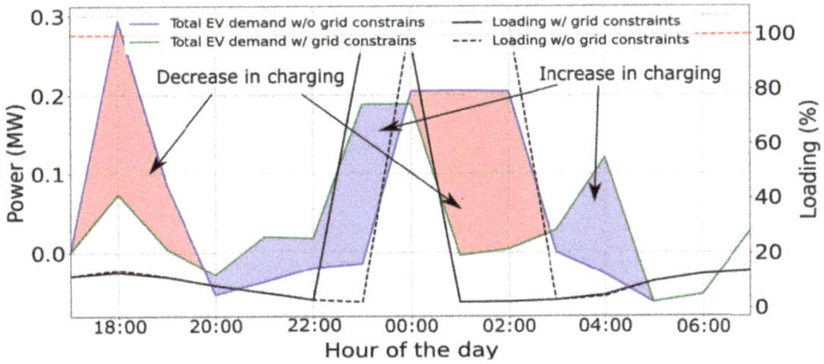

Figure 9. Cumulative EV charging profile comparison with and without grid constraints.

The proposed method successfully integrates the EV and the distance coverage constraint together with the grid constraints into the dispatch optimization problem of the energy system.

4. Discussion

In Section 3, it was found that the high share of PV at the low voltage level of the electrical system can lead to over-loads at the point of common coupling. This is caused by the residual energy fed into the grid. Neighborhoods with a PV capacity of 10 kW_p per household are prompted to lead to grid violations [8]. From the results, it may be inferred that a PV installed capacity of 5 kW_p per household can be led to feed-in levels exceeding the nominal voltage in a typical low voltage grid [26]. Nevertheless, over-voltages exceeding 10% of the nominal voltage are very unlikely to occur at this level of penetration. In the case of high share of PV without sector coupling or storage possibility, a curtailment up to ∼60 MWh per year is applied by the method to keep grid limits within acceptable ranges.

The integration of different sector coupling technologies may cause some grid events in the low voltage grids [28–30]. However, with a proper control strategy, such technologies can also provide some flexibility to the grid [31,32]. In the scenarios with flexibility options, the method re-schedules the power dispatch from these technologies. Table 2 provides a summary of the influence of the grid constraints in some indicators on each scenario. The summary demonstrates the relevance of the grid constraints in the optimization of an energy system. As it is shown in Table 2, this not only avoids grid limit violations, but it also saves energy from curtailment. From the simulation results, it was found that the methodology here proposed is capable of reducing the energy curtailment up to 45% by adapting to the economic dispatch, the technical grid constraints. Additionally, the self-consumption is increased up to ∼7% per year with the implementation of the method.

Table 2. Summary of the influence of grid constraints in the optimization parameters. Values are shown in MWh/year.

Scenario	Self-Consumption		Curtailment		Energy Saved
	Without Grid Constraints	With Grid Constraints	Without Grid Constraints	With Grid Constraints	
High PV Generation	288.60	288.60	270.04	270.04	0
Heat Pump	323.91	346.71	60.65	37.84	22.81
Electromobility	434.87	462.16	49.55	27.64	21.91

Therefore, the most economically and technically feasible dispatch of the energy system is achieved. The contribution of the *oemof-Solph*–PowerFactory toolchain results evident when flexible loads are present in the system. In this regard, the optimization decreases the overall energy that would be otherwise curtailed. Table 3 establishes a comparison between the proposed method and some of the most relevant researches found in the literature.

Table 3. Comparison with relevant works found in the literature.

Source	Considered Energy Carriers	Storage Technology	Grid Constraints	Approach
Cesena et al. [33]	Gas, heat, and power	Building heat inertia	Gas and power	MILP with heuristic penalization of grid constraints
Huang et al. [34]	Electromobility and power	EV battery	Power	Optimal EV charging scheduling through a genetic algorithm
Clegg et al. [35]	Gas, heat and power	—	Gas	Determination of flexibility according to gas availability and DC-OPF
Nolden et al. [17]	Power	—	Power	Unit commitment and power flow solver to achieve a techno-economic dispatch
This research	Electromobility, heat, gas, and power	Battery storage, EV battey, and heat storage	Power	MILP with iterative grid constraint generation

Due to the flexibility of energy systems provided by *oemof-Solph*, the proposed method surpasses the approaches found in the literature in terms of the complexity of the energy system itself. Whereas Martinez et al. [33] consider gas, heat, and power, this study additionally considers, the electromobility sector and the influence on the system by batteries and heat storage. On the other hand, this research only considers the influence of the electric power grid. In some other studies, the grid constraints are considered, but neglecting either storage or other energy sectors [17,34,35].

5. Summary and Outlook

The proposed method combines a linear optimizer such as *oemof-Solph* and a power analysis tool such as PowerFactory. This combination allows the optimization of an energy system with multiple energy carriers while considering the actual grid limits for sector coupling technologies integration. Such grid limits have an impact on the performance of the energy system:

- For scenarios with PV capacity above 5 kW_p per household, the voltage and line loading constraints are violated, therefore affecting the optimization results. In the case study, curtailment of ~60 MWh/year is needed to keep the grid within its limits. This due to the lack of storage or flexibilization options.

- If flexible technologies are present, parameters as curtailment and self consumption are affected by the grid constraints. In the case study, the self-consumption can increase 7% and the curtailment be reduced by 45% as compared with the typical optimization without grid constraints. The method successfully integrated the additional load from the heat pumps and EV without grid violations.
- In comparison to other approaches, the decoupled constraint generation in the optimization problem determines the techno-economical power dispatch for all the different energy sectors and storage technologies.

To adapt the method to large-scale grids, different open source tools might provide a faster interface and interaction with *oemof-Solph* compared to the PowerFactory API. Most of these tools are written in python or julia [36]. Another novel approach would be to consider a data-driven constraint generation. This might be especially relevant for online applications or analysis in the transmission system. Due to its high accuracy and run-time, this topic is yet to be exploited [37]. Furthermore, the influence of the heat network is a feature that could be added to this approach in future research.

Author Contributions: Conceptualization, A.R., P.K. and F.S.; investigation, A.R.; writing—original draft, A.R.; supervision, P.K.; writing—review and editing, A.R., P.K., F.S. and K.v.M. All authors have read and agreed to the published version of the manuscript.

Funding: This research received no external funding.

Institutional Review Board Statement: Not applicable.

Informed Consent Statement: Not applicable.

Data Availability Statement: Not applicable.

Conflicts of Interest: The authors declare no conflict of interest.

References

1. United Nations Framework Convention on Climate Change (UNFCCC). The Paris Agreement. 2015. Available online: https://unfccc.int/process-and-meetings/the-paris-agreement/the-paris-agreement (accessed on 19 May 2021)
2. Agora Energiewende. Energiewende: What do the new laws mean. In *Ten Questions and Answers about EEG*; Agora Energiewende: Berlin, Germany, 2017. Available Online: https://www.agora-energiewende.de/en/publications/energiewende-what-do-the-new-laws-mean/(accessed on 25 December 2021).
3. Bundesministerium für Umwelt, Naturschutz und nukleare Sicherheit. *Climate Action in Figures-Facts, Trends and Incentives for German Climate Policy*; Federal Ministry for the Environment, Nature Conservation and Nuclear Safety (BMU): Berlin, Germany, 2018.
4. David, A.; Mathiesen, B.V.; Averfalk, H.; Werner, S.; Lund, H. Heat roadmap Europe: Large-scale electric heat pumps in district heating systems. *Energies* **2017**, *10*, 578. [CrossRef]
5. Sensfuß, F.; Deac, G.; Bernath, C. *Vorabanalyse Langfristige Rolle und Modernisierung der Kraft-Wärme-Kopplung*; Kurzpapier. Hg. v. Bundesministerium für Wirtschaft und Energie BMWI: Berlin, Germany, 2017.
6. Jochem, P.; Babrowski, S.; Fichtner, W. Assessing CO2 emissions of electric vehicles in Germany in 2030. *Transp. Res. Part A Policy Pract.* **2015**, *78*, 68–83. [CrossRef]
7. Jain, P.; Jain, T. Assessment of electric vehicle charging load and its impact on electricity market price. In Proceedings of the 2014 International Conference on Connected Vehicles and Expo (ICCVE), Vienna, Austria, 3–7 November 2014; pp. 74–79. [CrossRef]
8. Arnold, M.; Friede, W.; Myrzik, J. Investigations in low voltage distribution grids with a high penetration of distributed generation and heat pumps. In Proceedings of the 2013 48th International Universities' Power Engineering Conference (UPEC), Dublin, Ireland, 2–5 September 2013; pp. 1–6. [CrossRef]
9. Bayer, B.; Matschoss, P.; Thomas, H.; Marian, A. The German experience with integrating photovoltaic systems into the low-voltage grids. *Renew. Energy* **2018**, *119*, 129–141. [CrossRef]
10. Energetisches Nachbarschaftsquartier Fliegerhorst Oldenburg. Available online: https://www.enaq-fliegerhorst.de/ (accessed on 28 April 2021).
11. Fortenbacher, P.; Ulbig, A.; Koch, S.; Andersson, G. Grid-constrained optimal predictive power dispatch in large multi-level power systems with renewable energy sources, and storage devices. In Proceedings of the IEEE PES Innovative Smart Grid Technologies, Europe, Istanbul, Turkey, 12–15 October 2014; pp. 1–6. [CrossRef]

12. Zargar, B.; Monti, A.; Ponci, F.; Martí, J.R. Linear Iterative Power Flow Approach Based on the Current Injection Model of Load and Generator. *IEEE Access* **2020**, *9*, 11543–11562. [CrossRef]
13. Handschin, E.; Kuhn, S.; Rehtanz, C.; Schultz, R.; Waniek, D. *Optimaler Kraftwerkseinsatz in Netzengpasssituationen. Innovative Modellierung und Optimierung von Energiesystemen*; LIT: Berlin, Germany, 2009; pp. 39–68. ISBN:978-3-8258-1359-8.
14. Liu, Y.; Zhang, N.; Wang, Y.; Yang, J.; Kang, C. Data-driven power flow linearization: A regression approach. *IEEE Trans. Smart Grid* **2018**, *10*, 2569–2580. [CrossRef]
15. de Novoa, M.; Martinez, L. Optimal Solar PV, Battery Storage, and Smart-Inverter Allocation in Zero-Net-Energy Microgrids Considering the Existing Power System Infrastructure. Ph.D. Thesis, UC Irvine, Irvine, CA, USA, 2020.
16. Jiang, M.; Guo, Q.; Sun, H.; Ge, H. Decoupled piecewise linear power flow and its application to under voltage load shedding. *CSEE J. Power Energy Syst.* **2020**, *7*, 976–985.
17. Nolden, C.; Schönfelder, M.; Eßer-Frey, A.; Bertsch, V.; Fichtner, W. Network constraints in techno-economic energy system models: Towards more accurate modeling of power flows in long-term energy system models. *Energy Syst.* **2013**, *4*, 267–287. [CrossRef]
18. Lohmeier, D.; Cronbach, D.; Drauz, S.R.; Braun, M.; Kneiske, T.M. Pandapipes: An Open-Source Piping Grid Calculation Package for Multi-Energy Grid Simulations. *Sustainability* **2020**, *12*, 9899. [CrossRef]
19. Geidl, M.; Andersson, G. Optimal power flow of multiple energy carriers. *IEEE Trans. Power Syst.* **2007**, *22*, 145–155. [CrossRef]
20. Levron, Y.; Guerrero, J.M.; Beck, Y. Optimal power flow in microgrids with energy storage. *IEEE Trans. Power Syst.* **2013**, *28*, 3226–3234. [CrossRef]
21. Acha, S.; Green, T.C.; Shah, N. Techno-economical tradeoffs from embedded technologies with storage capabilities on electric and gas distribution networks. In Proceedings of the IEEE PES General Meeting, Minneapolis, MN, USA, 25–29 July 2010; pp. 1–8. [CrossRef]
22. Qadrdan, M.; Ameli, H.; Strbac, G.; Jenkins, N. Efficacy of options to address balancing challenges: Integrated gas and electricity perspectives. *Appl. Energy* **2017**, *190*, 181–190. [CrossRef]
23. Hilpert, S.; Kaldemeyer, C.; Krien, U.; Günther, S.; Wingenbach, C.; Plessmann, G. The Open Energy Modelling Framework (oemof)-A new approach to facilitate open science in energy system modelling. *Energy Strategy Rev.* **2018**, *22*, 16–25. [CrossRef]
24. Gonzalez-Longatt, F.; Torres, J.L.R. *Advanced Smart Grid Functionalities Based on Powerfactory*; Springer: Berlin/Heidelberg, Germany, 2018. [CrossRef]
25. Schmeling, L.; Schönfeldt, P.; Klement, P.; Wehkamp, S.; Hanke, B.; Agert, C. Development of a decision-making framework for distributed energy systems in a German district. *Energies* **2020**, *13*, 552. [CrossRef]
26. Forschungsstelle für Energiewirtschaft e.V. (FfE) CC BY 4.0. Basisnetztopologien MONA 2030. Available online: https://www.ffe.de/themen-und-methoden/speicher-und-netze/752-ffe-stellt-rechenfaehige-basisnetztopologien-aus-projekt-mona-2030-zur-verfuegung (accessed on 28 April 2021).
27. Ben-Ameur, W.; Neto, J. A constraint generation algorithm for large scale linear programs using multiple-points separation. *Math. Program.* **2006**, *107*, 517–537. [CrossRef]
28. Clement-Nyns, K.; Haesen, E.; Driesen, J. The impact of charging plug-in hybrid electric vehicles on a residential distribution grid. *IEEE Trans. Power Syst.* **2009**, *25*, 371–380. [CrossRef]
29. Passey, R.; Spooner, T.; MacGill, I.; Watt, M.; Syngellakis, K. The potential impacts of grid-connected distributed generation and how to address them: A review of technical and non-technical factors. *Energy Policy* **2011**, *39*, 6280–6290. [CrossRef]
30. Mies, J.J.; Helmus, J.R.; Van den Hoed, R. Estimating the charging profile of individual charge sessions of Electric Vehicles in the Netherlands. *World Electr. Veh. J.* **2018**, *9*, 17. [CrossRef]
31. Hilpert, S. Effects of Decentral Heat Pump Operation on Electricity Storage Requirements in Germany. *Energies* **2020**, *13*, 2878. [CrossRef]
32. Rubio, A.; Behrends, H.; Geißendörfer, S.; Maydell, K.v.; Agert, C. Determination of the required power response of inverters to provide fast frequency support in power systems with low synchronous inertia. *Energies* **2020**, *13*, 816. [CrossRef]
33. Ceseña, E.A.M.; Loukarakis, E.; Good, N.; Mancarella, P. Integrated Electricity–Heat–Gas Systems: Techno–Economic Modeling, Optimization, and Application to Multienergy Districts. *Proc. IEEE* **2020**, *108*, 1392–1410. [CrossRef]
34. Huang, Z.; Fang, B.; Deng, J. Multi-objective optimization strategy for distribution network considering V2G-enabled electric vehicles in building integrated energy system. *Prot. Control Mod. Power Syst.* **2020**, *5*, 1–8. [CrossRef]
35. Clegg, S.; Mancarella, P. Integrated electrical and gas network flexibility assessment in low-carbon multi-energy systems. *IEEE Trans. Sustain. Energy* **2015**, *7*, 718–731. [CrossRef]
36. Auer, S.; Liße, J.; Mandha, S.R.; Horn, C. Power-Flow-Constrained Asset Optimization for Off-Grid Power Systems Using Selected Open-Source Frameworks. In Proceedings of the 4th International Hybrid Power Systems Workshop, Crete, Greece, 22–23 May 2019.
37. Pineda, S.; Morales, J.M.; Jiménez-Cordero, A. Data-driven screening of network constraints for unit commitment. *IEEE Trans. Power Syst.* **2020**, *35*, 3695–3705. [CrossRef]

energies

Article

Design of Nonlinear Backstepping Double-Integral Sliding Mode Controllers to Stabilize the DC-Bus Voltage for DC–DC Converters Feeding CPLs

Subarto Kumar Ghosh [1], Tushar Kanti Roy [2,*], Md. Abu Hanif Pramanik [3] and Md. Apel Mahmud [4]

1 Department of Electrical & Electronic Engineering, Rajshahi University of Engineering & Technology, Rajshahi 6204, Bangladesh; ghosh.sk@eee.ruet.ac.bd
2 Department of Electronics & Telecommunication Engineering, Rajshahi University of Engineering & Technology, Rajshahi 6204, Bangladesh
3 Department of Electrical & Computer Engineering, Rajshahi University of Engineering & Technology, Rajshahi 6204, Bangladesh; abuhanif@ece.ruet.ac.bd
4 School of Engineering, Deakin University, Geelong, VIC 3216, Australia; apel.mahmud@deakin.edu.au
* Correspondence: tkroy@ete.ruet.ac.bd or roy.kanti03@gmail.com; Tel.: +880-13-142-104-24

Abstract: This paper proposes a composite nonlinear controller combining backstepping and double-integral sliding mode controllers for DC–DC boost converter (DDBC) feeding by constant power loads (CPLs) to improve the DC-bus voltage stability under large disturbances in DC distribution systems. In this regard, an exact feedback linearization approach is first used to transform the nonlinear dynamical model into a simplified linear system with canonical form so that it becomes suitable for designing the proposed controller. Another important feature of applying the exact feedback linearization approach in this work is to utilize its capability to cancel nonlinearities appearing due to the incremental negative-impedance of CPLs and the non-minimum phase problem related to the DDBC. Second, the proposed backstepping double integral-sliding mode controller (BDI-SMC) is employed on the feedback linearized system to determine the control law. Afterwards, the Lyapunov stability theory is used to analyze the closed-loop stability of the overall system. Finally, a simulation study is conducted under various operating conditions of the system to validate the theoretical analysis of the proposed controller. The simulation results are also compared with existing sliding mode controller (ESMC) and proportional-integral (PI) control schemes to demonstrate the superiority of the proposed BDI-SMC.

Keywords: backstepping double-integral sliding mode control scheme; constant power load; exact feedback linearization approach; Lyapunov stability theory; negative-resistance characteristics; non-minimum phase; nonlinear dynamical model

Citation: Ghosh, S.K.; Roy, T.K.; Pramanik, M.A.H.; Mahmud, M.A. Design of Nonlinear Backstepping Double-Integral Sliding Mode Controllers to Stabilize the DC-Bus Voltage for DC–DC Converters Feeding CPLs. *Energies* **2021**, *14*, 6753. https://doi.org/10.3390/en14206753

Academic Editors: Victor Becerra and Ahmed Rachid

Received: 24 September 2021
Accepted: 14 October 2021
Published: 17 October 2021

1. Introduction

Over the past few decades, the power electronic converters (PECs) have been widely used in vehicular power systems (VPSs) (e.g., space vehicles, sea, land, etc.) and in renewable energy source (RES)-based systems (e.g., DC and AC microgrids) due to their voltage step-up, step-down, or conversion capabilities [1–5]. Recently, the power electronic-based DC distribution networks (DCDNs) are becoming more popular owing to their distinct advantages in terms of efficiency, controllability, flexibility, etc. [6,7]. However, two major challenges need to be addressed for a stable and reliable operation of power electronic-based DCDNs in conjunction of RESs, where the first challenge is the low terminal voltage of RESs while important to maintain a relatively high and stable terminal voltage of the DC-bus as all loads and RESs are either directly or indirectly connected to this bus [8–10]. As a results, DC–DC boost converters (DDBCs) are usually used in DCDNs as the interface between the DC-bus and RESs, which need to be appropriately controlled to achieve the desired DC-bus voltage [11,12]. On the other hand, the maintenance of stable and reliable operations becomes even more challenging when a large number of DC–DC converters are used to interface loads and RESs [13–15]. The second challenge is the instability issue caused by constant power loads (CPLs) in DCDNs with tightly coupled power electronic converters [16–18]. It is well-known that the constant power is drawn by CPLs, and consequently, it exhibits negative-impedance characteristics that destabilize as well as even cause blackouts [18,19]. Thus, it is essential to design an advance controller that not only deals with the destabilization but also provides quick dynamic response while guaranteeing the stability of the system.

Several control strategies have been proposed to resolve the instability issue of DDBCs caused by CPLs. Passive damping methods are proposed in [18,20] to neutralize the destabilization effect by increasing the damping into a system. However, these methods have limited applications due to their high costs and large sizes of passive elements such as capacitors and inductors. To alleviate these drawbacks, active damping methods such as virtual impedance [21], virtual resistance [22], and virtual capacitor [23] are proposed for stabilizing the system by adding extra control loops to reshape impedances. In these methods, the destructive effects of CPLs are mitigated by enforcing the unstable poles into the stable region through the modification in control loops. However, the satisfactory performance of these damping methods is highly dependent on the switching frequency. Thus, a voltage–current (V–I) droop-based dual-loop control scheme is proposed in [24] to stabilize the system by injecting sufficient damping torque into networks that overcomes the aforementioned drawbacks. However, a small variation in the droop gain leads to an inaccurate power sharing in networks [24]. To overcome this issue, an adaptive V–I droop control approach is proposed in [25] in which the virtual resistance is emulated at the output terminal of converters. However, these control approaches so far discussed here can only ensure the stability of the system near nominal operating points as these control schemes are developed by considering small disturbances around actual operating regions. Therefore, the system will be unstable when large disturbances due to variations in RESs and loads appear in the system.

A model predictive controller (MPC) is proposed in [26] to improve the stability of the system and extends the operating point. Though the MPC can handle the operating point issue, it cannot ensure better transient behaviors in the presence of large disturbances as it is designed without considering model uncertainties. This limitation is addressed in [27,28] by designing a hybrid MPC using the Takagi–Sugeno fuzzy-based scheme in conjunction with a traditional MPC. However, an accurate dynamical model is required to achieve the desired performance, which is always practically obstructed due to the presence of external disturbances and model uncertainties. Furthermore, the online computational complexity of MPCs limited their practical application in real-time platforms.

Nonlinear control schemes can be used to stabilize the system by considering the nonlinearity issue of CPLs and non-minimum phase problem of DDBCs. In [29], a passivity-based nonlinear controller is designed to resolve the instability issue due to CPLs while improving the transient and dynamic behaviors of the DC-bus voltage. However, the damping performance is severely affected with variations in the system parameters, which is overcome in [30] by proposing an adaptive passivity-based approach. However, the major drawback of these approaches is that the non-minimum phase problem of DDBCs cannot be avoided and further deteriorates the overall stability margin under large disturbances. Moreover, a time-scale model is required to design these controllers, which are quite hard to deal with in realistic applications.

Nonlinear feedback linearizing controllers (FBLCs) have an uncertainty/noise decoupling capability as discussed in [31–34], and these controllers are used in [35,36] for stabilizing the DC-bus voltage by overwhelming the non-minimum phase problem associated with DDBCs and the instability issue caused by CPLs. Though the approaches in [35,36] tackle the instability issue due to CPLs, the elimination of non-minimum phase problems is not completely achieved. Another FBLC is proposed in [37], where the non-minimum phase problem associated arising from DDBCs is avoided by considering the inductor current as an output function. However, the zero-dynamic stability is not presented in [37] though it is important and the control law in [37] is derived only for resistive loads though nonlinear dynamics of CPLs lead to unstable operation. Moreover, the performance of FBLC schemes so far discussed here highly rely on the precise parametric information of the system, which is quite impossible as these parameters change due to changes in operating points. On the other hand, the nonlinear sliding mode controller (SMC) is an effective approach for dealing with robustness against parameter uncertainties and external disturbances in both nonlinear and linear systems [38,39]. Hence, the parameter sensitivity problem of a FBLC can be resolved using SMC schemes and a robust pulse width modulation (PWM)-based SMC is used in [40] to achieve the desired transient and steady-state performance of DC microgrids where DDBCs feed CPLs. Similar approaches are presented in [41,42] to achieve the same control objective as that of [40]. Though SMCs effectively handle the effects of parameter variations, the utilization of a discontinuous function and fixed gain for the sliding surface results in unwanted chattering in control laws, which can even damage power electronic converters in practical applications.

The shortcomings of SMCs are overcome in [43–46] by proposing a nonlinear backstepping controller (BSC), and these controllers are explored in [47] for stabilizing DC-bus voltage in the DDBC feeding CPLs. However, the implementation of the BSC still requires knowing the accurate parameters of the system as neither parametric uncertainties nor external disturbances are incorporated during the design. The issues related to parametric uncertainties and external disturbances are alleviated in [48] using an adaptive BSC for ensuring the large signal stability of the DCDN with CPLs. However, an observer is used to estimate external disturbances, which is quite an expensive approach. Moreover, the non-minimum phase problem associated with DDBCs is not dealt with in this work. To avoid this drawback, a composite controller that is a combination of FBLC and adaptive BSC schemes is proposed in [49,50], where the non-minimum phases problem and impacts of the negative-impedance due to CPLs are canceled by transforming the nonlinear model into a linear canonical form. However, these controllers do not ensure better transient response when the system parameters change due to wide variations of operating points. Apart from this, the zero dynamic stability for the remaining state variables that are not transformed through the feedback linearization process is not discussed in these papers. To address the aforementioned problems, an SMC and adaptive BSC are combined in [51]. However, a conventional sliding surface is used in this controller ,which leads to the chattering phenomena in the control effort and makes it impossible to ensure the desired control performance.

Motivated from the limitations of existing literature, a composite scheme based on the backstepping double integral-sliding mode controller (BDI-SMC) is proposed in this work to enhance the stability of DDBC feeding CPLs in DCDNs. In this work, the issues related to the non-minimum phase arising from DDBCs and negative-impedance behaviors of CPLs are tackled by canceling nonlinearities and transforming the nonlinear dynamical model into a feedback linearized model using the exact feedback linearization approach. A double integral-sliding surface is then considered to derive the control law that can enhance the steady-state tracking performance while eliminating unwanted chattering effects. Afterwards, the actual control law is determined following design steps in both BSC and SMC. Finally, the feasibility of the actual control input in terms of maintaining the stability is analyzed using the Lyapunov function. A simulation study is conducted considering different operating scenarios of the system to verify its effectiveness. Additionally, its superiority is demonstrated by comparing the results with those of traditional nonlinear SMC and PI controllers.

2. System Description and Its Dynamical Model

The overall structure of a DCDN is depicted in Figure 1. It is well-known that the DC source generally consists of renewable energy-based distributed generation systems (DGS) such as solar photovoltaic (PV) system, rectifier in conjunction wind power generation systems, etc., which works as the main power supply. The DDBCs are commonly used for interfacing the DC source with the main DC-bus so that the voltage requirement of various loads can easily be met. Among these loads, some tightly regulated loads draw constant power from the DC-bus, which are also known as CPLs and exhibit negative-impedance characteristics. The electrical characteristic of CPLs can be described using the following voltage–current relation:

$$i_{CPL} = \frac{P_{CPL}}{v_{bus}} \tag{1}$$

where i_{CPL} denotes the current of a CPL, P_{CPL} represents the power of a CPL, and v_{bus} is the DC-bus voltage. The concept of the small disturbance analysis can be applied around an equilibrium point of Equation (1). Based on this analysis, the equivalent impedance of a CPL can be expressed as follows:

$$R_{CPL} = \frac{\partial v_{bus}}{\partial i_{CPL}} = -\frac{P_{CPL}}{I_{CPL}^2} \tag{2}$$

where I_{CPL} is the CPL steady-state value of the current. From Equation (2), it is obvious that the negative-impedance behavior of a CPL creates a negative impact on the performance, i.e., it decreases the damping of the system, which further destabilizes the system, especially when it is connected with a converter. Since the nonlinearity introduced into system is due to CPLs, traditional linear controllers cannot guarantee the stability against large disturbances. Hence, the employment of a nonlinear controller is essential to neutralize the nonlinear effect of CPLs. However, to achieve such an objective from a controller, it is necessary to have an appropriate dynamical model. Hence, a dynamical model for DDBCs feeding a CPL is developed in the following to meet the controller design requirement.

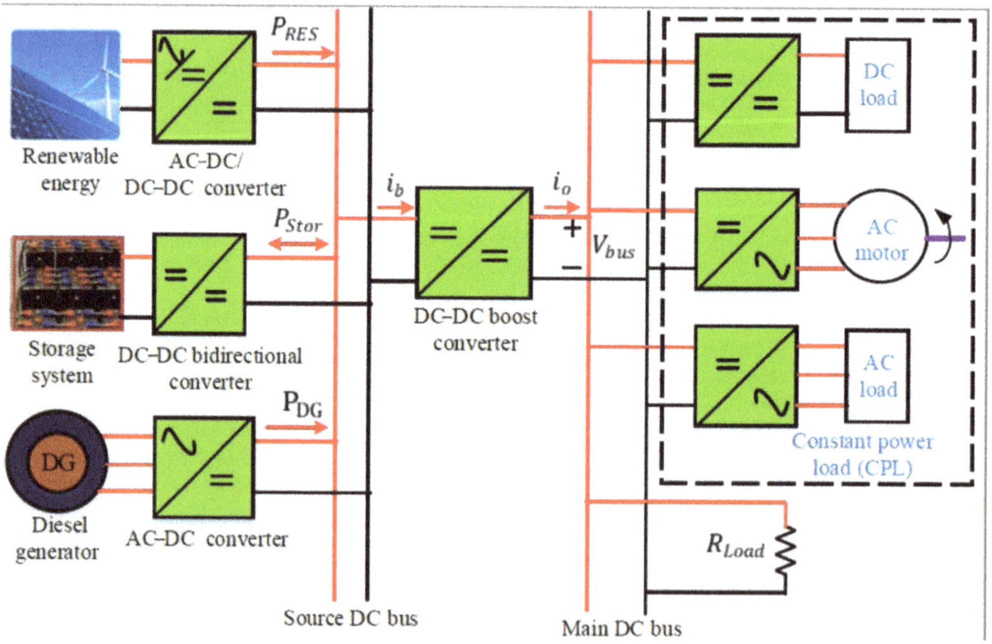

Figure 1. Typical layout of a DCDN.

The simplified structure of a DDBC feeding a CPL in DCDNs is shown in Figure 2. In this configuration, a current controlled source is used to represent a lumped CPL whereas the lumped resistive load is represented by R_{Load}. It is worth mentioning that the input DC voltage source of the DDBC will be obtained from RESs or another DC–DC converter. Now, according to the simplified structure as shown in Figure 2, the expression of its model while on (T_{on}) and off (T_{off}) can be obtained as follows:

$$\dot{x} = (A_{on}x + b_{on})u + (A_{off}x + b_{off})(1-u) \tag{3}$$

with

$$A_{on} = \begin{bmatrix} -\frac{r_b}{L_b} & 0 \\ 0 & -\frac{P_{CPL}}{v_{bus}} \end{bmatrix},\ b_{on} = \begin{bmatrix} \frac{V_{in}}{L_b} \\ 0 \end{bmatrix},\ A_{off} = \begin{bmatrix} -\frac{r_b}{L_b} & -\frac{1}{L_b} \\ \frac{1}{C_{bus}} & -\frac{P_{CPL}}{v_{bus}} \end{bmatrix},\ b_{off} = \begin{bmatrix} \frac{V_{in}}{L_b} \\ 0 \end{bmatrix},\ and\ x = \begin{bmatrix} i_b \\ v_{bus} \end{bmatrix}$$

where i_b represents the inductor current, V_{in} is the input voltage of the converter, v_{bus} is the main DC-bus voltage, L is the inductor, r_b is the parasitic resistance of an inductor, u is the duty cycle, and C_{bus} is the DC-bus capacitance. It worth noting that the inductor current, i_b, and the main DC-bus voltage, v_{bus}, are considered state variables. It is well-known that a discontinuous converter model can be approximated by a continuous model by selecting the converter switching frequency as being higher than the natural frequency. Hence, for continuous conduction mode (CCM) operations, the discontinuous control input u can be replaced by the duty ratio μ, which is a continuous function in the subinterval [0, 1]. Therefore, by considering the CCM operation, the model as represented by Equation (4) can be expressed as follows:

$$\begin{aligned} \frac{di_b}{dt} &= -\frac{r_b}{L_b}i_b + \frac{V_{in} - v_{bus}}{L_b} + \frac{v_{bus}}{L_b}\mu \\ \frac{dv_{bus}}{dt} &= \frac{1}{C_{bus}}(i_b - \frac{v_{bus}}{R_{Load}} - \frac{P_{CPL}}{v_{bus}}) - \frac{i_b}{C_{bus}}\mu \end{aligned} \tag{4}$$

The proposed BDI-SMC is designed based on the model as described by Equation (4) which is discussed in the next section.

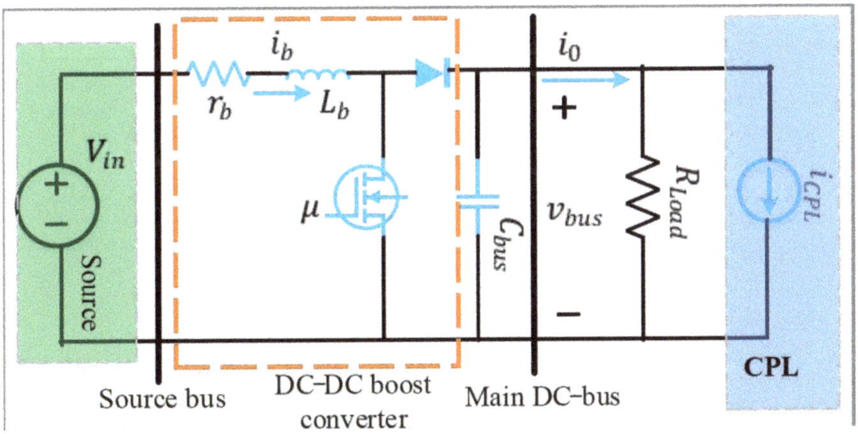

Figure 2. The simplified structure of a DDBC feeding a CPL.

3. Proposed Controller Design Approach

In this section, the proposed BDI-SMC is designed for DDBCs feeding CPLs. The key control objective is to achieve fast transient and desired steady-state tracking errors of the DC-bus voltage. However, the dynamical model represented by Equation (4) is not suitable for designing the BDI-SMC in its current form. Hence, the following steps need to be followed to make the model suitable for the controller design:

- Transformation of the model into an exactly linearized system using an exact feedback linearization approach and making it suitable to apply the backstepping control approach;
- Elimination of the non-minimum phase problem of DDBCs by considering the total stored energy and its rate of change as two new state variables;
- Derivation of the control law using the proposed scheme that satisfies the desired tracking performance to track the DC-bus voltage; and
- Stability analysis of the whole system with the newly derived control input performed using the Lyapunov stability theory.

All these steps are discussed in detail through the following subsections.

3.1. Transformation of the Model into an Exactly Linearized System Using the Exact Feedback Linearization

The transformation of the nonlinear model of the DDBC with a CPL in Equation (4) can be represented as the following generalized form:

$$\begin{aligned} \dot{x} &= f(x) + g(x)u \\ y &= h(x) \end{aligned} \tag{5}$$

where $x = [x_1, x_2, \ldots, x_n]^T$ is the state of the system, u is the input, y is the output of the system, and $f(x)$ and $g(x)$ are nonlinear functions of states and parameters with $g(x) \neq 0$. Fitting Equations (4) into (5), it can be written as follows:

$$f(x) = \begin{bmatrix} -\frac{r_b}{L_b}x_1 + (\frac{V_{in}-x_2}{L_b}) \\ \frac{1}{C_{bus}}(x_1 - \frac{x_2}{R_{Load}} - \frac{P_{CPL}}{x_2}) \end{bmatrix}, \ and \ g(x) = \begin{bmatrix} \frac{x_2}{L_b} \\ -\frac{x_1}{C_{bus}} \end{bmatrix}^T$$

where $x = [i_b \ \ v_{bus}]^T$ and $u = \mu$. The following steps are necessary to transform the nonlinear system into an exactly linearized system using the exact feedback linearization scheme.

Step 1: Selection of the output function

The feedback linearizability depends on the output of the system, which can be selected in a different way. For the DDBC with a CPL, the output function can be chosen as any of the states (i.e., the capacitor voltage or inductor current) and the combination of these states in the form of the total energy of the system. The system becomes partially linearized when the capacitor voltage is selected as the output. Hence, the system will be a non-minimum phase one as the internal dynamic stability cannot satisfy the stability criterion. This issue can be tackled by selecting the inductor current as the output, which also makes the system partially linearized but with the stable internal dynamic. However, the regulation of the output voltage by indirectly controlling the inductor current results in excessive overshoot and slower response. Under this situation, the output needs to be chosen in such a way that satisfies the requirement of the exact linearization approach, i.e., the system is exactly linearized for which the relative degree of the exactly linearized system is equal to the order of the system. Based on this discussion, the total stored energy is considered an output function that can be expressed as follows:

$$y = h(x) = \frac{1}{2}L_b i_b^2 + \frac{1}{2}C_{bus}v_{bus}^2 \tag{6}$$

Step 2: Relative degree calculation

In this step, the relative degree for the nonlinear system as presented by Equation (5) is calculated using the Lie derivative while considering the output function in Equation (6). To calculate the relative degree, the following condition should be satisfied:

$$L_g L_f^{r-1} h(x) \neq 0 \tag{7}$$

In Equation (7), r is the relative degree while L represents the Lie derivative along the vector field denoted by the subscript. To satisfy the condition for the exact linearization approach, the relative degree of the nth order system will be $r = n$, which satisfies the following expressions:

$$\begin{aligned} & L_g L_f^{1-1} h(x) = L_g L_f^{2-1} h(x) = \ldots = L_g L_f^{n-2} h(x) = 0 \\ & L_g L_f^{n-1} h(x) \neq 0 \end{aligned} \quad (8)$$

Based on the model and output function, the following expressions are obtained:

$$\begin{aligned} r = n = 2 \\ L_g L_f^{1-1} h(x) = 0 \end{aligned}$$

and

$$L_g L_f^{n-1} h(x) \neq 0$$

Hence, the system is exactly linearized.

Step 3: Nonlinear coordinate transformation and exact linearization

In this step, the coordinate transformation technique is adopted to convert the original x state variables into new z state variables, which can be discussed as follows:

$$z_1 = h(x) = L_f^{1-1} h(x) = \frac{1}{2} L_b i_b^2 + \frac{1}{2} C_{bus} v_{bus}^2 \quad (9)$$

in which the dynamic can be written as

$$\dot{z_1} = \frac{\partial h(x)}{\partial x} \dot{x} \quad (10)$$

Using Equation (5), Equation (10) can be written as follows:

$$\dot{z_1} = \frac{\partial h(x)}{\partial x} f(x) + \frac{\partial h(x)}{\partial x} g(x) u = L_f h(x) + L_g L_f^{1-1} h(x) u \quad (11)$$

Since $L_g L_f^{1-1} h(x) = 0$, Equation (11) can be written as follows:

$$\dot{z_1} = z_2 = L_f h(x) = V_{in} x_1 - \frac{x_2^2}{R_{Load}} - P_{CPL} - r_b x_1^2 \quad (12)$$

Equation (12) can be written as follows:

$$\dot{z_2} = L_f^2 h(x) + L_g L_f^{2-1} h(x) u = v \quad (13)$$

Since $L_g L_f^{2-1} h(x) \neq 0$, the original nonlinear system can be represented as the following exactly linearized form:

$$\begin{aligned} \dot{z_1} &= z_2 \\ \dot{z_2} &= v \end{aligned} \quad (14)$$

where z_1 and z_2 are the state variables of the exactly linearized system and v is the control variable of the transformed linear systems. The model in Equation (14) can be rewritten as follows:

$$\begin{aligned} \dot{z_1} &= z_2 \\ \dot{z_2} &= v = a(x) + b(x) \mu \end{aligned} \quad (15)$$

where

$$a(x) = L_f^2 h(x) = \frac{(V_{in} - 2x_1 r_b)(V_{in} - x_2 - r_b x_1)}{L_b} - \frac{2x_2}{R_{Load} C_{bus}} (x_1 - \frac{x_2}{R_{Load}} - \frac{P_{CPL}}{x_2})$$
$$b(x) = L_g L_f h(x) = \frac{(V_{in} - 2x_1 r_b) x_2}{L_b} + \frac{2x_1 x_2}{R_{Load} C_{bus}}$$

The proposed controller is designed based on this model, which is discussed in the following subsection.

3.2. Controller Design

In this subsection, the control law is determined using the combination of backstepping and double integral-sliding mode control schemes. Here, the main objectives are to track z_1 and z_2, which ultimately ensures the desired tracking of the DC-bus voltage under large disturbances. The following steps describe the design process for the proposed BDI-SMC.

Step 1: For fulfilling the design objective, the first tracking error (e_1) can be defined as follows:

$$e_1 = z_1 - z_{1(ref)} \tag{16}$$

where $z_{1(ref)}$ is the reference value of the state z_1, which can be calculated as follows:

$$z_{1(ref)} = \frac{1}{2} L_b i^2_{b(ref)} + \frac{1}{2} C_{bus} v^2_{bus(ref)} \tag{17}$$

where $i_{b(ref)} = \frac{P_{CPL}}{V_{in}}$ is the reference value of the inductor current and $v_{dc(ref)}$ is the reference value of the DC-bus voltage. Now, the dynamic of e_1 using Equation (16) can be obtained as follows:

$$\dot{e_1} = z_2 - \dot{z_1}_{(ref)} \tag{18}$$

As the actual control input does not appear in Equation (18), z_2 is assumed to be the virtual control input and γ is assumed to be the corresponding stabilizing function or virtual control law. Hence, the final error (e_2) can be defined as follows:

$$e_2 = z_2 - \gamma \tag{19}$$

Using Equation (19), Equation (18) can be rewritten as follows:

$$\dot{e_1} = e_2 + \gamma - \dot{z_1}_{(ref)} \tag{20}$$

At this stage, it is required to check the stability of the tracking error dynamic $\dot{e}_1$, and to do this, the following Lyapunov function (LF) is considered:

$$W_1 = \frac{1}{2} e_1^2 \tag{21}$$

in which the time derivative using Equation (20) can be written as follows:

$$\dot{W_1} = e_1 [e_2 + \gamma - \dot{z_1}_{(ref)}] \tag{22}$$

From Equation (22), γ can be selected as follows:

$$\gamma = -k_1 e_1 + \dot{z_1}_{(ref)} \tag{23}$$

where k_1 is a positive constant. Substituting Equations (23) into (22) yields the following:

$$\dot{W_1} = -k_1 e_1^2 + e_1 e_2 \tag{24}$$

From Equation (24), it can be seen that, if $e_2 = 0$, then $\dot{W}_1 \leq 0$. To achieve this, the next step is essential.

Step 2: The dynamic of e_2 can be written as follows:

$$\dot{e_2} = \dot{z_2} - \dot{\gamma} \tag{25}$$

Using Equation (15), the dynamic of e_2 can be expressed as follows:

$$\dot{e_2} = a(x) + b(x)\mu - \dot{\gamma} \tag{26}$$

At this instant, a double-integral sliding surface in term of e_2 can be selected as follows:

$$S = e_2 + \alpha_1 \int e_2 dt + \alpha_2 \int\int e_2 dtdt \tag{27}$$

where α_1 and α_2 are constant parameters of the sliding surface. The dynamic of Equation (27), using the value of $\dot{e}_2$, can be expressed as follows:

$$\dot{S} = a(x) + b(x)\mu - \dot{\gamma} + \alpha_1 e_2 + \alpha_2 \int e_2 dt \tag{28}$$

It is well-known that appropriately selecting a reaching law is very important to mitigate the chattering while improving the convergence time. Hence, a reaching law is selected as follows to meet the above objective:

$$\dot{S} = -\beta_1 sgn(S) - \beta_2 S \tag{29}$$

where β_1 and β_2 are positive constants and the chattering effect depends on these values. Combining Equations (28) and (29), it can be written as follows:

$$a(x) + b(x)\mu - \dot{\gamma} + \alpha_1 e_2 + \alpha_2 \int e_2 dt = -\beta_1 sgn(S) - \beta_2 S \tag{30}$$

From Equation (30), the actual control law can be determined as follows:

$$\mu = -\frac{1}{b(x)}[a(x) - \dot{\gamma} + \alpha_1 e_2 + \alpha_2 \int e_2 dt + \frac{1}{S} e_1 e_2 + \beta_1 sgn(S) + \beta_2 S] \tag{31}$$

Substituting Equation (31) into Equation (28) yields

$$\dot{S} = -\beta_1 sgn(S) - \beta_2 S - \frac{1}{S} e_1 e_2 \tag{32}$$

The following Lyapunov function is selected to ensure overall stability with the derived control law:

$$\dot{W}_2 = W_1 + \frac{1}{2} S^2 \tag{33}$$

Using Equations (24) and (32), $\dot{W}_2$ can be written as follows:

$$\dot{W}_2 = -k_1 e_1^2 - \beta_1 |S| - \beta_2 S^2 \tag{34}$$

Since $k_1 > 0$, $\beta_1 > 0$, and $\beta_2 > 0$, $\dot{W}_2 \leq 0$. Thus, the overall stability of the system is guaranteed with the designed control law as represented by Equation (31). The effectiveness of the designed BDI-SMC is demonstrated in the next section.

4. Simulation Results

To verify the effectiveness of the designed composite BDI-SMC strategy, a similar simulation model as that shown in Figure 2 is built on the MATLAB/Simulink platform.

The nominal voltage of the main-DC bus of the distribution network is considered as 110 V while the rated power of DC loads is considered as 2 kW. In order to capture the highest destabilization effect of the DC load on the DC distribution network, a pure CPL is considered the DC power demand rather than taking into account the composite load demand. The frequency of the power electronic interface boost converter is set as 5 kHz with a sampling frequency of 100 kHz to evaluate the performance of the designed BDI-SMC. It is well-known that, during the implementation of controllers, it is essential to know the system parameters. Hence, the system parameters used for the simulation study are listed in Table 1. The gain parameters of the designed nonlinear controller are listed as follows: $k_1 = 1000$, $\alpha_1 = 70$, $\alpha_2 = 0.45$, $\beta_1 = 100$, and $\beta_2 = 0.01$. It is worth mentioning that the gain parameters are selected based on the trial and error method to meet the desired control objective.

Table 1. Nominal system parameters of the system.

Parameters	Description	Value
V_{in}	Supply voltage	55 V
V_{bus}	Main DC-bus voltage	110 V
P_{CPL}	Nominal power in constant power load	2 kW
r_b	Parasitic resistance of an inductor	2 mΩ
L_b	Inductance of the converter	5 mH
C_{bus}	Capacitance of the main DC-bus	6 mF
R_{Load}	Resistive Load	∞

The performance of the BDI-SMC is verified by widely varying the operating region by considering the variations in the reference power of CPLs, main DC-bus reference voltage, and input power in terms of the input source voltage. Therefore, the following three case studies are considered to demonstrate the performance of the designed controller:

- **Case I:** Controller performance investigation with variations in the reference power of the CPL;
- **Case II:** Controller performance investigation with variations in the reference voltage of the DC-bus; and
- **Case III:** Controller performance investigation with variations in the input voltage.

To show the merits of the designed BDI-SMC, the performance is also compared with the existing SMC (ESMC), as proposed in [40], and with a conventional proportional-integral (PI) controller.

Case I: Controller performance investigation with variations in the reference power of the CPL

In this scenario, the variation in the reference power of CPLs is taken into consideration to evaluate the performance of the designed controller and existing controllers. At the beginning of the simulation, the rated power of the CPL is considered as 2 kW whereas it is increased to 4 kW at t = 1 s. On the other hand, other parameters are kept unchanged. Due to the transient power of CPLs, as shown in Figure 3a, the post-disturbance dynamic response of the main DC-bus and inductor current are affected, which can be seen in Figure 3b,c, respectively. From these responses, it can be observed that all three controllers can ensure stability. However, the designed BDI-SMC can provide faster dynamic responses along with less undershoot compared with the ESMC and PI. Again at t = 2 s, the CPL power is decreased from 4 kW to 500 W, which is considered a large disturbance in the system. Due to this large disturbance, the conventional PI controller is unable to provide sufficient damping torque into the networks, consequently leading to instability in both the main DC-bus voltage and inductor current, which can be clearly seen from Figure 3b,c. Meanwhile, both the designed BDI-SMC and the ESMC can ensure stability, which can be clearly seen from Figure 3b,c, respectively. However, though both controllers can maintain zero output voltage tracking errors, the designed BDI-SMC ensures a faster settling time and less overshoot. The corresponding control signal for all three controllers is depicted in

Figure 3d,e, where it is obvious that the designed control signal is more stable compared with the PI and ESMC controllers. Furthermore, it can be observed that the designed controller can effectively eliminate the chattering effect on the PWM whereas the ESMC cannot attenuate the chattering effect.

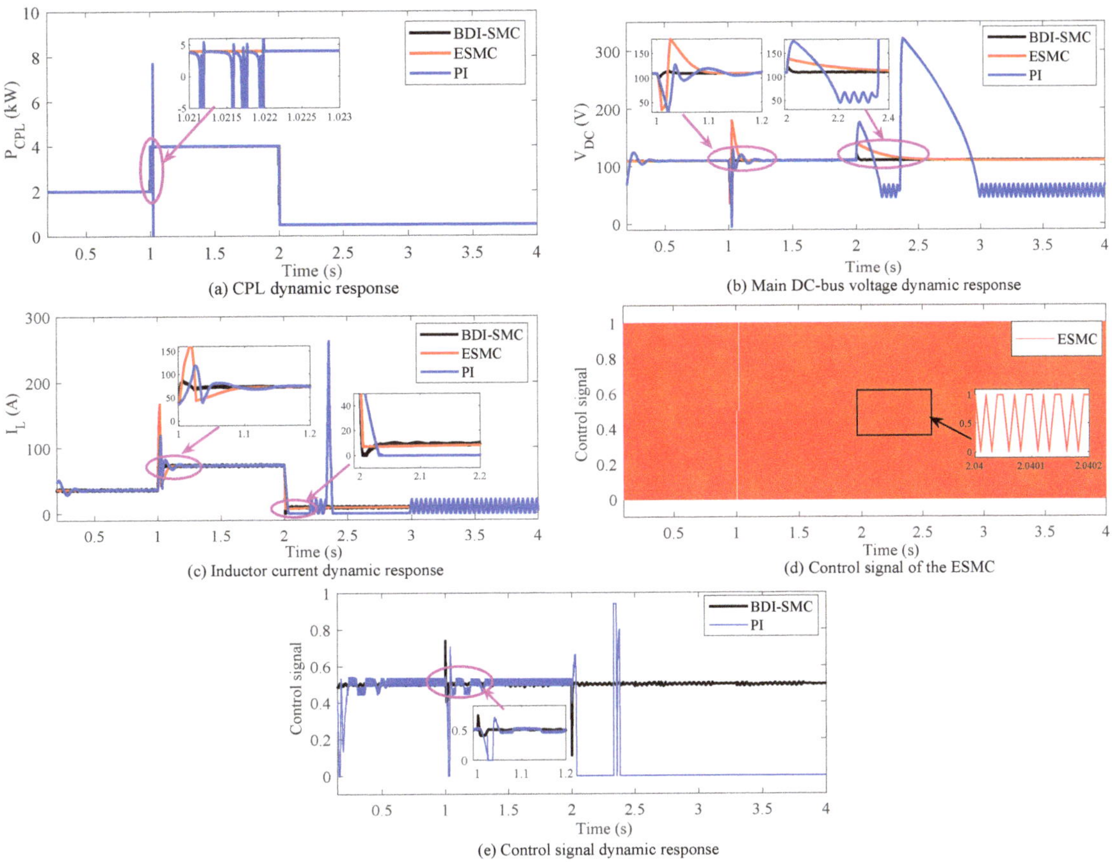

Figure 3. Dynamic response of the system with the variations in the CPL reference power.

Case II: Controller performance investigation with the variations in the reference voltage of the DC-bus

In this case study, the main DC-bus voltage increases from its equilibrium state 110 V to 160 V at t = 1.2 s and again increases from 160 V to 220 V at t = 2.2 s, whereas the CPL is set as 2 kW. The corresponding system responses are illustrated in Figure 4. From Figure 4a, it can be observed that the PI and ESMC cannot ensure a main DC-bus voltage tracking performance with a faster settling time and less overshoot/undershoot whilst the BDI-SMC can quickly track the new reference main DC-bus voltage as the settling time overshoot/undershoot are close to zero. Similarly, from Figure 4b, it can be observed that the designed BDI-SMC can provide a faster dynamic performance compared with the ESMC and PI. However, although the overshoot/undershoot in the inductor current is less with the ESMC, the control signal contains higher chattering, as illustrated in Figure 4d. As a result, the switching losses are higher, which reduces the overall system performance. On the other hand, the designed BDI-SMC is able to provide an oscillation-free control signal to the PWM compared with the ESMC, which can be clearly seen from

Figure 4d,e. Throughout this simulation study, the load power is constant due to the electrical characteristics of the CPL, which is shown in Figure 4c.

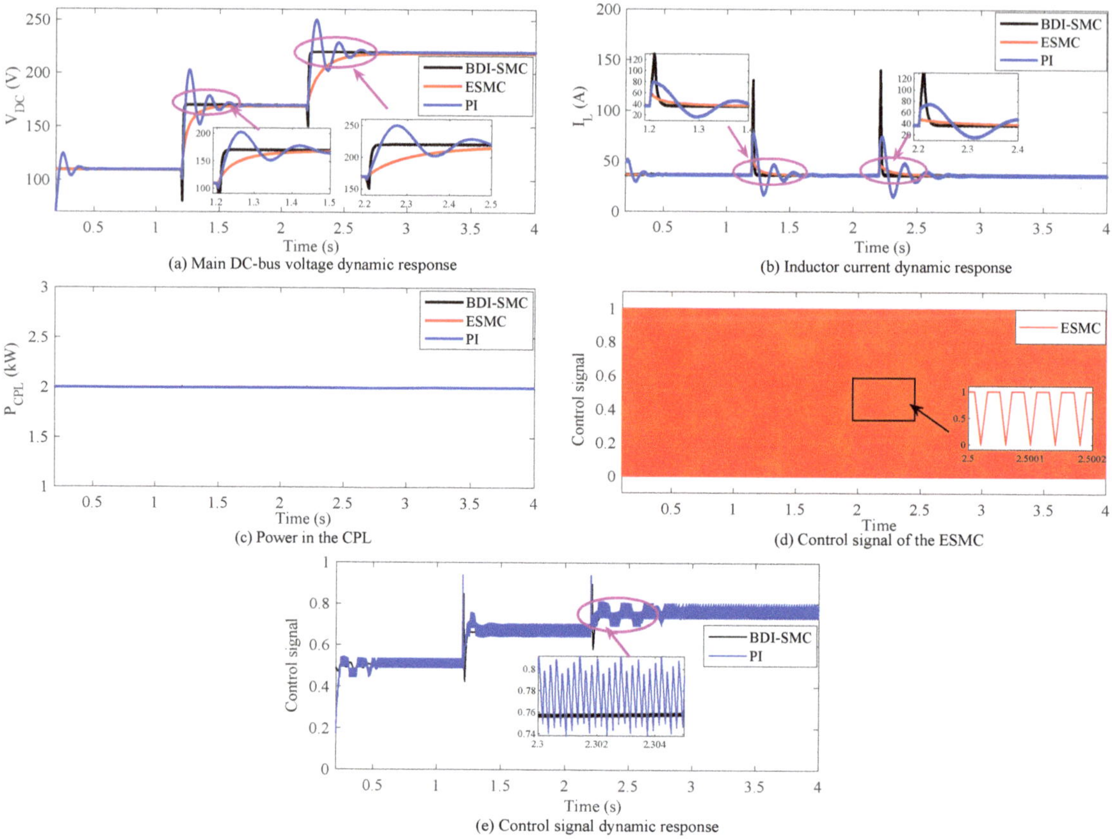

Figure 4. Dynamic response of the system with the variations of the main DC-bus reference voltage.

Case III: Controller performance investigation with the variations in the input voltage

Practically, the output power of RESs is continuously changeable due to their intermittent characteristics. To show that impact on the distribution network, in this case study, the variation in the input power is considered. For this purpose, the input voltage of the DC voltage increases from 55 V to 70 V and decreases from 70 V to 40 V at t = 1.4 s and t = 2.4 s, respectively. Figure 5 shows the corresponding dynamic responses of the system. From Figure 5b,c, it can be observed that all three controllers can accurately track the main DC-bus voltage and inductor current when input voltage variation occurs. However, the designed BDI-SMC has a faster dynamic response and a smaller overshoot compared with the ESMC and conventional PI control methods. Moreover, it can be seen that the CPL remained unchanged at the equilibrium state 2 kW, which is illustrated in Figure 5d.

From the above simulation results, it can be concluded that the designed BDI-SMC can provide fast dynamic convergence speed, accurate tracking of the DC-bus voltage, and good large disturbances stability for a DC–DC boost converter feeding a CPL. Moreover, it is obvious that the stability of the DC-bus voltage is not destroyed even after the large variations in the CPL power. Hence, the simulation result is consistent with the theoretical analysis.

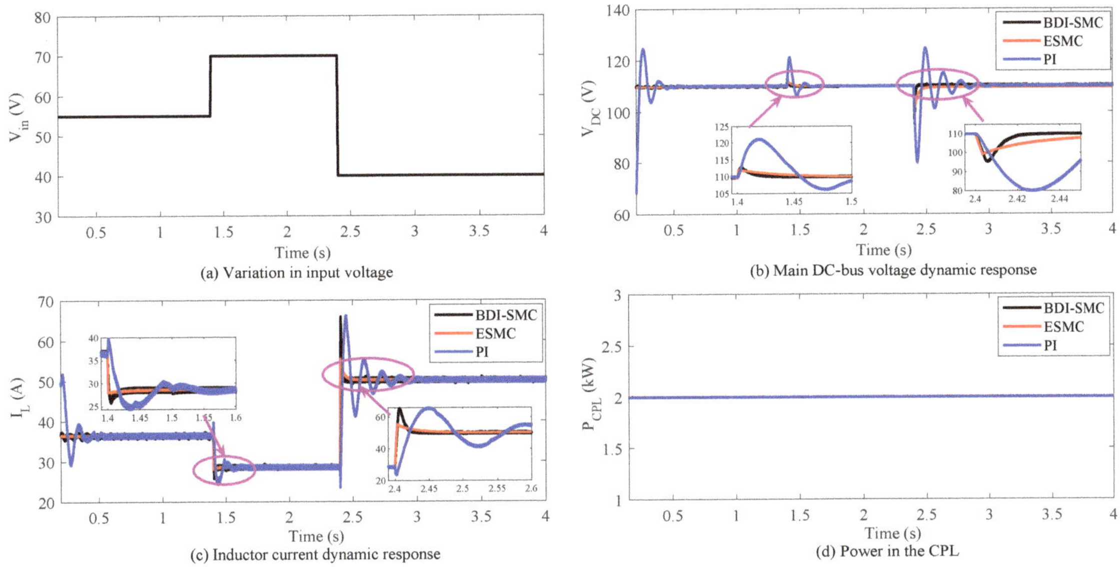

Figure 5. Dynamic response of the system with the variations in the input voltage.

5. Conclusions

A composite nonlinear controller combining the backstepping control theory and double integral-sliding mode control theory is proposed for a DC–DC boost converter feeding a CPL with distribution networks. In order to design the proposed controller, a canonical form is developed using the exact feedback linearization approach by considering the total stored energy as the output function. Based on the canonical form of the model, a composite nonlinear controller is designed to guarantee DC-bus voltage stability under large disturbances of a DC–DC boost converter feeding CPLs with negligible steady-state tracking error and fast transient responses. The effectiveness and theoretical analysis of the designed controller are verified through simulation results. In the future, a robust adaptive nonlinear composite controller will be designed by considering both parametric uncertainty and external disturbances.

Author Contributions: Conceptualization, S.K.G., T.K.R. and M.A.M.; methodology, S.K.G. and T.K.R.; software, S.K.G. and M.A.H.P.; validation, S.K.G. and T.K.R.; formal analysis, S.K.G., T.K.R. and M.A.H.P.; investigation, S.K.G. and T.K.R.; writing—original draft preparation, S.K.G. and T.K.R.; writing—review and editing, T.K.R. and M.A.M.; supervision, T.K.R. and M.A.M.; project administration, T.K.R. and M.A.M. All authors have read and agreed to the current version of the manuscript.

Funding: This research received no external funding.

Institutional Review Board Statement: Not applicable.

Informed Consent Statement: Not applicable.

Data Availability Statement: Not applicable.

Conflicts of Interest: The authors declare no conflicts of interest.

Abbreviations

The following abbreviations are used in this manuscript:

BDI-SMC	Nonlinear Backstepping Double Integral-Sliding Mode Controller
BSC	Nonlinear Backstepping Controller
CPL	Constant Power Load
CCM	Continuous Conduction Mode
DGSs	Distributed Generation Systems
DDBC	DC–DC Boost Converter
DCDNs	DC Distribution Networks
ESMC	Existing Sliding Mode Controller
FBLCs	Nonlinear Feedback Linearizing Controllers
MPC	Model Predictive Controller
PV	Solar Photovoltaic
PECs	Power Electronic Converters
PI	Proportional-Integral Controller
PWM	Pulse Width Modulation
RESs	Renewable Energy Sources
SMC	Sliding Mode Controller
VPSs	Vehicular Power Systems

References

1. Roy, T.K.; Mahmud, M.A.; Oo, A.M.T.; Haque, M.E.; Muttaqi, K.M.; Mendis, N. Nonlinear adaptive backstepping controller design for islanded DC microgrids. *IEEE Trans. Ind. Appl.* **2018**, *54*, 2857–2873. [CrossRef]
2. Orchi, T.F.; Mahmud, M.A.; Oo, A.M.T. Generalized dynamical modeling of multiple photovoltaic units in a grid-connected system for analyzing dynamic interactions. *Energies* **2018**, *11*, 296. [CrossRef]
3. Roy, T.K.; Mahmud, M.A. Active power control of three-phase grid-connected solar PV systems using a robust nonlinear adaptive backstepping approach. *Sol. Energy* **2017**, *153*, 64–76. [CrossRef]
4. Ghosh, S.; Roy, T.; Pramanik, M.; Mahmud, M.A. LMI-Based Optimal Linear Quadratic Controller Design for Multiple Solar PV Units Connected to Distribution Networks. In Proceedings of the 2021 IEEE Texas Power and Energy Conference (TPEC), College Station, TX, USA, 2–5 February 2021; pp. 1–6.
5. Roy, T.K.; Mahmud, M.A.; Oo, A.M.T.; Bansal, R.; Haque, M.E. Nonlinear adaptive backstepping controller design for three-phase grid-connected solar photovoltaic systems. *Electr. Power Compon. Syst.* **2017**, *45*, 2275–2292. [CrossRef]
6. Ghosh, S.K.; Roy, T.K.; Pramanik, M.A.H.; Sarkar, A.K.; Mahmud, M. An energy management system-based control strategy for DC microgrids with dual energy storage systems. *Energies* **2020**, *13*, 2992. [CrossRef]
7. Xiao, J.; Wang, P.; Setyawan, L.; Xu, Q. Multi-level energy management system for real-time scheduling of DC microgrids with multiple slack terminals. *IEEE Trans. Energy Convers.* **2015**, *31*, 392–400. [CrossRef]
8. Roy, T.K.; Mahmud, M.A. Dynamic stability analysis of hybrid islanded DC microgrids using a nonlinear backstepping approach. *IEEE Syst. J.* **2017**, *12*, 3120–3130. [CrossRef]
9. Nasir, M.; Khan, H.A.; Hussain, A.; Mateen, L.; Zaffar, N.A. Solar PV-based scalable DC microgrid for rural electrification in developing regions. *IEEE Trans. Sustain. Energy* **2017**, *9*, 390–399. [CrossRef]
10. Mendis, N.; Mahmud, M.A.; Roy, T.K.; Haque, M.E.; Muttaqi, K.M. Power management and control strategies for efficient operation of a solar power dominated hybrid DC microgrid for remote power applications. In Proceedings of the 2016 IEEE Industry Applications Society Annual Meeting, Portland, OR, USA, 2–6 October 2016; pp. 1–8.
11. Roy, T.K.; Mahmud, M.A.; Oo, A.M.T.; Haque, M.E.; Muttaqi, K.M.; Mendis, N. Nonlinear adaptive backstepping controller design for controlling bidirectional power flow of BESSs in DC microgrids. In Proceedings of the 2016 IEEE Industry Applications Society Annual Meeting, Portland, OR, USA, 2–6 October 2016; pp. 1–8.
12. Mahmud, M.A.; Roy, T.K.; Saha, S.; Haque, M.E.; Pota, H.R. Robust nonlinear adaptive feedback linearizing decentralized controller design for islanded DC microgrids. *IEEE Trans. Ind. Appl.* **2019**, *55*, 5343–5352. [CrossRef]
13. Ravada, B.R.; Tummuru, N.R.; Ande, B.N.L. Photovoltaic-Wind and Hybrid Energy Storage Integrated Multi-Source Converter Configuration for DC Microgrid Applications. *IEEE Trans. Sustain. Energy* **2020**, *12*, 83–91. [CrossRef]
14. Garg, A.; Tummuru, N.R.; Oruganti, R. Implementation of Energy Management Scenarios in a DC Microgrid using DC Bus Signaling. *IEEE Trans. Ind. Appl.* **2021**, *57*, 5306–5317. [CrossRef]
15. Ghosh, S.K.; Roy, T.K.; Pramanik, M.A.H.; Ali, M.S. Energy management techniques to enhance DC-bus voltage transient stability and power balancing issues for islanded DC microgrids. In *Advances in Clean Energy Technologies;* Elsevier: Amsterdam, The Netherlands, 2021; pp. 349–375.
16. Lucas, K.E.; Pagano, D.J.; Vaca-Benavides, D.A.; Garcia-Arcos, R.; Rocha, E.M.; Medeiros, R.L.; Rios, S.J. Robust Control of Interconnected Power Electronic Converters to Enhance Performance in dc distribution systems: A case of study. *IEEE Trans. Power Electron.* **2020**, *36*, 4851–4863. [CrossRef]

17. Chang, F.; Cui, X.; Wang, M.; Su, W. Region of Attraction Estimation for DC Microgrids with Constant Power Loads Using Potential Theory. *IEEE Trans. Smart Grid* **2021**, *12*, 3793–3808. [CrossRef]
18. Wang, M.; Tang, F.; Wu, X.; Niu, J.; Zhang, Y.; Wang, J. A Nonlinear Control Strategy for DC-DC Converter with Unknown Constant Power Load Using Damping and Interconnection Injecting. *Energies* **2021**, *14*, 3031. [CrossRef]
19. AL-Nussairi, M.K.; Bayindir, R.; Padmanaban, S.; Mihet-Popa, L.; Siano, P. Constant Power Loads (CPL) with Microgrids: Problem Definition, Stability Analysis and Compensation Techniques. *Energies* **2017**, *10*, 1656. [CrossRef]
20. Majstorovic, D.; Celanovic, I.; Teslic, N.D.; Celanovic, N.; Katic, V.A. Ultralow-latency hardware-in-the-loop platform for rapid validation of power electronics designs. *IEEE Trans. Ind. Electron.* **2011**, *58*, 4708–4716. [CrossRef]
21. Lu, X.; Sun, K.; Guerrero, J.M.; Vasquez, J.C.; Huang, L.; Wang, J. Stability enhancement based on virtual impedance for DC microgrids with constant power loads. *IEEE Trans. Smart Grid* **2015**, *6*, 2770–2783. [CrossRef]
22. Rahimi, A.M.; Emadi, A. Active damping in DC/DC power electronic converters: A novel method to overcome the problems of constant power loads. *IEEE Trans. Ind. Electron.* **2009**, *56*, 1428–1439. [CrossRef]
23. Magne, P.; Marx, D.; Nahid-Mobarakeh, B.; Pierfederici, S. Large-signal stabilization of a DC-link supplying a constant power load using a virtual capacitor: Impact on the domain of attraction. *IEEE Trans. Ind. Appl.* **2012**, *48*, 878–887. [CrossRef]
24. Gui, Y.; Han, R.; Guerrero, J.M.; Vasquez, J.C.; Wei, B.; Kim, W. Large-Signal Stability Improvement of DC-DC Converters in DC Microgrid. *IEEE Trans. Energy Convers.* **2021**, *36*, 2534–2544. [CrossRef]
25. Augustine, S.; Mishra, M.K.; Lakshminarasamma, N. Adaptive droop control strategy for load sharing and circulating current minimization in low-voltage standalone DC microgrid. *IEEE Trans. Sustain. Energy* **2014**, *6*, 132–141. [CrossRef]
26. Karami, Z.; Shafiee, Q.; Khayat, Y.; Yaribeygi, M.; Dragicevic, T.; Bevrani, H. Decentralized model predictive control of DC microgrids with constant power load. *IEEE J. Emerg. Sel. Top. Power Electron.* **2019**, *9*, 451–460. [CrossRef]
27. Karami, Z.; Shafiee, Q.; Sahoo, S.; Yaribeygi, M.; Bevrani, H.; Dragicevic, T. Hybrid Model Predictive Control of DC–DC Boost Converters With Constant Power Load. *IEEE Trans. Energy Convers.* **2020**, *36*, 1347–1356. [CrossRef]
28. Vafamand, N.; Yousefizadeh, S.; Khooban, M.H.; Bendtsen, J.D.; Dragičević, T. Adaptive TS fuzzy-based MPC for DC microgrids with dynamic CPLs: Nonlinear power observer approach. *IEEE Syst. J.* **2018**, *13*, 3203–3210. [CrossRef]
29. Zeng, J.; Zhang, Z.; Qiao, W. An interconnection and damping assignment passivity-based controller for a DC–DC boost converter with a constant power load. *IEEE Trans. Ind. Appl.* **2013**, *50*, 2314–2322. [CrossRef]
30. Hassan, M.A.; Li, E.P.; Li, X.; Li, T.; Duan, C.; Chi, S. Adaptive passivity-based control of DC–DC buck power converter with constant power load in DC microgrid systems. *IEEE J. Emerg. Sel. Top. Power Electron.* **2018**, *7*, 2029–2040. [CrossRef]
31. Mahmud, M.; Hossain, M.; Pota, H.; Oo, A.M. Robust partial feedback linearizing excitation controller design for multimachine power systems. *IEEE Trans. Power Syst.* **2016**, *32*, 3–16. [CrossRef]
32. Roy, T.K.; Mahmud, M.A.; Shen, W.; Oo, A.M. An adaptive partial feedback linearizing control scheme: An application to a single machine infinite bus system. *IEEE Trans. Circuits Syst. II Express Briefs* **2019**, *67*, 2557–2561. [CrossRef]
33. Roy, T.K.; Mahmud, M.A.; Shen, W.; Oo, A.M.T. A non-linear adaptive excitation control scheme for feedback linearized synchronous generations in multimachine power systems. *IET Gener. Transm. Distrib.* **2021**, *15*, 1501–1520. [CrossRef]
34. Pramanik, M.; Roy, T.; Ghosh, S.; Anower, M.; Mahmud, M.A. Robust Partial Feedback Linearizing Excitation Controller Design for Higher-Order Synchronous Generator in SMIB Systems to Improve the Transient Stability. In Proceedings of the 2021 IEEE Texas Power and Energy Conference (TPEC), College Station, TX, USA, 2–5 February 2021; pp. 1–6.
35. Rahimi, A.M.; Williamson, G.A.; Emadi, A. Loop-cancellation technique: A novel nonlinear feedback to overcome the destabilizing effect of constant-power loads. *IEEE Trans. Veh. Technol.* **2009**, *59*, 650–661. [CrossRef]
36. Sulligoi, G.; Bosich, D.; Giadrossi, G.; Zhu, L.; Cupelli, M.; Monti, A. Multiconverter medium voltage DC power systems on ships: Constant-power loads instability solution using linearization via state feedback control. *IEEE Trans. Smart Grid* **2014**, *5*, 2543–2552. [CrossRef]
37. Arora, S.; Balsara, P.; Bhatia, D. Input–output linearization of a boost converter with mixed load (constant voltage load and constant power load). *IEEE Trans. Power Electron.* **2018**, *34*, 815–825. [CrossRef]
38. Roy, T.; Mahmud, M.A.; Ghosh, S.; Pramanik, M.; Kumar, R.; Oo, A.M. Design of an Adaptive Sliding Mode Controller for Rapid Earth Fault Current Limiters in Resonant Grounded Distribution Networks to Mitigate Powerline Bushfires. In Proceedings of the 2021 IEEE Texas Power and Energy Conference (TPEC), College Station, TX, USA, 2–5 February 2021; pp. 1–6.
39. Roy, T.K.; Mahmud, M.A. Fault current compensations in resonant grounded distribution systems to mitigate powerline bushfires using a nonsingular terminal sliding model controller. *IET Gener. Transm. Distrib.* **2021**, 1–14. [CrossRef]
40. Singh, S.; Fulwani, D.; Kumar, V. Robust sliding-mode control of dc/dc boost converter feeding a constant power load. *IET Power Electron.* **2015**, *8*, 1230–1237. [CrossRef]
41. Zhang, M.; Li, Y.; Liu, F.; Luo, L.; Cao, Y.; Shahidehpour, M. Voltage stability analysis and sliding-mode control method for rectifier in DC systems with constant power loads. *IEEE J. Emerg. Sel. Top. Power Electron.* **2017**, *5*, 1621–1630. [CrossRef]
42. El Aroudi, A.; Martínez-Treviño, B.A.; Vidal-Idiarte, E.; Cid-Pastor, A. Fixed switching frequency digital sliding-mode control of DC-DC power supplies loaded by constant power loads with inrush current limitation capability. *Energies* **2019**, *12*, 1055. [CrossRef]
43. Roy, T.; Mahmud, M.; Shen, W.; Oo, A.; Haque, M. Robust nonlinear adaptive backstepping excitation controller design for rejecting external disturbances in multimachine power systems. *Int. J. Electr. Power Energy Syst.* **2017**, *84*, 76–86. [CrossRef]

44. Boutebba, O.; Semcheddine, S.; Krim, F.; Corti, F.; Reatti, A.; Grasso, F. A Nonlinear Back-stepping Controller of DC-DC Non Inverting Buck-Boost Converter for Maximizing Photovoltaic Power Extraction. In Proceedings of the 2020 IEEE International Conference on Environment and Electrical Engineering and 2020 IEEE Industrial and Commercial Power Systems Europe (EEEIC/I&CPS Europe), Madrid, Spain, 9–12 June 2020; pp. 1–6.
45. Roy, T.; Morshed, M.; Tumpa, F.; Pervej, M. Robust adaptive backstepping speed controller design for a series DC motor. In Proceedings of the 2015 IEEE International WIE Conference on Electrical and Computer Engineering (WIECON-ECE), Dhaka, Banglades, 19–20 December 2015; pp. 243–246.
46. Boutebba, O.; Laudani, A.; Lozito, G.M.; Corti, F.; Reatti, A.; Semcheddine, S. A Neural Adaptive Assisted Backstepping Controller for MPPT in Photovoltaic Applications. In Proceedings of the 2020 IEEE International Conference on Environment and Electrical Engineering and 2020 IEEE Industrial and Commercial Power Systems Europe (EEEIC/I&CPS Europe), Madrid, Spain, 9–12 June 2020; pp. 1–6.
47. Xu, Q.; Jiang, W.; Blaabjerg, F.; Zhang, C.; Zhang, X.; Fernando, T. Backstepping control for large signal stability of high boost ratio interleaved converter interfaced DC microgrids with constant power loads. *IEEE Trans. Power Electron.* **2019**, *35*, 5397–5407. [CrossRef]
48. Xu, Q.; Zhang, C.; Wen, C.; Wang, P.; Yeong, L.M. A novel adaptive backstepping controller for stabilization of DC/DC converter feeding constant power load. In Proceedings of the 2017 IEEE 26th International Symposium on Industrial Electronics (ISIE), Edinburgh, UK, 19–21 June 2017; pp. 570–575.
49. Xu, Q.; Zhang, C.; Wen, C.; Wang, P. A novel composite nonlinear controller for stabilization of constant power load in DC microgrid. *IEEE Trans. Smart Grid* **2017**, *10*, 752–761. [CrossRef]
50. Li, X.; Zhang, X.; Jiang, W.; Wang, J.; Wang, P.; Wu, X. A Novel Assorted Nonlinear Stabilizer for DC–DC Multilevel Boost Converter With Constant Power Load in DC Microgrid. *IEEE Trans. Power Electron.* **2020**, *35*, 11181–11192. [CrossRef]
51. Wu, J.; Lu, Y. Adaptive Backstepping Sliding Mode Control for Boost Converter With Constant Power Load. *IEEE Access* **2019**, *7*, 50797–50807. [CrossRef]

energies

Article

Reinforcement Learning for Energy-Storage Systems in Grid-Connected Microgrids: An Investigation of Online vs. Offline Implementation

Khawaja Haider Ali [1,2,*], Marvin Sigalo [1], Saptarshi Das [1], Enrico Anderlini [3], Asif Ali Tahir [1] and Mohammad Abusara [1,*]

1 Penryn Campus, College of Engineering, Mathematics and Physical Sciences, University of Exeter, Cornwall TR10 9FE, UK; ms924@exeter.ac.uk (M.S.); saptarshi.das@ieee.org (S.D.); a.tahir@exeter.ac.uk (A.A.T.)
2 Department of Electrical Engineering, Sukkur IBA University, Sukkur 65200, Pakistan
3 Department of Mechanical Engineering, Roberts Building, University College London, London WC1E 7JE, UK; e.anderlini@ucl.ac.uk
* Correspondence: ka400@exeter.ac.uk (K.H.A.); m.abusara@exeter.ac.uk (M.A.)

Abstract: Grid-connected microgrids consisting of renewable energy sources, battery storage, and load require an appropriate energy management system that controls the battery operation. Traditionally, the operation of the battery is optimised using 24 h of forecasted data of load demand and renewable energy sources (RES) generation using offline optimisation techniques, where the battery actions (charge/discharge/idle) are determined before the start of the day. Reinforcement Learning (RL) has recently been suggested as an alternative to these traditional techniques due to its ability to learn optimal policy online using real data. Two approaches of RL have been suggested in the literature viz. offline and online. In offline RL, the agent learns the optimum policy using predicted generation and load data. Once convergence is achieved, battery commands are dispatched in real time. This method is similar to traditional methods because it relies on forecasted data. In online RL, on the other hand, the agent learns the optimum policy by interacting with the system in real time using real data. This paper investigates the effectiveness of both the approaches. White Gaussian noise with different standard deviations was added to real data to create synthetic predicted data to validate the method. In the first approach, the predicted data were used by an offline RL algorithm. In the second approach, the online RL algorithm interacted with real streaming data in real time, and the agent was trained using real data. When the energy costs of the two approaches were compared, it was found that the online RL provides better results than the offline approach if the difference between real and predicted data is greater than 1.6%.

Keywords: reinforcement learning (RL); microgrid; battery management; offline and online RL; optimisation

Citation: Ali, K.H.; Sigalo, M.; Das, S.; Anderlini, E.; Tahir, A.A.; Abusara, M. Reinforcement Learning for Energy-Storage Systems in Grid-Connected Microgrids: An Investigation of Online vs. Offline Implementation. *Energies* **2021**, *14*, 5688. https://doi.org/10.3390/en14185688

Academic Editor: Seon-Ju Ahn

Received: 7 August 2021
Accepted: 7 September 2021
Published: 9 September 2021

1. Introduction

Grid-connected microgrids are becoming the main building blocks of smart grids. They facilitate the vast deployment and better utilisation of RES, reduce stress on the existing power grid, and provide consumers with uninterrupted power supply. The main aim for any Energy Management System (EMS) for grid-connected microgrids is to reduce operational costs by reducing the cost of power imported from the grid. This is achieved by controlling the Battery Energy Storage System (BESS) to store power when RES generation is higher than load demand and release power when it is less than the load demand. However, BESS capacity is finite and hence, depending on the battery size, imported power from the grid is likely to be used. Therefore, grid tariffs, which can vary during the day, must be taken into account when deciding charging and discharging commands. In addition, the feed-in tariff can also be available to enable consumers or prosumers to sell their excess power to the grid [1]. These factors demonstrate the need for an intelligent optimisation method. However, due to the lack of information regarding future generation and load profiles, this presents a challenge for EMS.

Inspired by the classical economic dispatch of power systems, various studies have suggested different optimisation techniques to plan, schedule, and control the BESS in grid-connected microgrids. These studies have formulated the operation of the BESS as a dispatch optimisation problem and solved it using Linear Programming (LP) [2,3], Mixed Integer Programming (MIP) [4], Mixed-Integer Linear Programming (MILP) [5–7], and Mixed-Integer Nonlinear Programming (MINLP) [8,9]. In Chen et al. [2], a general algebraic modelling system (GAMS) was developed and solved by CPLEX solver and then tested in a physical system based in Taiwan. The study compared two models to determine the impact of energy storage on optimal scheduling. The first model consisted of thermal and electrical loads and a CHP unit. The second model used additional thermal and electrical storage. Both models used in this work relied on the prediction of load profiles. In Luna et al. [5], the model reflects a deterministic problem that promotes self-consumption based on 24 h look-ahead forecast data. The microgrid consists of a supervisory control stage that compensates for any mismatch between the offline scheduling process and the real-time microgrid operation. In Li et al. [7], the microgrid optimisation problem is formalised using a general algebraic modelling system (GAMS) via a discretised step transformation (DST) approach and finally solved using the CPLEX solver. This paper proposes a new optimal scheduling mode by modelling the uncertainty of spinning reserves provided by energy storage with probabilistic constraints. These achievements are highly dependent on the proper estimation of spinning reserves, which is a big challenge while working on a real system. In Mosa and Ali [9], the MINLP algorithm was used to reduce the operational cost of a DC microgrid consisting of a photovoltaic (PV), fuel cell (FC), micro turbine (MT), diesel generator (DE), and battery/BESS. This study uses Egyptian grid load profiles over four seasons of the year based on the prediction.

The above traditional approaches require a detailed and accurate mathematical model of the system, while some of them require the linearisation of the system. In addition, previous knowledge of future RES generation and load demand over a period is required as an input to the optimisation problem. The accuracy of the prediction may affect the accuracy of the BESS operation. Therefore, different forecasting algorithms that can handle the stochastic nature of load demand and RES have been suggested in the literature. These algorithms are designed to forecast short (daily), medium (seasonal), and long (yearly) load demand and availability of RES. Most advanced forecasting algorithms include Artificial Neural Networks (ANN), dynamic programming (DP)-based optimisation, and fuzzy logic by considering the weather conditions. Although forecasting techniques vary within the vast amount of existing literature [10–18], the most common objective of these techniques is to decrease the forecasting error by better modelling the uncertainties in real time. Although forecasting algorithms have improved in recent years, it is still challenging to predict the future load demand and availability of RES with minimum error, especially if the decision making is implemented in real time.

Recent studies [19–26] have introduced reinforcement learning as a potential solution for the optimal operation of BESS due to its ability to develop an optimal policy online. In RL, an agent interacts with the surrounding environment and develops an optimal policy for taking the right action after exploring its state. The agent takes the action to maximise a future accumulative reward. The main advantage for RL over traditional methods is that it does not need any model of the environment and it can learn the optimum policy in real time. Yoldas et al. [27] used the MINLP technique guided by a Q-learning algorithm to decrease the daily energy cost and emission of harmful gases simultaneously. Performance comparisons were made using only conventional Q-learning. The result showed an approximately 1.2% reduction of the daily operational costs associated with the proposed technique over conventional Q-learning approaches.

There are two main types of online RL algorithms: off-policy and on-policy. In off-policy methods (e.g., Q-learning), the action-valued function is approximated independently of the policy being followed. Conversely, in on-policy approaches, e.g., in state–action–reward–state–action (SARSA), the action-valued function is continuously updated according to the developed policy, which makes it harder to converge [28]. Whereas with off-policy RL, the agent does not need to follow any specific policy and in fact could even act randomly, on-policy schemes rely on the policy that is being established. Despite the possibly random behaviour, off-policy methods, including the Q-learning algorithm, can converge onto the optimal policy independent on the policy employed during exploration [29]. On the other hand, offline, or data-driven RL develops the policy on pre-collected data. Once the policy is developed, it is deployed to control the system. The policy is not updated by interacting with the system in real time. Offline learning, such as batch RL and other conventional techniques such as MILP, MINLP, and LP algorithms, work with data in bulk. Therefore, the uncertainty of some unknown variables such as load demand and PV profiles make these offline methods more challenging because the training is done on forecasted and not real data. Conventionally, offline learning algorithms need to be re-run from scratch in order to learn from modified or new data.

In Mbuwir et al. [19], the authors proposed batch reinforcement learning, offline RL, to solve the optimisation of the microgrid problem in order to achieve a cost-efficient solution. The goal was to find or statistically learn the pattern of the best control policy from the training data (previous year's load and PV profiles) in the form of several smaller batches (sets) and then use this policy on the current environment in real time. When the batch RL was compared to the MILP approach, it was shown to be 19% less efficient than MILP. Kuznetsova et al. [30] developed a two-step-ahead RL algorithm to cut down utility bills by learning the stochastic behaviour of the environment using RL and then scheduling the battery two hours ahead from the current time. RL trains the agent and produces the optimal actions of the battery using forecasted wind and load demand power profiles. Liu et al. [20] proposed a cooperative RL algorithm for distributed economic dispatch in microgrids. However, the challenge of using forecasted PV and load data, which can affect function approximation, is not addressed in this paper. Jiang and Fei [21], suggested a Q-learning based, economical smart microgrid with two-level hierarchical agents with flexible demand response and distributed energy resource management. The authors claim that the suggested scheme is very effective while satisfying the user's preference. However, the suggested work requires load demand from the user before scheduling its distributed units and batteries. This can affect the cost optimisation adversely if the user's demand request changes during real-time operations.

In [22,25,31–43], shallow and deep neural networks have been suggested to approximate the Q-value function to achieve better optimisation results with shorter convergence time. Low convergence time is also desirable for online applications. Lu et al. [22] used deep RL to develop an energy-trading scheme according to the predicted future renewable energy generation, estimated future power demand, and battery level. This work also depended on forecasted renewable energy power generation. In Bui et al. [25], a double deep neural network (DDQN) was proposed for function approximation of Q-values. The authors claim that this method trains the model faster as compared with a single deep (having one network) RL algorithm. This work also depends on the estimation of future load demand and PV generation. Zhou et al. [24] suggested an algorithm to train the agent in real time using real data profiles instead of forecasted datasets. A fuzzy Q-learning approach is adopted for a system consisting of household users and a local energy pool in which customers are free to trade with the local energy pool and enjoy low-cost renewable energy while avoiding the installation of new energy generation equipment. Another online approach was proposed in Kim et al. [26] in which real-time pricing is used to reduce the system cost. Both Zhou et al. [24] and Kim et al. [26] do not provide information regarding the efficiency of their algorithms with respect to other offline Q-learning techniques. Another study by Kim and Lim [44] applied Q-learning directly in real time.

Optimal cost was achieved for the whole year rather than for a single day. In contrast to other offline approaches, this direct online approach trains the agent in real time using real data by moving from one day to another. In the beginning, the agent experience is low; however, as days progress, the agent starts exploiting more actions, and an optimal policy is developed.

The literature reviewed here suggests that offline policies (including RL) require predicted data in order to produce optimal results if the real data is the same as the predicted data, i.e., zero prediction error. The RL online approach, on the other hand, does not rely on predictions and uses real streaming data. However, it is not clear how effective the online RL algorithm is as compared to the offline approaches when the prediction error increases. Motivated by this shortcoming in the existing literature, this paper provides a comprehensive comparison between the two approaches. Using one year of real PV generation and load data obtained from [45], different profiles for predicted data were created by generating random noise with different standard deviations. The noise was added to the real data to create synthetic predicted data. Then, the 24 h of predicted data were used to train the RL agent, and then, the optimised battery command achieved in this process was applied to the real data (offline RL approach). The online RL, on the other hand, interacted with real data in real time. Then, the energy costs of the two approaches were compared to help users make decisions on the most appropriate approach given the accuracy of the available forecasted data. Finally, the case with zero prediction error was considered in the comparison of MINLP versus RL to establish a benchmark between conventional offline approaches versus the offline RL.

2. Energy Management System

The aim of the EMS is to reduce the cost of the power imported from the grid as shown in Figure 1. Thus, the grid supplies energy only if the RES and BESS do not fulfil the demand, and the time of transfer is chosen in order to minimise the cost. Export of energy to the main grid might also be possible if a feed-in tariff is available. Renewable energy, PV in this study, has a priority to fulfil the load requirement first. If it is not sufficient, then the battery, main grid, or combination of both are used to fulfil the demand. The BESS may charge from the PV directly or charge from or discharge into the main grid if needed. The EMS can make use of different tariff rates within the day by charging the BESS during low-tariff periods and discharging it during high-tariff periods. In this work, a fixed feed-in tariff is assumed, and there are three different tariffs (peak, mid, low) which are assumed to import the energy from the main grid depending on time of the day.

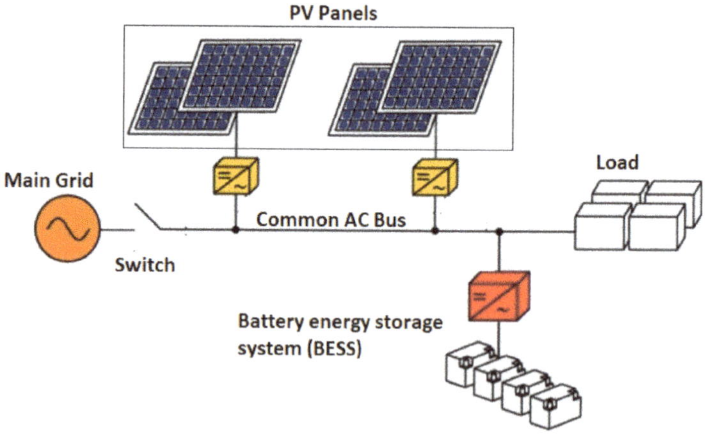

Figure 1. Illustration of microgrid model with BESS.

In this paper, an RL algorithm is used to interact with the microgrid and take automated decisions to control the BESS, taking into account a dynamically changing environment characterised by the available PV output, load demand, and the level of battery charge (SOC). The decisions are in the form of actions for the battery to charge, discharge, or remain idle. The recommended actions for the BESS are developed through Q-learning.

2.1. System States

The state space (S) is discretised at $\Delta t = 30$ min, which suggests that the learning agent captures the information related to the dynamics of the microgrid after the time interval of 30 min. In Equation (1), t represents the time period, which has 48 states in 24 h of a day due to its discretisation every 30 min.

$$s_t = [SOC, e_t^{Net}, t] \in S, \tag{1}$$

where SOC, e_t^{Net} are the battery state of charge and net power demand, respectively. The e_t^{Net} is the difference between the load demand and the energy generated by PV such that:

$$e_t^{Net} = e_t^{demand} - e_t^{PV}. \tag{2}$$

The SOC should be bounded by maximum and minimum limits such that:

$$SOC_{\min} \leq SOC(t) \leq SOC_{\max}. \tag{3}$$

We discretise the state space as shown in Equation (4) below in which the i, j, k indices represent the SOC, e_t^{Net} and t, respectively as:

$$S_{discrete} = \left\{ S_{i,j,k} \right\}, \tag{4}$$

where each index in the state space has the following levels: $i = 3$ levels, $j = 2$ levels, i.e., positive ($e_t^{Net} \geq e_t^{PV}$) or negative ($e_t^{Net} < e_t^{PV}$), and $k = 48$ levels. Thus, the total number of states is $48 \times 3 \times 2 = 288$.

2.2. Action Space

The action space consists of the charge, discharge, and idle command of the battery such as:

$$A = \{ a | (\text{Discharge, Idle, Charge}) \}. \tag{5}$$

At each time step t, one action is selected from the action space A. If the action "Discharge" is chosen, the battery discharges into the main grid, supplies the load, or both. In case of the action "Idle", the load demand is fulfilled by the PV source, main grid, or both. If the "Charge" action is selected, the battery is charged from the PV, the grid, or both.

2.3. Backup Controller

In this work, we used a backup controller, which acts as a filter for every control action resulting from the policy π to take care of the practical constraint, such as the inability of the battery to charge or discharge beyond its maximum and minimum SOC level, respectively. In addition, there is a certain limit of battery charging or discharging at time t. For example, if the "Charge" action is selected by the RL agent at time t and one of the discrete states is ($e_t^{Net} > e_t^{PV}$), then the battery should charge from the main grid up to a certain limit (Δe) defined in Table 1 even if the capacity of the battery is more than Δe. Moreover, if at time t, one of the discrete states is ($e_t^{Net} < e_t^{PV}$) and the RL agent selects the action "Charge", the battery will charge from the extra PV available (after fulfilling the load demand) by respecting the charging rate parameter of the battery. If for example the current PV power is more or less than the charging rate of the battery, the battery is charged up to the maximum charging rate (Δe) or current PV power, respectively.

Table 1. Chosen parameters of the microgrid used in this paper.

Name	Values
Total capacity of the battery	12,000 KWh
Max/Min charging rate of battery (Δe)	(2300/2) KW
SOCmax	100%
SOCmin	70%
Initial SOC of the battery (SOC(0))	[min, max]
Time step lenght (Δt)	30 min
α, γ, ε	0.5, 0.5, 0.6
Total iterations (Offline)	10,000–15,000

2.4. Reward

The reward function $r(s_t,a_t)$ is the immediate incentive gained by taking a specific action a at time t in states. The reward function is chosen to minimise the running cost of importing power from the grid and maximise the revenue of selling power to the grid. The cost is calculated every 30 min (as Δt = 30) by multiplying the respective tariff rates, as mentioned in Section 2.5. The reward function is the negative of the cost of imported energy or the cost of exported energy. Hence, the reward function can be formulated as follows:

$$r(s_t, a_t) = \left\{ \begin{array}{ll} -P_t^{grid} \times \Delta t \times Tariff_{imp}, & P_{grid} \geq 0 \\ P_t^{grid} \times \Delta t \times Tariff_{exp}, & P_{grid} < 0 \end{array} \right\}, \tag{6}$$

where $Tariff_{imp}$ and $Tarriff_{exp}$ are the import and export tariffs, respectively. P_t^{grid} is the grid power and is given by:

$$P_t{}^{grid} = e_t{}^{Net} + P_t{}^{batt}, \tag{7}$$

where P_t^{batt} is the power used to charge the battery.

2.5. Tariff

The import tariff has three different values depending on the time of use:

$$Tariff_{imp} = \left\{ \begin{array}{ll} 0.05£/\text{kWh} & \text{low peak, } 22:00 \text{ to } 8:00 \\ 0.08£/\text{kWh} & \text{medium peak, } 9:00 \text{ to } 12:00 \\ 0.171£/\text{kWh} & \text{high peak, } 19:00 \text{ to } 21:00 \end{array} \right\}. \tag{8}$$

The export tariff does not vary and it is $Tariff_{exp} = 0.033£/\text{kWh}$.

3. Q-Learning Algorithm

The backbone of the Q-learning algorithm is based on the two components described in Equation (9) as:

$$R_t{}^{\pi} = r(s_t, a_t) + \sum_{i=1}^{\infty} \gamma^i . r(s_{t+i} . a_{t+i}). \tag{9}$$

The first component shows the impact of the current action on future rewards, and the second component is the total discounted rewards at time step t under a given policy π. Therefore, R_t^{π} is defined as the sum of the instant reward at time step t plus the future discounted rewards. The parameter γ is the discount factor used to determine the importance of future rewards from the next time step (t + 1) up to infinity. If γ = 0, the algorithm considers the current reward only, while if γ = 1, both current and future rewards have equal weight. In Q-learning, the policy is learned implicitly without any prior knowledge. This is done by approximating the action-value function by repeatedly updating the $Q(s_t, a_t)$ through experience such as:

$$Q(s_t, a_t) = Q(s_t, a_t) + \alpha[r(s_t, a_t) + \gamma \text{max} Q(s_{t+1}, a_{t+1}) - Q(s_t, a_t)]. \tag{10}$$

To approximate the Q-table, it is important to estimate all state–action pairs of the table. Parameter α is the learning rate that determines how much the newly obtained value overrides the old value of $Q(s_t, a_t)$. If α = 0, the newly obtained information is ignored during training, whereas if α = 1, only the latest information is used. Therefore, in Q-learning, the selection of α (ranging between 0 and 1) is very important to keep a balance between the old and new information. The RL agent takes random actions in the beginning if it follows an ε-greedy policy, which is adopted in this work. The idea behind the purely ε-greedy approach is to try every decision once and then keep picking the one that results in the highest reward as learning progresses. After certain iterations and by performing different actions in each state of the Q-table, the agent learns to maximise the value (state–action) of the Q-table by taking greedy actions. Random and greedy actions correspond to exploration (ε) versus exploitation, respectively. In this work, taking a greedy action, the decision is based on:

$$\varepsilon \leftarrow \varepsilon / \sqrt{M(s) - M_{\max}} \tag{11}$$

where $M(s)$ is the number of times a certain action is taken in a specific state. $M_{\max}$ is the maximum constant value selected after which greedy actions are selected by the Q-learning algorithm.

3.1. *Offline RL Implementation*

In this section, the implementation of offline RL to control BESS in microgrids will be discussed. At the beginning of each day, the forecasted PV and load data are gathered as inputs to RL. Then, Q-learning is run using the same input data until convergence is achieved. The policy developed at the end of this phase is used to generate the charging, discharging and idle commands for the next 24 h. This strategy is repeated for each day. The backup controller monitors the control parameters of the battery. After selecting the control actions of the battery from Q-learning, the backup controller ensures that all physical constraints and limitations are met before actually applying the battery actions on the physical system. Figure 2 illustrates the offline implementation of RL-based EMS.

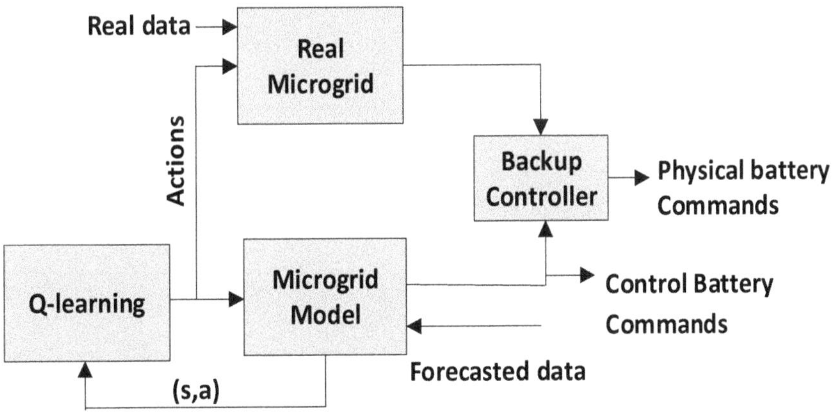

Figure 2. Offline implementation of the RL algorithm.

This offline RL implementation is similar to the traditional EMS approaches such as MILP, where estimated data are used by the optimiser to produce decision variables, such as charging/discharging/idle commands of the battery. The estimated synthetic data for PV generation and load consumption for the next 24 h are used by RL to schedule the battery command. Each episode of one day consists of 48 steps (30 min time interval). The RL keeps on using the same data until convergence is achieved. A total of 10,000–15,000 iterations were employed for better convergence. The optimised battery commands are dispatched at the start of the following day for the battery to operate in real time. The same process is repeated for every day of the year. As we are interested to find the total cost of a complete year (365 days), for the initialization of offline Q-learning, the Q-table simply initialises the action-value function at time step 0 with the value of 0 or ∞.

3.2. Online RL Implementation

Online RL is applied directly to real data in real time. Therefore, the agent learns the optimal policy by interacting with the real system. There is no pre-training in this online approach, unlike offline techniques. Figure 3 shows the online RL for EMS. The online RL algorithm updates the actions of the battery and dispatches them every 30 min in real time regardless of the status of convergence. Learning can be very slow, especially in first few days. Before convergence, the performance would be suboptimal. With time, the agent develops an optimal policy. The function of the backup controller in the online RL implementation is the same as for offline RL, as described in Section 2.3.

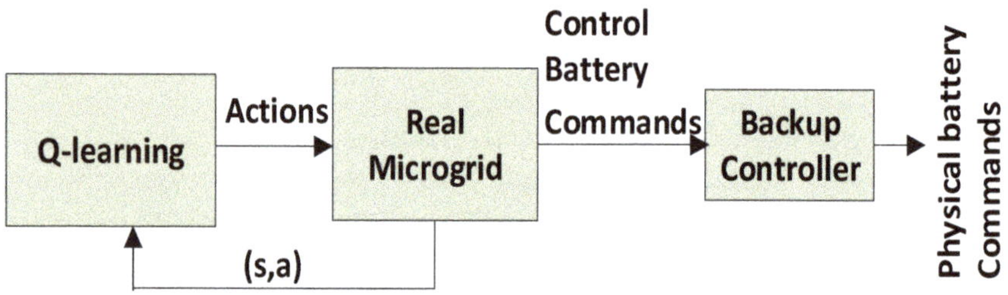

Figure 3. Online implementation of the RL algorithm.

There is no separate training stage in online RL; rather, the agent gathers experiences during real-time interaction with the environment. Therefore, in the beginning, the Q-table is initialised with a shortsighted future reward with the algorithm hyper-parameter γ set to 0. Then, the table will be updated in real time by interacting with the real system. This simple initialisation step reduces the convergence time substantially as per [44].

3.3. Prediction Error Generation

To compare the performance of both offline and online Q-learning, there is a need to create a difference between forecasted data (PV and load) and real data to represent the prediction error. We use an algorithm to add random noise to the real net power demand given by $e_t^{Net} = e_t^{demand} - e_t^{PV}$. The real PV and load profiles for a complete year are obtained from [45]. The noise is generated using normally distributed white Gaussian noise having different standard deviation (σ) values. For each σ, five noise profiles are generated, averaged, and then added to the real net power demand data to produce the forecasted net power demand, as shown in Figure 4. The increase in the σ value will increase the standard deviation error. Thus, the forecasted net demand with higher sigma values represents an increasing trend of deviation with respect to the real power demand and vice versa.

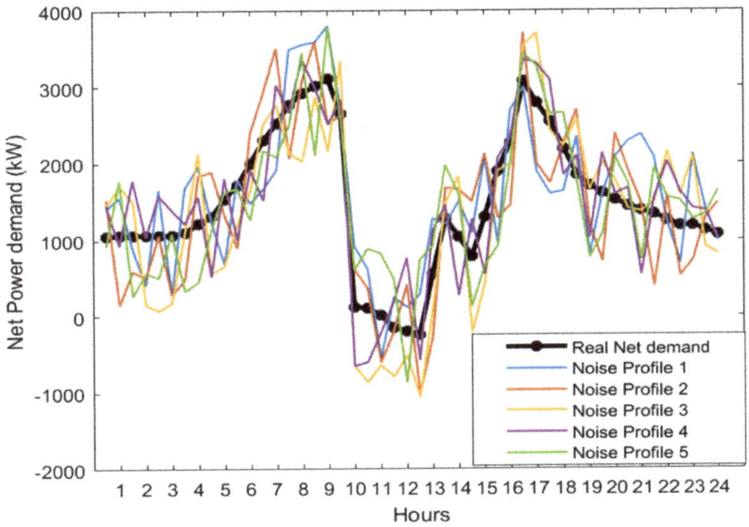

Figure 4. Real and forecasted net demand per hour for one day after noise addition (at σ =100).

4. Simulation Results

This section evaluates the performance of the offline and online optimisation techniques. The offline and online RL were both applied on a daily basis with an interval of 30 min. Firstly, we compared the results of offline RL with MILP to analyse the efficiency of both algorithms in terms of cost saving in a grid-tied microgrid system. Then, we compared offline RL with online RL after establishing a benchmark between the offline RL and MILP techniques. In this regard, both the offline and online RL optimisation techniques need to be investigated in terms of cost saving per year. This work compares the behaviour of both the approaches when there is a different percentage of errors present between forecasted and real data profiles (PV and load). This information can be used to decide between offline and online RL when the real data (PV and Load) profiles deviate from the forecasted data. The real data assumed in this work are gathered from the online open-source data platform in reference [45]. The chosen parameters used to simulate the behaviour of the microgrid are provided in Table 1.

Below, Figure 5 shows the average net forecasted $e_{forecast}^{sum/year}/5$ demand per year using all five samples at each σ. Then, Equation (12) is used to find the percentage error between the real and forecasted power demand per year:

$$Net^{error/year} = \frac{1}{n}\sum_{1}^{n}\left(\frac{e_{Real}^{sum/year} - e_{forecast}^{sum/year}}{e_{Real}^{sum/year}}\right). \tag{12}$$

The error bars in Figure 5 above show the $Net^{error/year}$ for each sigma using the highest and lowest sample of $e_{forecast}^{sum/year}$. The arrows indicating 1 and 2 in Figure 5 indicate the constant $e_{real}^{sum/year}$ and varying $Net^{error/year}$ in Equation (12), respectively.

Figure 6 depicts the difference between the generated $e_{forecast}^{sum/year}/5$ and the real power demand. The difference of the total real power demand per year with the highest and lowest sum of the generated forecasted power demand per year out of five samples at each sigma is described using the error bars on the secondary axis of Figure 6.

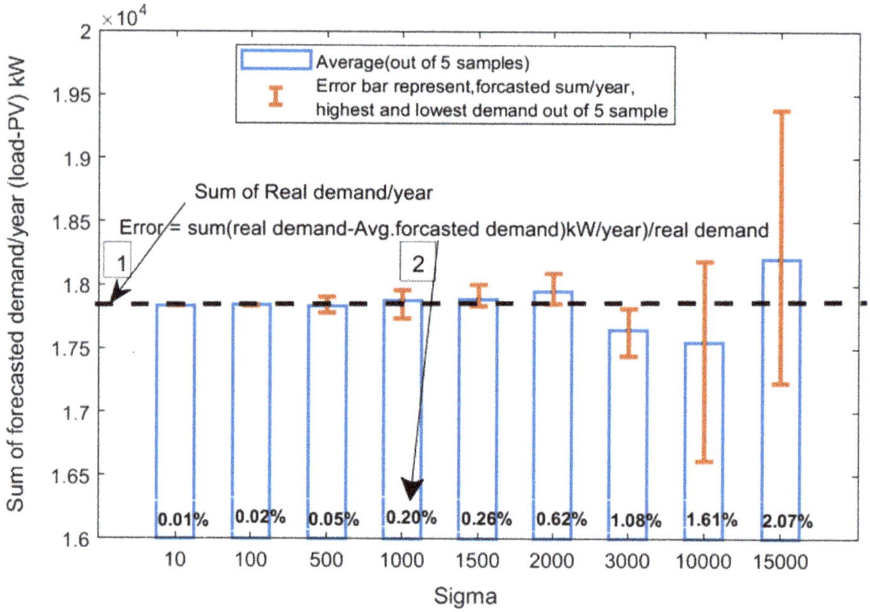

Figure 5. Average sum of the net forecasted demand per year out of five random samples.

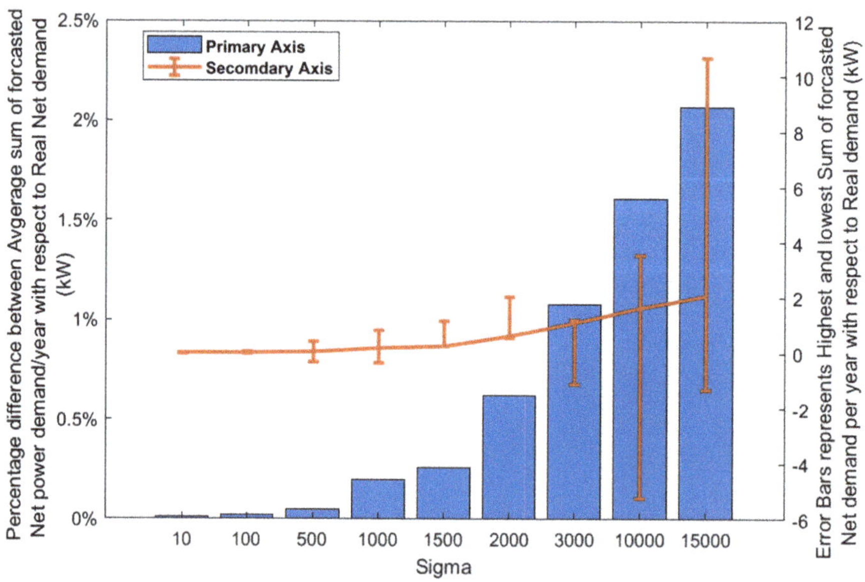

Figure 6. Average sum of forecasted demand with respect to real demand per year.

The two datasets, real (original) and forecasted (synthetic), are related to the net power demand per year and used to compare the performance of offline and online RL. Firstly, we compare offline MILP with offline RL by considering both the forecasted and real data profiles (PV, load demand) to be the same for both approaches. The results (Figure 7) show that both approaches have almost identical performance in terms of cost optimisation of the microgrid, with negligible difference. Over that small difference, MILP behaves slightly better due to the convergence requirements of RL. Therefore, the offline approaches, such as MILP and RL, which use the same forecasted and real data (having 0% error) for training, are equally good in real time.

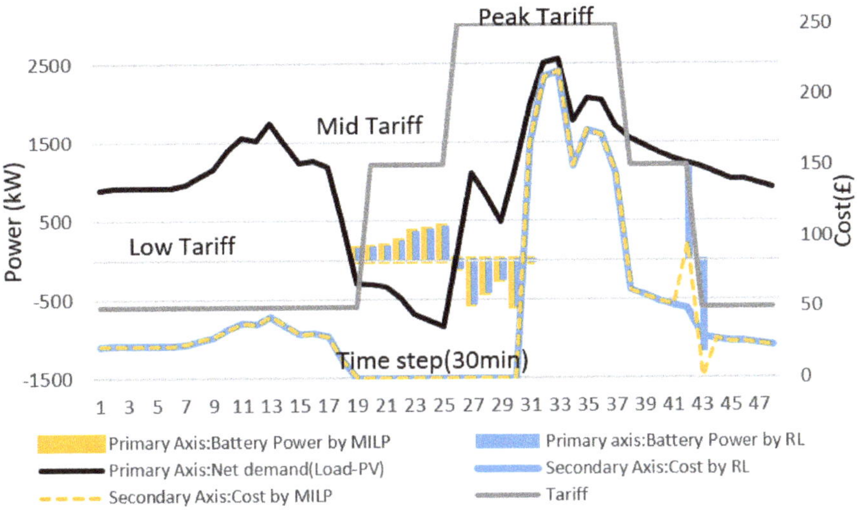

Figure 7. Comparison between MILP and RL in terms of cost per hour.

Figure 8 shows the convergence in terms of cost using offline RL. In this figure, only real data (PV and load) profiles were employed to analyse the convergence pattern of the offline RL as an ideal case. In the beginning, the cost is high and the curve shows random behaviour. As the number of episodes increases, the learning ability of the agent improves until the Q-table converges to show the optimal cost.

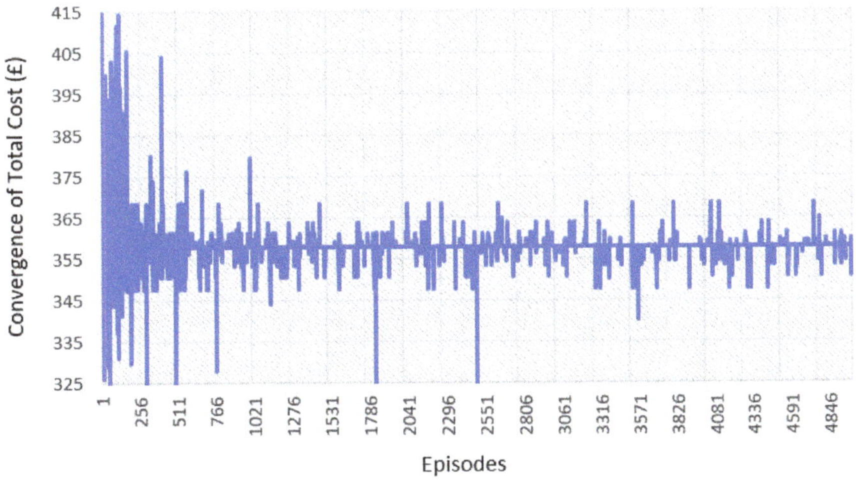

Figure 8. Convergence of the cost during the offline RL algorithm training.

4.1. Offline vs. Online RL with No Forecast Error

The performance of the two RL approaches is first compared when there is no forecast error; i.e., the forecast data are the same as real data. Of course, this ideal situation does not exist in practice, but it provides an initial benchmark for the results. Both RL approaches are implemented as explained earlier in Section 3.1 and Section 3.2. In addition, MILP was also used to optimise the battery operation. Figures 9 and 10 show the energy imported from the grid with respective cost on a daily basis, respectively. It can be seen that when the forecast error is zero, offline RL produces superior results. In Figures 9 and 10, it can be seen that before the convergence of the online RL algorithm, a higher amount of grid energy is imported; therefore, the cost is also higher with respect to offline RL in the initial days. However, as the days progress and online RL learning converges to the optimal policy, the controller follows the same pattern as offline RL in terms of cost saving and reducing the imported energy. In this work, convergence was achieved in between 75 and 90 days in the case of online RL.

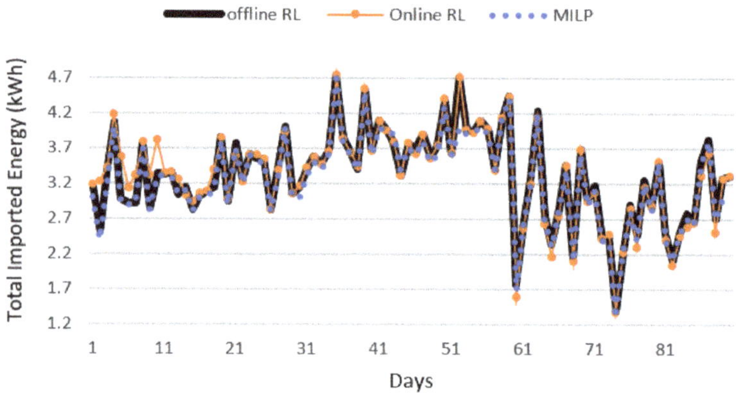

Figure 9. Performance of offline (ideal scenario) vs. online RL in terms of imported energy per day.

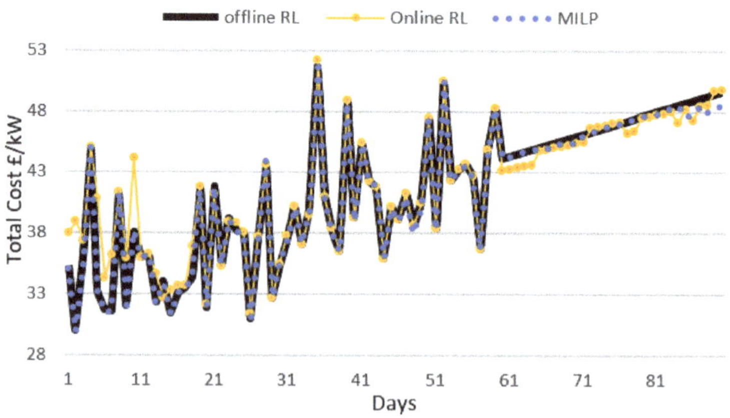

Figure 10. Performance of offline (ideal scenario) vs. online RL in terms of cost per day.

Figure 11 shows the overview of the battery actions generated in an average day after applying the Q-learning algorithm. We used the offline RL for a single day in Figure 11 to show the different states of the battery after convergence during different time intervals of the day with respect to load demand and PV availability. The difference between the load and PV (e_t^{Net}) is shown in the graph below at each time step. The e_t^{Net} can be negative if the PV power generated is greater than the load demand at time t. The suggested battery actions by the RL agent pass through the backup controller to accommodate all physical constraints, as described in Section 2.3. As shown in Figure 11, when $e_t^{PV} > e_t^{Net}$ at time t, the battery charges from the current PV power up to the maximum level of Δe after fulfilling the load demand. The outstanding PV power is sold to the main grid. During discharging, the battery discharges up to maximum level of Δe either to fulfil the load demand or sell power to the utility grid.

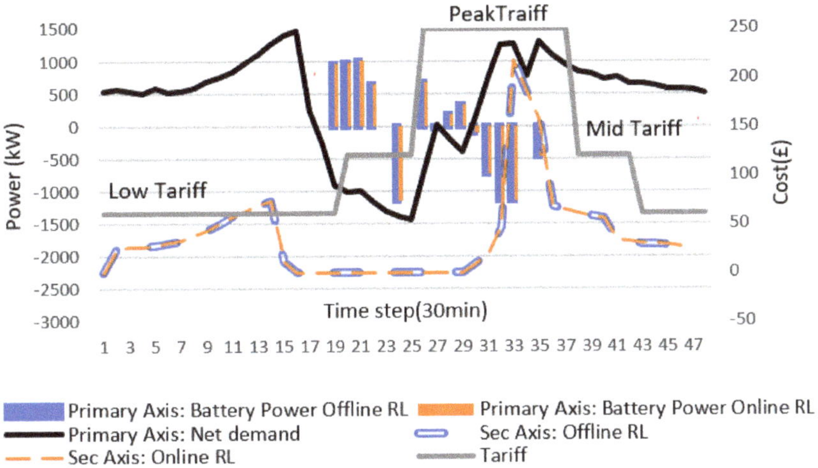

Figure 11. Comparison of offline vs. online RL after convergence.

Figure 11 is the randomly selected day of the year when both online and offline RL converge. The plot shows the same battery actions for both offline and online RL. Therefore, the cost achieved in both approaches are also same.

4.2. *Offline vs. Online with Varying Forecast Error*

The forecasted profiles generated in Section 3.3 are used by the offline RL to create the battery charge/discharge/idle commands, which are then applied to the real data as was explained in Section 3.1 and Figure 2. For online RL, real data are used to generate the battery commands that are directly applied to the physical microgrid system. Figure 12 shows the overview of the battery actions (kW) generated in an average day after introducing 1.6% forecasted error with respect to the real net demand. Figure 12 also showed the daily cost of offline and online RL. The offline RL cost is higher than that of online RL at the time steps 31 to 35. Therefore, the overall average cost of offline RL in a day is higher than that of online RL (after convergence).

Figure 13 shows the optimal average cost achieved per year for both offline and online RL.

The results show that the average cost and imported energy of the offline Q-learning increase as the relative error between the forecasted and real power demand grows or vice versa. The error between real and synthetic predicted profiles are calculated using Equation (12). A rise in the value of the standard deviation reflects the increase in the error (in percent), as shown in Figure 5. Therefore, the noise level (σ) and the relative error are proportional to each other. At the start, when the error of the forecasted demand with respect to the real demand is low, the offline Q-learning performs better in comparison with the online RL in terms of cost optimisation per year. However, as the error increases, for example at 1.61% (between forecasted and real net demand), the online Q-learning begins to perform better and results in a lower cost than the offline Q-leaning.

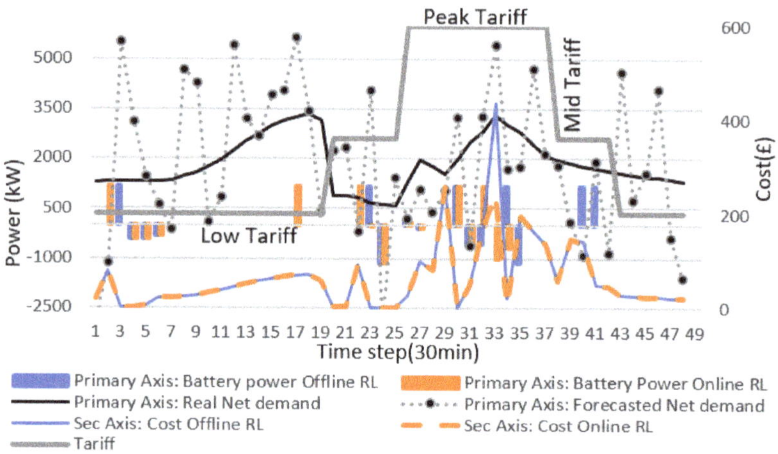

Figure 12. Comparison of offline vs. online RL (with forecast error) after convergence.

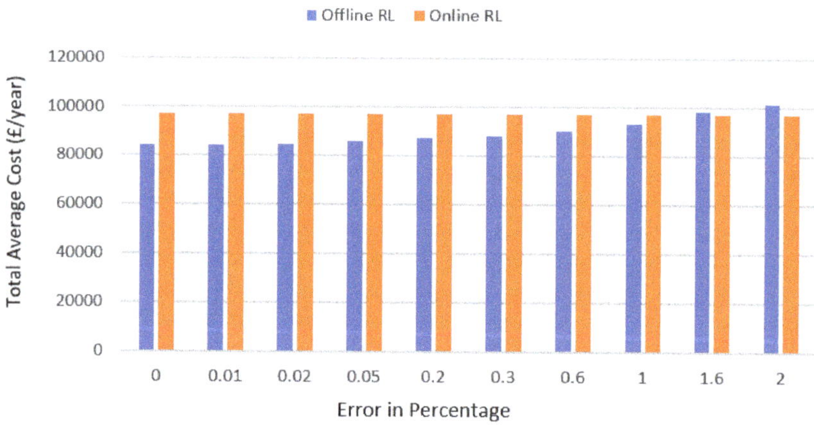

Figure 13. Performance of online vs. offline RL in terms of average total cost per year.

5. Discussion

In this study, we report a comparison between the offline and online optimisation approaches to reduce the operational cost of a grid-connected microgrid by optimally managing the BESS in the presence of forecasting error. A variation between the forecasted and real-time demand may occur due to a change in the weather, ultimately affecting the suggested real-time optimal battery actions obtained through offline training. Hence, the results may not be optimal in terms of reducing cost. To avoid the complication of using forecasted data profiles (PV, load) offline, the literature suggested the online Q-learning [44]. However, online Q-learning needs some days to converge, as its training happens online after exploring real-time data. This knowledge gap in the existing research indicated a need for thorough analysis and comparison of both offline and online approaches. Therefore, we proposed the comparison between offline versus online algorithms on an annual basis. Conventional MILP and offline RL show approximately similar behaviour in terms of saving cost in microgrid operation. Then, the average annual costs of offline and online RL approaches were analysed using different forecasted data profiles. The synthetic forecasted data with respect to real data were produced by adding random white Gaussian noise with specified standard deviation.

Table 2 suggested the possible implementation of offline or online RL on modern research to optimise the economy of power systems. For example, different type of optimisation problems were solved by using different methods in the past, as mentioned in Section 1. These solutions faced a range of challenges, including convergence, inefficient optimisation (cost, Co2 Emission), and computational time under different or varied conditions. The scope of this article suggests that reinforcement learning (offline/online) has full potential to deal with the different types of optimisation problems and challenges.

Table 2. Proposed offline and online RL on the current application scenario.

Reference	Application	Used Method	Future Approach and Strategy
[3]	Sizing large-scale thermal energy storage (TES)	MILP	Apply offline day beforeApply online in real time
[4]	Minimise the use of fossil fuels	LP and MILP	Online RL
[7]	Reduce the cost	GAMSCPLEX	Online RL
[27]	Multi-objective Optimisation	MINLP guided by Q-learning	Multi-objective RL (online/offline)
[46]	Control load shedding	IOT, mathematical modelling	Online RL
[47]	Energy trading and security	Blockchain based	Compare block-chain based mechanism with RL (online/offline)
[48]	Power management	Fuzzy logic controller	Offline RL
This work	BESS management in MG	Offline and online RL	Apply offline RL for training, online RL at real time

6. Conclusions

The following are the key findings of this paper:

- When the error is in between 0 and 1.5%, the offline RL algorithm performs better in terms of cost with respect to the online RL.
- In the first few days, RL performs better as it converges from day 1, while online RL shows better results in terms of cost after a few days. The number of days may vary depending upon the difference between the real and forecasted demand.
- The operating cost of the microgrid is proportional to the imported energy from the main grid by considering PV and battery operating cost equal to zero.
- The computational cost and time of offline RL is higher than that of online RL.
- In the literature [10–18], it was evident that the forecasting of PV has less accuracy than load forecasting.

Therefore, a higher difference between forecasted and real PV and load profiles suggests adopting an online Q-learning approach. For example, there are certain countries and areas where the forecasting of PV and load demand are not certain due to abrupt changes in the weather condition or the user behaviour. The actual energy demand at run times may change a lot in contrast to the predictions. The online Q-learning for cost optimisation provides a better solution in these regions. While operating in favourable prediction conditions, the offline Q-learning performs better.

In the future, there may be other quantification methods such as root mean square error (RMSE) that can be used to introduce errors in forecasted data with respect to real data to compare these two approaches for other type of noise distributions. In the future, both offline and online Q-learning approaches may be employed as a two-layer structure. This can provide a better and more efficient solution for the cost optimisation in a real microgrid.

Author Contributions: Conceptualisation, M.A. and K.H.A.; methodology, K.H.A.; validation, M.A. and S.D.; formal analysis, E.A.; investigation, K.H.A. and M.S.; resources, A.A.T.; writing—

original draft preparation, K.H.A.; writing—review and editing, M.A., E.A., S.D., A.A.T. and K.H.A.; supervision, M.A. and A.A.T. All authors have read and agreed to the published version of the manuscript.

Funding: This research was funded by the Engineering and Physical Sciences Research Council (EPSRC) grant number EP/T025875/1 and The APC was partially funded by the editor allowance.

Data Availability Statement: The data is available from the lead or the corresponding author upon reasonable requests.

Acknowledgments: The financial support was provided by the Engineering and Physical Sciences Research Council (EPSRC), U.K. under research Grant No. EP/T025875/1. K.H.A. would also like to thank Sukkur Institute of Business Administration University, Pakistan for fully funded Ph D studentship.

Conflicts of Interest: The authors declare no conflict of interest.

Nomenclature

ANN	Artificial Neural Networks
BESS	Battery Energy Storage System
CHP	Combined Heat and Power
DDQN	Double Deep Neural Network
DG	Diesel Generator
DP	Dynamic Programming
DSP	Discretised Step Transformation
EMS	Energy Management System
FC	Fuel Cell
GAMS	General Algebraic Modelling System
LP	Linear Programming
MC	Micro Turbine
MILNP	Mixed-Integer Nonlinear Programming
MILP	Mixed-Integer Linear Programming
MIP	Mixed Integer Programming
PV	Photovoltaic
RES	Renewable Energy Sources
RL	Reinforcement Learning
SOC	State of Charge

References

1. Sutton, R.S.; Barto, A.G. *Reinforcement Learning: An Introduction*, 2nd ed.; MIT Press: Cambridge, UK, 2018.
2. Chen, Y.-H.; Lu, S.-Y.; Chang, Y.-R.; Lee, T.-T.; Hu, M.-C. Economic analysis and optimal energy management models for microgrid systems: A case study in Taiwan. *Appl. Energy* **2013**, *103*, 145–154. [CrossRef]
3. Benalcazar, P. Optimal sizing of thermal energy storage systems for CHP plants considering specific investment costs: A case study. *Energy* **2021**, *234*, 121323. [CrossRef]
4. Dolara, A.; Grimaccia, F.; Magistrati, G.; Marchegiani, G. Optimization Models for Islanded Micro-Grids: A Comparative Analysis between Linear Programming and Mixed Integer Programming. *Energies* **2017**, *10*, 241. [CrossRef]
5. Luna, A.C.; Diaz, N.L.; Graells, M.; Vasquez, J.C.; Guerrero, J. Mixed-Integer-Linear-Programming-Based Energy Management System for Hybrid PV-Wind-Battery Microgrids: Modeling, Design, and Experimental Verification. *IEEE Trans. Power Electron.* **2016**, *32*, 2769–2783. [CrossRef]
6. Cosic, A.; Stadler, M.; Mansoor, M.; Zellinger, M. Mixed-integer linear programming based optimization strategies for renewable energy communities. *Energy* **2021**, *237*, 121559. [CrossRef]
7. Li, Y.; Yang, Z.; Li, G.; Zhao, D.; Tian, W. Optimal Scheduling of an Isolated Microgrid With Battery Storage Considering Load and Renewable Generation Uncertainties. *IEEE Trans. Ind. Electron.* **2018**, *66*, 1565–1575. [CrossRef]
8. Stluka, P.; Godbole, D.; Samad, T. Energy management for buildings and microgrids. In Proceedings of the 2011 50th IEEE Conference on Decision and Control and European Control Conference, Orlando, FL, USA, 12–15 December 2011; pp. 5150–5157. [CrossRef]
9. Mosa, M.A.; Ali, A. Energy management system of low voltage dc microgrid using mixed-integer nonlinear programing and a global optimization technique. *Electr. Power Syst. Res.* **2020**, *192*, 106971. [CrossRef]

10. Dong, J.; Olama, M.M.; Kuruganti, T.; Melin, A.M.; Djouadi, S.M.; Zhang, Y.; Xue, Y. Novel stochastic methods to predict short-term solar radiation and photovoltaic power. *Renew. Energy* **2019**, *145*, 333–346. [CrossRef]
11. Soubdhan, T.; Ndong, J.; Ould-Baba, H.; Do, M.-T. A robust forecasting framework based on the Kalman filtering approach with a twofold parameter tuning procedure: Application to solar and photovoltaic prediction. *Sol. Energy* **2016**, *131*, 246–259. [CrossRef]
12. Pedro, H.; Coimbra, C.F. Assessment of forecasting techniques for solar power production with no exogenous inputs. *Sol. Energy* **2012**, *86*, 2017–2028. [CrossRef]
13. Kushwaha, V.; Pindoriya, N.M. A SARIMA-RVFL hybrid model assisted by wavelet decomposition for very short-term solar PV power generation forecast. *Renew. Energy* **2019**, *140*, 124–139. [CrossRef]
14. Lv, Z.; Wang, L.; Guan, Z.; Wu, J.; Du, X.; Zhao, H.; Guizani, M. An Optimizing and Differentially Private Clustering Algorithm for Mixed Data in SDN-Based Smart Grid. *IEEE Access* **2019**, *7*, 45773–45782. [CrossRef]
15. Voyant, C.; Notton, G.; Kalogirou, S.; Nivet, M.-L.; Paoli, C.; Motte, F.; Fouilloy, A. Machine learning methods for solar radiation forecasting: A review. *Renew. Energy* **2017**, *105*, 569–582. [CrossRef]
16. Nti, I.K.; Teimeh, M.; Nyarko-Boateng, O.; Adekoya, A.F. Electricity load forecasting: A systematic review. *J. Electr. Syst. Inf. Technol.* **2020**, *7*, 1–19. [CrossRef]
17. Kuster, C.; Rezgui, Y.; Mourshed, M. Electrical load forecasting models: A critical systematic review. *Sustain. Cities Soc.* **2017**, *35*, 257–270. [CrossRef]
18. Jahan, I.S.; Snasel, V.; Misak, S. Intelligent Systems for Power Load Forecasting: A Study Review. *Energies* **2020**, *13*, 6105. [CrossRef]
19. Mbuwir, B.V.; Ruelens, F.; Spiessens, F.; Deconinck, G. Battery Energy Management in a Microgrid Using Batch Reinforcement Learning. *Energies* **2017**, *10*, 1846. [CrossRef]
20. Liu, W.; Zhuang, P.; Liang, H.; Peng, J. Distributed Economic Dispatch in Microgrids Based on Cooperative Reinforcement Learning. *IEEE Trans. Neural Netw. Learn. Syst.* **2018**, *29*, 2192–2203. [CrossRef] [PubMed]
21. Jiang, B.; Fei, Y. Smart Home in Smart Microgrid: A Cost-Effective Energy Ecosystem with Intelligent Hierarchical Agents. *IEEE Trans. Smart Grid* **2014**, *6*, 3–13. [CrossRef]
22. Lu, X.; Xiao, X.; Xiao, L.; Dai, C.; Peng, M.; Poor, H.V. Reinforcement Learning-Based Microgrid Energy Trading With a Reduced Power Plant Schedule. *IEEE Internet Things J.* **2019**, *6*, 10728–10737. [CrossRef]
23. Foruzan, E.; Soh, L.-K.; Asgarpoor, S. Reinforcement Learning Approach for Optimal Distributed Energy Management in a Microgrid. *IEEE Trans. Power Syst.* **2018**, *33*, 5749–5758. [CrossRef]
24. Zhou, S.; Hu, Z.; Gu, W.; Jiang, M.; Zhang, X.-P. Artificial intelligence based smart energy community management: A reinforcement learning approach. *CSEE J. Power Energy Syst.* **2019**, *5*, 1–10. [CrossRef]
25. Bui, Y.-H.; Hussain, A.; Kim, H.-M. Double Deep *Q* -Learning-Based Distributed Operation of Battery Energy Storage System Considering Uncertainties. *IEEE Trans. Smart Grid* **2019**, *11*, 457–469. [CrossRef]
26. Kim, B.-G.; Zhang, Y.; van der Schaar, M.; Lee, J.-W. Dynamic Pricing and Energy Consumption Scheduling With Reinforcement Learning. *IEEE Trans. Smart Grid* **2015**, *7*, 2187–2198. [CrossRef]
27. Yoldas, Y.; Goren, S.; Onen, A. Optimal Control of Microgrids with Multi-stage Mixed-integer Nonlinear Programming Guided Q-learning Algorithm. *J. Mod. Power Syst. Clean Energy* **2020**, *8*, 1151–1159. [CrossRef]
28. Watkins, C.J.C.H.; Dayan, P. Q-learning. *Mach. Learn.* **1992**, *8*, 279–292. [CrossRef]
29. Perera, A.; Kamalaruban, P. Applications of reinforcement learning in energy systems. *Renew. Sustain. Energy Rev.* **2020**, *137*, 110618. [CrossRef]
30. Kuznetsova, E.; Li, Y.-F.; Ruiz, C.; Zio, E.; Ault, G.; Bell, K. Reinforcement learning for microgrid energy management. *Energy* **2013**, *59*, 133–146. [CrossRef]
31. Ji, Y.; Wang, J.; Xu, J.; Fang, X.; Zhang, H. Real-Time Energy Management of a Microgrid Using Deep Reinforcement Learning. *Energies* **2019**, *12*, 2291. [CrossRef]
32. Fujimoto, S.; Meger, D.; Precup, D. Off-policy deep reinforcement learning without exploration. *Proc. Mach. Learn. Res.* **2019**, *97*, 2052–2062.
33. Wang, Y.; Jin, H. A Boosting-based Deep Neural Networks Algorithm for Reinforcement Learning. In Proceedings of the 2018 Annual American Control Conference (ACC), Milwaukee, WI, USA, 27–29 June 2018; pp. 1065–1071. [CrossRef]
34. Zhang, K.; Yang, Z.; Başar, T. Multi-agent reinforcement learning: A selective overview of theories and algorithms. In *Handbook of Reinforcement Learning and Control*; Springer: Berlin/Heidelberg, Germany, 2021; pp. 321–384.
35. Staddon, J.E.R. The dynamics of behavior: Review of Sutton and Barto: Reinforcement Learning: An Introduction (2 nd ed.). *J. Exp. Anal. Behav.* **2020**, *113*, 485–491. [CrossRef]
36. Das, A.; Ni, Z. A Computationally Efficient Optimization Approach for Battery Systems in Islanded Microgrid. *IEEE Trans. Smart Grid* **2017**, *9*, 6489–6499. [CrossRef]
37. Vinyals, O.; Babuschkin, I.; Czarnecki, W.M.; Mathieu, M.; Dudzik, A.; Chung, J.; Choi, D.H.; Powell, R.; Ewalds, T.; Georgiev, P.; et al. Grandmaster level in StarCraft II using multi-agent reinforcement learning. *Nature* **2019**, *575*, 350–354. [CrossRef]
38. Boait, P.; Advani, V.; Gammon, R. Estimation of demand diversity and daily demand profile for off-grid electrification in developing countries. *Energy Sustain. Dev.* **2015**, *29*, 135–141. [CrossRef]
39. Hernandez-Aramburo, C.; Green, T.; Mugniot, N. Fuel Consumption Minimization of a Microgrid. *IEEE Trans. Ind. Appl.* **2005**, *41*, 673–681. [CrossRef]

40. Bui, V.-H.; Hussain, A.; Kim, H.-M. Q-Learning-Based Operation Strategy for Community Battery Energy Storage System (CBESS) in Microgrid System. *Energies* **2019**, *12*, 1789. [CrossRef]
41. Rancilio, G.; Lucas, A.; Kotsakis, E.; Fulli, G.; Merlo, M.; Delfanti, M.; Masera, M. Modeling a Large-Scale Battery Energy Storage System for Power Grid Application Analysis. *Energies* **2019**, *12*, 3312. [CrossRef]
42. Hernández, L.; Baladrón, C.; Aguiar, J.M.; Carro, B.; Sánchez-Esguevillas, A.; Lloret, J. Artificial neural networks for short-term load forecasting in microgrids environment. *Energy* **2014**, *75*, 252–264. [CrossRef]
43. Castronovo, M.; François-Lavet, V.; Fonteneau, R.; Ernst, D.; Couëtoux, A. Approximate bayes optimal policy search using neural networks. In Proceedings of the ICAART 2017—9th International Conference on Agents and Artificial Intelligence, Porto, Portugal, 24–26 February 2017; Volume 2, pp. 142–153. [CrossRef]
44. Kim, S.; Lim, H. Reinforcement Learning Based Energy Management Algorithm for Smart Energy Buildings. *Energies* **2018**, *11*, 2010. [CrossRef]
45. Data Platform—Open Power System Data. Available online: https://data.open-power-system-data.org/ (accessed on 8 February 2021).
46. Hussain, M.M.; Siddique, M.; Raees, A.; Nouman, M.; Javed, W.; Razaq, A. Power Management through Smart Grids and Advance Metering Infrastructure. In Proceedings of the 2020 6th IEEE International Energy Conference (ENERGYCon), Gammarth, Tunisia, 28 September–1 October 2020; pp. 767–772. [CrossRef]
47. Ahmad, R.F.; Siddique, M.; Riaz, K.; Hussain, M.M.; Bhatti, M. Blockchain based Secure Energy Trading Mechanism for Smart Grid. *Pak. J. Eng. Technol.* **2021**, *4*, 100–107. [CrossRef]
48. Al Badwawi, R.; Issa, W.R.; Mallick, T.K.; Abusara, M. Supervisory Control for Power Management of an Islanded AC Microgrid Using a Frequency Signalling-Based Fuzzy Logic Controller. *IEEE Trans. Sustain. Energy* **2018**, *10*, 94–104. [CrossRef]

energies

MDPI

Article

Modified Beetle Annealing Search (BAS) Optimization Strategy for Maxing Wind Farm Power through an Adaptive Wake Digraph Clustering Approach

Yanfang Chen [1,2], Young-Hoon Joo [2,*] and Dongran Song [3]

[1] School of Electronics and Information Engineering, Jiujiang University, Jiujiang 332005, China; 6110011@jju.edu.cn
[2] School of IT Information and Control Engineering, Kunsan National University, Kunsan 54150, Korea
[3] School of Automation, Central South University, Changsha 410083, China; songdongran@csu.edu.cn
* Correspondence: yhjoo@kunsan.ac.kr

Abstract: Owing to scale-up and complex wake effects, the centralized control that processes the command from turbines may be unsuitable, as it incurs high communication overhead and computational complexity for a large offshore wind farm (OWF). This paper proposes a novel decentralized non-convex optimization strategy for maxing power conversion of a large OWF based on a modified beetle antennae search (BAS) algorithm. First, an adaptive threshold algorithm which to establish a pruned wake direction graph while preserving the most critical wake propagation relationship among wind turbines are presented. The adaptive graph constraints were used to create wake sub-digraphs that split the wind farm into nearly uncoupled clustering communication subsets. On this basis, a Monte Carlo-based beetle annealing search (MC-BAS) nonlinear optimization strategy was secondly designed to adjust the yaw angles and axial factors for the maximum power conversion of each turbine subgroup. Finally, the simulation results demonstrated that a similar gain could be achieved as a centralized control method at power conversion and reduces the computational cost, allowing it to solve the nonlinear problem and real-time operations of the OWF.

Keywords: beetle antennae search optimization; wake propagation; direct graph; offshore wind farm; clustering subset; graph adaptive pruning

Citation: Chen, Y.; Joo, Y.-H.; Song, D. Modified Beetle Annealing Search (BAS) Optimization Strategy for Maxing Wind Farm Power through an Adaptive Wake Digraph Clustering Approach. *Energies* **2021**, *14*, 7326. https://doi.org/10.3390/en14217326

Academic Editor: Frede Blaabjerg

Received: 21 September 2021
Accepted: 30 October 2021
Published: 4 November 2021

1. Introduction

With increasing environmental problems, such as the greenhouse effect, clean energy has become a critical issue that needs to be solved worldwide. Wind energy has become very competitive in comparison with other green energy production technologies for the mature control technology, such as the control approach to improve the performance of wind turbines in different conditions [1], model predictive control [2], etc. Moreover, the exploitation of wind energy is mainly fulfilled by wind turbines in arrays or groups to reduce the cost of energy [3,4]. Additionally, energy storage is also a very important branch [5].

To increase the total output power, the optimization methodology [6] plays a vital role by considering the topography, prevailing wind direction environment [7], and the space of turbines. For a designed wind farm, there are also some other methodologies for improving the total output power [8–10], decreasing the thrust load [11], improving the lifetime of turbines [12], and tracking power reference signals to improve wind integration into the power grid [13–16]. Researchers have recently proposed some control methods for optimizing power conversion in large wind farms without considering wake infection [17,18]. Some kinds of literature [19–21] propose novel graph theories. The production of generating an interaction matrix is treated in the same way. However, the difference is the kinetic energy deficits with considering the wind speed and direction probabilities, which is explained in detail in Equation (7) in paper [18]. Additionally, in this paper, we focus on the variations of wind velocity between the upstream and downstream turbines. However, the constraints of the offshore land limit the space of turbines, not infinitely long. Therefore, the wake effects in a wind farm are inevitable, and the closely spaced turbines will produce little output power. The reason is that wake affects the downstream wind speed. Moreover, the relationship between the upstream and downstream turbines is inseparable from the basic turbine model, and the existing turbine models currently contain Jensen's model [22], Frandsen's model [23], CFD-wake model [24], etc. The Jensen and Frandsen models have assumed "top-hat" shape distributions for the wake velocity deficit. In fact, the actual distribution is Gaussian as in [21,25], which is based on self-similarity theory and is often used in free shear flows. It has become a hot topic that the optimization algorithm performs the search of the values of optimal axial and yaw angle to maximize the output power and mitigate the wake interference.

For large-scale wind farms, the burden of communication with the central unit is significantly large, and the speed of centralized implementation is no longer suited for real-time control [15,16]. More recent studies have investigated decentralized control, distributed control [26], and cooperate control [27] which are proposed to decrease the communication burden and improve control speed by dividing whole turbines into several decoupled subsets. Centralized control only suits a small-scale wind farm because there is much large information sharing between turbines and the central controller, and every turbine needs no other turbines' information. Distributed control method was proposed in [26]; there are several controllers in local groups instead of the central controller, and some key information of shared turbines will be transmitted between them. Moreover, every turbine considers the power information of itself and the neighbors to set the control actions. Additionally, a cooperative optimization problem is designed in [27], the information of the whole wind farm output power is transmitted to every turbine, and the control variable is axial induction factors.

For example, an equivalent model for wind farm reduction was proposed in [28] to cluster the same-feature wind turbines into a group. However, for a complex wake, the same feature presented a difficult problem to find the k-means clustering algorithm [29,30], k-median clustering algorithm [31], and k-nearest physical neighbor [32], which were proposed to uncover the clustered index. Moreover, the disadvantage of [33,34] was the clustering index with one-dimensional data, which only focuses on wind speed without considering the wake effect. However, for a large-scale offshore wind farms, one-dimensional data cannot explain the wake relationship matrix of the wake effect, and two-dimensional data, considering the wind speed and the wake effect are more suitable. Therefore, based on a clustering index with two-dimensional data, another possible solution was to propose a novel clustering approach that can construct decouple communication architecture. In [35], the turbines were clustered into some groups through the singular value decomposition (SVD) clustering algorithm. However, those clustering algorithms were assumed to behave similarly and were only valid for limited wind directions without considering the differences in the incoming wind.

As mentioned previously, there is a lack of efficient cluster methods to cluster the subset of communication neighbors of the turbines considering complex wake effects. To solve this problem, in this paper, we propose an adaptive pruning wake digraph to divide the large-scale offshore wind farm (OWF) into decoupled groups, then cluster the neighbors into the same subset by setting the k-median of two turbines' wake imperfect weights as the clustering index. The wind farm control and the optimization problem in every subset are non-convex, so it is critical to devise an efficient optimization algorithm. By combining the wake model and the single objective function, there are some optimization methods proposed in the literature [26,36–39]: in [38], a gradient-based method on wind farm layout optimization is presented; however, a gradient-based optimization algorithm is based on an analytical wind farm function which is simplified, so it cannot precisely reflect the wind farm control. Therefore, without the wind farm function, only some necessary measurement data for data-driven optimization algorithms are proposed, for example, in [26,37]. However, the high model complexity of the measurable data will become significantly large. Moreover, the performance may not be good with noisy measurement data unavoidable in real applications. The authors of [36] explored a cooperative greedy algorithm optimizing energy production of wind farms; because of the wake effects, the downstream turbine could not generate much energy, so the total output power was not maximum.

Due to the large amount of large-scale wind farm data and nonlinear and non-convex characteristics, the above-mentioned control methods become unsuitable for obtaining the global optimum. Evolutionary algorithms (EAs) are one class of novel nature-inspired global optimization algorithms that are proposed in the literature [40,41]. For maximizing the total output power of wind farms, other algorithms such as modified grey wolf (GW) [42,43], particle swarm optimization (PSO) [44,45], and genetic algorithm (GA) [46] have been proposed. The authors of [44] present PSO intelligent algorithms for power conversion maximization using a nonlinear wake model. Importantly, no one algorithm can fit all systems. The above-mentioned algorithms are prone to premature convergence leading to a local optimum, not a global optimum, because of some unsuitable parameters. Therefore, we use a novel intelligent algorithm dubbed the beetle antennae search (BAS) [47] to improve calculation performance and searching ability to maximize wind farm power conversion. The BAS may generally converge early and fall into the optimal local solution for the unsuitable step size. In addition, the Monte Carlo (MC) method can be used to prevent the evolutionary algorithm from stagnating at a local optimum.

To summarize, in this paper, we first define an adaptive wake digraph to cluster a subset of communication neighbors of turbines. Secondly, we propose an MC-BAS optimization algorithm based on adaptive communication network topology to solve the non-convex power optimization problem. Finally, the yaw angles and axial factors are optimized to increase the power output by simulating a wind farm with 2×2 turbines, 3×3 turbines, and 5×5 turbines. The main contributions are summarized as follows:

- A decentralized coordination control scheme is achieved by controlling the yaw angles and axial factors to maximize power conversion on the wind farm. Large-scale wind farms are divided into several decoupled subsets, and then the local controller only controls the local subset's data. The proposed control scheme enables efficiency in the real-time application by optimizing the decentralized coordination to reduce computational burden and information exchange.
- A wake-based graph adaptive pruning approach is presented to split a large wind farm into several clustering subsets. This approach aims to find a decoupled sub-graph that can preserve essential distribution characteristics of the original wake direction graph. We adopt a graph clustering algorithm to divide turbines via wake graphs adaptive pruning constraint, and threshold ε_k which is a vital point parameter to control the number of groups of the pruned wake digraph.
- We develop a modified BSA optimization algorithm based on adaptive pruned communication architectures. The Monte Carlo (MC) law of Simulate Anneal (SA) is introduced to improve the BAS, which significantly improves the reproducibility and stability of the algorithm. Finally, the improved algorithm is applied for wake steering control and maximum power conversion on the wind farm.

This study is organized as follows: Section 2 introduces the Gaussian-based wake model considering yaw angle. Section 3 introduces the algorithm to cluster turbine via adaptive pruning wake digraph through setting the suitable global threshold ε_p. In Section 4, the new MC-BAS control strategy in OWF is proposed to optimize the axial vale and yaw angle in every subset. Section 5 presents the simulated result of the proposed algorithm, the efficiency of power optimization and minimizing the calculating time is verified, and some important look-up tables are constructed. Finally, some important conclusions and summaries are presented in Section 6.

2. Gaussian-Based Wake Model Considering Yaw Angle

This section describes the wind turbine wake model for total output power optimization through axial induction and yaw angle control. The three-dimensional wind velocity deficit behind the upstream turbine i is defined as Gaussian shape, which was derived through Navier–Stokes equation [25]:

$$\frac{V(x,y,z)}{V_\infty} = 1 - Ce^{-(y-\delta)^2/2\sigma_y^2}e^{-(z-z_h)^2/2\sigma_z^2} \tag{1}$$

where V denotes the velocity in the wake, V_∞ denotes the free-stream inflow wind velocity of the wind farm, x, y, z is the direction of streamwise, horizontal spanwise, and vertical spanwise, a decoupled topology, δ is the wake centerline, z_h is the hub height, σ_y, σ_z is the wake expansion in y, z direction, and C is the velocity deficit at the wake center. The main parameters are shown in Appendix A.

The relationship of α, and γ, can be defined as [21,25]:

$$\alpha \approx \frac{0.3\gamma}{\cos\gamma}\left(1 - \sqrt{1 - C_T\cos\gamma}\right) \tag{2}$$

where C_T denotes the thrust coefficient.

The relationship between the initial lateral deflection of wake deflection δ_0 as denied as in [25], x_0 is the length of the near wake as defined in [25], and the wake deflection angle α can be defined as the following equation [21]:

$$\delta_0 = \delta_0 \tan\alpha \tag{3}$$

$$\delta = \delta_0 + \frac{\theta E_0}{5.2}\sqrt{\frac{\sigma_{y0}\sigma_{z0}}{k_y k_z M_0}}\ln\left[\frac{\left(1.6 + \sqrt{M_0}\right)\left(1.6\sqrt{\frac{\sigma_y\sigma_z}{\sigma_{y0}\sigma_{z0}}} - \sqrt{M_0}\right)}{\left(1.6 - \sqrt{M_0}\right)\left(1.6\sqrt{\frac{\sigma_y\sigma_z}{\sigma_{y0}\sigma_{z0}}} + \sqrt{M_0}\right)}\right] \tag{4}$$

After describing the atmospheric wake model of the wind field, the turbine model used in this paper is introduced. The turbine model consists of a power coefficient C_p and thrust coefficient C_T which are all based on wind speed, tip speed ratio, and a blade pitch angle. In this paper, the C_p and C_T curves are used to form the fitting data of FAST and the National Renewable Energy Laboratory's (NREL's) 5 MW turbine [48].

To calculate the output power P of a wind turbine, the formula can be shown as follows [30]:

$$P_j(\alpha_j, \gamma_j; V_j) = \frac{1}{2}\eta\rho A_j C_P \cos(\alpha_j, \gamma_j)^{1.88} V_j^3 \tag{5}$$

where η denotes generator efficiency; ρ is the air density; A_j is the rotor swept area; $\cos(\alpha_j, \gamma_j)^{1.88}$ represents the correction factor of axial factors α_j and yaw misalignment angle γ_j; wind velocity V_j can be calculated from Equation (1).

As depicted in Equations (1)–(5), the power conversion can be optimized by adjusting the axial factors α_j and yaw angle γ_j. The interested reader can read a more detailed description of wake deflection in [19].

3. Clustering Turbine via Pruning Wake Digraph

This section will partition the large-scale OWF into several decoupled subsets based on the weights calculated from the k-Median clustering algorithm [31].

The adaptive pruning wake digraph process can be summarized in Figure 1, including the wake farm model, original digraph generation, digraph pruning, and turbines clustering. The decision variable is the magnitude of the strength between turbines w_{ij} as shown in Figure 1c,e,g, then the large-scale OWF can be defined into several decoupled subsets. The section is focused on how to obtain adaptive wake digraphs of the large-scale OWF with wake interaction.

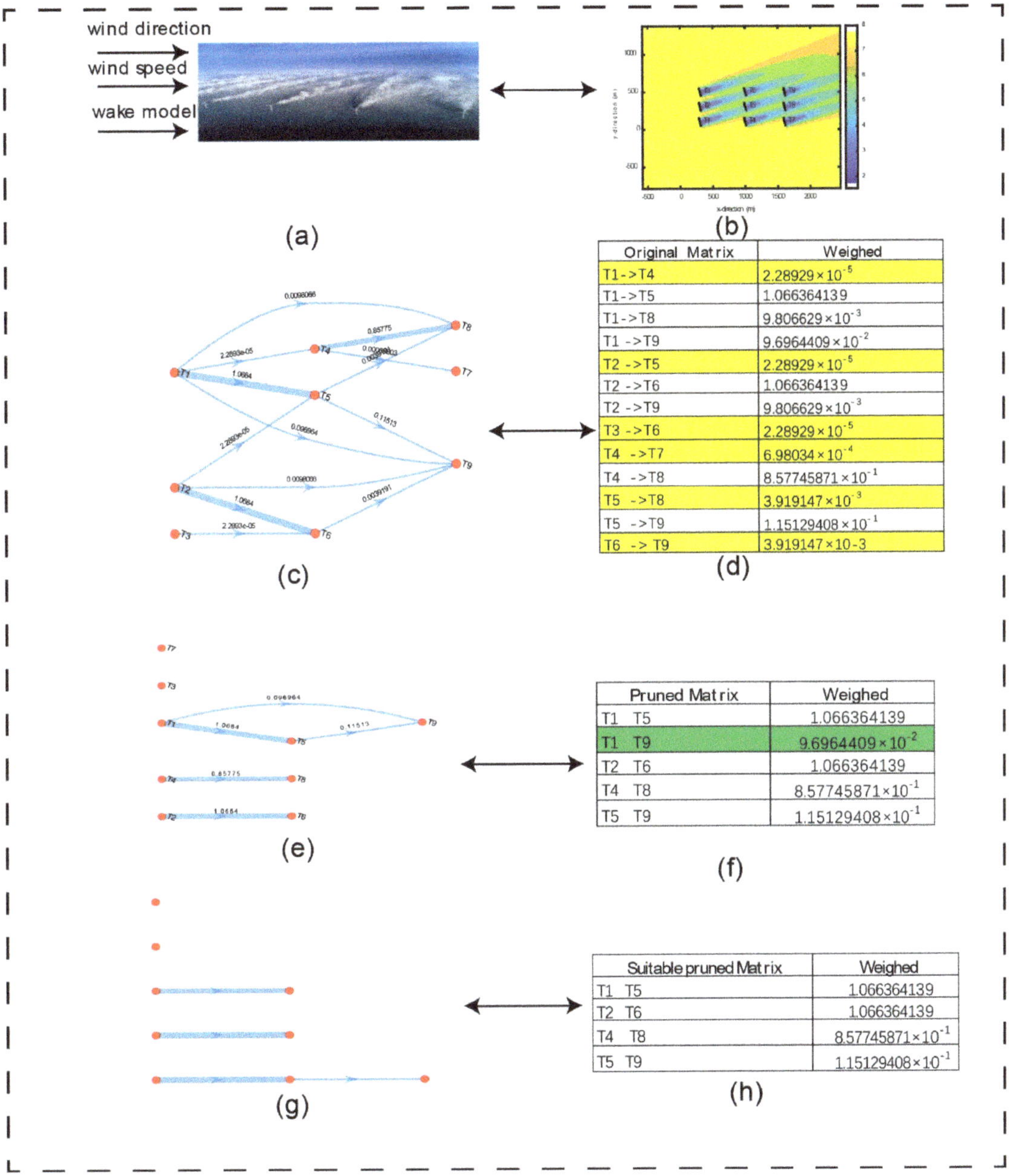

Original Matrix	Weighed
T1->T4	2.28929×10^{-5}
T1->T5	1.066364139
T1->T8	9.806629×10^{-3}
T1 ->T9	9.6964409×10^{-2}
T2 ->T5	2.28929×10^{-5}
T2 ->T6	1.066364139
T2 ->T9	9.806629×10^{-3}
T3 ->T6	2.28929×10^{-5}
T4 ->T7	6.98034×10^{-4}
T4 ->T8	$8.57745871 \times 10^{-1}$
T5 ->T8	3.919147×10^{-3}
T5 ->T9	$1.15129408 \times 10^{-1}$
T6 -> T9	3.919147×10-3

Pruned Matrix	Weighed
T1 T5	1.066364139
T1 T9	9.6964409×10^{-2}
T2 T6	1.066364139
T4 T8	$8.57745871 \times 10^{-1}$
T5 T9	$1.15129408 \times 10^{-1}$

Suitable pruned Matrix	Weighed
T1 T5	1.066364139
T2 T6	1.066364139
T4 T8	$8.57745871 \times 10^{-1}$
T5 T9	$1.15129408 \times 10^{-1}$

Figure 1. The proposed whole decentralized control scheme on a 3 × 3 matrix OWF with $V_\infty = 8$ m/s, $\varphi = 15°$: (**a**) Wind Farm; (**b**) wake field; (**c**) original wake digraph; (**d**) turbine original matrix; (**e**) pruned wake digraph with $k = 3.2$; (**f**) turbine pruned matrix with $k = 3.2$; (**g**) pruned wake digraph with $k = 5.1$; (**h**) turbine pruned matrix with $k = 5.1$.

3.1. The Original Wake Graph of Wind Farm

An autonomous wind farm can be modeled as a weighted directed graph with wake impact from upstream turbines i to downstream turbines j [49]. Under this paradigm, the digraph is determined by the wind speed direction and wind farm layout. Assuming that there are N turbines in the OWF, we will explain how to construct the original wake digraph $\mathcal{G}$ in detail.

To demonstrate the advanced approach, some definitions as follows are necessary:

Definition 1. *The original wake digraph $\mathcal{G} = (\mathcal{V}, \mathcal{E})$ where vertices $\mathcal{V}$, $\mathcal{V} = \{v_i i = 1, 2, \ldots, N\}$ denote the turbines. Edges $\mathcal{E}$, $\mathcal{E} \subset \mathcal{V} \times \mathcal{V}$ represents the wake distribution between every pair of upstream turbine and downstream turbine.*

$\mathcal{E}$ can also be used the weight to indicate the strength of interactions between two nodes [30]:

$$\mathcal{E} = \{w_{ij} : i, j \in \mathcal{V}\} \tag{6}$$

where w_{ij} is a non-negative value. When the upstream turbine v_i exerts a wake effect on the downstream turbine v_j, the following equation can describe it:

$$w_{ij} = \begin{cases} \frac{A_{overlap\ i,j} * V_{wake}}{x/D}, & \text{shadowing,} \\ 0, & \text{no shadowing.} \end{cases} \tag{7}$$

where $V_{wake} = \frac{V_\infty - V_j}{V_\infty}$, the wake overlap effect area $A_{overlap\ i,j}$ is described in Figure 2, x represents the physical distance between the upstream turbine V_i and downstream turbine V_j; D represents the turbine rotor diameter of all the turbines. It is critical to note that the wake distribution should remain constant during the control period so that the control speed is high enough to counteract the changing wake distribution.

- The communication neighbors of vertex (turbine) v_i are denoted by $N_i = \{v_j | ((v_i, v_j) \in \mathcal{E})\}$.
- The set of shared turbine S_i in communication neighbors between the subset N_i and subset N_j, are denoted as $S_i = \{T_i | T_i \in N_i \cap N_j\}$ where T_i represents the shared turbine numbers in different subsets.

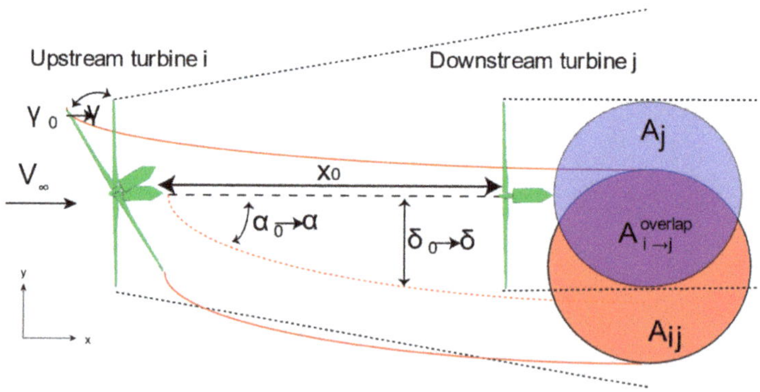

Figure 2. An example of a two-turbine wake redirection control through setting the yaw misalignment angle. γ is the upstream turbine i's yaw angle, α is the deflection angle, and δ denotes the wake deflection. The black dashed lines represent the wake of the upstream turbine i with no yaw control, and the red lines indicate the wake of the upstream turbine i with yaw control. $A_{i\to j}^{overlap}$ represents the area overlap ratio that the wake effect area $A(Ti|Tj)$ of the upstream turbine i to the downstream turbine j and the rotor area $A(Tj)$ of downstream turbine j.

3.2. Decoupled Communication Scheme of Wind Farm Based on Adaptive Pruned Algorithm

For most turbines, the coupling degree of one turbine with other turbines may be high or low. Therefore, the pruned wake digraph $\mathcal{G}_p$ can be pruned from the original wake digraph $\mathcal{G}$.

Definition 2. *The pruned wake digraph* $\mathcal{G}_p = (\mathcal{V}, \mathcal{E}_p,)$, *pruning edges* $\mathcal{E}_p$ *set* $\mathcal{E}_p \subset \mathcal{V} \times \mathcal{V}$, $\mathcal{E}_p = \{w_{ijp} : i, j \in \mathcal{V}\}$
where

$$w_{ijp} = \begin{cases} w_{ij} & w_{ij} \geq \varepsilon_k, \\ 0, & w_{ij} < \varepsilon_k \end{cases} \tag{8}$$

- The adaptive threshold $\varepsilon_k = k * \varepsilon$, k is one hyper-parameter, and ε is the basic threshold.
- Basic threshold ε is defined as the geometric median of the whole wake weight coefficients. The central idea of the geometric median is as follows: given the set of n^2 points $w_{11}, w_{12} \ldots, w_{ij}, \ldots, w_{nn}$ find a value ε that minimizes the sum of Euclidean distance:

$$f(x) \stackrel{\text{def}}{=} \sum_{i=1}^{N} \sum_{j=1}^{N} \|x - W_{ij}\|_2 \tag{9}$$

where $\varepsilon \epsilon \ argmin(f(x))$.

A decoupled topology in this section will be achieved by using the adaptive pruned wake digraph algorithm. However, determining the degree of pruning digraph has not been deeply studied, and there is little literature discussing it. This paper proposes an adaptive pruned algorithm to find a suitable threshold ε_k to solve the grouped problem by obtaining more reasonable decoupled subsets.

Based on the pruned wake digraph $\mathcal{G}_p = (\mathcal{V}, \mathcal{E}_p)$, for each angle $\varphi \in \{\varphi_1, \ldots, \varphi_w\}$, there are clustering subsets $N_l, l \geq 1$. A given direction φ has a corresponding cluster subset $N_i \in \{N_1, \ldots, N_M\}$, M is the number of subsets. Then, according to the wind farm layout, we build the original wake digraph $\mathcal{G}$ and calculate the weight coefficient matrix w_{ij} to find the communication neighborhood of turbines. The algorithm for turbine clustering via adaptive pruned wake digraph is shown as follows (Algorithm 1):

Algorithm 1: The method of clustering turbine via pruning wake digraph (Adaptive pruned wake digraph algorithm)

Step 1: Based on the layout of the position of the wind farm $(\boldsymbol{X}, \boldsymbol{Y})$, collect all relevant parameters, including wind direction $\boldsymbol{\Phi}$, wind speed $\boldsymbol{V_\infty}$. Additionally, the parameters of the wind turbines, for example, the rotor diameter $\boldsymbol{D}$, the physical distance $\boldsymbol{x}$ between WTs, and the overlap wake area $\boldsymbol{A_{i \to j}^{overlap}}$, etc.
Step 2: Calculate the threshold $\mathcal{E}$, and set the initial hyper-parameter $\boldsymbol{k}$, step hyper-parameter $\boldsymbol{\Delta k}$.
Step 3: Obtain the pruned digraph $\mathcal{G}_p$ from the original wake digraph $\mathcal{G}$ according to the global threshold ε_k according to the global threshold ε_k.
Step 4: Digraph clustering. Firstly, define the leading turbines for each subset that is experiencing free-stream velocity $\boldsymbol{V_\infty}$. Secondly, each leading turbine decides the communication neighbors through the connectivity information of the digraph $\mathcal{G}_p$ by a depth-first tree search (BFS) algorithm into the same subset $\boldsymbol{N_i}$.
Step 5: If there is a set of shared turbines $\boldsymbol{S_i}$, we need to continue to tune the value of k by set $\boldsymbol{k} = \boldsymbol{k} + \boldsymbol{\Delta k}$, go back to Step 3. If not, go directly to step 6.
Step 6: Calculate the output power and calculating time with the $\boldsymbol{k}$ value from step 5, save the parameters.
Step 7: If the coefficients of $\mathcal{G}_p$ are not all 0, continue to tune the $\boldsymbol{k}$ value by setting $\boldsymbol{k} = \boldsymbol{k} + \boldsymbol{\Delta k}$, go back to Step 2. If the coefficients of $\mathcal{G}_p$ are all 0, go to step 8.
Step 8: Based on the adaptive pruned wake digraph $\mathcal{G}_p$, we can establish turbine clustering subsets $\boldsymbol{N_i}$ and analyze all the parameters with different $\boldsymbol{k}$, and select the suitable value $\boldsymbol{k_2}$.

Overall, the adaptive pruned digraph and the contribution can be simplified as follows:

Firstly: for one, considered wind speed and wind direction, to satisfy the pruned wake digraph are decoupled (no shared wind turbines between all the subsets), the range of k can be found out to be $k \in [k_1, k_3]$, k_1 is the minimum value, and k_3 is the maximum value of k. In most literature, the k is one random value between the range so that the result may be suboptimal.

Secondly: by comparing the performance of output power and calculating time, deduce a suitable value k_2.

Thirdly: change the wind speed and direction by the above method and obtain another corresponding suitably k_2.

Lastly: create a line-off query look-up table using the obtained result, which can be used for quick reference with input winds velocity V_∞ and wind direction φ, the output is k_2.

The contributions are mainly to select suitable parameter k_2 with the proposed algorithm. Based on this condition, the control optimization can obtain the best result. The procedure will be demonstrated in Section 5 by one simulation example.

4. Wind Farm Control Strategy

Upon establishing the pruned wake digraph and clustering subsets, the overall OWF is controlled by multiple individual independents rather than by a single controller. The decentralized optimization process is described in this section. In order to achieve the control objectives of real-time output power optimization, a decentralized control scheme is proposed for the large-scale OWF, as shown in Figure 3. This control scheme is divided into two steps. First, wind farms are decoupled into several independent wind turbine clusters, and their communication neighbors are determined by the network topology of the adaptive pruned wake digraph. Therefore, a decentralized control strategy for OWF is proposed to realize the power control of the host computer. Second, we use the beetle antennae search (BAS) algorithm approach to optimize the yaw setting and the axial induction factor setting in the OWF to maximize the total wind plant power conversion.

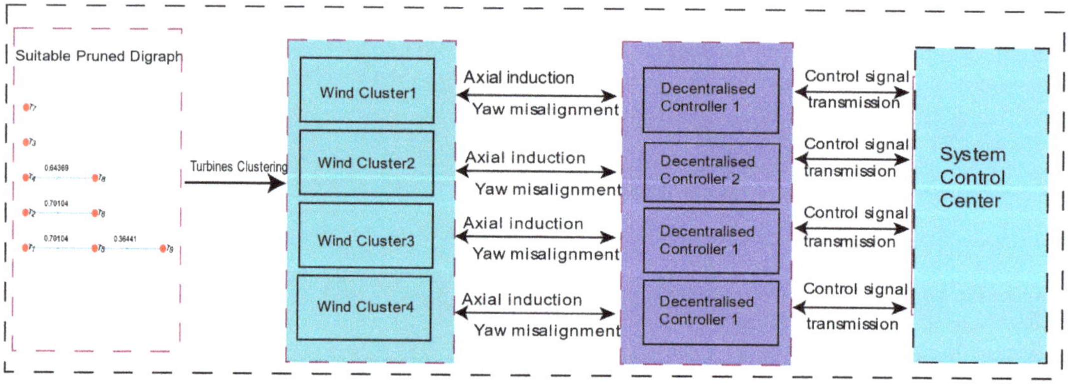

Figure 3. The proposed whole decentralized control scheme on a 3 × 3 turbines offshore wind farm.

4.1. The Output Power Optimization Problem

For a given decoupled clustering subsets N_i, the corresponding decentralized power control is based on the optimization objective: minimizing the power output of wind farms in Equation (4). Each wind turbine cluster is composed of several neighbor wind turbines that are decentralized on different communication network topologies of adaptive pruned wake digraphs. The proposed decentralized control scheme is shown in Figure 3. The corresponding decentralized power control for a given clustering subset N_i is based on the following single-objective optimization problem. The control variables for the optimization are the axial factor α and yaw angle γ of the wind turbines. The whole OWF decentralized optimization function $f(x)$ can be expressed as follows:

$$\begin{aligned} \min_{x} f(\boldsymbol{x}) &= -\sum_{i=1}^{M}\sum_{j=1}^{K}(P_{i,j}(\gamma_{i,j},\alpha_{i,j};V_{i,j})) \\ \text{s. t.} &\begin{cases} -\gamma_{min} \le \gamma_{i,j} \le \gamma_{max} \\ \alpha_{min} \le \alpha_{i,j} \le \alpha_{max} \end{cases} \end{aligned} \tag{10}$$

where i indicates subsets, j indicates turbines in subsets, M indicates the number of uncoupled subsets on the wind farm, and K indicates the number of turbines in every subset. The yaw angle $\gamma_{i,j}$ is between the upper bound γ_{max} and the lower bound $-\gamma_{min}$; α_{min} and α_{max} represent the lower bound and upper bound range of axial factor $\alpha_{i,j}$ and the power $P_{i,j}$ is between the minimum power P_{low} and the rated power P_{rate}.

4.2. Monte Carlo Law with the BAS(MC-BAS) Controller for Wind Turbines

The OWF cost function is given in Equation (9), which is a nonlinear and non-convex optimization problem. Many methods do not guarantee to find the optimal global solution. Herein, we adopt the BAS algorithm to solve optimization problems. Like genetic algorithms (GA), particle swarm optimization (PSO), and other intelligent algorithms, BAS does not require prior knowledge of the specific shape of the function or gradient information to optimize efficiency. The main two steps are the searching process and the result detection, tuning the adaptive step size until the optimization value is reached. The advantage of BAS is simple and fast speed to get the optimization object than other intelligence algorithms. Moreover, the high-speed advantage of BAS over the particle swarm optimization algorithm is that it requires only one individual, a longicorn beetle.

The Monte Carlo (MC) law of the annealing algorithm (SA) is introduced to improve the repeatability and stability of the algorithm. The improved algorithm is then applied to wake steering control so as to maximize power conversion on the wind farm. The optimal target value of the object is determined by simulating the annealing process, and the lowest energy of the target and simulated annealing incorporates random variables during the search process. For example, it embraces a worse solution than the current solution with a certain probability, increasing the possibility of exiting local optimization. The Modified MC-BAS algorithm is shown below (Algorithm 2).

1. Random direction vector
   To simulate the search behavior of longicorn, its direction vector is defined as [47]:

$$\vec{\boldsymbol{b}} = \frac{\text{rand}(k,1)}{\| \text{rand}(k,1) \|} \tag{11}$$

   where rand$(k,1)$ denotes a random function, and k represents position dimensions.
2. The coordinate of both right-hand and left-hand sides of the antennae of beetles are presented as [47]:

$$\begin{aligned} \boldsymbol{x}_r &= \boldsymbol{x}^t + d^t \vec{\boldsymbol{b}} \\ \boldsymbol{x}_l &= \boldsymbol{x}^t - d^t \vec{\boldsymbol{b}} \end{aligned} \tag{12}$$

   where t represents the number of iterations; x_r and x_l denote the spatial position of the right and left beetles of longicorn beetles in the t iteration, respectively; d^t represents the exploitability of antennae sensing length in the t iteration.
3. Fitness value:

$$\begin{cases} f_{right} = f(\boldsymbol{x}_r) \\ f_{left} = f(\boldsymbol{x}_l) \end{cases} \tag{13}$$

where f_{right} and f_{left} denote the fitness value of the right beetle and the left beard in the current spatial position; $f(\cdot)$ is the fitness function as Equation (9).

Algorithm 2: The grouped MC-BAS methodology for wind farm power production (MC-BAS Algorithm)

Result: The best yaw angles and the best axial factors x_n^{bst} and the best output power f_n^{bst}.
Input: Establish output objective function $f(x_n^t)$, where variable $x_n^t = [x_1^t, x_2^t, \ldots, x_N^t]$ and initialize the parameters $x^0, d^0, \delta^0, \eta_d, \eta_\delta, M_T, N, \alpha, \delta_{\text{criterion}}, t_{max}$.

```
While (n < N) do
    While (t < t_max) or (δ < δ_criterion) do
        Generate the direction vector b via Equation (9);
        Search in variable space with two kinds of antennae x_r and x_l via Equation (12);
        Update fitness value f(x_r), f(x_l) via Equation (13);
        Update the state variable x^t via Equation (14);
        if f(x_n^t) < f_n^bst then
            x_n^bst = x_n^t,
            f_n^bst = f(x_n^t),
        else
            delta=(f(x_n^t) - f_n^bst)/f_n^bst;
            L_P= exp (-delta / M_T);
            if rand (1) <= L_p,then
                x_n^bst = sample(x_n^t),
                f_n^bst = (f(x_sample^t)),
                M_T = α * M_T
            End
        End
        Update sensing diameter d,step δ via Equation (16)
    End
    Update t: t = t + 1
    Update n: n = n + 1
End
```

4. Pre-update position:

$$x^t = x^{t-1} + \delta^t \overrightarrow{b} sign\left(f_{right} - f_{left}\right) \tag{14}$$

Pre-update the position of the beetles based on the iteration, and the $sign(\cdot)$ is a symbol function; δ^t is the step size factor of the algorithm in the t iteration.

5. Accepted solution using the Monte Carlo law
The Monte Carlo law of the SA algorithm is embedding into BAS. In the iterative process, the probability P is used to accept the inferior solution to improve the global optimization ability of BAS:

$$L_P = \begin{cases} 1, & f(x^t) < f(x^{t-1}) \\ \exp\left(-\frac{f(x^t) - f(x^{t-1})}{M_T}\right), & f(x^t) \geq f(x^{t-1}) \end{cases} \tag{15}$$

where $f(x^t)$ denotes a pre-update position, $f(x^{t-1})$ denotes the best position in the last iteration; exp (.) represents the exponential function; M_T is the higher temperature.

6. Step size:

$$\begin{aligned} d^t &= \eta_d d^{t-1} + d^0 \\ \delta^t &= \eta_\delta \delta^{t-1} + \delta^0 \end{aligned} \tag{16}$$

where d and δ denote antennae length and step size, d^0 and δ^0 is the initial value, where d^t and δ^t is the step size factor of the algorithm in the t iteration, and the two parameters η_d and η_δ are set by the user.

The Monte Carlo law is mixed with the BAS algorithm. The basic steps of the grouped MC-BAS algorithm can be summarized as the pseudo-code shown in MC-BAS algorithm.

5. Validation and Discussion

Due to the randomness and intermittence, there are no constant wind speed and constant wind direction. For simplification, based on the probability of the known wind rose, we can calculate the average value of them in 10 min, then obtain an approximately constant value to describe them. In this experiment, assume that the average wind speed $V_\infty = 8$ m/s, the range of wind direction is $\varphi = \{0°, 15°, 30°, \ldots, 180°\}$, with which the baseline direction is the x-axis direction and under the assumption that it is constant within one control cycle.

The reason is that the wake is affected not much by the wind speed V_∞ but by the wind direction φ, so we only study the one wind speed $V_\infty = 8$ m/s, however, the whole wind direction is $\varphi \in [0°, 360°]$, with considering the symmetricity in the square wind farm, we only need to study the wind direction $\varphi \in [0°, 180°]$, and in this Simulink, we choose 5° as the step size in wind direction.

The wake digraph is the basic digraph from the wake field, as shown in Figure 4. The performance of proposed optimization approaches will be verified in this section by simulation results with the same NREL-5 MW Type III WT [48] turbine. The main parameters are given in Table A1 (Appendix A). The layout structure of OWF with lateral distance X = 500 m, longitudinal distance Y = 200 m, and rotor diameter D = 126 m, and nominal power P = 5 MW. The test is conducted for a 10-minute average of free wind speed and the direction range belongs to $\varphi \in [0°, 180°]$ at 15° increment. Furthermore, to verify the scalability and the feasibility of the proposed algorithm, in this paper, we study three different scales of wind farms with the different numbers of turbines N = 4, N = 9, and N = 25, respectively. The initial yaw angles γ are set to 0 with a range of $\gamma \in [-30°, 30°]$, and the initial factors are set to 1/3 with a range of $\alpha \in [0, 1/3]$. It was observed that the numerical results showed that the proposed control method could reach an improved increase rate with a larger wind farm by comparing the result of 2×2, 3×3, 5×5 matrix turbine wind farms. In other words, verification of the proposed method means that it is suited for a large-scale wind farm.

5.1. Processing the Adaptive Pruning Wake Redirect Digraph

In this section, the cluster method splits the large OWF into several independent subsets using the proposed pruned wake digraph clustering approach (see Section 3). The 5×5 wind farm location under different wind directions is shown in the proposed wake digraph. In this case, we consider a wind direction of $\Phi = 45°, 90°$, the wake original digraph $\mathcal{G}$, pruned digraph $\mathcal{G}_p$ as illustrated in Figures 5 and 6, respectively. In addition, the decoupled communication topology comes from the suitably pruned wake digraph.

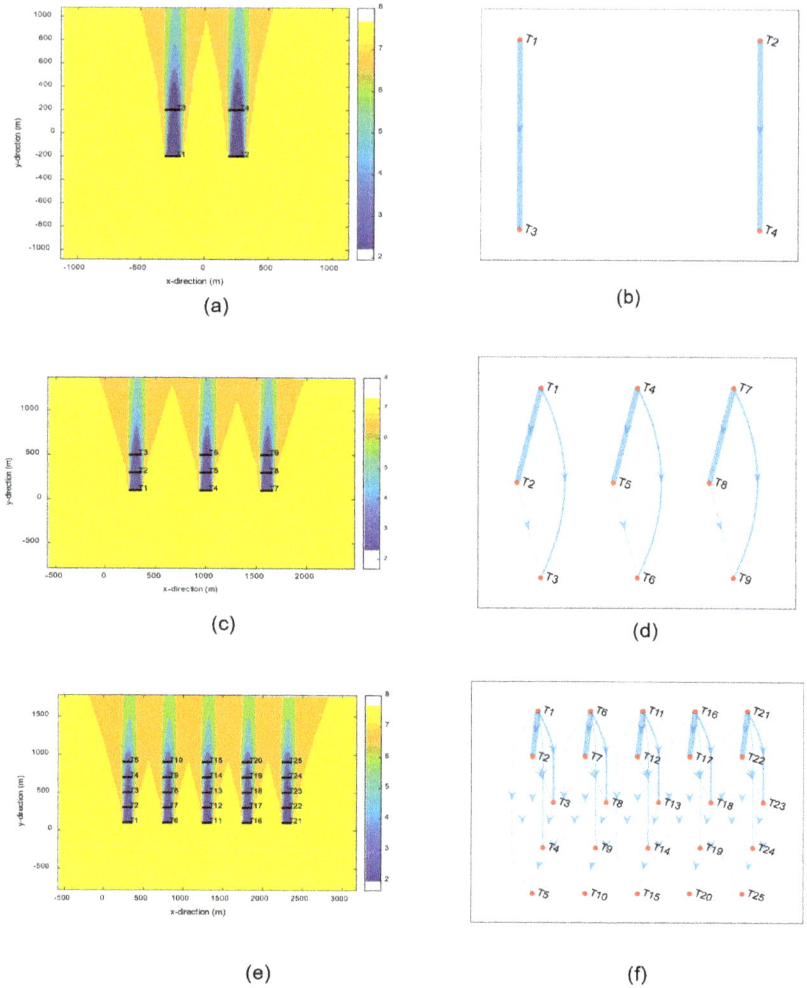

Figure 4. The original wake digraph $\mathcal{G}$ under different wind turbines $\varphi = 90°$: (**a**) 2×2 turbines wake field; (**b**) 2×2 turbines original wake digraph; (**c**) 3×3 turbines wake field; (**d**) 3×3 turbines original wake digraph; (**e**) 5×5 turbines wake field; (**f**) 5×5 turbines original wake digraph.

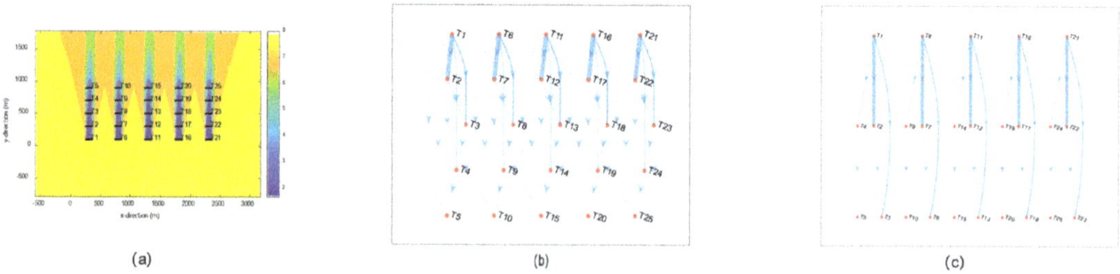

Figure 5. The wake digraph ($\varepsilon = 0.157481474$) with wind direction $\varphi = 90°$: (**a**) wake field; (**b**) original wake digraph ($k = 0.1$); (**c**) pruned wake digraph ($k = 3.2$).

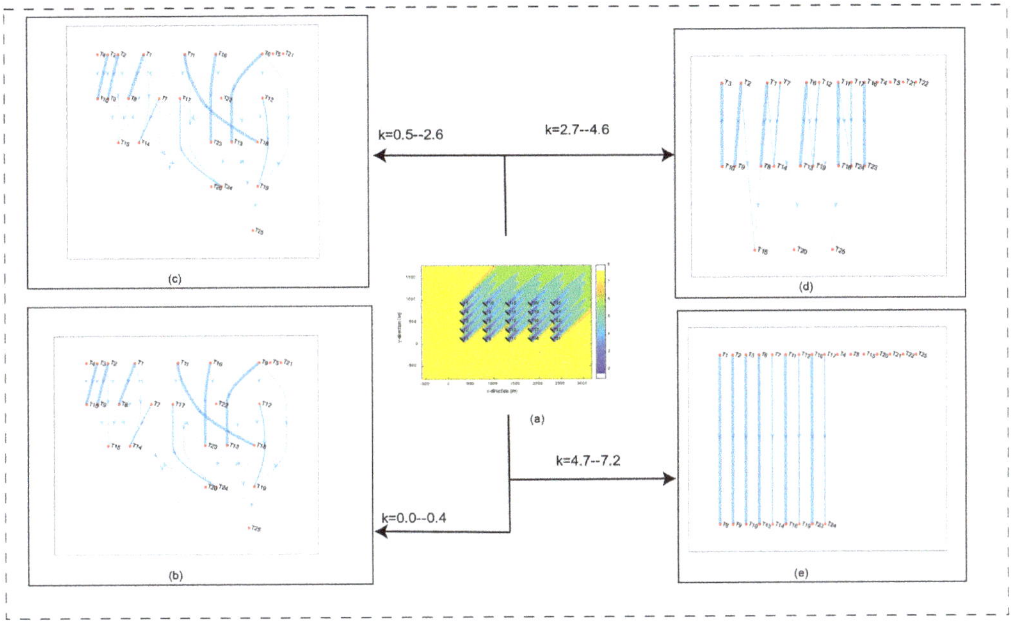

Figure 6. The process of adaptive pruning wake digraph method with varying k for the wake digraph ($\varepsilon = 0.0122$) with wind direction $\varphi = 45°$: (**a**) wake field; (**b–e**) pruned wake digraph $\mathcal{G}_p$ with different k.

Figure 5 shows that the original wake digraph is the same as the pruned wake digraph. This is because when the wind direction is $\varphi = 90°$, the entire wake effect is concentrated on the downstream turbines without any diffused. However, when the wind direction changes to $\varphi = 45°$, the power conversion of the OWF is increased since the influence of wake interaction becomes low.

As shown in Figure 6, the original wake digraph $\mathcal{G}$ differs in the adaptive pruned wake digraph $\mathcal{G}_p$. The pruned wake digraph $\mathcal{G}_p$ is different with varying k. From Figure 6b to Figure 6c, the threshold $\varepsilon_p = k*\varepsilon$, $k \in [0.5, 2.6]$, so the edges which are smaller than ε_p are cut off, such as the edges $\mathcal{E}_{11,25}$, etc. From Figure 6c to Figure 6d, the global threshold ε_p become bigger k, with the range of $k \in [2.7, 4.6]$, then the edges which are smaller than ε_p are cut off, for example, the edge $\mathcal{E}_{1,15}$, $\mathcal{E}_{16,25}$, $\mathcal{E}_{22,25}$, etc. Lastly, from Figure 6d to Figure 6e, get bigger k at the range of $k \in [4.7, 7.2]$, the more edges are cut off, for example, the edges $\mathcal{E}_{13,20}$, $\mathcal{E}_{18,25}$, etc. and then the wake field is divided into 13 subsets with no shared turbine.

Figure 6 illustrates that the wake topology is parameter-dependent, as the external relevant variables (φ, k) influence the wake effect. For a given φ, the suitable k is vital for pruning the wake digraph to obtain the optimization decoupled subsets. The method regarding tuning the hyperparameters k will be discussed in the next step, which is shown in Tables 1 and 2 as follows.

Table 1. The relationship of subsets and shared turbines (ST) with different k.

k	No of Subsets	With ST or Not
0–0.4	9	Yes
0.5–2.6	13	Yes
2.7–4.6	13	No
4.7–7.2	16	No
7.3	21	NO

Table 2. Comparing the results of different pruning wakes under different k.

k	Baseline Power	Total Power (W)	Groups	T(s)	ΔP
2.7	3.23E+07	3.62×10^7	13	2.84×10^2	12.19%
2.8	3.23E+07	3.62×10^7	13	2.44×10^2	12.19%
3.3	3.23E+07	3.62×10^7	13	2.42×10^2	12.19%
3.8	3.23E+07	3.62×10^7	13	2.36×10^2	12.19%
4.3	3.23E+07	3.58×10^7	13	2.35×10^2	10.95%
4.8	3.23E+07	3.58×10^7	13	2.32×10^2	10.95%
5.3	3.23E+07	3.51×10^7	13	2.30×10^2	8.78%
5.7	3.23E+07	3.43×10^7	13	2.26×10^2	6.30%
5.8	3.23E+07	3.03×10^7	16	3.57×10^2	−6.10%
6.3	3.23E+07	3.03×10^7	16	3.59×10^2	−6.10%
6.8	3.23E+07	3.01×10^7	16	3.66×10^2	−6.72%
7.3	3.23E+07	3.01×10^7	21	3.66×10^2	−6.72%

Table 1 shows that there are shared turbines in the subsets depending on the value of k. In this paper, we focus on the range of $2.7 \leq k \leq 7.3$ because of no shared wind turbine. In other words, the subsets are all decoupled. Then, there is another problem of how to set the suitable value k. In this paper, the proposed adaptive pruning algorithm can solve this problem. An adaptive threshold ε_p can be proposed by comparing the output power and calculating time, and the comparison results are displayed in Table 2. It is essential to note that in Table 2, considering the objective of the real-time control, we choose the suitable value k_2 that focuses more on computational efficiency and an increased power rate by more than 4%. Moreover, when k is set as 5.3 as in Table 2, the control time is 226 s which is smaller than others. The high control speed is a very vital parameter during the control process. Therefore, we can find the suit $k_2 = 5.3$.

From Table 2, we chose $k = 5.7$ as a suitable value. The reason is that when $k > 5.7$, the output power is smaller than the baseline value, which is not permitted in this paper. Moreover, the calculation time is the smallest at the range of $0 \leq k \leq 5.7$. The controller speed is important for the objection of real-time control. In this condition, the pruned wake digraph will be divided into 13 decoupled subsets, and the clustered turbines' neighbor turbines of every subset are $\text{N1} = \{\text{T1}, \text{T8}, \text{T15}\}$, $\text{N2} = \{\text{T2}, \text{T9}\}$, $\text{N3} = \{\text{T3}, \text{T10}\}$, $\text{N4} = \{\text{T4}\}$, $\text{N5} = \{\text{T5}\}$, $\text{N6} = \{\text{T6}, \text{T13}, \text{T20}\}$, $\text{N7} = \{\text{T7}, \text{T14}\}$, $\text{N8} = \{\text{T11}, \text{T18}, \text{T25}\}$, $\text{N9} = \{\text{T12}, \text{T19}\}$, $\text{N10} = \{\text{T16}, \text{T23}\}$, $\text{N11} = \{\text{T17}, \text{T24}\}$, $\text{N12} = \{\text{T21}\}$, $\text{N13} = \{\text{T22}\}$.

Using the above-mentioned method, the range of wind direction extends to $0^\circ \leq \varphi \leq 90^\circ$ with an increment of 15°. Under different wind directions, to obtain decoupled communication topology by pruning the wake digraph, the experimental range of k is $k_1 \leq k \leq k_3$ and k_2 is the suitable value, which can be obtained from the proposed adaptive pruning algorithm.

In this way, wind speed keeps $V_\infty = 8\ \text{m/s}$, we can also obtain the suitable value k_2 when wind direction φ varies from the range of $[0^\circ, 90^\circ]$ which is shown in Table 3 as follows. When wind speed V_∞ and wind direction φ changed, a look-up table of k_2 can be obtained by the proposed adaptive algorithm, which is shown in Appendix B—Table A2.

5.2. The Combined Evaluation of the Decentralized MC-BAS Algorithm

When $V_\infty = 8\ \text{m/s}$, $\varphi = 45^\circ$, and $k = 5.7$, as shown in Tables 1 and 2, the OWF wake digraph can be divided into 13 decoupled subsets. Taking the subset N2, for example, it concludes two neighbor turbines in cluster N2, the upstream turbine WT2 and the downstream turbine WT9. In this paper, the control actions and wake infection only work in the same subset. To maximize the output power of OWF, the yaw angles γ and the axial factors α are activated in an optimally decentralized manner. We will explain the sensitive relationship between the control actions α, γ, the output power P, and the consequent wind speed direction φ of the neighbor wind turbine in one subset as shown in Figure 7.

Taking subset N2, for example, in the range of wind direction $\varphi \in [90°, 180°]$, the upstream turbine WT2, the output power of MC-BAS is larger than that with the greedy method as shown in Figure 7b,c. Moreover, for the downstream WT9, the output power of MC-BAS is larger than that of the greedy method. However, the reason is not because of controlling the axial factors α and the yaw angle γ of WT2 but the decreasing wake effect of WT2, which is clearly shown in Figure 7e,f. For WT9 having no downstream turbines, the control parameters do not need to change the value significantly. This method is also applicable to other wind direction ranges and some other subsets. For brevity, we will not repeat the description in this paper.

When the wind direction is in the range of wind direction $\varphi \in [20°, 50°]$, WT2 and WT9 are in the same subset. In other words, WT2 and WT9 are neighbor turbines. Essentially, when wind direction φ changes significantly, the communication topology will also vary. The wind turbine will infect each other for the same subset, and for different subsets, the wind turbines are all independent. In this way, every wind turbine in all subsets with varying wind direction is optimized, allowing the total output power to reach the maximum value. The result is shown in Figure 8 and Table 4.

Table 3. The different adaptive k with the varying wind direction φ.

φ	k	P(W)	ΔP	T(s)
$\varphi = 0^{\circ}$	Baseline	2.4556×10^7	0%	0.1896
	$k_1 = 0.1$	2.8528×10^7	16.18%	276.245
	$k_2 = 5.6$	2.8526×10^7	16.17%	211.5501
	$k_3 = 11.7$	2.0173×10^7	−17.85%	243.3781
$\varphi = 15^{\circ}$	Baseline	2.4973×10^7	0%	0.1659
	$k_1 = 1.6$	2.8554×10^7	14.34%	268.4627
	$k_2 = 2.4$	2.8152×10^7	12.73%	138.1018
	$k_3 = 2.7$	2.1252×10^7	−14.90%	189.6079s
$\varphi = 30^{\circ}$	Baseline	3.1375×10^7	0%	0.1595
	$k_1 = 1.9$	3.3643×10^7	7.23%	276.8732
	$k_2 = 3.5$	3.2142×10^7	2.45%	239.3284
	$k_3 = 4.1$	2.9763×10^7	−5.14%	293.1692
$\varphi = 45°$	Baseline	3.9268×10^7	0%	0.1402
	$k_1 = 2.5$	4.1118×10^7	4.57%	284.4385
	$k_2 = 5.7$	4.0926×10^7	4.16%	226.5321
	$k_3 = 7.3$	3.8126×10^7	−2.97%	366.5429
$\varphi = 60°$	Baseline	3.0271×10^7	0.00%	0.1385
	$k_1 = 0.9$	3.2014×10^7	5.76%	259.6893
	$k_2 = 6.8$	3.1139×10^7	2.87%	271.8649
	$k_3 = 7.9$	2.7853×10^7	−7.99%	350.6543
$\varphi = 75^{\circ}$	Baseline	2.3257×10^7	0%	0.1243
	$k_1 = 0.6$	2.5473×10^7	9.53%	174.9643
	$k_2 = 16.5$	2.4385×10^7	4.85%	136.9856
	$k_3 = 27.3$	2.1072×10^7	−9.39%	181.6532
$\varphi = 90°$	Baseline	1.8731×10^7	0%	0.1133
	$k_1 = 0.0$	2.2795×10^7	21.70%	112.5742 s
	$k_2 = 31.6$	2.1596×10^7	15.30%	98.7756 s
	$k_3 = 83.7$	1.6765×10^7	−12.00%	117.329 s

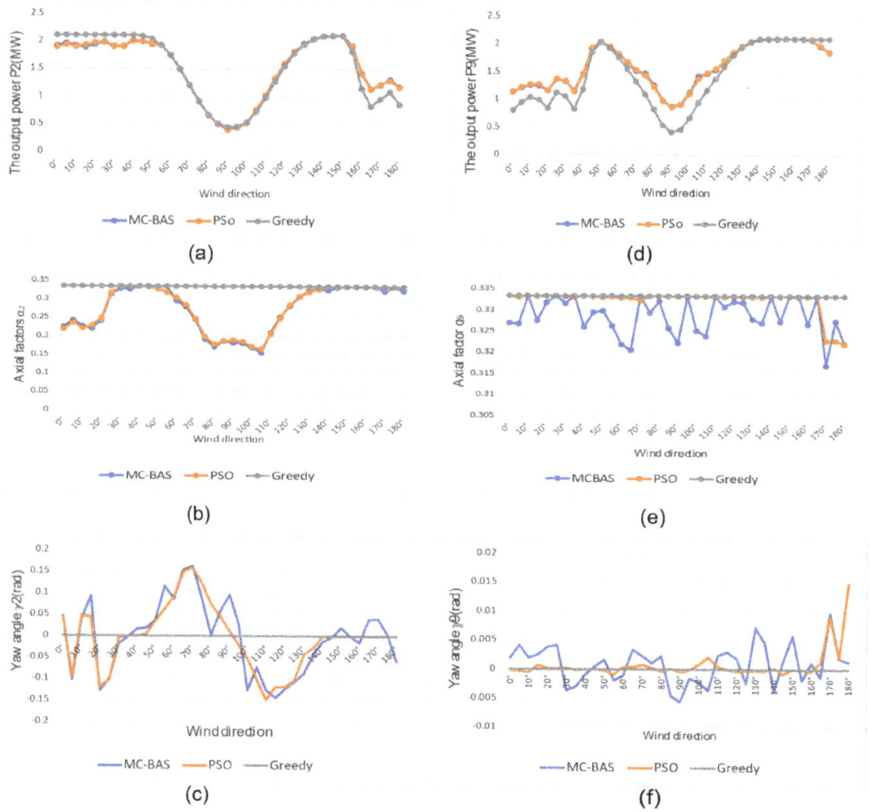

Figure 7. Comparison of control parameters of subset N2 between MC-BAS, PSO, and Greedy control with different wind direction: (**a**) The output power of WT2; (**b**) The axial factors of WT2; (**c**) The yaw angle of WT2; (**d**) The output power of WT9; (**e**) The axial factors of WT9; (**f**) The yaw angle of WT9.

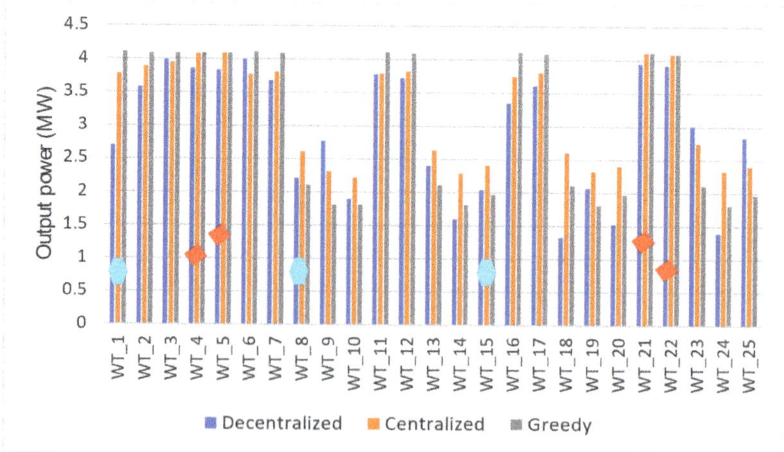

Figure 8. The respective power output of wind turbine: $\varphi = 45°$.

Table 4. Comparison of power output and calculating time on OWF using centralized and decentralized algorithms on 5 × 5 matrix wind farm.

Wind Direction	Control Method	P_Total (w)	ΔP_Total	T_Total (s)
$\varphi = 0°$	Centralized Greedy	2.4556×10^7	0.00%	0.1896
	Centralized MC-BAS	2.8529×10^7	16.18%	636.5711
	Decentralized MC-BAS (k_1 = 0.1)	2.8528×10^7	16.17%	276.245
	Decentralized MC-BAS (k_2 = 5.6)	2.8526×10^7	16.17%	211.5501
$\varphi = 25°$	Centralized Greedy	3.1281×10^7	0.00%	0.1279
	Centralized MC-BAS	3.4255×10^7	9.51%	466.0449
	Decentralized MC-BAS (k_1 = 1.6)	3.4254×10^7	9.50%	218.4627
	Decentralized MC-BAS (k_2 = 2.4)	3.4252×10^7	9.50%	138.1018
$\varphi = 45°$	Centralized Greedy	3.9268×10^7	0.00%	0.1102
	Centralized MC-BAS	4.2164×10^7	7.37%	286.72145
	Decentralized MC-BAS (k_1 = 2.5)	4.1118×10^7	4.57%	284.4385
	Decentralized MC-BAS (k_2 = 5.7)	4.0926×10^7	4.16%	226.5321
$\varphi = 90°$	Centralized Greedy	1.8731×10^7	0.00%	0.1133
	Centralized MC-BAS	2.3853×10^7	27.35%	399.0926
	Decentralized MC-BAS (k_1 = 0.0)	2.2795×10^7	21.70%	112.5742
	Decentralized MC-BAS (k_2 = 31.6)	2.1596×10^7	15.24%	98.7756

Figure 8 presents the results of the respective power output, while Table 4 shows the comparison of total power output. Generally, when the MC-BAS control method is used to implement a centralized and decentralized approach, the total produced power increases compared to the greedy control method. The decoupled cluster subset with wind direction $\varphi = 45°$ can be obtained from Table 4. For example, one cluster subset N1 includes turbines T1, T8, and T15, symbolled as blue hexagon lines. Figure 8 shows that, with the greedy control algorithm, the lead wind turbine T1 produces the maximum power output, while the communication neighbors T8 and T15 only produce minimal power output without regulating the wake effect. Furthermore, the upstream wind turbine can cause significant wake disruption, reducing wind speed and lowering power conversion of the downstream wind turbine [17]. The wake effect is taken into account in the MC-BAS decentralized and centralized control scheme for optimizing overall power output. The majority of upstream turbines reduce output power, whereas downstream turbines increase power conversion, thereby increasing the entire power conversion. Furthermore, subsets N4, N5, N12, N13 have only one turbine, which is symbolized as a little red diamond on WT4, WT5, WT21, and WT22, and, respectively, the output power has no significant difference in the three different control methods since they are unconcerned about the downstream turbine. Other wind directions can be analyzed in the same way. However, the methodologies were not described in this paper to maintain brevity.

The calculated time differs between the decentralized and centralized methods. The control speed of decentralized control is higher than the centralized control because there are fewer turbines to solve, as shown in Table 4. The rate of power (ΔP_total) is the increased power at the baseline of *P*_total of the greedy centralized algorithm. Table 4 shows that ΔP_total increases at varying degrees in the decentralized MC-BAS control and centralized MC-BAS method under different wind directions. Moreover, the decentralized MC-BAS computation time (T_total) is reduced to less than 1/3 times of the centralized approach. The mean total power generated by the centralized MC-BAS algorithm and decentralized MC-BAS algorithm improves by 14.4% and 11.3676%, respectively, compared to the baseline. This indicates about 3.0324% power loss in the decentralized MC-BAS compared to the centralized MC-BAS method. Thus, the proposed control strategy is practical for increasing power output and improving calculation speed from the perspective of real-time control and the profit of the large-scale OWF. For different wind directions, we set the appropriate value k_2 so as to improve the calculation efficiency. Moreover, the importance of the adaptive pruned algorithm is also verified in Table 4.

5.3. The Advantage of MC-BAS over Other Algorithms

Generally, the higher the number of iterations, the more accurate the computation. This section aims to take the least number of iterations possible to reach the optimum control actions, resulting in improved total power conversion and communication burden.

According to Figure 9, the calculating time will increase significantly as the number of iterations increases by 100 to 300 in 20 increments, verifying the statements made in Section 3.1. The number of iterations plays a crucial role in reducing the calculation time, therefore reducing the communication burden. Consequently, the exact iteration value is a significant tuning value for optimization algorithms. The MC-BAS algorithm takes far less calculating time than the other three control algorithms and is about 1/9 time of the PSO method and 1/4 time of the GA method. Thus, the proposed centralized MC-BAS method outperforms other intelligent methods (GA and PSO) in terms of calculating speed.

Figure 10 shows the power conversion depending on wind directions and iterations with four different control algorithms. The PSO method can produce more power than others in most iterations. However, when the number of iterations exceeds 140, the total production in the PSO algorithm is equal to that of the MC-BAS algorithm. Consequently, we set the number of iterations to 140 to obtain better total power conversions with the proposed MC-BAS. The main drawback of the GA algorithm is the unstable output power, which varies at different iterations, as shown in Figure 10b,d. Therefore, the GA algorithm is not a suitable choice for OWF. The convergence of the algorithm can be measured by error variation and the number of iterations. In this paper, in order to test the influence of the number of iterations on the results, the number of iterations is used as the condition for the end of the simulation.

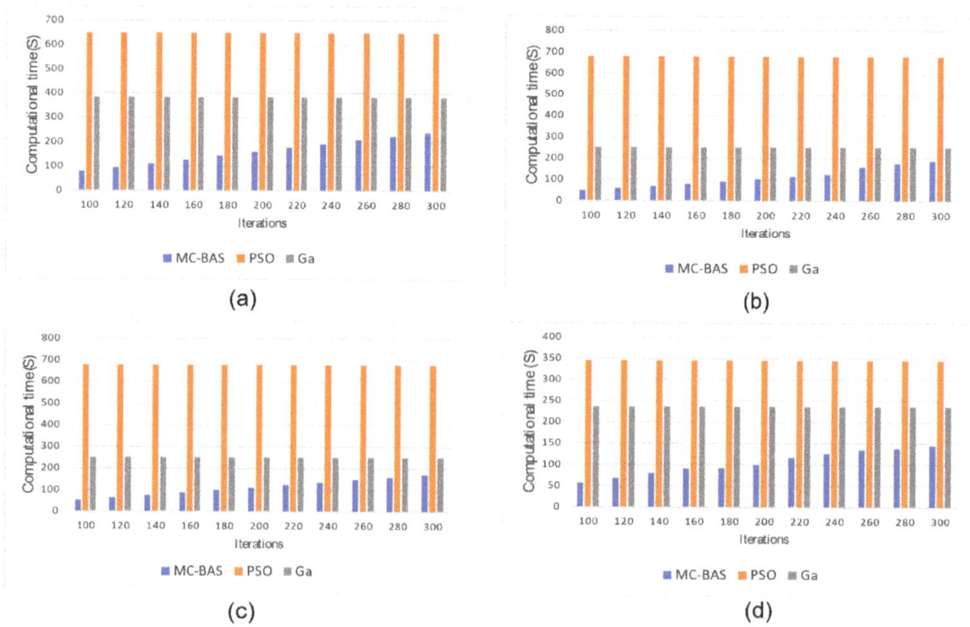

Figure 9. The computational time versus iteration of different algorithms in 3 × 3 matrix OWF: (**a**) $\varphi = 0°$; (**b**) $\varphi = 25°$; (**c**) $\varphi = 45°$; (**d**) $\varphi = 90°$.

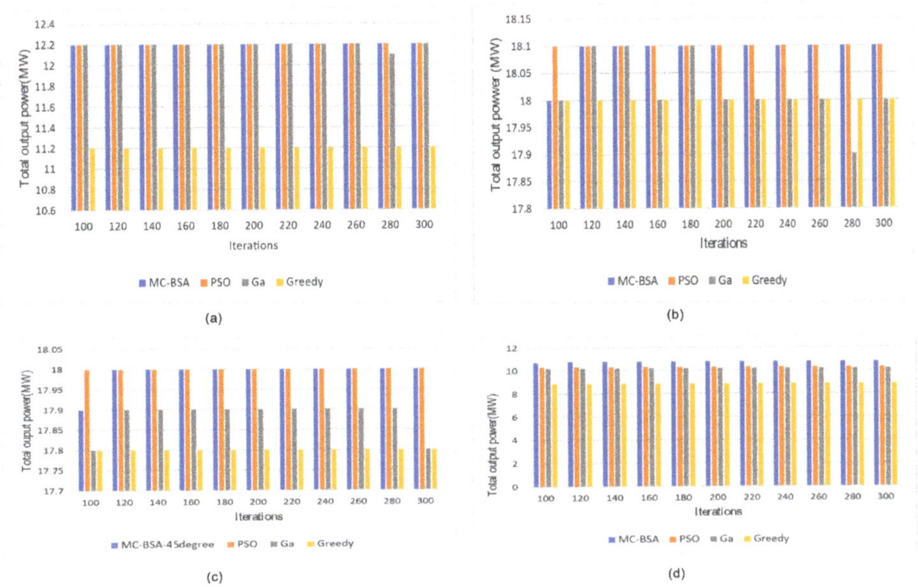

Figure 10. The total power output of 3 × 3 matrix OWF on MC-BAS algorithm compared to the iteration algorithms: (**a**) $\varphi = 0°$; (**b**) $\varphi = 25°$; (**c**) $\varphi = 45°$; (**d**) $\varphi = 90°$.

Moreover, we demonstrate the effectiveness and scalability of the proposed methodology. The range of wind direction is $\varphi \in [0°, 10°, \ldots, 180°]$ and the number of turbines is a 2 × 2 matrix and 3 × 3 matrix as shown in Figure 11.

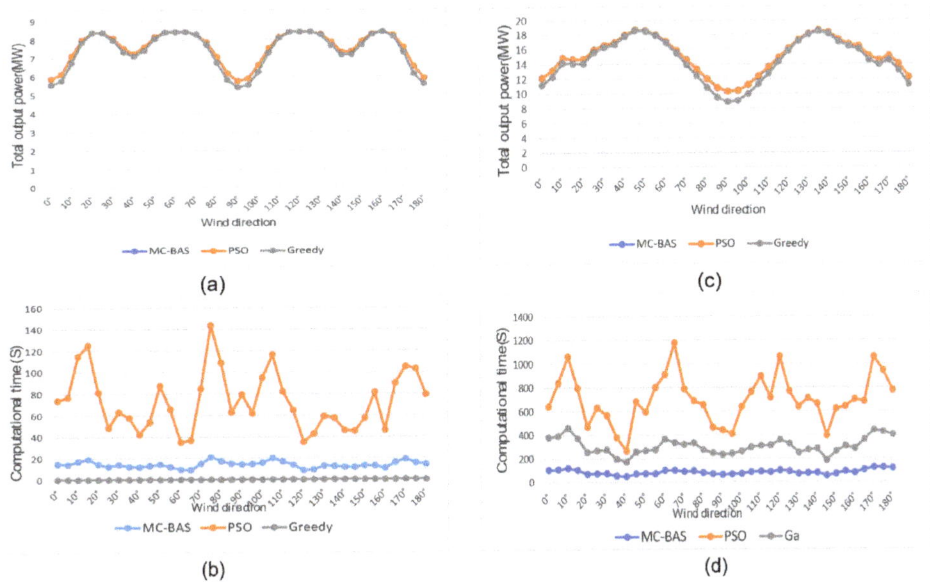

Figure 11. Comparison of the centralized MC-BAS algorithms with other iteration methods: (**a**) the total power conversion with 2 × 2 matrix turbines; (**b**) the calculating time with 2 × 2 matrix turbines; (**c**) the total power conversion with 3 × 3 matrix turbines; (**d**) the calculating time with 3 × 3 matrix turbines.

Figure 11a,c shows that the total power conversion of MC-BAS is exactly the same as that of PSO, which is better than the value produced by the greedy algorithm in most wind directions. In addition, the MC-BAS algorithm takes less calculation time than the PSO algorithm on the 2×2 matrix and 3×3 matrix wind farms (Figure 11b,d). Notably, the ordinate scale of Figure 11c is 10 times that of Figure 11a,d and is 10 times that of Figure 11b. A more detailed analysis with the increasing output is displayed in Tables 5 and 6.

Table 5. The total power conversion improvement rate compared to the baseline with the 3×3 matrix OWF.

φ	0°	5°	10°	15°	20°	25°	30°	35°	40°	45°	50°	55°	60°
▲MCBAS	9%	7%	5%	4%	4%	3%	2%	2%	1%	1%	1%	1%	1%
▲PSO	9%	7%	5%	4%	4%	3%	2%	2%	1%	1%	1%	1%	1%
φ	65°	70°	75°	80°	85°	90°	95°	100°	105°	110°	115°	120°	125°
▲MCBAS	3%	5%	7%	11%	14%	16%	15%	13%	9%	7%	3%	2%	1%
▲PSO	3%	5%	7%	11%	14%	16%	15%	13%	9%	7%	3%	2%	1%
φ	130°	135°	140°	145°	150°	155°	160°	165°	170°	175°	180°		
▲MCBAS	1%	0%	1%	2%	2%	3%	4%	4%	3%	7%	9%		
▲PSO	1%	0%	1%	2%	2%	3%	4%	4%	3%	7%	9%		

Table 6. The total power conversion improvement rate compared to the baseline in 2×2 matrix WF.

φ	0°	5°	10°	15°	20°	25°	30°	35°	40°	45°	50°	55°	60°
▲MCBAS	6%	6%	6%	2%	0%	0%	2%	3%	2%	3%	1%	0%	0%
▲PSO	6%	6%	6%	2%	0%	0%	2%	3%	2%	3%	1%	0%	0%
φ	65°	70°	75°	80°	85°	90°	95°	100°	105°	110°	115°	120°	125°
▲MCBAS	0%	0%	2%	5%	7%	7%	6%	6%	3%	1%	0%	0%	0%
▲PSO	0%	0%	2%	5%	6%	6%	6%	6%	3%	1%	0%	0%	0%
φ	130°	135°	140°	145°	150°	155°	160°	165°	170°	175°	180°		
▲MCBAS	1%	3%	2%	2%	2%	1%	0%	1%	4%	7%	6%		
▲PSO	1%	3%	2%	2%	2%	1%	0%	1%	4%	7%	6%		

Tables 5 and 6 show the total production improvement rate is relative to the baseline of greedy control with 2×2 matrix turbines and 3×3 matrix turbines. It can be observed in some wind directions that the increased power rate is zero, such as $\varphi = 20°, 25°$, etc., in Table 6 and $\varphi = 135°$ in Table 5. The advantage of the proposed MC-BAS algorithm was verified by comparing it with several other algorithms, especially in increasing the output power and decreasing the calculating time. Additionally, in a large-scale OWF, an adaptive pruned wake digraph is proposed to divide it into several decoupled subsets. Then, the same controller works on every subset to ensure real-time control. By analyzing the data in Tables 3–6, we can conclude that the ΔP with the MC-BAS algorithm increases with a greater number of wind turbines on the wind farm. For example, when $\varphi = 15°$, $\Delta P = 2\%, 4\%, 12.73\%$ on the 2×2, 3×3, and 5×5 matrix wind farms, respectively. We anticipate that the proposed algorithm will demonstrate a good performance of large-scale wind farms.

6. Conclusions

This paper proposed a decentralized real-time power optimization for large-scale OWFs using an adaptive pruned wake digraph approach. The results of this paper can be summarized as follows:

1. The proposed adaptive pruning algorithm fully considers the real-time power optimization control goals, providing a suitable method of grouping to avoid obtaining a sub-optimization result due to the unsuitable communication topology. The vital point of the adaptive pruned digraph is to uncover the accurate global threshold ε_p corresponding to the different wind by setting the suitable parameter k_2. Moreover, the proposed method was verified to be efficient by the Simulink result, and the off-line look-up table was constructed in Appendix B.
2. This work presents a modified BAS algorithm to raise BAS's ability and efficiency for dealing with high-dimensional nonlinear problems. The BAS can use fewer iterations to rapidly search for the fitness function maximum in the parameter selection space. Meanwhile, the Monte Carlo (MC) law of Simulate Anneal (SA) was introduced to improve the reproducibility and stability of the algorithm by avoiding blind searching and escaping the local traps minima.
3. For a large-scale wind farm, real-time state information may be excessive for the high communication and computational burden—centralized control approaches might fail. However, the adaptive pruned digraph decentralized operation can solve this problem by dividing the large-scale wind farm into several decoupled subsets; the local controller only deals with the local subset.

Future work will focus on increasing the control parameters and control objectives of the large-scale OWF, considering the infection of nonlinear turbulent flow [50,51]. Moreover, optimizing the wind farm layout with irregularly shaped wind farms will be studied by decreasing the wake effect.

Author Contributions: Conceptualization, Y.-H.J.; methodology, Y.C.; software, Y.C.; validation, Y.-H.J.; investigation, D.S.; data curation, Y.C.; writing—original draft preparation, Y.C.; writing—review and editing, D.S.; visualization, D.S.; supervision, Y.-H.J. All authors have read and agreed to the published version of the manuscript.

Funding: This work was partially supported by the Basic Science Research Program through the National Research Foundation of Korea (NRF), funded by the Ministry of Education (NRF-2016R1A6A1A03013567, NRF-2021R1A2B5B01001484) and the Innovation-Driven Project of Central South University (2020CX031).

Institutional Review Board Statement: Not applicable.

Conflicts of Interest: The authors declare no conflict of interest.

Appendix A

Table A1. The main parameters of the 5 MW wind turbine for an offshore wind farm.

P_rate	5 MW
D	126 m
ω^{min}	6.9 rpm
ω^{max}	12.1 rpm
β^{max}	90°
Gearbox ratio	97:1
Rated wind speed	11.4 m/s
C_p^{max}	0.485
Hub height	90 m

Appendix B

Table A2. The look-up table of different k_2 with the varying wind.

V_∞	φ	0°	15°	30°	45°	60°	75°	90°	105°	120°	135°	150°	165°	180°
8 m/s	k_2	5.6	2.4	3.5	4.8	6.8	16.5	31.6	19.9	7.5	5.3	4.3	3.8	6.2
9 m/s	k_2	5.9	2.8	3.9	5.1	7.4	16.9	31.8	21.3	8.0	5.7	4.5	5.3	6.7
10 m/s	k_2	6.2	3.3	4.2	5.7	7.6	17.6	41.8	27.3	9.6	6.7	5.8	6.9	7.6
11 m/s	k_2	6.8	3.9	6.4	8.4	18.9	43.4	29.8	10.9	7.9	6.9	7.3	7.6	9.4

References

1. Song, D.; Zheng, S.; Yang, S.; Yang, J.; Dong, M.; Su, M.; Hoon Joo, Y. Annual Energy Production Estimation for Variable-Speed Wind Turbine at High-Altitude Site. *J. Mod. Power Syst. Clean Energy* **2021**, *9*, 684–687. [CrossRef]
2. Song, D.; Chang, Q.; Zheng, S.; Yang, S.; Yang, J.; Hoon Joo, Y. Adaptive Model Predictive Control for Yaw System of Variable-Speed Wind Turbines. *J. Mod. Power Syst. Clean Energy* **2021**, *9*, 219–224. [CrossRef]
3. Wang, L. *Numerical Optimization of Wind Farm Layout and Control Strategy*; Queensland University of Technology: Brisbane City, Australia, 2017.
4. Serrano González, J.; Burgos Payán, M.; Santos, J.M.R.; González-Longatt, F. A Review and Recent Developments in the Optimal Wind-Turbine Micro-Siting Problem. *Renew. Sustain. Energy Rev.* **2014**, *30*, 133–144. [CrossRef]
5. Chen, X.; Huang, L.; Liu, J.; Song, D.; Yang, S. Peak Shaving Benefit Assessment Considering the Joint Operation of Nuclear and Battery Energy Storage Power Stations: Hainan Case Study. *Energy* **2022**, *239*, 121897. [CrossRef]
6. Kirchner-Bossi, N.; Porté-Agel, F. Wind Farm Area Shape Optimization Using Newly Developed Multi-Objective Evolutionary Algorithms. *Energies* **2021**, *14*, 4185. [CrossRef]
7. Ahmad, T.; Basit, A.; Anwar, J.; Coupiac, O.; Kazemtabrizi, B.; Matthews, P. Fast Processing Intelligent Wind Farm Controller for Production Maximisation. *Energies* **2019**, *12*, 544. [CrossRef]
8. Yang, J.; Wang, L.; Song, D.; Huang, C.; Huang, L.; Wang, J. Incorporating Environmental Impacts into Zero-Point Shifting Diagnosis of Wind Turbines Yaw Angle. *Energy* **2022**, *238*, 121762. [CrossRef]
9. Song, D.; Yang, J.; Su, M.; Liu, A.; Liu, Y.; Joo, Y. A Comparison Study between Two MPPT Control Methods for a Large Variable-Speed Wind Turbine under Different Wind Speed Characteristics. *Energies* **2017**, *10*, 613. [CrossRef]
10. Yang, J.; Fang, L.; Song, D.; Su, M.; Yang, X.; Huang, L.; Joo, Y.H. Review of Control Strategy of Large Horizontal-Axis Wind Turbines Yaw System. *Wind Energy* **2021**, *24*, 97–115. [CrossRef]
11. Harrison, M.; Bossanyi, E.; Ruisi, R.; Skeen, N. An Initial Study into the Potential of Wind Farm Control to Reduce Fatigue Loads and Extend Asset Life. *J. Phys. Conf. Ser.* **2020**, *1618*, 022007. [CrossRef]
12. Kanev, S.K.; Savenije, F.J.; Engels, W.P. Active Wake Control: An Approach to Optimize the Lifetime Operation of Wind Farms: Active Wake Control: An Approach to Optimize the Lifetime Operation of Wind Farms. *Wind Energy* **2018**, *21*, 488–501. [CrossRef]
13. Xiao, X.-Y.; Yang, R.-H.; Zheng, Z.-X.; Wang, Y. Cooperative Rotor-Side SMES and Transient Control for Improving the LVRT Capability of Grid-Connected DFIG-Based Wind Farm. *IEEE Trans. Appl. Supercond.* **2019**, *29*, 1–5. [CrossRef]
14. Ahmed, S.D.; Al-Ismail, F.S.M.; Shafiullah, M.; Al-Sulaiman, F.A.; El-Amin, I.M. Grid Integration Challenges of Wind Energy: A Review. *IEEE Access* **2020**, *8*, 10857–10878. [CrossRef]
15. Shafiullah, G.M.; Amanuallh, M.T.O.; Shawkat Ali, A.B.M.; Wolfs, P. Potential Challenges of Integrating Large-Scale Wind Energy into the Power Grid–A Review. *Renew. Sustain. Energy Rev.* **2013**, *20*, 306–321. [CrossRef]
16. Wang, L.; Li, H.-W.; Wu, C.-T. Stability Analysis of an Integrated Offshore Wind and Seashore Wave Farm Fed to a Power Grid Using a Unified Power Flow Controller. *IEEE Trans. Power Syst.* **2013**, *28*, 2211–2221. [CrossRef]
17. Zha, X.; Liao, S.; Huang, M.; Yang, Z.; Sun, J. Dynamic Aggregation Modeling of Grid-Connected Inverters Using Hamilton's-Action-Based Coherent Equivalence. *IEEE Trans. Ind. Electron.* **2019**, *66*, 6437–6448. [CrossRef]
18. Chowdhury, M.A.; Shen, W.X.; Hosseinzadeh, N.; Pota, H.R. A Novel Aggregated DFIG Wind Farm Model Using Mechanical Torque Compensating Factor. *Energy Convers. Manag.* **2013**, *67*, 265–274. [CrossRef]
19. Dhoot, A.; Antonini, E.G.A.; Romero, D.A.; Amon, C.H. Optimizing Wind Farms Layouts for Maximum Energy Production Using Probabilistic Inference: Benchmarking Reveals Superior Computational Efficiency and Scalability. *Energy* **2021**, *223*, 120035. [CrossRef]
20. Wakasa, Y.; Yamasaki, S. Distributed Particle Swarm Optimization Based on Primal-Dual Decomposition Architectures. *Stoch. Syst. Theory Appl.* **2015**, *2015*, 97–101. [CrossRef]
21. Annoni, J.; Dall'Anese, E.; Hong, M.; Bay, C.J. Efficient Distributed Optimization of Wind Farms Using Proximal Primal-Dual Algorithms. In Proceedings of the 2019 American Control Conference (ACC), Philadelphia, PA, USA, 10–12 July 2019; pp. 4173–4178.
22. Katic, I.; Højstrup, J.; Jensen, N.O. A Simple Model for Cluster Efficiency. In *EWEC'86. Proceedings*; Palz, W., Sesto, E., Eds.; A. Raguzzi: Rome, Italy, 1987; Volume 1, pp. 407–410.

23. Sizhuang, L.; Youtong, F. Analysis of the Jensen's Model, the Frandsen's Model and Their Gaussian Variations. In Proceedings of the 2014 17th International Conference on Electrical Machines and Systems (ICEMS), Hangzhou, China, 22–25 October 2014; pp. 3213–3219.
24. Kuo, J.Y.J.; Romero, D.A.; Beck, J.C.; Amon, C.H. Wind Farm Layout Optimization on Complex Terrains—Integrating a CFD Wake Model with Mixed-Integer Programming. *Appl. Energy* **2016**, *178*, 404–414. [CrossRef]
25. Bastankhah, M.; Porté-Agel, F. Experimental and Theoretical Study of Wind Turbine Wakes in Yawed Conditions. *J. Fluid Mech.* **2016**, *806*, 506–541. [CrossRef]
26. Gebraad, P.M.O.; van Dam, F.C.; van Wingerden, J.-W. A Model-Free Distributed Approach for Wind Plant Control. In Proceedings of the 2013 American Control Conference, Washington, DC, USA, 17–19 June 2013; pp. 628–633.
27. Deljouyi, N.; Nobakhti, A.; Abdolahi, A. Wind Farm Power Output Optimization Using Cooperative Control Methods. *Wind Energy* **2021**, *24*, 502–514. [CrossRef]
28. Ni, Y.; Li, C.; Du, Z.; Zhang, G. Model order reduction based dynamic equivalence of a wind farm. *Int. J. Electr. Power Energy Syst.* **2016**, *83*, 96–103. [CrossRef]
29. Fang, R.; Shang, R.; Wu, M.; Peng, C.; Guo, X. Application of Gray Relational Analysis to K-Means Clustering for Dynamic Equivalent Modeling of Wind Farm. *Int. J. Hydrogen Energy* **2017**, *42*, 20154–20163. [CrossRef]
30. Shu, T.; Song, D.; Hoon Joo, Y. Decentralised Optimisation for Large Offshore Wind Farms Using a Sparsified Wake Directed Graph. *Appl. Energy* **2022**, *306*, 117986. [CrossRef]
31. Charikar, M.; Guha, S.; Tardos, É.; Shmoys, D.B. A Constant-Factor Approximation Algorithm for the k-Median problem. *J. Comput. Syst. Sci.* **2002**, *65*, 129–149. [CrossRef]
32. Quiroga-Novoa, P.; Cuevas-Figueroa, G.; Preciado, J.L.; Floors, R.; Peña, A.; Probst, O. Towards Better Wind Resource Modeling in Complex Terrain: A k-Nearest Neighbors Approach. *Energies* **2021**, *14*, 4364. [CrossRef]
33. Ali, M.; Ilie, I.-S.; Milanovic, J.V.; Chicco, G. Wind Farm Model Aggregation Using Probabilistic Clustering. *IEEE Trans. Power Syst.* **2013**, *28*, 309–316. [CrossRef]
34. Ma, S.; Geng, H.; Yang, G.; Pal, B.C. Clustering-Based Coordinated Control of Large-Scale Wind Farm for Power System Frequency Support. *IEEE Trans. Sustain. Energy* **2018**, *9*, 1555–1564. [CrossRef]
35. Zhang, B.; Liu, J. Wind Turbine Clustering Algorithm of Large Offshore Wind Farms Considering Wake Effects. *Math. Probl. Eng.* **2019**, *2019*, 6874693. [CrossRef]
36. Marden, J.R.; Ruben, S.D.; Pao, L.Y. A Model-Free Approach to Wind Farm Control Using Game Theoretic Methods. *IEEE Trans. Control Syst. Technol.* **2013**, *21*, 1207–1214. [CrossRef]
37. Park, J.; Law, K.H. Bayesian Ascent: A Data-Driven Optimization Scheme for Real-Time Control With Application to Wind Farm Power Maximization. *IEEE Trans. Contr. Syst. Technol.* **2016**, *24*, 1655–1668. [CrossRef]
38. Thomas, J.J.; Annoni, J.; Fleming, P.A.; Ning, A. Comparison of Wind Farm Layout Optimization Results Using a Simple Wake Model and Gradient-Based Optimization to Large Eddy Simulations. In *Proceedings of the AIAA Scitech 2019 Forum*; American Institute of Aeronautics and Astronautics: San Diego, CA, USA, 2019.
39. Gebraad, P.M.O.; van Wingerden, J.W. Maximum Power-Point Tracking Control for Wind Farms. *Wind Energy* **2015**, *18*, 429–447. [CrossRef]
40. Lee, J.; Son, E.; Hwang, B.; Lee, S. Blade Pitch Angle Control for Aerodynamic Performance Optimization of a Wind Farm. *Renew. Energy* **2013**, *54*, 124–130. [CrossRef]
41. Li, W.; Özcan, E.; John, R. Multi-Objective Evolutionary Algorithms and Hyper-Heuristics for Wind Farm Layout Optimisation. *Renew. Energy* **2017**, *105*, 473–482. [CrossRef]
42. Tumari, M.Z.M.; Suid, M.H.; Ahmad, M.A. A Modified Grey Wolf Optimizer for Improving Wind Plant Energy Production. *Indones. J. Electical Eng. Comput. Sci.* **2020**, *18*, 1123. [CrossRef]
43. Dongran, S.; Liu, J.; Yang, Y.; Yang, J.; Su, M.; Wang, Y.; Gui, N.; Xuebing, Y.; Huang, L.; Joo, Y.H. Maximum Wind Energy Extraction of Large-Scale Wind Turbines Using Nonlinear Model Predictive Control via Yin-Yang Grey Wolf Optimization Algorithm. *Energy* **2021**, *221*, 119866. [CrossRef]
44. Tian, J.; Su, C.; Soltani, M.; Chen, Z. Active Power Dispatch Method for a Wind Farm Central Controller Considering Wake Effect. In Proceedings of the IECON 2014—40th Annual Conference of the IEEE Industrial Electronics Society, Dallas, TX, USA, 29 October–1 November 2014; pp. 5450–5456.
45. Song, D.; Fan, X.; Yang, J.; Liu, A.; Chen, S.; Joo, Y.H. Power Extraction Efficiency Optimization of Horizontal-Axis Wind Turbines through Optimizing Control Parameters of Yaw Control Systems Using an Intelligent Method. *Appl. Energy* **2018**, *224*, 267–279. [CrossRef]
46. Rezaei, N.; Uddin, M.N.; Amin, I.K.; Othman, M.L.; Marsadek, M. Genetic Algorithm-Based Optimization of Overcurrent Relay Coordination for Improved Protection of DFIG Operated Wind Farms. *IEEE Trans. Ind. Applicat.* **2019**, *55*, 5727–5736. [CrossRef]
47. Jiang, X.; Li, S. BAS: Beetle Antennae Search Algorithm for Optimization Problems. *arXiv* **2017**, arXiv:1710.10724. [CrossRef]
48. Jonkman, J.; Butterfield, S.; Musial, W.; Scott, G. *Definition of a 5-MW Reference Wind Turbine for Offshore System Development*; NREL/TP-500-38060; National Renewable Energy Laboratory: Golden, CO, USA, 2009; p. 947422.
49. Siniscalchi Minna, S. Advanced Wind Farm Control Strategies for Enhancing Grid Support. Ph.D. Thesis, Polytechnic University of Catalonia, Barcelona, Spain, 2019.

50. King, R.; Hamlington, P.; Dykes, K.; Graf, P. Adjoint Optimization of Wind Farm Layouts for Systems Engineering Analysis. In Proceedings of the 34th Wind Energy Symposium, San Diego, CA, USA, 4 January 2016.
51. King, R.N.; Dykes, K.; Graf, P.; Hamlington, P.E. Optimization of Wind Plant Layouts Using an Adjoint Approach. *Wind Energy Sci.* **2017**, *2*, 115–131. [CrossRef]

energies

MDPI

Article

Home Energy Management Considering Renewable Resources, Energy Storage, and an Electric Vehicle as a Backup

Oussama Ouramdane, Elhoussin Elbouchikhi *, Yassine Amirat, Franck Le Gall and Ehsan Sedgh Gooya

ISEN Yncréa Ouest Brest, LABISEN, 20, Rue Cuirassé Bretagne, 29200 Brest, France; oussama.ouramdane@isen-ouest.yncrea.fr (O.O.); yassine.amirat@isen-ouest.yncrea.fr (Y.A.); franck.le-gall@isen-ouest.yncrea.fr (F.L.G.); ehsan.sedgh-gooya@isen-ouest.yncrea.fr (E.S.G.)
* Correspondence: elhoussin.elbouchikhi@isen-ouest.yncrea.fr; Tel.: +33-06-44-25-24-08

Abstract: The vehicle-to-grid concept emerged very quickly after the integration of renewable energy resources because of their intermittency and to support the grid during on-peak periods, consequently preventing congestion and any subsequent grid instability. Renewable energies offer a large source of clean energy, but they are not controllable, as they depend on weather conditions. This problem is solved by adding energy storage elements, implementing a demand response through shiftable loads, and the vehicle-to-grid/vehicle-to-home technologies. Indeed, an electric vehicle is equipped with a high-capacity battery, which can be used to store a certain amount of energy and give it back again later when required to fulfill the electricity demand and prevent an energy shortage when the main-grid power is limited for security reasons. In this context, this paper presents a comparative study between two home microgrids, in one of which the concept of vehicle-to-home is integrated to provide a case study to demonstrate the interest of this technology at the home level. The considered microgrid is composed of renewable energy resources, battery energy storage, and is connected to the main grid. As the vehicle is not available all day, in order to have consistent results, its intervention is considered in the evening, night, and early morning hours. Two case studies are carried out. In the first one, the vehicle-to-home concept is not taken into account. In this case, the system depends only on renewable resources and the energy storage system. Subsequently, the electric vehicle is considered as an additional energy storage device over a few hours. Electric vehicle integration brings an economic contribution by reducing the cost, supporting the other MG components, and relieving the main grid. Simulation results using real weather data for two cities in France, namely Brest and Toulon, show the effectiveness of the vehicle-to-home concept in terms of cost, energy self-sufficiency, and continuity of electrical service.

Keywords: vehicle-to-grid; vehicle-to-home; microgrids; distributed energy generation; energy storage; optimal sizing; energy management systems; optimization; energy dispatching

Citation: Ouramdane, O.; Elbouchikhi, E.; Amirat, Y.; Le Gall, F.; Sedgh Gooya, E. Home Energy Management Considering Renewable Resources, Energy Storage, and an Electric Vehicle as a Backup. *Energies* **2022**, *15*, 2830. https://doi.org/10.3390/en15082830

Academic Editor: Angelo Zarrella

Received: 14 February 2022
Accepted: 4 April 2022
Published: 13 April 2022

1. Introduction

Vehicle-to-grid (*V2G*) could be considered as the biggest technological advancement since renewable energy resources (RER) became commercially viable [1]. With more than 3 million EVs worldwide [2], these EVs are used only about 5% of the time; the rest of the time they are parked when the owners are working or at home [3]. EVs can be used for a secondary role during the 95% of available time. Indeed, electric vehicles (EV) with a large battery can recover energy produced by RER, which would otherwise be lost. In this case, the EV functions as a secondary storage system. In the opposite case, where production is low, the storage systems and the main grid intervene by sending the missing energy, to achieve the production consumption balance. This demand for energy can overload the grid and cause blackouts or voltage and frequency drops. To relieve the grid, the energy it supplies can be limited to a certain amount, the rest being supplied by the EV as it discharges. In some cases, for same precedents reasons, it can be possible to sell the energy stored in the EV to the grid. In this case, the EV functions as a secondary energy source; hence, the concept of *V2G* [4,5].

The microgrid (MG) concept has emerged for the optimal integration of renewable energy resources, energy storage systems (ESS), and shedding loads, which may include EVs. The integration of EVs must not change the roles of the MG elements, both in grid-connected or islanded modes, i.e., to supply a given load at a given time. RER must be exploited to the maximum, by implementing maximum power point tracking (MPPT) and excess energy is stored in ESS. In the *V2G* concept, EVs represent a secondary production and storage system in the MG. One of the advantages of bi-directional energy in the *V2G* concept is the exchange of energy between EVs [6]. However, it is mandatory to consider the degradation of the EVs batteries while implementing *V2G* technology. Optimal MG sizing is one of the studies to be conducted for *V2G* concept integration. It involves many aspects that need to be addressed, including the appropriate RER technologies with the optimal siting and the environment considered. Moreover, other considerations should be taken into account, such as improving conversion efficiency at charging stations, transmission efficiency and loss minimization, optimal control of operation, the need for adequate loads that better match the generation elements, and, finally, the type, capacity, and number of EVs needed for safe and reliable operation [7]. As a result, interest in EV technologies has increased considerably in recent years, resulting in several scientific publications on the subject [8,9]. Moreover, the market share of EVs is expected to grow exponentially. Most of the research deals with very important topics such as energy efficiency [10], EV charger topologies and the impact of charging on the network [11], communication architecture [12], degradation of the vehicle battery [13], frequency regulation [14], and the support provided to the integration of RERs [15], mode of payment [16], etc.

Several technical aspects have been investigated in the literature. Indeed, an EV can be considered as a storage medium for RERs depending on the availability and states of charge of the EVs. In [17], Hasan and Elyas aim to optimise the operation of EVs in order to mitigate the intermittency of RERs and to reduce energy costs by minimising the charge/discharge cycles of vehicle batteries to avoid their degradation. In [18], Abdul et al. proposed a new formulation for the design and management of grid-connected EV charging stations with integrated solar panels. In [19], Mehdi et al. examined the value of the *V2H* concept at the building level, and whether it can operate in an islanded mode by introducing PV as an energy source to relieve the grid. One of the cited advantages of *V2H* is the support it provides to the grid. In fact, an EV parked at home during peak hours can power appliances. In [20], Philipp et al. assessed the ability of the expanded German transmission network to cope with the additional demand due to the integration of electric vehicles using a transmission problem formulation. In [21], authors presented an EMS of an energy system composed of main grid, load, and a fleet of EVs. The scheduling of EV charging/discharging can be done 24 h in advance, which allows for voltage and frequency regulation in the grid. The EV can sometimes be used as a back-up when the house is disconnected from the grid or when a blackout has occurred. Critical loads will be connected directly to the vehicle, to ensure the continuity of supply for the user's comfort. In [22], Vítor et al. designed a *V2H*, such that the EV intervenes during power outages. In [23], Hasan presented a model that significantly improves the energy resilience of a building based on PV, a battery for storage, and an EV for *V2H* operating in connected and islanded mode over a few hours. The coordination of the EV and the battery allows one to respond to the load after failures in the grid and PV. Furthermore, an EV can be added to increase battery lifetime. Since the vehicle's storage capacity is greater than the battery's, the vehicle will experience fewer charge/discharge cycles, and, therefore, reduce the charge/discharge cycles of the battery without degrading its battery too much [24]. Finally, an EV can allow for an efficient management of load shedding [25].

The economic perspective has been investigated in some studies. In [26], Sangyoon and Dae-Hyun presented a hierarchical deep reinforcement learning method for energy management of household appliances considering RER, ESS, and EVs. They consider four study cases: two during the week with different weather data, and two during the weekend, with different vehicle states of charge. They find that the energy cost decreases significantly in the cases where the RERs are productive. The presence of the vehicle at home during the weekend is very cost effective due to its high storage capacity. In [27], Harun and Seddik developed seven optimal strategies studied on 1000 use cases and with four different daily energy price profiles in order to reduce the vehicle charging cost. A $V2G$ algorithm named Optimal Logic Control based on series of logic commands has been developed. The results show the advantage of this algorithm over the traditional optimal charging strategies available in the literature. The algorithm is mainly made to be very efficient when the selling price/purchase ratio is greater than one, allowing the user to earn money. In [28], Sausen et al. proposed a model for coupling an EV to a smart home containing PV and batteries, having the objective of achieving cost minimization that was conditional on maximizing vehicle efficiency. Four scenarios are considered: in the first one, uncontrolled EV charging was adopted; in the second one, economic charging was applied; in the third one, economic EV charging was adopted, including distributed generation; and in the fourth one, economic EV charging was considered, with distributed generation and $V2H$ operation. The results show that the peak consumption has been reduced thanks to the intervention of the EV and the bill has been decreased. In [29], Xuan et al. proposed a strategy to extend the lifetime of the batteries, such that the batteries operate when there is a surplus of RER production and when electricity prices increase during peak hours. The results show the effectiveness of this strategy, as when the battery discharge cost is higher than the market energy price, no power is supplied by the batteries, and conversely, when the discharge cost is lower than the market energy price, the power demanded is first supplied by the batteries, then, if the load is very high, the grid is allowed to supply the remaining load.

There is a lot of literature on $V2G$, but the $V2H$ concept is not thoroughly addressed, as most research studies focus on the interaction of electric vehicles with the electric grid. Additional research is needed to address the cost of implementing the $V2H$ concept and its profitability. In this context, this paper presents a sizing and optimization study of a MG operating in a grid-connected mode at the home level. The MG is composed of PVs and WT as RER, a battery as ESS, and a domestic load. The EV is used as a secondary energy storage device, which allows for the enhancement of the overall MG performance. This study concerns two aspects that are related to MG optimal sizing and smart day-ahead energy management. Two case studies are considered in this paper: a first grid-connected MG with only RERS and ESS is investigated; next, a second grid-connected MG with the integration of RERs, ESS, and an EV is implemented and the results are compared. An economic study is performed to investigate the advantages offered by the use of the EV and the interest of $V2H$ technology in terms of energy supply reliability, cost reduction, and grid support during congestion periods. Simulation results are presented using actual weather data from two cities, which are Brest (cold climate) and Toulon (hot climate) in France. Generally speaking, the contributions targeted in this work are threefold:

- Optimal sizing of a home MG considering actual weather data, energy price, EV batteries as a secondary ESS and limiting the grid power for enhancing the energy self-consumption;
- Energy management system taking into account the availability of the EV;
- Investigate the economic profitability of $V2H$ for the users by considering the electric vehicle as backup in order to mitigate the grid congestion.

This article is organised as follows: Section 2 presents the proposed MG architecture, presenting the used renewable energy resources and energy storage devices. Section 3 describes the MG optimal sizing and energy management considering two aspects, which are optimal energy dispatching and economic viability. Section 4 discusses the simulation results for the two considered case studies for two different cities in France. Finally, Section 5 concludes this paper and gives some perspectives for future work.

2. Microgrid Architecture

A home microgrid, which implements the *V2H* concept, is designed for the relief of main grid constraints. The MG is operated in grid-connected mode and consists of PV, WT, battery energy storage, and an EV. The architecture of the MG is presented in Figure 1. All components are linked to the DC bus through *DC*/*DC* converters (PV, batteries, and EV) or *AC*/*DC* converters (WT and grid). The PV and WT produce a quantity of energy depending on weather conditions, which does not always match the load demand. There are two possible scenarios: energy production is insufficient or, conversely, there is surplus energy generation. Energy storage and the main grid allow for the buffering of insufficient or excess power generation. Indeed, the battery and grid can restore the production/consumption balance by supplying the lack of energy or, conversely, by absorbing excess energy. Since the energy absorbed/fed by the battery is limited by its SOC and its maximum power, and since grid power can also limited, an EV can be used as an additional backup energy storage. Indeed, an EV can be charged through public charging stations or when parked at home. In order to support the grid and ensure the continuity of electricity supply, it can have a bidirectional energy exchange with the home MG.

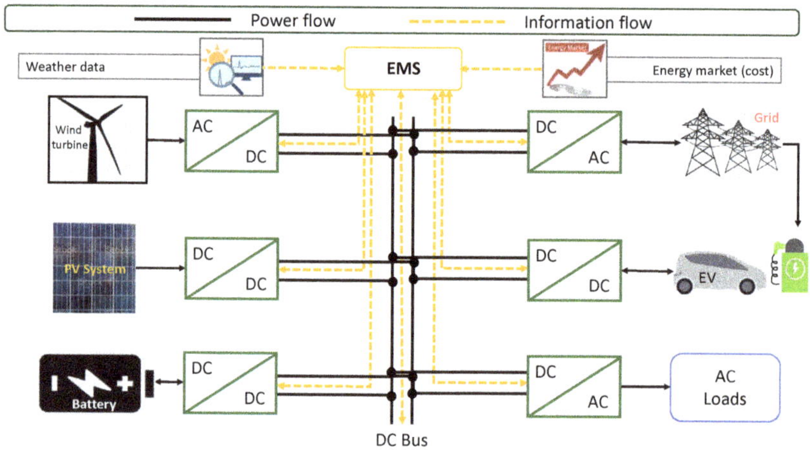

Figure 1. Structure of the home MG considering an EV as backup energy storage.

2.1. Home MG Components

The choice of MG components is based on the maturity of the distributed energy resources (DER) and the efficiency of the ESS. Currently, the most mature RER is PV, which is the most mastered and used by researchers in projects aiming to achieve autonomous systems, relieve the grid, or produce clean energy. The PV used in this project is monocrystalline panels, which are suitable for cold climates, while polycrystalline panels are more suitable for warm climates [30]. A domestic wind turbine is the second mature renewable technology, known as an individual wind turbine [30]. WT is mainly installed at small-scale residential customers. It can also be used to power larger loads, such as small businesses and agricultural fields. The interest of this installation is to satisfy a part of the load, reducing the dependence on the power grid. Two classes of these wind turbines can be distinguished: vertical-axis wind turbines and horizontal-axis wind turbines. In the considered MG, the installed wind turbine is a horizontal-axis configuration with three blades, a drive train consisting of a gearbox and a generator, and a tower to support the rotor.

The considered battery in this study is a lithium-ion battery. It is lighter than a lead-acid battery (five times lighter), has a long lifespan (1500 cycles), and is more resistant to deep discharges. New trends in the use of batteries for energy systems focus on integration with multiple RERs (PV, WT, etc.) and also on the integration with other energy storage systems that complement them [31]. This trend fits perfectly with the proposed MG, which includes PV and WT as well as an EV as an additional energy storage device.

The installed PV and WT maximum power and battery capacity are unknown and are computed by the optimisation program based on the load data, weather data, grid energy cost, renewables and battery investment cost, battery power limits, grid power limits, and battery capacity of the EV. Subsequently, an optimal energy dispatching is computed, which determines the power flows between all components within the battery power and capacity limits, grid limits, RER availability, and EV displacement, considering the economic aspects.

2.2. Modeling of the Renewable Resources and Energy Storage Devices

PV panels and wind turbines are considered as energy resources in this study. The following subsections describe these renewable resources modeling for optimal sizing and smart energy management.

2.2.1. PV Panels

The power generated by the PV panels depends on the installed maximum power and weather conditions. The PV output power depends on the irradiance, the efficiency of the generation, the area of the panels and the optimal orientation depending on the location. The chosen PV technology has an efficiency $\eta_{pv} = 15\%$. The power produced by PV over a day is given as follows:

$$P_{PV}(t) = \eta_{PV} \cdot S_{PV} \cdot P_{Irr}(t); \quad 1 < t < 24\ \text{h} \tag{1}$$

where:

- S_{PV} is the PV panels area in m^2;
- $P_{Irr}(t)$ corresponds to the irradiance in the considered location W/m^2;
- η_{PV} stand for the chosen PV technology efficiency.

2.2.2. Wind Turbine

A wind turbine has an output power that mainly depends on its radius and the wind speed in the considered area. The other variables are constants, such as air density (ρ = 1.225 kg/m^3), or can be considered as constants by setting them to a value given by the control algorithm by implementing maximum power point algorithm (MPPT), such as the power coefficient ($C_p = 0.4$). The electrical power extracted by a wind turbine is as follows:

$$P_{WT}(t) = \frac{1}{2} \cdot \rho \cdot C_p \cdot \pi \cdot R_{WT}^2 \cdot V_{Wind}(t)^3; \quad 1 < t < 24\ \text{h} \tag{2}$$

where:

- R_{WT} is the wind turbine radius in m;
- V_{Wind} corresponds to wind velocity, in m/s.

2.2.3. Energy Storage Devices: Battery and EV

The energy supplied by the battery and the EV depend on their capacity and their state of charge (SOC). Moreover, the output power is limited to enhance the lifespan of energy storage devices. The SOC represents the ratio between the energy contained in the battery (respectively, EV battery) and its maximum capacity (respectively, EV battery capacity). The SOC of the battery (respectively, EV battery) depends on the power supplied or recovered by the battery (respectively, EV battery), the time of application of this power (δt = 1 h), and its maximum capacity (respectively, EV battery capacity). The SOC varies from one hour to another and can increase during charging and decrease during discharging. The SOC for both the battery and the EV battery is given by the following formula:

$$SOC_{bat}(t) = SOC_{bat}(t-1) - \frac{P_{bat}(t) \cdot dt}{E_{bat}^{max}}; \quad 1 < t < 24\text{ h} \quad (3)$$

$$SOC_{EV}(t) = SOC_{EV}(t-1) - \frac{P_{EV}(t) \cdot dt}{E_{EV}^{max}}; \quad 1 < t < 24\text{ h} \quad (4)$$

where:

- $SOC_{bat}(t)$ and $SOC_{EV}(t)$ correspond to battery and EV battery state of charge at time t;
- $P_{bat}(t)$ and $P_{EV}(t)$ stand for power of battery and EV battery. This power can be either positive or negative depending on the operation conditions;
- E_{bat}^{max} and E_{EV}^{max} are the maximum battery and EV energy capacity.

3. Optimal Sizing and Planning of a Home MG with *V2H* Technology

A domestic MG with *V2H* concept optimization is carried out in two steps, which are the optimal sizing and the energy management system (EMS), as shown in Figure 2. MG sizing is performed based on the weather data, the market energy cost, and the investment cost, all while fulfilling the technical constraints. Next, an optimal energy management algorithm is proposed considering that the home MG is optimally designed. In this section, the objective of the proposed energy management system is presented, which is mainly related to the sizing and the minimisation of the energy cost over 20 years of operation of the MG while integrating the *V2H* technology.

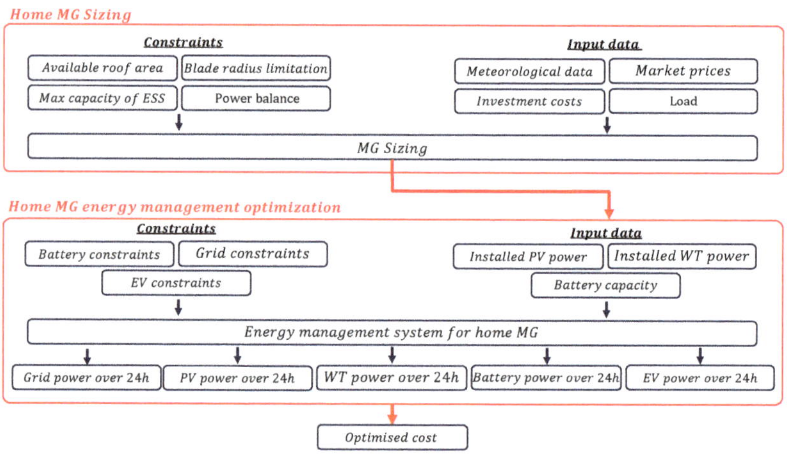

Figure 2. Scheme describing the proposed MG sizing and energy management system.

3.1. Energy Management Strategy

The objective in this work is to achieve a *V2H* system with maximum autonomy and minimum energy cost. Figure 3 shows the scheme of the proposed energy management algorithm. Five different layers can be distinguished in this energy management strategy. Indeed, in order to ensure maximum autonomy and favour the use of renewable energy, the load has to be fed by RERs at any time. On-maximum power point tracking (On-MPPT) mode is prioritized. Depending on the value of net power $P_{net} = P_{Load} - (P_{PV} + P_{WT})$, there are three possible scenarios as follows:

- $P_{net} = 0$: Renewables fully satisfy the load and there is no need for energy from battery, power grid, and EV;
- $P_{net} > 0$: In this case, the battery, the grid, and the EV are required the complement renewables underproduction. The use of each of these resources depends on batteries state of charge, power limitations, and EV availability;

- $P_{net} < 0$: Battery free capacity is used to store the overproduction. If the battery is fully charged or the required power is higher than the maximum power, the excess power is supplied to the grid. However, if the EV is available or the grid power is not within its power limitations, the EV battery is used as a backup energy supply.

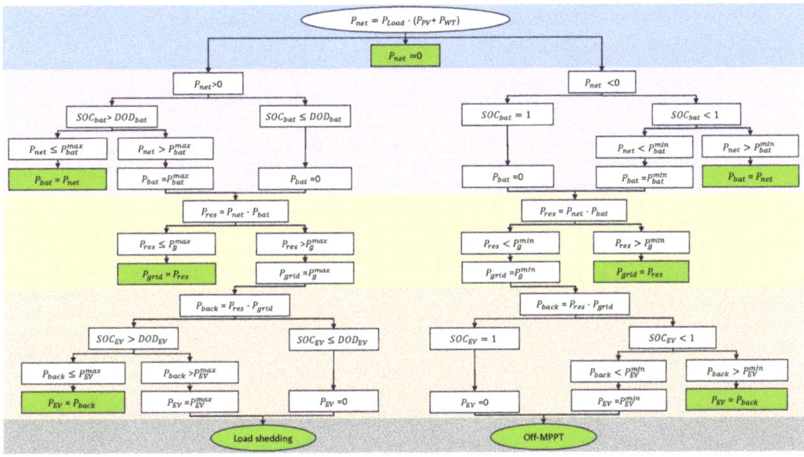

Figure 3. Energy management strategy (P_{net} is the net power, P_{res} stands for residual power, and P_{back} corresponds to backup power).

In the renewables underproduction situation, the first element that will intervene is the battery. This step is shown in purple on the algorithm diagram. On one hand, if the battery is sufficiently charged and the power demand does not exceed the maximum discharge power specified by the manufacturer, the battery provides the power required to ensure the generation/consumption power balance. On the other hand, if the discharge power exceeds the maximum acceptable battery discharge power, battery provides its maximum power while the remaining power is provided by the grid, as shown by the yellow stage in the EMS scheme. Even though the grid can be considered as a large energy resource, its power can be limited for security and reliability reasons (to prevent overload, for example). If the power demand does not exceed grid power limitations, the load is supplied by the grid. Otherwise, the EV is used as a backup to ensure the electricity continuity of service, as shown in the pink section of the algorithm. A final situation can occur for which the EV maximum power is exceeded or the EV state of charge does not allow the use of the EV battery. In this case, a load shedding is required as depicted by the grey part of the EMS scheme.

Conversely, in an overproduction situation, the battery is first used to store the excess energy generation, as shown by the purple section of the algorithm. If the battery is fully charged or the required power exceeds the maximum charge power given by the manufacturer, the main grid intervenes to guarantee the generation/consumption balance (yellow part of the diagram). As previously specified, the grid power is limited to enhance the system reliability and improve autonomy, which is the reason why the EV is used to store a part of the produced energy if the grid maximum power is achieved, as described in the pink part of the energy management strategy. Finally, if the EV is not available or the EV battery is fully charged, renewables are operated in off-MPPT mode in order to reduce energy production (grey part).

3.2. Objective Function

The objective of this work is to design an optimal home MG, which minimizes the investment and energy costs while ensuring the integration of $V2H$ technology. This is done by maximizing home autonomy and reducing the energy issued from the grid. The optimization is performed over 20 years, taking into account weather data, grid energy price, and investment costs. The investment cost includes all equipment, namely PV panels, a wind turbine, batteries, power electronics, and cables, purchase and installation, and the operating and maintenances costs. The energy cost, however, mainly consists of the energy bills from the grid operator. The objective function is to ensure the optimal operation of the microgrid by minimizing its operating cost. Indeed, the area of the PV panels and the wind turbine power should be minimized to satisfy the load while minimizing the energy cost. Moreover, battery installed capacity should be optimized by taking into consideration its degradation cost. Consequently, the cost function to be optimized can be expressed as follows:

$$Cost = C_{inv} + C_{ene} \tag{5}$$

where:

- C_{inv} corresponds to the investment cost, which is given by:

$$C_{inv} = C_{PV} \cdot P_{PV}^{ins} + C_{WT} \cdot P_{WT}^{ins} + C_{Bat} \cdot E_{bat}^{max} \tag{6}$$

  with:

  - C_{PV}, C_{WT}, and C_{Bat} are PV panels cost in (€/W), WT cost in (€/W) and battery cost in (€/J), respectively;
  - P_{PV}^{ins}, P_{WT}^{ins}, and E_{bat}^{max} are the installed PV power, wind power, and battery capacity, respectively.

- C_{ene} stand for the energy cost of the power provided by the main grid. It can be expressed as follows:

$$C_{ene} = C_g \cdot \sum_{i=1}^{20\text{ years}} \left(\sum_{t=1}^{24\text{ h}} (P_g^i(t))\right) \tag{7}$$

  with:

  - C_g corresponds to the grid energy cost, which is variable during the day: On peak periods have higher price than off-peak periods;
  - $P_g^i(t)$ is the power fed or supplied by the power grid, which is minimized in this study.

3.3. System Constraints

3.3.1. Generation/Consumption Balance and Power Limitations

Minimizing the previously presented cost function is performed while respecting constraints related to each component of the home MG. The production/consumption balance is the main constraint of the system. To ensure a continuous and reliable power supply, the energy supplied by the RERs, the batteries, the grid, and the EV must be equal to the power required by the load at any time. This constraint is provided in Equation (8). Energy conversion efficiency is considered to account for the losses that occur when charging or discharging batteries or the EV battery. To express these losses in the power balance, the charging power is divided by the charging energy conversion efficiency ($\eta_{bat}^{cha} = \eta_{EV}^{cha} = 0.9$) and the discharging power is multiplied by the discharging energy conversion efficiency ($\eta_{bat}^{disch} = \eta_{EV}^{disch} = 0.9$). Equation (9) represents the new power balance constraint. Where $P_{bat}^{+}(t)$, $P_g^{+}(t)$, and $P_{EV}^{+}(t)$ correspond to power supplied to the load and $P_{bat}^{-}(t)$, $P_g^{-}(t)$, and $P_{EV}^{-}(t)$ stand for power fed to battery, power grid, and EV, respectively. Equations (10)–(12) describe the fact that these energy resources cannot charge and discharge at the same time. Moreover, to ensure that the battery, main grid, and EV do not exchange energy with each other, the constraint in Equation (13) is added. This is mainly set to minimize losses introduced by the energy conversion between these elements.

Equations (14) and (15) are set to ensure that the battery and EV battery recover its initial state of charge at the end of each day. To ensure this, the battery and EV provide a quantity of energy during underproduction periods, which is recovered during overproduction periods. Since the EV is not available all the day, it can be used only for a certain period of the day as specified by Equation (16). Finally, the power supplied (absorbed) by the battery is limited because of the characteristics of the conductors and switches, as shown by Equation (17). In addition of that, grid power should be limited to increase autonomy and avoid overloading. Its power is limited between minimum and maximum values as depicted by Equation (18).

$$P_{Load}(t) = P_{PV}(t) + P_{WT}(t) + P_{bat}(t) + P_g(t) + P_{EV}(t) \quad ; \quad 1 < t < 24\text{ h} \tag{8}$$

$$\begin{aligned} P_{Load}(t) = P_{PV}(t) + P_{WT}(t) + \eta_{bat}^{disch} \cdot P_{bat}^{+}(t) - \frac{1}{\eta_{bat}^{cha}} \cdot P_{bat}^{-}(t) + P_g^{+}(t) - \\ P_g^{-}(t) + \eta_{EV}^{disch} \cdot P_{EV}^{+}(t) - \frac{1}{\eta_{EV}^{cha}} \cdot P_{EV}^{-}(t) \quad ; \quad 1 < t < 24\text{ h} \end{aligned} \tag{9}$$

$$P_{bat}^{+}(t) \cdot P_{bat}^{-}(t) = 0 \quad ; \quad 1 < t < 24\text{ h} \tag{10}$$

$$P_g^{+}(t) \cdot P_g^{-}(t) = 0 \quad ; \quad 1 < t < 24\text{ h} \tag{11}$$

$$P_{EV}^{+}(t) \cdot P_{EV}^{-}(t) = 0 \quad ; \quad 1 < t < 24\text{ h} \tag{12}$$

$$|P_{EV}(t)| + |P_{bat}(t)| + |P_g(t)| = |P_{EV}(t) + P_{bat}(t) + P_g(t)| \; ; \; 1 < t < 24\text{ h} \tag{13}$$

$$\sum_{t=1}^{24\text{ h}} P_{bat}(t) = 0 \tag{14}$$

$$\sum_{t=1}^{24\text{ h}} P_{EV}(t) = 0 \tag{15}$$

$$P_{EV}^{+}(t) = P_{EV}^{-}(t) = 0 \quad ; \quad 8 < t < 18\text{ h} \tag{16}$$

$$P_{bat}^{min} < P_{bat}(t) < P_{bat}^{max} \quad ; \quad 1 < t < 24\text{ h} \tag{17}$$

$$P_g^{min} < P_g(t) < P_g^{max} \quad ; \quad 1 < t < 24\text{ h} \tag{18}$$

3.3.2. Renewable Resources and ESS Constraints

The installed PV power is determined by the optimisation algorithm. This power depends on the peak irradiation and is limited by the available area on the roof of the house. Equation (19) ensures that the area is within the tolerated limits. The installed wind power depends on the peak value of the wind speed. The radius of the turbine is also limited by a maximum value. This maximum value should not exceed the value tolerated for domestic installations in urban areas and it cannot be negative, which is expressed by the constraint in Equation (20).

$$0 < S_{PV} < S_{limit} \tag{19}$$

$$0 < R_{WT} < R_{limit} \tag{20}$$

To ensure maximum lifespan (maximum cycles) for the battery and the EV battery, minimum SOC should be limited. On one hand, to ensure a safe operation of batteries (respectively EV battery), the SOC is limited by an upper limit, equal to 1. On the other hand, batteries (respectively, EV battery) usually have a maximum DOD set by the designer, indicating that by respecting this DOD the battery (respectively, EV battery) will be able to be operated at the maximum of charge/discharge cycles. SOCs for both the battery ESS and EV battery are then limited by a lower limit equal to the DOD for the whole day. These constraints are presented in Equations (21) and (22). Maximum battery capacity to be installed is optimized and is consequently limited to a reasonable capacity, denoted C_{bat}^{max}, as expressed by Equation (23).

$$DOD_{bat} < SOC_{bat}(t) < 1; \quad 1 < t < 24\text{ h} \tag{21}$$

$$DOD_{EV} < SOC_{EV}(t) < 1; \quad 1 < t < 24\,\text{h} \tag{22}$$

$$0 < E_{bat}^{max} < C_{bat}^{max} \tag{23}$$

with:

- $SOC_{bat}(t)$ and $SOC_{EV}(t)$ are state of charge of battery ESS and EV battery;
- E_{bat}^{max} is the installed battery capacity that need to be optimized, which is limited by a maximum capacity C_{bat}^{max};
- DOD_{bat} and DOD_{EV} are depth of discharge of battery ESS anf EV battery.

4. Simulation Results

Numerical studies have been carried out to investigate the benefits of integrating EVs in terms of financial benefits, optimal sizing, and guaranteed energy autonomy by limiting power issued from main grid. From Equations (6) and (9), it is possible to distinguish decision variables, which must have the optimal values to minimize the total cost while respecting constraints; mainly the power balance constraint. The decision variables are:

- Grid power: $P_g^+(t)$ and $P_g^-(t)$;
- Charging and discharging power of the battery: $P_{bat}^+(t)$ and $P_{bat}^-(t)$;
- Battery capacity: E_{bat}^{max};
- PV and WT dimensions: S_{PV} and R_{WT};
- The evolution of renewable powers during a day: $P_{PV}(t)$ and $P_{WT}(t)$;
- Vehicle charging and discharging powers: $P_{EV}^+(t)$ and $P_{EV}^-(t)$.

The dealt with problem is a constrained non-linear optimization problem, which is described by Figure 4. The model is characterised by non-linear equality constraints and non-linear inequality constraints. In addition to these constraints, it is mandatory to define a set of lower and upper bounds on the design variables so that the solution is always in the specified range, such as the limits of the PV area and the radius of the wind turbine.

The sizing and energy management system is formulated as an optimization problem and is solved using the following method based on the Interior-point algorithm [32] using MATLAB software:

$$\min_x f(x), \text{s.t.:} \begin{cases} C(x) \leq 0 \\ C_{eq}(x) = 0 \\ A \times x \leq b \\ A_{eq} \times x = b_{eq} \\ l_b \leq x \leq u_b \end{cases} \tag{24}$$

With:

- x: Decision variables;
- $A \times x \leq b$: Linear inequality constraints;
- $A_{eq} \times x = b_{eq}$: Linear equality constraints;
- $C(x)$: Non-linear inequality constraints;
- $C_{eq}(x)$: Non-linear equality constraints.

This method fits the problem as it minimises a multivariate function; some variables are scalars and some are vectors. It allows one to compute the minimum of the objective function under non-linear equality and non-linear inequality constraints, with upper and lower bounds of the decision variables.

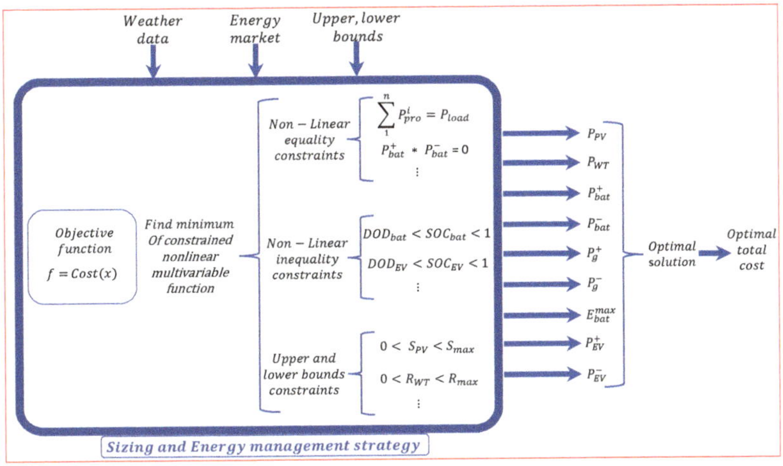

Figure 4. Scheme of the optimization problem.

In the following, two case studies are presented and the results are discussed as follows:

- Case study 1: Numerical studies for a grid-connected renewables-based MG without EV integration for Brest and Toulon;
- Case study 2: Numerical studies for a grid-connected renewables-based MG with EV integration for Brest and Toulon.

4.1. Optimization Input Data

The investment, maintenance, and operating costs of PV, WT, battery, and grid energy costs are taken from [33], considering a lifetime of 20 years for PV and wind turbine, and 5 years for batteries. These data are provided in Table 1. Variable energy consumption for 24 h is considered in this study. Indeed, the used load profile is equivalent to the energy consumption of a small house with variable consumed power. Figure 5b shows the hourly load pattern of 24 h. Three load peaks can be noticed in the provided load profile: in the morning, at noon, and in the evening. The same consumer profile is considered for the two case studies and for the two cities for a fair and reliable comparison.

The weather data used for home MG optimal sizing and energy management system design are actual weather data issued from [34,35]. Numerical studies have been performed for two different cities in France considering a cold and hot climate for the sake of comparison and to prove the effectiveness of the proposed approaches for different climates and weather conditions. First, a cold city, namely Brest, is considered, which is characterised by high speed wind during a significant period of the year and low sunshine with high solar irradiance variability during the year. Second, Toulon, which is a city in the south of France, is considered. Toulon is characterised by higher constant irradiance compared to Brest, but it has less wind due to its geographical location. Weather conditions of these two cities are different, which is of great interest to show differences in terms of MG sizing, autonomy of the installation, and the economical benefits. Since the designed MG is connected to the main grid, the worst case is not presented hereafter. Indeed, yearly mean values are considered for both PV panels and wind turbine optimal sizing. The irradiance and wind curves for the considered cities are shown in Figure 5b and Figure 5c, respectively.

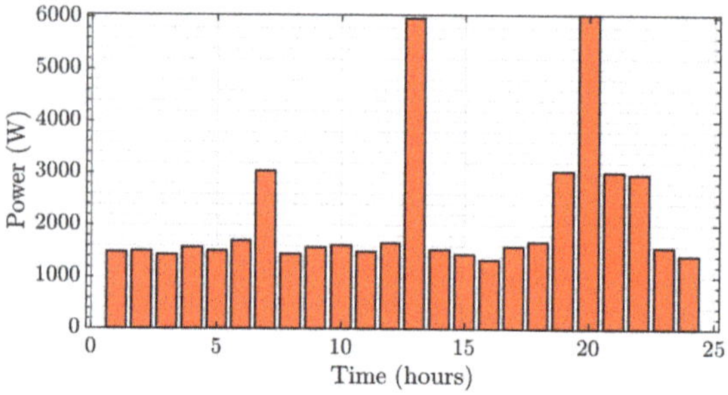

(**a**) Considered load power.

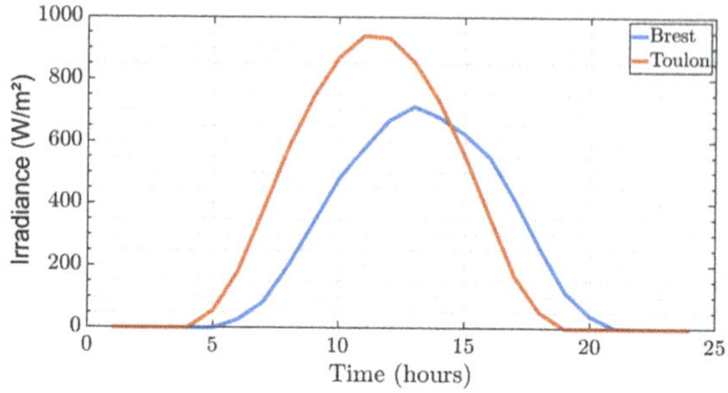

(**b**) Solar irradiance.

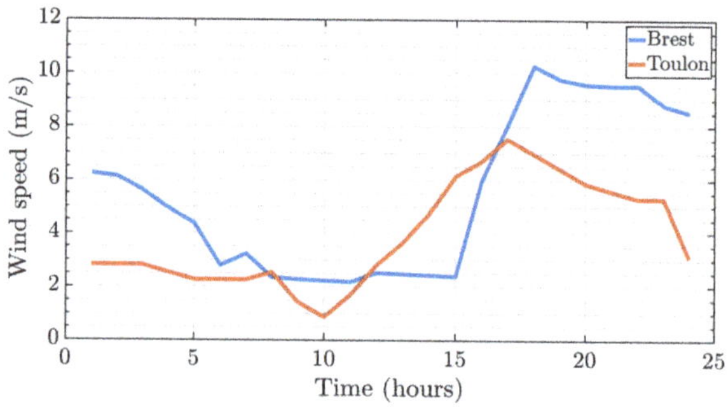

(**c**) Wind speed.

Figure 5. Load power and weather data for typical day in Brest and Toulon.

Table 1. Investment, maintenance, and lifetime costs of the used energy resources, with a lifetime of 20 years for PV and wind turbine, and 5 years for batteries.

	Solar Panels (€/kW)	Battery ESS (€/kWh)	Wind Turbine (€/kW)	Grid Energy Price (€/kWh)
Investment cost	2835.00	148.00	5832.00	On-peak periods: 0.216
Maintenance and operating costs	56.7	2.96	38.08	Off-peak periods: 0.108

In the simulations carried out, all parameters to be optimized are bounded in order to get more realistic installation corresponding to urban usage. All decision variables' boundaries are provided in Table 2, as described in the optimization problem constraints.

Table 2. Decision variables boundaries.

Variables	C_{bat}^{max} (MJ)	P_{bat}^{max} (W)	P_{bat}^{min} (W)	P_{g}^{max} (W)	P_{g}^{min} (W)	S_{limit} (m^2)	R_{limit} (m)	DOD_{bat}	DOD_{EV}
Bounds	100	2500	−2500	1000	−1000	30	2	0.4	0.85

4.2. Simulation Results for Home MG without EV Integration

Simulations have been performed first for a home MG without considering EV usage. Hot and cold weather climate regions in France are selected to assess the renewable resources-based MG optimal design and, consequently, to study the output power and the economic benefits. The design results give the optimal sizing of the energy generation system and is given in Table 3.

Table 3. Case study 1: PV, wind turbine, and battery optimal sizing.

	PV Power (kW)	PV Surface (m^2)	Wind Turbine Power (kW)	Wind Turbine Radius (m)	Battery Energy (kWh)
Brest	2.58	26.66	2.46	1.88	5.59
Toulon	3.84	27.26	0.89	1.86	16.51

These results are coherent since the solar energy potential in Toulon is higher than in Brest and wind power generation is more available in Brest. On one hand, the area of PV panels is larger in Toulon than in Brest, which is due to Toulon's high solar energy potential compared to Brest. The installed power in Toulon is 3.84 kW, whereas in Brest it is equal to 2.58 kW. On the other hand, due to high wind speed in Brest, the installed wind power is higher than in Toulon. Indeed, wind power reaches 2.46 kW against 0.89 kW in Toulon. Battery capacity in Toulon is much higher than in Brest. This is mainly due to the fact that the net power, which represents the difference between energy consumption and the actual PV and WT energy generation is higher in Toulon. Indeed, the battery is the first component to intervene in the energy management algorithm in order to maintain the consumption/generation balance.

Once the optimal sizing is performed, a one-day simulation is performed in order to show the behaviour of the MG components. Hence, PV and WT energy generation in Brest and Toulon are provided in Figure 6. Figure 7 shows the total RER energy production and net power (difference between RER production and load consumption). Net power can be provided using battery ESS if the SOC and the required power are within the specified limits or using the main grid, as shown by Figure 8. The sum of the battery and grid powers is equal to net power at any time, which shows that the main constraint related to power balance is achieved. The constraint related to the prohibited exchange of power between the battery and the grid (in both directions) is also fulfilled during the day.

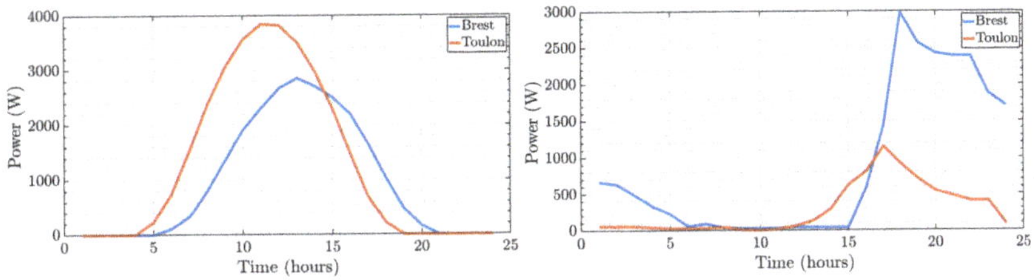

(**a**) PV energy generation. (**b**) WT energy generation.

Figure 6. Case study 1: Renewable energy resources energy generation in Brest and Toulon.

The variation of battery SOC is provided in Figure 8b. It can be noticed that the battery charges and discharges during the day while respecting the the specified DOD. The battery is safely operated, its SOC varies between the limits indicated in the constraints, which makes it possible to consider a lifetime of 5 years. It is also shown that the battery SOC at the end of the day is equal to the initial value.

The total energy from the grid before the integration of the RERs was 377.24 MWh during 20 years. This energy consumption issued from the grid is reduced to 92.2 MWh in Brest and 126.95 MWh in Toulon after the integration of the RERs, allowing an autonomy of 75.56% in Brest and 66.35% in Toulon. These results show the profitability of the system after 20 years of operation. Indeed, the total energy cost before the integration of the RER was 70,703 €. This energy cost is reduced to 40,449 € in Brest and 46,105 € in Toulon. The investment cost, the grid energy cost, and the autonomy in the considered cities are given in Table 4.

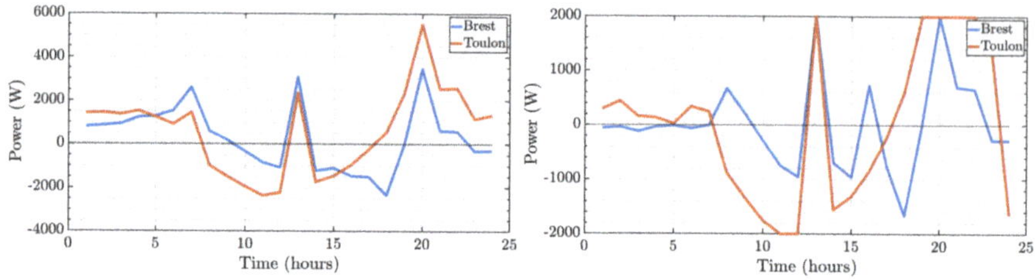

(a)

(b)

Figure 7. Case study 1: Net power and battery power for Brest and Toulon. (**a**) Net power highlighting production/consumption unbalance during the day. (**b**) Battery power for 24 h.

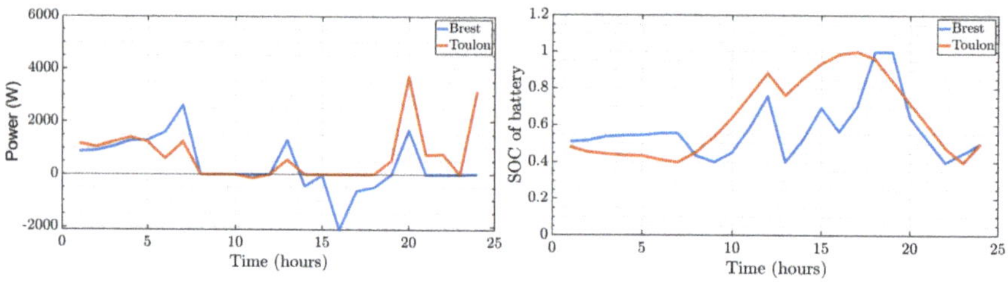

(**a**) Grid power for 24 h.

(**b**) State of charge of battery ESS.

Figure 8. Case study 1: Grid power and state of charge of the battery for Brest and Toulon.

Table 4. Case study 1: Energy cost and autonomy achieved without EV.

	Total Energy Consumption (MWh)	Cost without RER (€)	Local Generation (MWh)	Investment Cost (€)	Grid Energy (MWh)	Grid Energy Cost (€)	Battery Energy (MWh)	Total Cost (€)	Achieved Autonomy (%)
Brest	377.24	70,703	322.89	28,152	92.2	12,296	50.771	40,449	75.56
Toulon	377.24	70,703	272.19	27,424	126.95	18,680	98.984	46,105	66.35

4.3. Simulation Results for a Home MG with EV Integration

A second case study has been investigated to assess the interest of integrating the *V2H* concept in a home MG in terms of energy self-sufficiency and total cost comprising the investment, operating and maintenance costs of the MG, and the energy bill from the main grid. Two operating conditions may arise during operation, which are the vehicle is parked at home and can be used as an additional ESS or the vehicle is in normal every-day use. In the last operating situation, the home MG is operated as in case study 1 operating conditions. Numerical studies have been conducted for the same domestic MG, taking into account the use of EV batteries and limiting the available main grid power. Indeed, another constraint is considered in this second case study in order to relieve the grid by limiting the power exchange with the main grid to ±1000 W. The same cities are considered for the sake of comparison. This allows for the studying of the power outputs of renewable resources, battery ESS, EV, and grid, as well as the economic benefits of the integration of EVs at domestic MGs.

Domestic MG optimal sizing results with the integration of the EV are provided in Table 5 for Brest and Toulon. The same conclusions can be drawn from these results as the results discussed in case study 1. Indeed, the solar energy potential of Toulon remains higher than that of Brest and vice versa for the wind energy generation. The battery ESS to be installed has increased as the grid power is severely limited to enhance the use of renewables and EV batteries, and, consequently, to limit the stress on the grid, especially during on-peak periods.

Table 5. Case study 2: PV, battery, and wind turbine sizing results.

	PV Power (kW)	PV Surface (m^2)	Wind Turbine Power (kW)	Wind Turbine Radius (m)	Battery Energy (kWh)
Brest	3.2	29.78	1.11	1.26	8.74
Toulon	4.1	29	0.967	1.9	17.8

Renewable resources generated powers for 24 h are shown in Figure 9. It can be noticed that the integration of the EV as an additional ESS does not significantly change the sizing and the installed power of the RERs. It can be considered that renewable energy production mainly depends on meteorological data and these results allow for the extraction of the maximum energy from the RERs. The imbalance between renewable energy production and load consumption is shown in Figure 10. This imbalance is compensated for using battery-based ESS, the main grid, and the EV when the EV is parked at home. Otherwise, during EV daily use, only the battery-based ESS and the grid intervene to guarantee the production/consumption balance. System constraints are respected and power balance is achieved all the time. Moreover, there is no energy exchange between the battery, the grid, and the EV, as can be seen from Figure 11. Finally, the EV power exchange with a domestic MG and state of charge of the EV batteries are given by Figure 12. It can be seen that the EV is mainly used during periods where renewables energy generation is lower and load consumption is higher. This usage relieves the grid and decreases the energy bill.

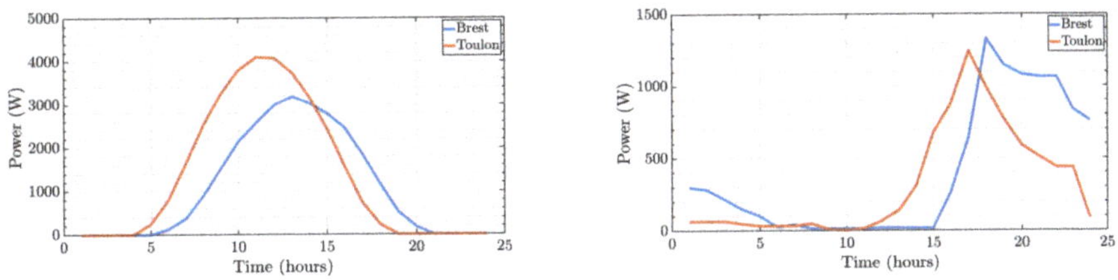

(**a**) PV energy generation for 24 h. (**b**) WT energy generation for 24 h.

Figure 9. Case study 2: Renewable energy resources generation for Brest and Toulon.

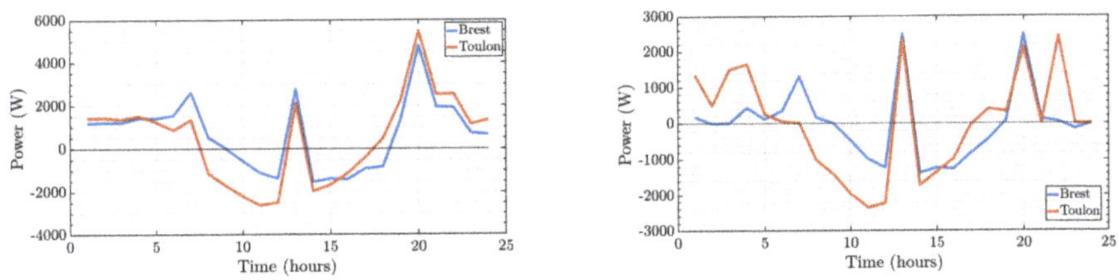

(**a**) Net power for 24 h. (**b**) Battery power during 24 h.

Figure 10. Case study 2: Net power representing the difference between renewables energy generation and load consumption and batteries power for Brest and Toulon.

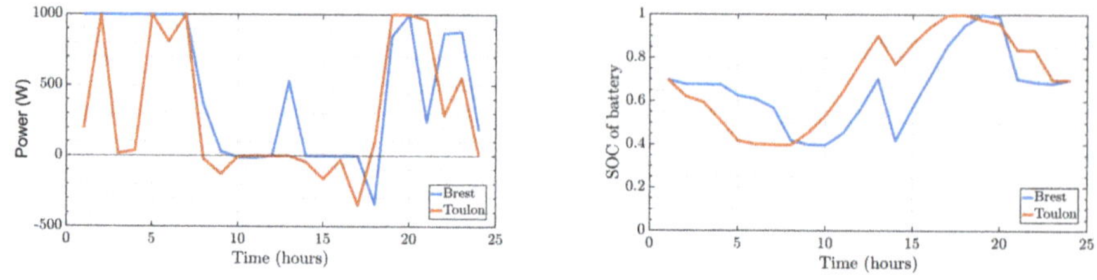

(**a**) Grid power during 24 h. (**b**) State of charge of battery.

Figure 11. Case study 2: Grid power and batteries state of charge for Brest and Toulon.

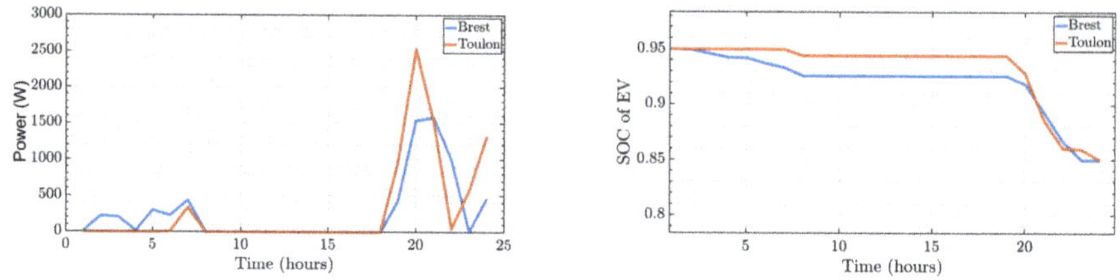

(**a**) EV power during 24 h. (**b**) State of charge of EV batteries.

Figure 12. Case study 2: EV batteries power and state of charge for Brest and Toulon.

EV usage in a home MG, even if limited to few hours, is still profitable because the home MG self-sufficiency is considerably increased in Toulon, which raises to 84.61%, and slightly increases in Brest, being raised to 76.87%, reducing the high electric network utilisation. The total cost for 20 years operation is 35,063 € in Brest and 38,827 € in Toulon. This allows financial savings of 5000 € and 8000 € in both cities, respectively compared to case study 1. The investment cost, battery energy, main grid energy, total energy cost, and achieved energy self-efficiency in the two considered cities are given in Table 6. It can be seen that the total cost is divided by almost 2 for the two cities. Moreover, the grid is relieved due to the use of the *V2H* concept and the energy bill is significantly decreased while achieving a very good autonomy.

Table 6. Case study 2: Energy cost and achieved autonomy with EV.

	Total Energy Consumption (MWh)	Cost without RER (€)	Local Generation (MWh)	Investment Cost (€)	Grid Energy (MWh)	Grid Energy Cost (€)	Battery Energy (MWh)	EV Energy (MWh)	Total Cost (€)	Achieved Autonomy (%)
Brest	377.24	70,703	257.520	21,880	87.238	13,182	57.642	47.289	35,063	76.87
Toulon	377.24	70,703	291.170	29,476	58.073	9350	95.126	53.494	38,827	84.61

4.4. Discussion

Numerical studies have proved that the integration of an EV in home microgrids has brought several benefits, starting with the economic gains, which is of interest for householders. Such financial savings encourage customers to switch to the *V2H* concept, given the reduction in the electricity bill and in the capacity of the installed battery ESS. The major disadvantage is the investment cost, which is still a bit expensive, but considering the duration of the project, this cost is quickly paid back in few years.

The second advantage to mention is the reduction of the energy provided by the main grid. Simulation results showed that the integration of an EV reduces the grid power and, consequently, lowers the stress on the grid, especially during congestion periods. The energy decreases from 92.2 MWh to 87.238 MWh in Brest, and it decreases by almost 50% in Toulon as the energy issued from the main grid decreases from 126.95 MWh to 58.073 MWh. The EV battery provides a certain amount of energy when it is connected to the home

MG, and, therefore, allows grid power limitation. If the *V2H* and *V2G* concept is further expanded, there will be a decrease in grid power and therefore fewer outages, fewer voltage and frequency regulation problems, and also fewer CO_2 emissions thanks to the massive use of local renewable resources.

In a home MG, battery power decreases after the integration of an EV as an additional ESS. This leads to the reduction of the number of cycles, and, therefore, the lifetime of the battery increases. The increase in lifetime induces financial savings as the operating and maintenance costs will straightforwardly decrease. Moreover, the installed battery capacity decreases, which allows for a decrease in the initial investment cost. Indeed, a part of the required energy storage system to handle the intermittency of RER is ensured by using EV batteries.

Despite the large amount of energy supplied by the EV, its SOC does not significantly vary, which can be explained by the large storage capacity of the EV to ensure certain autonomy. Indeed, the EV will have enough energy to support the home MG while limiting main grid power. In this work, the EV SOC is limited to a DOD of 85% to avoid its battery being significantly discharged, but in reality, the EV can be discharged even more. Indeed, the Lithium-ion batteries used in EVs can be discharged even more than 50% of their maximum capacity to support the home MG.

Table 7 summarizes the two conducted case studies. It is clear from this table and the previously discussed results that the integration of the EV brings great benefits to a home MG. Indeed, the energy self-sufficiency has increased in both cities, with a large improvement for Toulon (hot climate with huge PV potential). In terms of financial benefits, it is obvious from the given results that the use of renewable resources and the EV energy storage capacity lowers the energy bill for 20 years of operation, even if the investment cost is quite high, mainly due the price of battery energy storage devices.

The use of the electric vehicle as a RER energy storage device leads to the degradation of its battery over time [13], as the lifespan of the battery depends on the number of charge/discharge cycles. It would be more interesting to integrate an EV battery degradation cost in future work. Moreover, more aspects related to home MG EMS can be investigated, such as demand response and energy trading in the community integrating an EV fleet for enhancing the overall operation of the community MG [36].

Table 7. Summary of the achieved benefits for 20 years of operation of the home MG in Brest and Toulon.

	MG without EV	MG with EV
Brest	• Total Cost = 40,449 (€) • Financial savings = 30,254 (€) • Achieved autonomy = 75.56%	• Total Cost = 35,063 (€) • Financial savings = 35,640 (€) • Achieved autonomy = 76.87%
Toulon	• Total Cost = 46,105 (€) • Financial savings = 24,598 (€) • Achieved autonomy = 66.35%	• Total Cost = 38,827 (€) • Financial savings = 31,876 (€) • Achieved autonomy = 84.61%

5. Conclusions and Perspectives

This paper has presented an optimal sizing algorithm and energy management scheme for a home MG integrating the *V2H* technology for both hot and cold climates. The presented case studies show the advantage of the *V2H* concept, especially with the high integration of renewable energy resources. The electric vehicle presents a support for renewable energies by offering a larger storage capacity, thus efficiently increasing renewable energy generation by increasing on-MPPT mode usage. This increase in renewable production reduces the dependence on the grid, as shown in the simulation results. Therefore, the energy bill is reduced and the use of clean energy is empowered for sustainable development. It is mainly demonstrated that the use of *V2H* presents a financial interest for homeowners, supports the integration of RER for sustainable development, and relieves the main grid during on-peak periods.

The economic advantage of $V2H$ and its cost-effectiveness may encourage homeowners to switch from thermal to electric vehicles, thus reducing the pollution rate, knowing that the total amount of CO_2 emitted by the design of the electric vehicle is half that emitted by the manufacturing and operation of the combustion vehicle. This greatly reduces the level of CO_2 emissions over a period of 20 years. Moreover, in a house MG, the electric vehicle reduces the size of the storage system and avoids its degradation by storing surplus production, which lowers the total investment cost and reduces the rate of manufacturing and changing of storage systems, in this case the battery ESS.

In the near future, it is interesting to scale up to a district, using a fleet of electric vehicles, to show the benefits of the $V2G$ concept. It is foreseeable that the integration of electric vehicles will significantly reduce the pollution rate. However, the economic profitability needs to be demonstrated through further investigations. The move towards a larger scale will lead to more constraints, as vehicles will not have the same displacement and availability over a day, and this constraint will further complicate the optimisation of the objective function. It is not certain that, by increasing the number of electric vehicles, the network will be relieved, it may happen the other way round, where the charging stations require a huge amount of energy, and if the number of stations increases, the energy required increases accordingly. One solution to this problem is to power the charging stations with renewable energy.

Other aspects linked to artificial intelligence must be used to forecast weather data for optimal energy production and also to predict the displacement and consumption of vehicles, as well as their availability. Other parameters need to be considered in the modelling of the battery degradation cost, solar panels, and wind turbines to achieve more accuracy, such as the temperature, the change of the energy cost in the energy market, and load shedding.

Author Contributions: Conceptualization, O.O. and E.E.; methodology, O.O. and E.E.; validation, E.E.; formal analysis, O.O., E.E.; investigation, O.O. and E.E.; resources, O.O. and E.E.; writing—original draft preparation, O.O.; writing—review and editing, E.E., Y.A., F.L.G. and E.S.G.; visualization, O.O.; supervision, E.E.; project administration, E.E.; funding acquisition, E.E. and Y.A. All authors have read and agreed to the published version of the manuscript.

Funding: This research was funded by Brest Métropole and ISEN Yncréa Ouest, France.

Conflicts of Interest: The authors declare no conflict of interest. The funders had no role in the design of the study; in the collection, analyses, or interpretation of data; in the writing of the manuscript, or in the decision to publish the results.

Nomenclature

The following variables are used in this manuscript:

P_{Irr}	Irradiance of the day W/m^2
P_{PV}^{ins}	Installed PV power (computed considering maximum irradiance)
$P_{PV}(t)$	Power produced by PV at time t
P_{WT}^{ins}	Installed wind power (computed considering maximum wind speed)
$P_{WT}(t)$	Power produced by WT at time t
$P_{bat}^{+}(t)$	Battery discharge power at time t
$P_{bat}^{-}(t)$	Battery charging power at time t
$P_{g}^{+}(t)$	Power supplied by the grid at time t
$P_{g}^{-}(t)$	Power sent to the grid at time t
$P_{g}^{i}(t)$	Grid power at time t at for the ith year
$P_{EV}^{+}(t)$	Electric Vehicle discharge power at time t
$P_{EV}^{-}(t)$	Electric Vehicle charging power at time t
P_{bat}^{max}	Maximum discharge power of the battery
P_{bat}^{min}	Maximum charging power of the battery
P_{g}^{max}	Maximum power supplied by the grid
P_{g}^{min}	Maximum power sent to the grid

E_{bat}^{max}	Capacity of the installed battery (J)
E_{EV}^{max}	Capacity of the electric vehicle (J)
$SOC_{bat}(t)$	Battery state of charge at time t
$SOC_{EV}(t)$	Electric Vehicle state of charge at time t
DOD_{bat}	Battery depth of discharge
DOD_{EV}	Electric vehicle depth of discharge
S_{PV}	PV surface m^2
S_{limit}	Surface limit of the PV
R_{WT}	Radius of the wind turbine
R_{limit}	Radius limit of the WT
η_{bat}^{cha}	Battery charging efficiency
η_{bat}^{disch}	Battery discharging efficiency
η_{EV}^{cha}	Electric vehicle charging efficiency
η_{EV}^{disch}	Electric vehicle discharging efficiency
η_{PV}	PV efficiency
ρ	Air density
C_p	Power coefficient
V_{Wind}	Wind velocity (m/s)
dt	Power application time (1 h)
C_{PV}	Total PV investment, maintenance, and operating costs in €/kW
C_{WT}	Total wind turbine investment, maintenance, and operating costs in €/kW
C_{bat}	Total batteries investment, maintenance, and operating costs in in €/kWh
C_g	Cost of energy purchased from the grid in €/kWh
ESS	Energy storage system
MG	Microgrid
PV	Photovoltaic
WT	WInd turbine
$V2H$	Vehicle to home
$V2G$	Vehicle to grid
RER	Renewable energy resources
SOC	State of charge
DOD	Depth of charge

References

1. Tan, K.M.; Ramachandaramurthy, V.K.; Yong, J.Y. Integration of electric vehicles in smart grid: A review on vehicle to grid technologies and optimization techniques. *Renew. Sustain. Energy Rev.* **2016**, *53*, 720–732. [CrossRef]
2. Zeng, X.; Li, M.; Abd El-Hady, D.; Alshitari, W.; Al-Bogami, A.S.; Lu, J.; Amine, K. Commercialization of lithium battery technologies for electric vehicles. *Adv. Energy Mater.* **2019**, *9*, 1900161. [CrossRef]
3. Sharma, A.; Sharma, S. Review of power electronics in vehicle-to-grid systems. *J. Energy Storage* **2019**, *21*, 337–361. [CrossRef]
4. Ouramdane, O.; Elbouchikhi, E.; Amirat, Y.; Gooya, E.S. Optimal Sizing and Energy Management of Microgrids with Vehicle-to-Grid Technology: A Critical Review and Future Trends. *Energies* **2021**, *14*, 4166. [CrossRef]
5. Corchero, C.; Cruz-Zambrano, M.; Heredia, F.J. Optimal energy management for a residential microgrid including a vehicle-to-grid system. *IEEE Trans. Smart Grid* **2014**, *5*, 2163–2172.
6. Borge-Diez, D.; Icaza, D.; Açıkkalp, E.; Amaris, H. Combined vehicle to building (V2B) and vehicle to home (V2H) strategy to increase electric vehicle market share. *Energy* **2021**, *237*, 121608. [CrossRef]
7. Kataoka, R.; Shichi, A.; Yamada, H.; Iwafune, Y.; Ogimoto, K. Comparison of the economic and environmental performance of V2H and residential stationary battery: Development of a multi-objective optimization method for homes of EV owners. *World Electr. Veh. J.* **2019**, *10*, 78. [CrossRef]
8. Khaligh, A.; Li, Z. Battery, ultracapacitor, fuel cell, and hybrid energy storage systems for electric, hybrid electric, fuel cell, and plug-in hybrid electric vehicles: State of the art. *IEEE Trans. Veh. Technol.* **2010**, *59*, 2806–2814. [CrossRef]
9. Chan, C.C.; Bouscayrol, A.; Chen, K. Electric, hybrid, and fuel-cell vehicles: Architectures and modeling. *IEEE Trans. Veh. Technol.* **2009**, *59*, 589–598. [CrossRef]
10. Mwasilu, F.; Justo, J.J.; Kim, E.K.; Do, T.D.; Jung, J.W. Electric vehicles and smart grid interaction: A review on vehicle to grid and renewable energy sources integration. *Renew. Sustain. Energy Rev.* **2014**, *34*, 501–516. [CrossRef]
11. Yilmaz, M.; Krein, P.T. Review of the impact of vehicle-to-grid technologies on distribution systems and utility interfaces. *IEEE Trans. Power Electron.* **2012**, *28*, 5673–5689. [CrossRef]
12. Gope, P.; Sikdar, B. An efficient privacy-preserving authentication scheme for energy internet-based vehicle-to-grid communication. *IEEE Trans. Smart Grid* **2019**, *10*, 6607–6618. [CrossRef]

13. Farzin, H.; Fotuhi-Firuzabad, M.; Moeini-Aghtaie, M. A practical scheme to involve degradation cost of lithium-ion batteries in vehicle-to-grid applications. *IEEE Trans. Sustain. Energy* **2016**, *7*, 1730–1738. [CrossRef]
14. Liu, H.; Huang, K.; Yang, Y.; Wei, H.; Ma, S. Real-time vehicle-to-grid control for frequency regulation with high frequency regulating signal. *Prot. Control Mod. Power Syst.* **2018**, *3*, 13. [CrossRef]
15. Colmenar-Santos, A.; Muñoz-Gómez, A.M.; Rosales-Asensio, E.; López-Rey, Á. Electric vehicle charging strategy to support renewable energy sources in Europe 2050 low-carbon scenario. *Energy* **2019**, *183*, 61–74. [CrossRef]
16. Gao, F.; Zhu, L.; Shen, M.; Sharif, K.; Wan, Z.; Ren, K. A blockchain-based privacy-preserving payment mechanism for vehicle-to-grid networks. *IEEE Netw.* **2018**, *32*, 184–192. [CrossRef]
17. Mehrjerdi, H.; Rakhshani, E. Vehicle-to-grid technology for cost reduction and uncertainty management integrated with solar power. *J. Clean. Prod.* **2019**, *229*, 463–469. [CrossRef]
18. Quddus, M.A.; Kabli, M.; Marufuzzaman, M. Modeling electric vehicle charging station expansion with an integration of renewable energy and Vehicle-to-Grid sources. *Transp. Res. Part E Logist. Transp. Rev.* **2019**, *128*, 251–279. [CrossRef]
19. Alirezaei, M.; Noori, M.; Tatari, O. Getting to net zero energy building: Investigating the role of vehicle to home technology. *Energy Build.* **2016**, *130*, 465–476. [CrossRef]
20. Staudt, P.; Schmidt, M.; Gärttner, J.; Weinhardt, C. A decentralized approach towards resolving transmission grid congestion in Germany using vehicle-to-grid technology. *Appl. Energy* **2018**, *230*, 1435–1446. [CrossRef]
21. Amamra, S.A.; Marco, J. Vehicle-to-grid aggregator to support power grid and reduce electric vehicle charging cost. *IEEE Access* **2019**, *7*, 178528–178538. [CrossRef]
22. Monteiro, V.; Exposto, B.; Ferreira, J.C.; Afonso, J.L. Improved vehicle-to-home (iV2H) operation mode: experimental analysis of the electric vehicle as off-line UPS. *IEEE Trans. Smart Grid* **2016**, *8*, 2702–2711. [CrossRef]
23. Mehrjerdi, H. Resilience oriented vehicle-to-home operation based on battery swapping mechanism. *Energy* **2021**, *218*, 119528. [CrossRef]
24. Zeynali, S.; Rostami, N.; Ahmadian, A.; Elkamel, A. Two-stage stochastic home energy management strategy considering electric vehicle and battery energy storage system: An ANN-based scenario generation methodology. *Sustain. Energy Technol. Assess.* **2020**, *39*, 100722. [CrossRef]
25. von Bonin, M.; Dörre, E.; Al-Khzouz, H.; Braun, M.; Zhou, X. Impact of Dynamic Electricity Tariff and Home PV System Incentives on Electric Vehicle Charging Behavior: Study on Potential Grid Implications and Economic Effects for Households. *Energies* **2022**, *15*, 1079. [CrossRef]
26. Lee, S.; Choi, D.H. Energy management of smart home with home appliances, energy storage system and electric vehicle: A hierarchical deep reinforcement learning approach. *Sensors* **2020**, *20*, 2157. [CrossRef]
27. Turker, H.; Bacha, S. Optimal minimization of plug-in electric vehicle charging cost with vehicle-to-home and vehicle-to-grid concepts. *IEEE Trans. Veh. Technol.* **2018**, *67*, 10281–10292. [CrossRef]
28. Sausen, J.P.; Binelo, M.d.F.B.; Campos, M.; Sausen, A.Z.R.; Sausen, P.S. Economic feasibility study of using an electric vehicle and photovoltaic microgeneration in a smart home. *IEEE Lat. Am. Trans.* **2018**, *16*, 1907–1913. [CrossRef]
29. Hou, X.; Wang, J.; Huang, T.; Wang, T.; Wang, P. Smart home energy management optimization method considering energy storage and electric vehicle. *IEEE Access* **2019**, *7*, 144010–144020. [CrossRef]
30. Bull, S.R. Renewable energy today and tomorrow. *Proc. IEEE* **2001**, *89*, 1216–1226. [CrossRef]
31. Ota, Y.; Taniguchi, H.; Nakajima, T.; Liyanage, K.M.; Baba, J.; Yokoyama, A. Autonomous distributed V2G (vehicle-to-grid) satisfying scheduled charging. *IEEE Trans. Smart Grid* **2011**, *3*, 559–564. [CrossRef]
32. Roos, C.; Terlaky, T.; Vial, J.P. *Interior Point Methods for Linear Optimization*; Springer: Berlin/Heidelberg, Germany, 2005.
33. Hazem Mohammed, O.; Amirat, Y.; Benbouzid, M. Economical evaluation and optimal energy management of a stand-alone hybrid energy system handling in genetic algorithm strategies. *Electronics* **2018**, *7*, 233. [CrossRef]
34. European Commission. Photovoltaic Geographical Information System. Available online: https://re.jrc.ec.europa.eu/pvg_tools/en/tools.html#PVP (accessed on 4 April 2022).
35. EDF. Open Data Wind Measurements—île de Sein. Available online: https://opendata-iles-ponant.edf.fr/explore/dataset/mesures-de-vent-ile-de-sein/information/?sort=date_heure (accessed on 4 April 2022).
36. Liu, H.; Wang, B.; Wang, N.; Wu, Q.; Yang, Y.; Wei, H.; Li, C. Enabling strategies of electric vehicles for under frequency load shedding. *Appl. Energy* **2018**, *228*, 843–851. [CrossRef]

energies
MDPI

Article

Solar-Based DG Allocation Using Harris Hawks Optimization While Considering Practical Aspects

Suprava Chakraborty [1], Sumit Verma [2,*], Aprajita Salgotra [2], Rajvikram Madurai Elavarasan [3,*], Devaraj Elangovan [1] and Lucian Mihet-Popa [4,*]

1 TIFAC-CORE Research Center, Vellore Institute of Technology, Vellore 632014, India; suprava1008@gmail.com (S.C.); elangovan.devaraj@vit.ac.in (D.E.)
2 Department of Industrial and Management Engineering, Indian Institute of Technology Kanpur, Kanpur 208016, India; aprajita.salgotra3@gmail.com
3 Research & Development Division (Power and Energy), Nestlives Private Limited, Chennai 600091, India
4 Faculty of Electrical Engineering, Ostfold University College, 1757 Halden, Norway
* Correspondence: drsumitverma007@gmail.com (S.V.); rajvikram787@gmail.com (R.M.E.); lucian.mihet@hiof.no (L.M.-P.)

Abstract: The restructuring of power systems and the ever-increasing demand for electricity have given rise to congestion in power networks. The use of distributed generators (DGs) may play a significant role in tackling such issues. DGs may be integrated with electrical power networks to regulate the drift of power in the transmission lines, thereby increasing the power transfer capabilities of lines and improving the overall performance of electrical networks. In this article, an effective method based on the Harris hawks optimization (HHO) algorithm is used to select the optimum capacity, number, and site of solar-based DGs to reduce real power losses and voltage deviation. The proposed HHO has been tested with a complex benchmark function then applied to the IEEE 33 and IEEE 69 bus radial distribution systems. The single and multiple solar-based DGs are optimized for the optimum size and site with a unity power factor. It is observed that the overall performance of the systems is enhanced when additional DGs are installed. Moreover, considering the stochastic and sporadic nature of solar irradiance, the practical size of DG has been suggested based on analysis that may be adopted while designing the actual photovoltaic (PV) plant for usage. The obtained simulation outcomes are compared with the latest state-of-the-art literature and suggest that the proposed HHO is capable of processing complex high dimensional benchmark functions and has capability to handle problems pertaining to electrical distribution in an effective manner.

Keywords: RDS; Harris hawks; optimal power flow; optimization; solar PV; transmission loss

Citation: Chakraborty, S.; Verma, S.; Salgotra, A.; Elavarasan, R.M.; Elangovan, D.; Mihet-Popa, L. Solar-Based DG Allocation Using Harris Hawks Optimization While Considering Practical Aspects. *Energies* **2021**, *14*, 5206. https://doi.org/10.3390/en14165206

Academic Editor: Carlo Renno

Received: 22 July 2021
Accepted: 17 August 2021
Published: 23 August 2021

1. Introduction

Installing distributed generation (DG) sources in the distribution network system has been standard practice in recent years to minimize overall power losses and enhance the power quality [1,2]. The optimum sizing and positioning of DGs in power system networks are essential to maximize the benefits from those installations. The incorrect allocation and unreasonable sizing of DG units in the power system networks may increase voltage sags, voltage flickering, harmonic distortion, fault current, and power losses. With the application of DG units, the power system losses may be reduced by 13% [3,4]. In the functioning of power systems, economic damage and voltage collapse may be avoided through the reduction in power loss and voltage stability enhancement, respectively [5]. Thus, the investigation in optimal location selection and sizing of DG units in the distribution network is a step towards a profitable electricity supply [6,7]. Among all DG systems, solar photovoltaic DG systems seek attention worldwide for their abundant availability, easy installation, maintenance, and environment-friendly features.

The major goals of most techniques to determine the best location and size of DG units are to reduce power loss and improve voltage profile. The various techniques such as analytical methods, ant bee colony (ABC), genetic algorithm (GA), tabu search (TS), particle swarm optimization (PSO), fuzzy system, evolutionary programming, dynamic programming, etc., have been utilized to achieve the aforesaid objectives in a distribution network through the proper allocation and sizing of DG units. In the literature, GA is used to estimate the placement and size of DG units to improve the voltage profile and reduce power loss. Once DG units are appropriately placed in the distribution system network, voltage stability and loss reduction are improved significantly. The GA is utilized as the most applied optimization technique in resolving the problem of DGs allocation and sizing [8,9]. The multiobjective genetic optimization method is used in radial distribution systems to determine the best position and size for renewable-based DG units [10]. For site determination of DGs planning and performance index-based size, a GA-based multiobjective optimization is utilized to minimize the actual power loss in distribution systems with constant power, current, and impedance models [11]. Almabsout et al. [12] suggest an improved GA to determine the best placement and capacity of the simultaneous allocation of DGs/SCs in radial systems by combining the benefits of genetic algorithms and local search [12]. To minimize system losses, a mix of analytical and genetic algorithm approaches is utilized to optimize the allocation of numerous DGs in a distribution network [13].

To reduce real power losses and improve the voltage profile, Madhusudhan et al. [14] proposed the GA to identify the optimum location, as well as the size of the distribution network's DG units. Ayodele et al. [15] used GA to find the best DG technology for optimal power system functioning, as well as the best position and size of the DG to reduce network power loss. GA is applied to reduce the cost of system expansion and improves system stability [16,17]. However, GA convergence time is high, especially, when applied in the solution of complex problems, and may suggest inaccurate solutions. When compared to GA and TS techniques, Hassan et al. and Fan et al. [18,19] employed simulated annealing (SA) to find and specify the capacity of DGs while lowering computation time. However, the SA method has disadvantages such as termination at a local minimum, significant computational time, no information regarding the divergence of the local minimum from the global minimum, and no upper constraint for the calculation time. Using the TS approach, Liu et al. and Azam et al. [20,21] concentrated on DG optimum planning with the goal of minimizing both losses and line loadings. The TS technique, on the other hand, has the drawback of requiring a large number of iterations and parameter calculations. PSO was used to determine the best scale and distribution of DG units in the power system, together with its benefits [22].

One of the most effective and widely used optimization strategies is the PSO [23–25]. Barik et al. [26] presented a multiobjective PSO method for determining the best location and size of DG units while taking economic and technical factors into account. The advanced versions of PSO methods, such as improved PSO [27], binary PSO [28], social learning PSO [29], PSO with inertia weight, and PSO with constriction factor [30], are also applied in the DG allocation and sizing problems. However, the PSO technique has some disadvantages, such as difficulty in initializing the design parameters and inapplicability to scattering problems. Tolabi et al. and Oloulade et al. [31,32] introduced the ant colony optimization (ACO) technique to tackle the allocation and size problem of renewable energy source-based DGs in radial distribution networks with the goal of minimizing overall system losses. Their analysis showed that ACO gives a better solution, and computational time is less than GA. However, ACO takes more time to converge due to the complex nature of the problem but is still shorter than analytical methods. The major disadvantage of the ACO technique lies with its uncertainty in time to convergence. Das et al. [33] and Seker and Hocaoglu [34] used the artificial bee colony (ABC) method to compare results to the PSO technique and discovered that ABC provides a higher-quality solution with a faster convergence rate. The cuckoo search algorithm was used by Yuvaraj and Ravi [35] to improve the voltage profile and reduce power losses in biomass and solar–thermal DG units. To optimize the system voltage profile and decrease line losses, Arya and Koshti [36] used a shuffled frog leaping algorithm.

Rajaram et al. [37] used a plant growth simulation algorithm with objectives such as decreasing the losses and improving the voltage profile. To reduce energy losses in a distribution network system, Othman et al. [38] used the big bang–big crunch approach to find appropriate DG units. The bat algorithm was suggested by Sudabattula and Kowsalya [39] for the efficient allocation of solar-based DGs in the distribution network. To decrease power losses while preserving voltage profile, Duong et al. [40] developed an efficient biogeography-based optimization for optimal location and size of solar photovoltaic distributed generating units.

Harris hawks optimization (HHO) is a new metaheuristic optimization algorithm used in various applications, as tabulated in Table 1.

After a thorough search in credible academic publications, as shown in Table 1, to-date, the efficient newly invented HHO method has not been utilized to optimal solar-based DG allocation in a radial distribution system. As a result, this study compares and contrasts the suggested work with well-known optimization techniques. Suitable DG unit placement may bring significant benefits, including cost saving through a reduction in power loss and increasing the purchasing power capacity.

Table 1. Application of HHO in different literatures.

Year	Area of Application	Research Objectives	Research Findings	Reference No.
2021	Design of truss structures	The use of HHO to solve planar and spatial trusses with discrete design variables was investigated in this paper. Five benchmark structural issues were used to assess HHO's performance, and the resultant designs were compared to 10 state-of-the-art algorithms.	The statistical results demonstrate that HHO is quite consistent and reliable when related to truss structure optimization.	[41]
2021	Prediction of slope stability	The study's major goal is to develop a new metaheuristic optimization approach HHO for improving the accuracy of the traditional multilayer perceptron technique in estimating the factor of safety in the presence of inflexible foundations. Four slope stability conditioning elements are taken into account in this method: slope angle, rigid foundation position, soil strength, and applied surcharge.	The findings revealed that employing the HHO improves the ANN's prediction accuracy while analyzing slopes with unknown circumstances.	[42]
2021	Power flow controller	To reduce oscillations in single and multimachine power systems, a HHO tuned dual interval type-2 fuzzy lead–lag (Dual-IT2FLL)-based universal power flow controller (UPFC) is suggested. The suggested damping controller uses speed deviation, a distant input signal for stability enhancement, to coordinate between the modulation index (MI) and phase angle of series and shunt converters of UPFC at the same time.	Different performance indicators (PIs) such as mean, standard deviation, overshoots, and settling time are used to demonstrate that the proposed HHO-tuned dual-IT2FLL-based UPFC outperforms others under various operating circumstances.	[43]

Table 1. *Cont.*

Year	Area of Application	Research Objectives	Research Findings	Reference No.
2021	Shear strength estimation of reinforced concrete walls	The authors suggested three novel models for estimating peak shear strength using a mix of support vector regression and metaheuristic optimization techniques including teaching–learning-based optimization (TLBO), PSO, and HHO. The authors compiled a huge database with 228 RC shear wall experimental data and eight input parameters.	The suggested models may be used to estimate the shear strength of RC shear walls, potentially improving the accuracy of forecasting the structure's behavior and lowering construction costs.	[44]
2021	Screening of COVID-19 CT-scans	For the identification of COVID-19 from CT scan images, they suggested a two-stage pipeline consisting of feature extraction followed by feature selection (FS). A state-of-the-art convolutional neural network (CNN) model based on the DenseNet architecture was used for feature extraction. The HHO method was used in conjunction with SA and Chaotic initialization to remove noninformative and redundant features. The SARS-COV-2 CT-Scan dataset, which contains 2482 CT-scans, was used to test the suggested method.	The technique has an accuracy of about 98.42% without the chaotic initialization and the SA, which improves to 98.85% when the two are included, and therefore outperforms several state-of-the-art methods including other metaheuristic-based feature selection (FS) algorithms. The suggested approach reduces the number of characteristics chosen by around 75%, which is significantly better than most existing algorithms.	[45]
2021	Drug design and discovery	The authors presented a modified Henry gas solubility optimization (HGSO) based on heavy-tailed distributions (HTDs) utilizing improved HHO. A dynamical exchange between five HTDs were employed in this work to increase the HHO, which alters the exploitation phase in HGSO.	According to the values of accuracy, fitness value, and the number of selected characteristics, the results show that dynamic modified HGSO based on improved HHO has a high quality.	[46]
2021	Prediction of meteorological drought	In this study, the SVR (support vector regression) model was combined with two distinct optimization methods, PSO and HHO, to forecast the effective drought index (EDI) one month in advance in various sites across Uttarakhand, India.	The SVR-HHO model beat the SVR-PSO model in forecasting EDI, according to the results. SVR-HHO performed better than SVR-PSO in recreating the median, interquartile range, dispersion, and pattern of the EDI calculated from observed rainfall, according to visual assessment of model.	[47]
2021	Wireless sensor networks	The authors applied the HHO method to sensor node localization and compared their findings to other well-known optimization techniques that had just become available.	The suggested work's simulation results revealed that it outperforms existing computational intelligence methods in terms of average localization error, number of localized sensor nodes, and computational cost.	[48]

Table 1. *Cont.*

Year	Area of Application	Research Objectives	Research Findings	Reference No.
2021	Groundwater	The HHO method was used to minimize the sum of absolute deviation between observed and simulated water-table levels in order to optimize hydraulic conductivity and specific yield parameters of a modular three-dimensional finite-difference (MODFLOW) groundwater model.	According to the findings, the Pareto parameter sets gave appropriate results when the maximum and minimum aquifer drawdown were defined in the range of −40 to +40 cm/year.	[49]
2020	Parameter optimization of support vector regression	The goal of this research is to look at the SVR approach that is optimized using HHO, also known as HHO-SVR. To establish the performance of the HHO-SVR, five benchmark datasets were used to assess it. The HHO method is also compared to various metaheuristic algorithms and kernel types.	The findings revealed that the HHO-SVR has almost the same performance as other techniques, but is less time efficient.	[50]
2020	MPPT control	This study offers a new MPPT controller based on HHO that successfully tracks maximum power in all weather situations.	The suggested HHO outperforms the competition in terms of maximum power point tracking (MPPT) and convergence at the global maximum power point. The HHO-based MPPT approach provides faster maximum power point (MPP) tracking, decreased computing burden, and increased efficiency.	[51]
2020	Data dissemination for the Internet of Things	This study offers reliable data dissemination for the Internet of Things using HHO technique, which is a safe data diffusion mechanism for wireless sensor networks (WSN)-based IoT that accoutered a fuzzy hierarchical network model.	Simulation results show that RDDI delivers a more dependable approach and a better result than the other three disposals.	[52]
2020	Image segmentation	The HHO algorithm and the lowest cross-entropy as a fitness function are used to provide an efficient approach for multilevel segmentation in this work.	This HHO-based method outperforms other segmentation methods currently in use in the literature.	[53]
2020	Modeling of rainfall–runoff	To simulate the rainfall–runoff connection, data-driven approaches such as a multilayer perceptron (MLP) neural network and least squares support vector machine (LSSVM) are combined with a sophisticated nature-inspired optimizer, namely HHO.	All of the enhanced models with HHO outperformed other integrated models with PSO in predicting runoff changes, according to the findings. Furthermore, when HHO was combined with LSSVM, a high degree of accuracy in forecasting runoff levels was attained.	[54]
2020	Image segmentation	The HHO technique is used in this study to find reduced pulse coupled neural network settings.	The results of the experiments show that the HHO method is superior in image segmentation.	[55]

Table 1. *Cont.*

Year	Area of Application	Research Objectives	Research Findings	Reference No.
2020	Prediction of scour depth downstream of the ski-jump spillway	To forecast scour depth (SD) downstream of the ski-jump spillway, an alternative to standard techniques was used in this study. To improve the performance of an artificial neural network (ANN) to predict the SD, a novel optimization technique HHO was suggested.	The ANN-HHO model beat other existing models during the testing period, according to the findings. Furthermore, graphical evaluation reveals that the ANN-HHO model is more accurate than other models in predicting SD near the ski-jump spillway.	[56]
2020	Optimal power flow	By addressing single and multiobjective Optimal Power Flow (OPF) problems, this study provides a unique nature-inspired and population-based HHO approach for reducing emissions from thermal producing sources.	The findings are compared to artificial intelligence (AI), whale optimization algorithm (WOA), salp swarm algorithm (SSA), moth flame (MF), and glow warm optimization (GWO). Furthermore, according to the study on DG deployment, system losses and emissions are decreased by 9.83% percent and 26.2%, respectively.	[57]
2020	Water distribution network	A model based on the HHO was created to optimize the water distribution network for a one-month period, in Homashahr, Iran.	The findings showed that the HHO algorithm performed effectively in the challenge of optimal water supply network design. This method was equivalent to approximately 12% of the optimization in the end.	[58]
2020	Design of load frequency control	The best settings of the proportional-integral (PI) controller modeling load frequency control (LFC) in a multi-interconnected system with renewable energy sources are evaluated using a reliable technique-based HHO.	The collected findings proved the validity and superiority of the suggested HHO-based strategy for developing LFC for the systems under consideration.	[59]
2019	Design of microchannel heat sinks	For the reduction of entropy production, a unique Harris hawks optimization technique is used to microchannel heat sinks. The slip flow velocity and temperature jump boundary conditions were taken into account when creating the microchannel heat transfer model.	The Harris hawks method outperforms the other algorithms in terms of reducing microchannel entropy production.	[60]

Motivation and Contributions

The primary motivation behind this work is to design a novel technique for appropriate allocation and sizing of solar photovoltaic DGs to reduce power losses and enhance the voltage profile. Worldwide sustainable development is possible through the generation of electricity from renewable energy resources. The Indian government has taken a number of steps to stimulate the use of renewable energy (RE) resources, including setting state-specific RE objectives in the form of solar purchase obligations (SPO) and renewable purchase obligations (RPO). Every state has set a goal to fulfill a significant part of its overall energy demand from renewable resources under the provisions of RPO and SPO. The solar photovoltaic DGs (PV DG) considered in this paper are among all the renewable energy resources; solar energy has received major importance due to its abundant availability worldwide. Although researchers have previously used a variety of approaches to tackle the problem of DG allocation and size in the power system network, the authors have not considered the actual field installation capacity of PV DG; instead, they have considered only the actual power to be injected. Whereas the output of the solar PV DG is a meteorological parameter and PV module parameter-dependent system, thus, it is imperative to calculate the actual size of the PV DG to be installed to inject the targeted power into the grid. The contribution of this work is presented below:

- The proposed HHO has been tested with complex benchmark functions;
- Assign a novel approach for appropriate allocation and sizing of PV DGs in IEEE 33 bus and IEEE 69 bus power system network using HHO to minimize the power losses and improve the voltage profile;
- Compare the simulation outcomes of the proposed technique together with the recently available methods such as the teaching–learning-based optimization (TLBO), genetic algorithm (GA), particle swarm optimization (PSO), quasi-oppositional TLBO (QOTLBO), comprehensive teaching learning-based optimization (CTLBO), CTLBO ε-method, improved multiobjective elephant herding optimization (IMOEHO), improved decomposition-based evolutionary algorithm (I-DBEA), bat algorithm (BA), simulated annealing (SA), invasive weed optimization (IWO), bacterial foraging optimization algorithm (BFOA), and moth–flame optimization (MFO) to determine the effectiveness of the proposed algorithm over the exciting ones;
- Calculate the actual/practical size of the solar PV DG units to be installed to inject the targeted power into the power system grid.

The remainder of this article is structured in the following order. The mathematical formulation of the problem with various constraints is detailed in Section 2. The detail of the proposed HHO and the solution approach for the considered problem is presented in Section 3. In Section 4, the problem is tested with a benchmark function and with standard test systems. Section 5 deals with the practical calculation of solar PV DG. Section 6 concludes with some final observations together the breadth of future development.

2. Formulation of the Mathematical Problem

2.1. Loss Minimization

The objective of the present work is to relax the congestion in power lines along with determining the proper size and optimal location of DGs while keeping the losses (202.67 kW and 224.9 kW for IEEE 33 and IEEE 69 bus RDS, respectively) to the minimum. The major objective function (OF) is framed in the form of total system losses. Therefore, the OF may be stated by Equation (1).

$$OF = Minimize(P_{Loss}) \tag{1}$$

where

$$P_{Loss} = \sum_{k=1}^{n} g_k \left(V_i^2 + V_j^2 - 2V_i \times V_j \times cos(\delta_i - \delta_j) \right) \tag{2}$$

The various constraints of the proposed optimization problem are as stated in Equations (3)–(7).

$$V_i^{min} \leq V_i \leq V_i^{max} \tag{3}$$

$$P_{DG}^{min} \leq P_{DG} \leq P_{DG}^{max} \tag{4}$$

$$Q_{DG}^{min} \leq Q_{DG} \leq Q_{DG}^{max} \tag{5}$$

$$Q_{G_i}^{min} \leq Q_{G_i} \leq Q_{G_i}^{max} \tag{6}$$

$$P_{G_i}^{min} \leq P_{G_i} \leq P_{G_i}^{max} \tag{7}$$

where g_k is the conductance of branch k; V_i and V_j are the magnitude of voltages at sending and receiving bus, respectively; P_{DG} and Q_{DG} represent active and reactive power generation by DG; δ_i is the phase angle at ith and jth bus, respectively; and P_{G_i} represents active power generation at ith bus.

In Equations (3)–(7), the superscripts max and min represent the upper and the lower limits of the respective variables. The major objective here is to reduce congestion in lines and minimize losses.

2.2. Practical Sizing of PV DG

The power output of the PV module depends on meteorological parameters (such as ambient temperature and solar irradiance at the particular location) and on the parameters of the PV modules. To address the dependence on solar irradiation, the beta probability density function was used to model the uncertain nature of solar irradiance. The distribution of solar irradiance may be written as Equation (8) [61].

$$f_b(s) = \begin{cases} \frac{\Gamma(\alpha+\beta)}{\Gamma(\alpha)\Gamma(\beta)} s^{(\alpha-1)} & (1-s)^{(\beta-1)} \quad 0 \le s \le 1 \ \& \ \alpha, \beta \ge 0 \\ 0 & \end{cases} \tag{8}$$

$$\beta = (1-\mu)\left(\frac{\mu(1-\mu)}{\sigma^2} - 1\right) \tag{9}$$

$$\alpha = \frac{\mu\beta}{(1-\mu)} \tag{10}$$

where $\Gamma(^\circ)$ is defined as the gamma function, s is defined as the random variable of solar irradiance, $f_b(s)$ is defined as the beta distribution function of s, α and β are defined as the parameters of the beta distribution function, μ and σ are defined as the mean and standard deviation of s.

Equations (11)–(13) have been used to address the effect of ambient temperature on the output of the PV module. The temperature of the PV module is influenced by the nominal operating module temperature (NOMT), solar irradiance, and ambient temperature, as shown by Equation (11) [62].

$$T_M = T_A + s\left(\frac{\text{NOMT} - 20}{0.8}\right) \tag{11}$$

The output current of the PV module is a function of the solar irradiance, short-circuit current, temperature coefficient of current, and temperature of PV module, shown by Equation (12) [63].

$$I_M = s[I_{SC} + \varepsilon_i(T_M - 25)] \tag{12}$$

The voltage of the PV module is a function of the open-circuit voltage, voltage temperature coefficient of the module, and its temperature, as shown by Equation (13).

$$V_M = [V_{OC} - \varepsilon_v(T_M - 25)] \tag{13}$$

$$FF_M = \frac{I_{mpp} \times V_{mpp}}{I_{SC} \times V_{OC}} \tag{14}$$

Considering aforesaid environmental and PV module parameters correction factors, Equation (14) will be modified to Equation (15).

$$FF_M = \frac{I_{mpp} \times V_{mpp}}{I_M \times V_M} \tag{15}$$

The output power of PV module, operating at maximum power point at solar irradiance s, may be estimated using Equation (16).

$$P_o(s) = I_{mpp} \times V_{mpp} = FF_M \times I_M \times V_M \tag{16}$$

The output power of the PV plant, operating at maximum power point at solar irradiance s may be estimated using Equation (17) [64].

$$P_o(s) = N_M \times FF_M \times I_M \times V_M \quad (17)$$

The power output from the PV module considering maximum power point may be obtained by Equations (14) and (15). The variables used are defined as follows: T_M is the temperature of the PV module; T_A is the ambient temperature; NOMT is the nominal operating module temperature; I_M is the current of the PV module; V_M is the voltage of the PV module; I_{SC} is the short-circuit current of PV module; V_{OC} is the open-circuit voltage of PV module; ε_i is the temperature coefficient of current; ε_v is the temperature coefficient of voltage; I_{mpp} is the current at maximum power point at standard test condition (STC); V_{mpp} is the voltage at maximum power point at STC; FF_M is defined as the fill factor; N_M denotes the number of PV modules used in the PV plant; and $P_o(s)$ is the power output from the PV module ($N_M = 1$)/plant at solar irradiance s.

The expected output power from the PV module considering the effect of solar irradiance s and ambient temperature T_A may be calculated using Equation (18), and the expected total output power for a specific time period may be calculated using Equation (19) [64].

$$EOP(s) = P_o(s) \times f_b(s) \quad (18)$$

$$ETOP = \int_0^1 EOP(s)\, ds \quad (19)$$

Monocrystalline silicon PERC PV module of the following specifications as presented in Table 2, was used for calculation [65].

$$\varepsilon_i = T_i \times I_{mpp} \quad (20)$$

$$\varepsilon_v = T_v \times V_{mpp} \quad (21)$$

Table 2. PV module parameters.

Parameter	Specification
Nominal power—P_{mpp} (Wp)	350
V_{mpp} (V)	38.9
I_{mpp} (A)	9.0
V_{OC} (V)	46.7
I_{SC} (A)	9.72
T_v (Temperature coefficient of voltage)	−0.30 %/°C
T_i (Temperature coefficient of current)	0.066 %/°C
NOMT	44.6 °C
Area	2.01 m^2

To convert the temperature coefficient of voltage and current from %/°C (T_i and T_v) to A/°C (ε_i) and V/°C (ε_v), Equations (20) and (21) are used, respectively. Impact of NOMT and irradiance on the temperature, voltage, and current of the PV module are depicted in Figure 1.

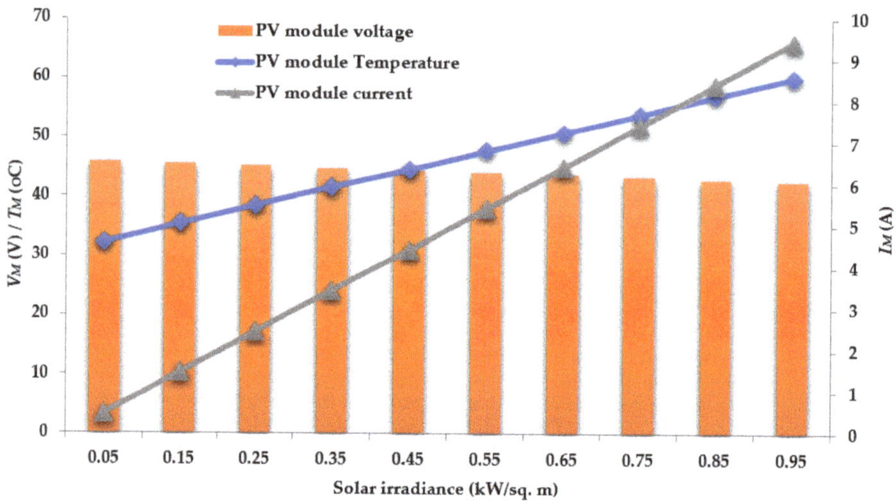

Figure 1. Different PV module parameters considering the NOMT and solar irradiance.

Considering the parameters effecting the output of PV module, it is observed that the voltage and the current of the PV module varies from 45.85 to 42.62 V and 0.49 to 9.43 A, respectively, with the variation of solar irradiance. The temperature of the PV module varies from 32.3 to 60 °C as irradiance changes, as shown in Figure 1.

The ambient temperature, mean, and standard derivation of solar irradiance during a specified time period are considered as 30.76°, 0.52 kW/m^2, and 0.21 kW/m^2, respectively [61]. The expected output power from the PV module considering the effect of solar irradiance and ambient temperature, associated environmental parameters, PV module parameters and modeling parameters are tabulated in Table 3.

Table 3. The expected output power from the PV module considering correction factors.

Environmental Parameters		PV Module Parameters Considering Correction Factors		Modeling Parameters		Output	
s	T_A	I_M	V_M	$f_b(s)$	β	$P_o(s)$	$EOP(s)$
0.05	30.76	0.49	45.85	2.16	0.14	17.26	2.38
0.15	30.76	1.47	45.49	2.16	0.56	51.48	28.66
0.25	30.76	2.45	45.13	2.16	0.98	85.28	83.44
0.35	30.76	3.44	44.77	2.16	1.32	118.66	156.88
0.45	30.76	4.43	44.41	2.16	1.54	151.62	234.05
0.55	30.76	5.42	44.05	2.16	1.62	184.16	297.55
0.65	30.76	6.42	43.70	2.16	1.52	216.27	329.15
0.75	30.76	7.42	43.34	2.16	1.26	247.95	311.49
0.85	30.76	8.42	42.98	2.16	0.83	279.20	230.65
0.95	30.76	9.43	42.62	2.16	0.27	310.02	83.51
						Average	175.78

The power output from the PV module $P_o(s)$ and $f_b(s)$ with respect to solar irradiance s is presented in Figure 2.

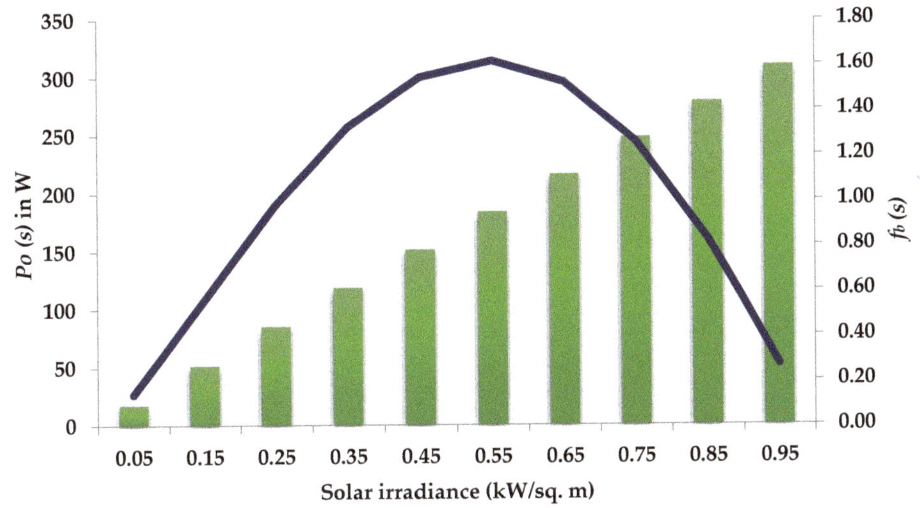

Figure 2. Output power from the PV module at different solar irradiance s.

The expected total power obtained from a single PV module is the average of $EOP(s)$, which is shown in Table 3 and depicted in Figure 3, i.e., 175.78 W.

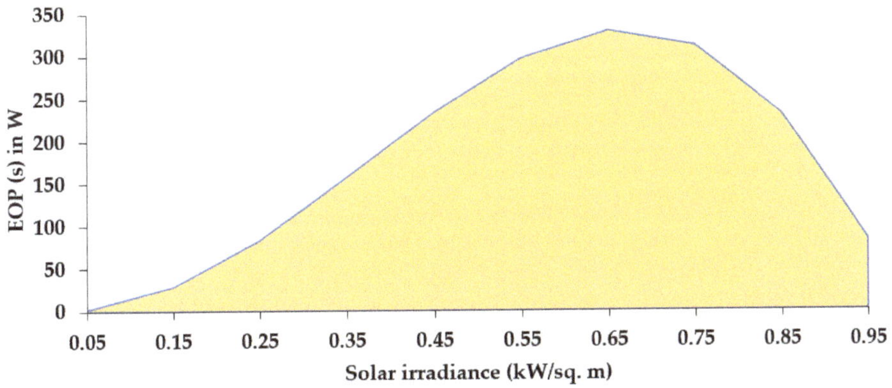

Figure 3. Expected output power from the PV module.

3. Proposed HHO and Solution Approach

3.1. HHO: Features

The Harris hawks is a recent population-based and gradient-free metaheuristic [66], hence, equally applicable to all optimization models or problems. The different phases of Harris hawks formulation are described in the next subsections.

3.2. Exploration Phase

In this phase, Harris hawks randomly search on locations and adopt a wait and watch strategy to catch the prey, as per Equation (22) [66].

$$X(t+1) = \begin{cases} X_{rand}(t) - r_1|X_{rand}(t) - 2r_2X(t)| & q \geq 0.5 \\ (X_{rabit}(t) - X_m(t)) - r_3(LB + r_4(UB - LB)) & q < 0.5 \end{cases} \quad (22)$$

where Xrabbit is the rabbit position and *X (t+1)* is the hawks' position for next iteration; *x (t)* shows the current position of the hawks. The *LB* and *UB* are maximum and minimum of decision variables. $X_{rand}(t)$ is a randomly selected hawk from the current position. The random number in the range (0,1) is shown by the $r_{(1\text{–}4)}$.

$$X_m(t) = \frac{1}{N}\sum_{i=1}^{N} X_i(t) \tag{23}$$

where X_m is the mean of the current population of hawks while N indicates the hawks' total population.

Assuming the energy of the rabbit is given by

$$E = 2E_o\left(1 - \frac{t}{T}\right) \tag{24}$$

E, E_o, and T represent the escaping energy of prey, primary energy, and the maximum number of iterations taken, respectively.

3.3. Exploitation Phase

3.3.1. Soft Besiege

This behavior is demonstrated by Equation (25) [66].

$$X(t+1) = \Delta X(t) - E|JX_{rabit}(t) - X(t)| \tag{25}$$

$$\Delta X(t) = X_{rabit}(t) - X(t) \tag{26}$$

where $X(t)$ and J represent the difference between the position vector of the rabbit and the current location in iteration and the random jump strength of the rabbit, respectively.

3.3.2. Hard Besiege

This behavior is showcased by (27).

$$(t+1) = X_{rabit}(t) - E|\Delta X(t)| \tag{27}$$

3.3.3. Soft Besiege along with Rapid Drives

In this behavior, it is assumed that hawks may choose their next step provided by the rule given in Equation (28) [66].

$$Y = X_{rabit}(t) - E|JX_{rabit}(t) - X(t)| \tag{28}$$

$$Z = Y + S \times LF(D) \tag{29}$$

D, S, and LF are problem dimensions, a random number of order (1 × D), and levy flight function, respectively. In addition, u and v are random numbers (0 to 1 range), while beta is the default constant value (assuming 1.5).

$$LF(x) = 0.01 \times \frac{\mathcal{U} \times \sigma}{|\vartheta|^{\frac{1}{\beta}}},\ \sigma = \left(\frac{\Gamma(1+\beta) \times \sin\left(\frac{\pi\beta}{2}\right)}{\Gamma\left(\frac{1+\beta}{2}\right) \times \beta \times 2^{\left(\frac{\beta-1}{2}\right)}}\right)^{\frac{1}{\beta}} \tag{30}$$

Soft besiege updates the position of the hawks by

$$X(t+1) = \begin{cases} Y, & if\ F(Y) < F(X(t)) \\ Z, & if\ F(Z) < F(X(t)) \end{cases} \tag{31}$$

3.3.4. Hard Besiege along with Rapid Drives

Hard besiege condition given by the following rule:

$$X(t+1) = \begin{cases} Y, & if\ F(Y) < F(X(t)) \\ Z, & if\ F(Z) < F(X(t)) \end{cases} \quad (32)$$

$$Y = X_{rabit}(t) - E|JX_{rabit}(t) - X_m(t)| \quad (33)$$

$$Z = Y + S \times LF(D) \quad (34)$$

The step-by-step procedure of HHO is summarized to the pseudocode, as shown in Figure 4 [66].

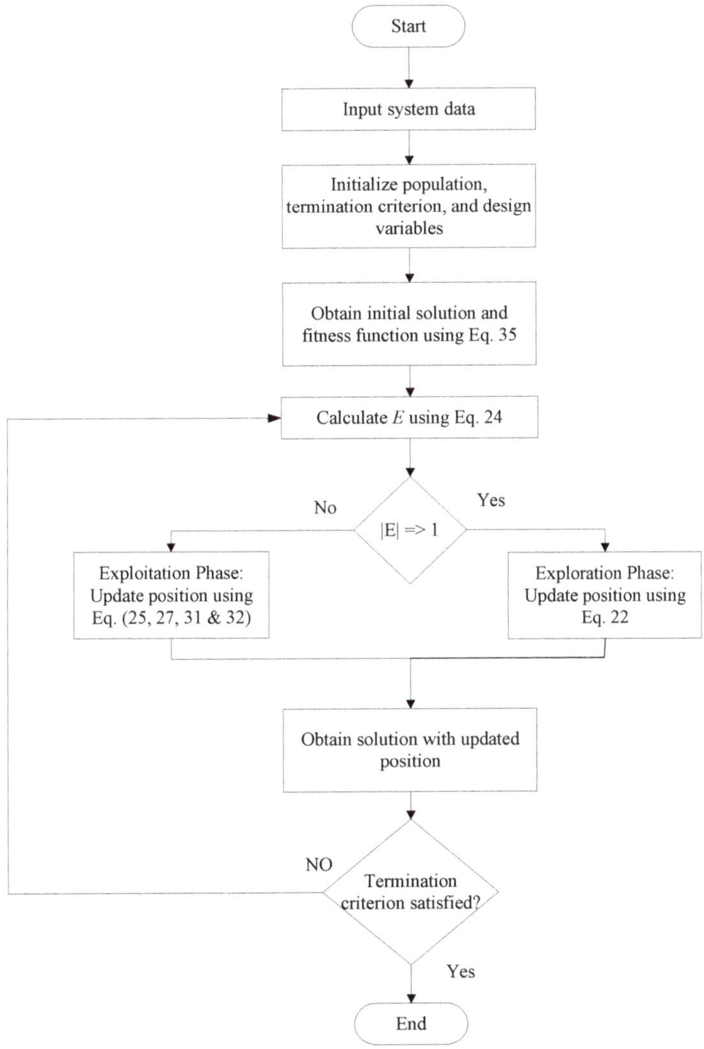

Figure 4. Flow chart of HHO. Reproduced from [66], Elsevier: 2019.

3.4. Solution Approch

3.4.1. HHO for PV DG Placement and Location

The major goal of this research is to determine the best placement and size for numerous PV DGs with the least amount of network power loss and a better voltage profile. In this work, the inequality constraints are converted to the penalty functions (*PFs*), and these *PFs* are added to the *OF* to construct the fitness function (*FF*) defined in Equation (35).

$$Minimum\ FF = OF + PF \times \sum_{j=1}^{VB} (\Delta V_j)^2 \tag{35}$$

Here, *FF* is essential to be minimized in order to get minimum loss value, *VB* represents the set of overloaded lines and voltage violated load buses, and *PF* represents the penalty factor. The violation in inequality constraints such as load bus voltage and line power flows was handled using the penalty function approach. *PF* that represents penalty factor was taken as 10,000 throughout the simulation process.

3.4.2. Computational Practice of HHO for DG Location and Values

Step 1 Read the input data of the system, such as the maximum number of iterations, number of PV DG units, and population size.
Step 2 Generate the value of the size of PV DG within their upper (DG_{max}) and lower limits (DG_{min}). The same is shown in Equation (36).

$$DG_i = DG_i^{min} + rand \times \left(DG_i^{max} - DG_i^{min}\right) \tag{36}$$

Here, DG_i represents the size of *i*th *DG* unit. Now, constitute a vector X_j, that contains the possible locations (*LOC*) and size of *DG*s as mentioned in Equation (37).

$$X_j = \left[DG_{j,1}, DG_{j,2}, \ldots\ldots\ldots, DG_{j,n}, LOC_{j,1}, LOC_{j,2}, \ldots\ldots\ldots, LOC_{j,n}\right] \tag{37}$$

The *LOC* is generated randomly. Initial solution set *X* is then formulated as shown in Equation (38).

$$X = [X_1, X_2, \ldots\ldots\ldots\ldots, X_N] \tag{38}$$

Step 3 Evaluation of the fitness function is processed using Equation (35) for individual Harris hawks, and the best hawk location is acknowledged.
Step 4 Calculate *E* using Equation (24).
Step 5 Exploration phase: Update the location of Harris hawks using Equation (22).
Step 6 Exploitation phase: Update the position using Equation (25), (27), (31), and (32).
Step 7 Once the number of iterations reaches the maximum value, then terminate. Else, go back to Step 3.

4. Simulation Results and Discussions

4.1. Testing Strategies

The simulations were run on a MATLAB 9.9 computer with an Intel i3 CPU running at 2.4 GHz and 4 GB of RAM. The software utilized is MATPOWER 7.2, which is a well-known power modeling tool.

4.2. Case 1

In order to establish an algorithm, the proposed HHO was tested with selected extremely complex benchmark functions taken from CEC-2014 (see Table 4). The results obtained are tabulated in Table 5. The HHO seems to provide very competitive results as compared to other recent metaheuristic optimization techniques.

Table 4. Summary of the CEC-2014 benchmark functions considered.

Type	ID	Functions	Fi = Fi(x)
Unimodal	F1	Rotated High Conditioned Elliptic	100
	F2	Rotated Bent Cigar	200
Simple Multimodal	F3	Shifted and Rotated Rastrigin's	900
	F4	Shifted Schwefel's	1000
Hybrid	F5	Hybrid Function 3 (N = 4)	1900
	F6	Hybrid Function 4 (N = 4)	2000
Composition	F7	Composition Function 8 (N = 3)	3000

Table 5. Comparative experimental outcomes on selected benchmark functions.

ID	Parameters	PSO	TLBO	CS	GSA	SFS	HHO
F1	max	4.56×10^8	8.93×10^8	5.51×10^8	5.31×10^7	1.17×10^6	3.01×10^5
	min	2.47×10^8	4.39×10^7	1.18×10^8	4.56×10^6	1.54×10^5	1.43×10^4
	median	3.31×10^8	3.42×10^8	3.10×10^8	8.37×10^6	6.16×10^5	1.52×10^5
	std	7.92×10^7	3.42×10^8	1.05×10^8	1.32×10^7	2.35×10^5	1.23×10^5
F2	max	3.63×10^{10}	4.06×10^4	2.42×10^4	1.61×10^4	2.00×10^2	2.00×10^2
	min	6.00×10^7	6.00×10^3	3.09×10^2	3.47×10^3	2.00×10^2	$\mathbf{2.00 \times 10^2}$
	median	1.55×10^{10}	1.52×10^4	8.08×10^3	8.38×10^3	2.00×10^2	2.00×10^2
	std	1.43×10^{10}	8.65×10^3	6.00×10^3	2.90×10^3	7.89×10^{-9}	0. 00
F3	max	1.24×10^3	1.12×10^3	1.34×10^3	1.10×10^3	9.84×10^2	9.03×10^2
	min	1.13×10^3	1.06×10^3	1.15×10^3	1.02×10^3	9.35×10^2	$\mathbf{9.20 \times 10^2}$
	median	1.18×10^3	1.09×10^3	1.25×10^3	1.06×10^3	9.61×10^2	9.19×10^2
	std	4.33×10	0.25×10^2	4.41×10	1.74×10	1.11×10	1.017×10
F4	max	7.90×10^3	5.92×10^3	3.21×10^3	5.25×10^3	2.71×10^3	1.05×10^3
	min	6.26×10^3	4.14×10^3	1.36×10^3	3.45×10^3	1.02×10^3	$\mathbf{1.00 \times 10^3}$
	median	7.18×10^3	5.06×10^3	2.17×10^3	4.37×10^3	1.49×10^3	1.01×10^3
	std	5.98×10^2	7.89×10^2	4.33×10^2	3.61×10^2	3.62×10^2	1.45×10
F5	max	2.10×10^3	1.91×10^3	2.04×10^3	2.00×10^3	1.91×10^3	1.92×10^3
	min	1.91×10^3	1.90×10^3	1.91×10^3	1.91×10^3	1.90×10^3	$\mathbf{1.90 \times 10^3}$
	median	1.97×10^3	1.91×10^3	1.92×10^3	2.00×10^3	1.91×10^3	1.91×10^3
	std	7.07×10	1.65	3.30×10	3.43×10	1.47	1.46
F6	max	4.37×10^3	5.34×10^3	6.02×10^4	6.82×10^4	2.10×10^3	2.75×10^3
	min	2.55×10^3	2.30×10^3	2.22×10^4	2.32×10^3	2.02×10^3	$\mathbf{2.00 \times 10^3}$
	median	3.00×10^3	2.74×10^3	3.68×10^4	1.77×10^4	2.06×10^3	2.26×10^3
	std	5.32×10^2	7.00×10^2	8.42×10^4	1.39×10^4	2.60×10	2.06×10^2
F7	max	9.70×10^5	1.56×10^6	5.08×10^5	1.14×10^5	7.66×10^3	5.62×10^3
	min	6.90×10^4	2.08×10^4	6.26×10^4	1.22×10^4	4.25×10^3	$\mathbf{3.56 \times 10^3}$
	median	3.35×10^5	6.56×10^5	1.77×10^5	1.46×10^4	5.63×10^3	4.71×10^3
	std	3.63×10^5	5.64×10^5	9.11×10^4	1.84×10^4	7.38×10^2	1.30×10^3

4.3. Case 2

The proposed HHO-based approach is applied to find the suitable location and capacity of the DGs in the IEEE 33-bus RDS test system where the network and load data may be obtained from [67]. The single line diagram of the IEEE 33-bus RDS is shown in Figure 5. IEEE 33-bus RDS has a total of 33 buses, among which 32 are load buses and 1 is a generator bus. It can be visualized from Figure 5 that at bus no. 1 generator is connected; the other buses may have any type of load connected, as per the requirement. The total active power demand is 3.72 MW while reactive is 2.3 MVAR. Total power loss of the system is 202.67 kW.

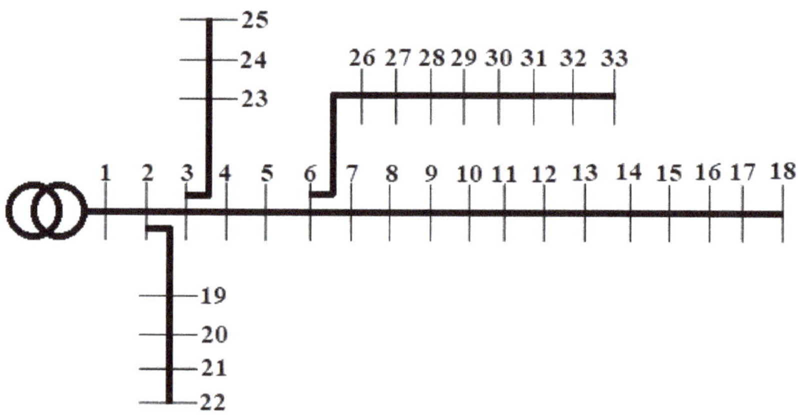

Figure 5. Single line diagram of IEEE 33-bus RDS.

In order to find candidate buses for locating a PV DG using this approach for each individual bus, it is assumed that there is a PV DG at that bus at a time. For optimal sizing of a PV DG at this stage, it is assumed that the PV DG may produce electric power in all possible ranges (e.g., 0–1 MW). The proposed HHO algorithm is applied for the minimization of overall loss as the objective function of the problem. First, only one PV DG is used to relax the congestion in lines, and the results obtained are tabulated in Table 6. With the application of proposed HHO on distribution problem, the losses are reduced to 129.2 from 202.67 kW with only one DG in installation of size 0.95 MW.

Table 6. Variation in power loss with change in an optimal allocation of PV DGs.

Test System	Buses Count	Array Location	P_{loss} (kW)	Loss Reduction (%)
33 bus system	1	30	129.20	38.76
	2	12, 30	86.90	58.81
	3	13, 24, 30	72.10	64.42

For further improvement, the problem is tested by installing two and three PV DGs in the power network. The results obtained are presented in Table 6. The overall active power losses decreased to 86.9 and 72.10 kW with the application of two and three PV DGs, respectively, using HHO. The comparative results are portrayed in Table 7 in terms of the best location and size of PV DGs. The locations suggested by HHO to install PV plants in IEEE 33 bus are depicted in Figure 6.

Table 7. Comparative results for optimal location and values of PV DGs corresponding to case 2.

Optimization Method	Bus Count	Array Location	DG Size (MW)	Total DG Size (MW)	P_{loss} (kW)	Loss Reduction (%)
Base case	-	-	-	-	202.67	0.00
TLBO [68]	3	12	1.1826	3.560	124.70	38.47
		28	1.1913			
		30	1.1863			
GA [69]	3	11	1.5000	2.994	106.30	47.55
		29	0.4230			
		30	1.0710			
PSO [69]	3	8	1.1770	2.989	105.30	48.04
		13	0.9820			
		32	0.8300			

Table 7. *Cont.*

Optimization Method	Bus Count	Array Location	DG Size (MW)	Total DG Size (MW)	P_{loss} (kW)	Loss Reduction (%)
GA/PSO [69]	3	11 16 32	0.9250 0.8630 1.2000	2.998	103.40	48.98
QOTLBO [68]	3	13 26 30	1.0834 1.1876 1.1992	3.470	103.40	48.98
CTLBO ε-method [70]	3	13 25 30	1.1926 0.8706 1.6296	3.693	96.17	52.55
IMOEHO [71]	3	14 24 30	1.0570 1.0540 1.7410	3.852	95.00	53.13
I-DBEA [72]	3	13 24 30	1.0980 1.0970 1.7150	3.913	94.85	53.20
CTLBO [70]	3	13 24 30	1.0364 1.1630 1.5217	3.721	85.96	57.59
BA [39]	3	15 25 30	0.81630 0.95235 0.95235	2.721	75.05	62.97
HHO [Proposed]	3	13 24 30	0.8311 0.9500 0.9500	2.731	72.10	**64.42**

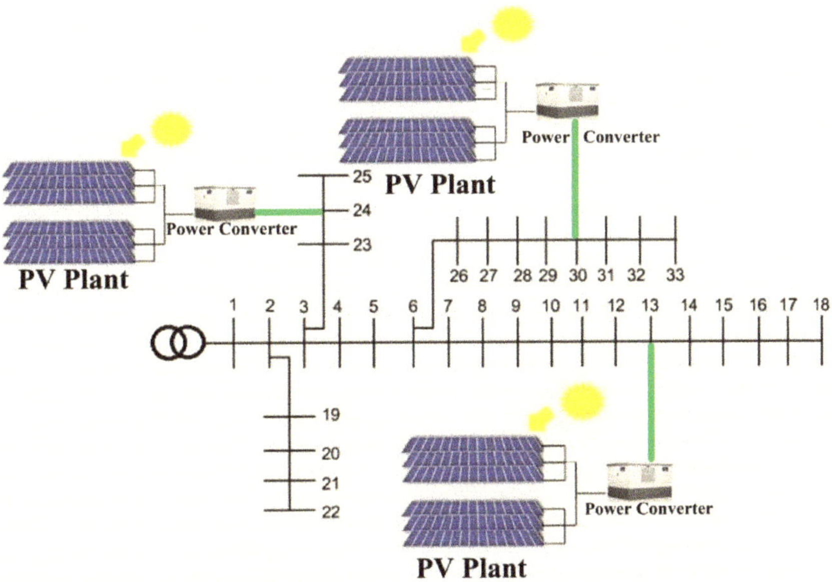

Figure 6. IEEE 33 bus line diagram with PV plants at locations suggested by HHO.

The size and location suggested by HHO (refer to Table 7) provide maximum reduction in losses as compared to TLBO [68], GA [69], PSO [69], GA/PSO [69], QOTLBO [68], CTLBO ε-method [70], IMOEHO [71], I-DBEA [72], CTLBO [70], and BA [39]. In addition, the voltage graph of all the buses obtained after utilization of PV DGs is showcased in Figure 7.

The bus voltages are obtained from the load flow analysis. The bus voltage profile improves significantly under the application of three PV DGs at their respective optimal locations. The variation of fitness function against the number of iterations for installation with three PV DGs using HHO is showcased in Figure 8. The iterative graph shows that the HHO converges to an optimal solution value with very few iterations.

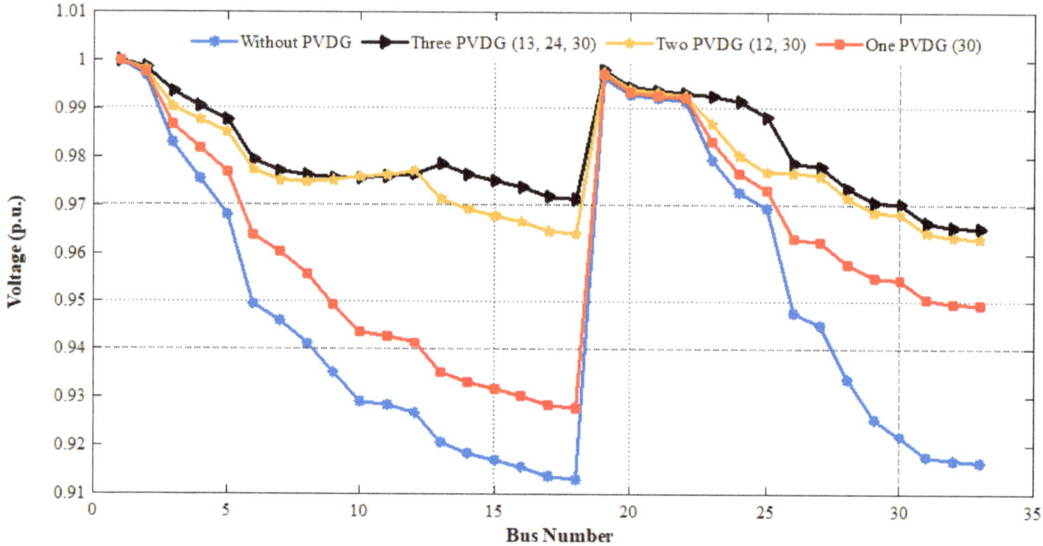

Figure 7. Voltage graph for IEEE 33 bus system without and with PV DGs.

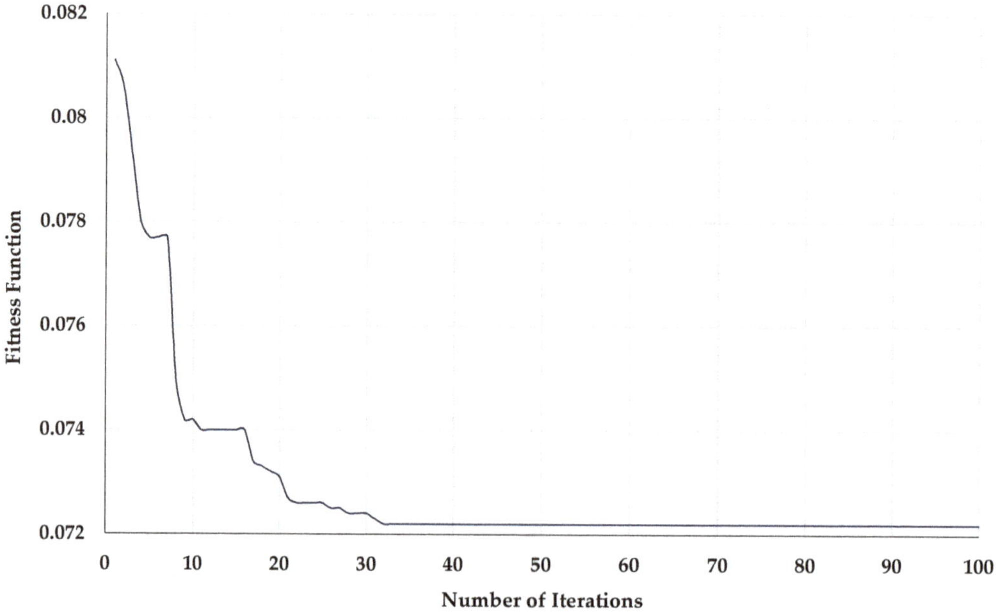

Figure 8. Convergence characteristics of fitness function pertaining to case 2 (with three PV DG).

4.4. Case 3

To test the effectiveness of the HHO on a larger system, the proposed approach is tested to find the suitable location and capacity (size) of the DGs in the IEEE 69 bus RDS test system where the load and branch data values may be obtained from [73]. The single line diagram of the IEEE 69-bus RDS is shown in Figure 9. IEEE 69-bus RDS consists of 69 buses, including 68 load buses and 1 generator bus. The generator is connected at bus no. 1 and a load of the required amount can be connected to the other buses. The total active power demand is 3.80 MW while reactive is 2.69 MVAR. Total power loss of the system is 224.9 kW.

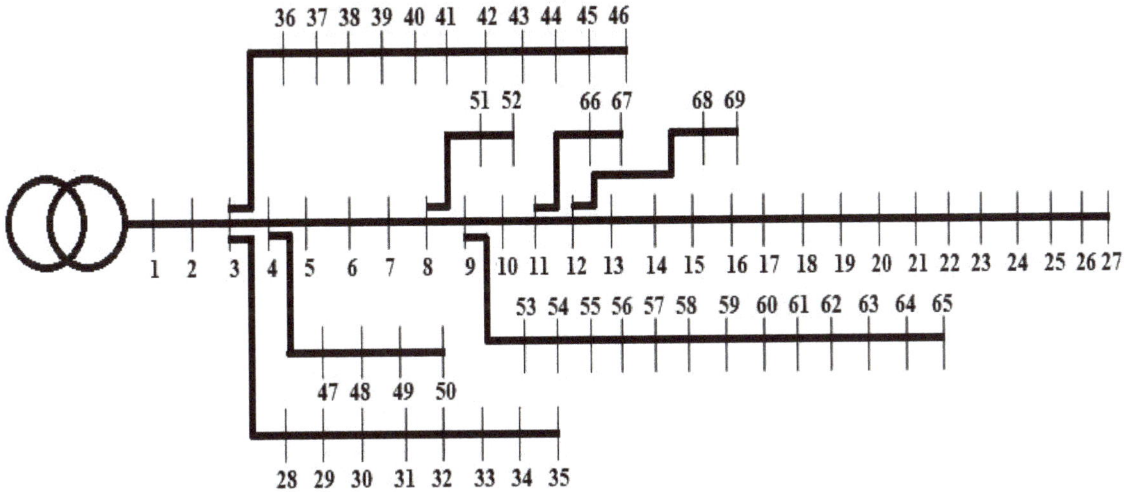

Figure 9. Single line diagram of IEEE 69-bus RDS.

For addressing the most suitable candidate buses for locating a PV DG using this approach for each individual bus, it is assumed that there is a PV DG at that bus at a time. For optimal sizing of a PV DG at this stage, it is assumed that the PV DG may produce electric power in all possible ranges (e.g., 0–1 MW). The proposed HHO algorithm is applied for the reduction/minimization of overall loss as the objective function of the problem. First, only one PV DG is used to relax the congestion and reduce losses in lines, and the results obtained are tabulated in Table 8. The installation of one optimum PV DG the line losses reduced by 48.86% with DG size of 0.95 MW.

Table 8. Variation in power loss with change in an optimal allocation of PV DGs.

Test System	Bus Count	Array Location	DG Size (MW)	P_{loss} (kW)	Loss Reduction (%)
69 bus system	1	61	0.95	115	48.866
	2	61, 62	0.95, 0.9118	83.4	62.916
	3	17, 61, 62	0.5329, 0.95, 0.822	71.8	68.074

For further improvement, the problem is tested by installing two and three PV DGs in the power network. The results obtained are presented in Table 8. The overall active power losses decreased to 71.80 kW with the application of three PV DGs using HHO. The comparative results are portrayed in Table 9 in terms of the best location and size of PV DGs. The locations suggested by HHO to install PV plants in the IEEE 69 bus are depicted in Figure 10.

Table 9. Comparative results for optimal location and values of PV DGs corresponding to case 3.

Optimization Method	Bus Count	Array Location	DG Size (MW)	Total DG Size (MW)	P_{loss} (kW)	Loss Reduction (%)
Base case	-	-	-	-	224.9	0.00
GA [69]	3	21 62 64	0.9297 1.0752 0.9848	2.9897	89	60.43
PSO [69]	3	61 63 17	1.1998 0.7956 0.9925	2.9879	83.2	60.43
TLBO [68]	3	13 61 62	1.0134 0.9901 1.1601	3.1636	82.172	63.46
GA/PSO [69]	3	63 61 21	0.8849 1.1926 0.9105	2.988	81.1	63.94
QOTLBO [68]	3	15 61 63	0.8114 1.1470 1.0022	2.9606	80.585	64.17
CTLBO ε-method [70]	3	12 25 61	0.9658 0.2307 2.1336	3.3301	79.66	64.58
I-DBEA [72]	3	61 19 11	2.1487 0.4717 0.7126	3.32	78.347	65.16
SA [74]	3	18 60 65	0.4204 1.3311 0.4298	2.1813	77.09	65.72
CTLBO [70]	3	11 18 61	0.5603 0.4274 2.1534	3.1411	76.372	66.04
IWO [75]	3	27 65 61	0.2381 0.4334 1.3266	1.9981	76.12	66.15
BFOA [76]	3	27 65 61	0.2954 0.4476 1.3451	2.0881	75.21	66.56
MFO [77]	3	61 18 11	2.0000 0.3803 0.5822	2.9625	72.37	67.82
HHO [Proposed]	3	17 61 62	0.5329 0.9500 0.8220	2.3049	71.8	**68.07**

The size and location suggested by HHO (refer to Table 9) provides maximum reduction in losses as compared to GA [69], PSO [69], TLBO [68], GA/PSO [69], QOTLBO [68], CTLBO ε-method [70], I-DBEA [72], SA [74], CTLBO [70], IWO [75], BFOA [76], and MFO [77]. In addition, the voltage graph of all the buses obtained after utilization of PV DGs is showcased in Figure 11. The bus voltages are obtained from the load flow analysis. The bus voltage profile improves significantly under the application of three PV DGs at their respective optimal locations. The variation of fitness function against the number of iterations for installation with three PV DGs using HHO is showcased in Figure 12. The iterative graph shows that the HHO converges to an optimal solution value with very few iterations.

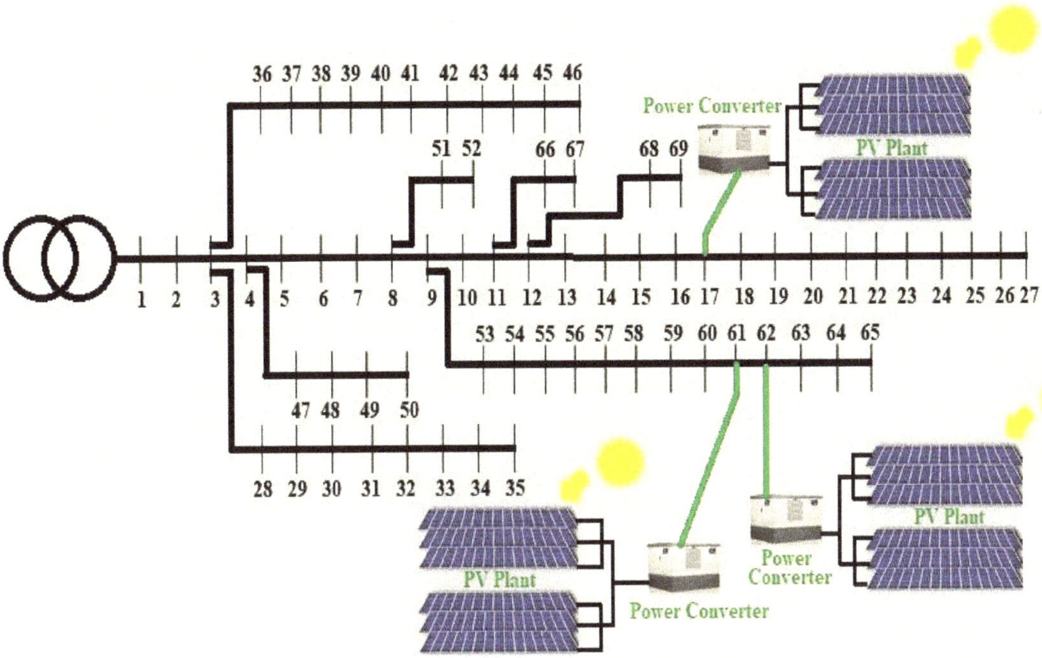

Figure 10. IEEE 69 bus line diagram with PV plants at locations suggested by HHO.

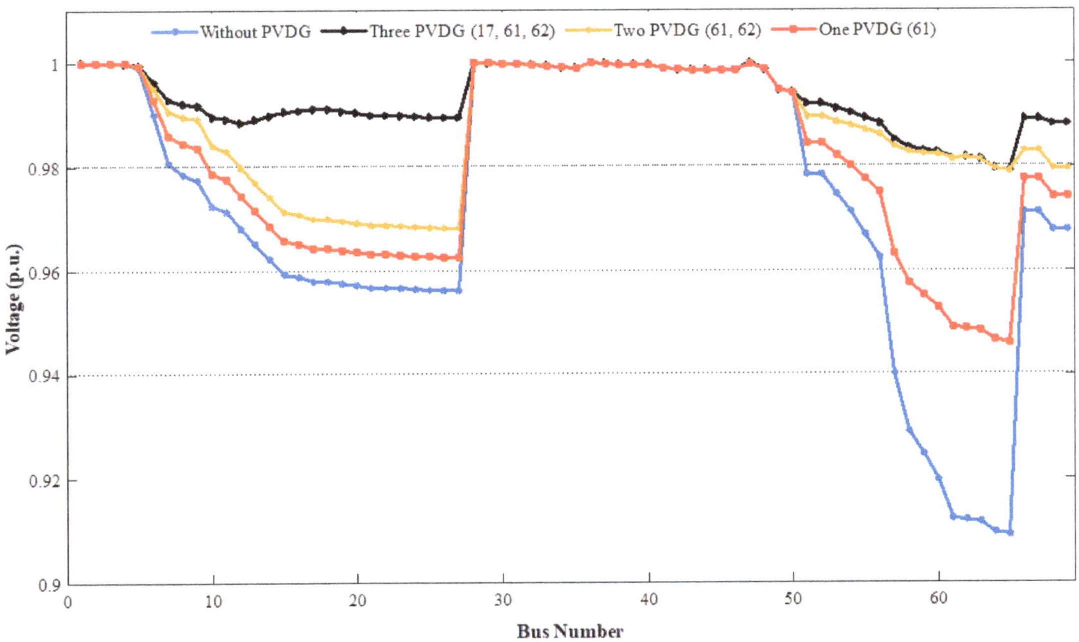

Figure 11. Voltage graph for IEEE 69 bus system without and with PV DGs.

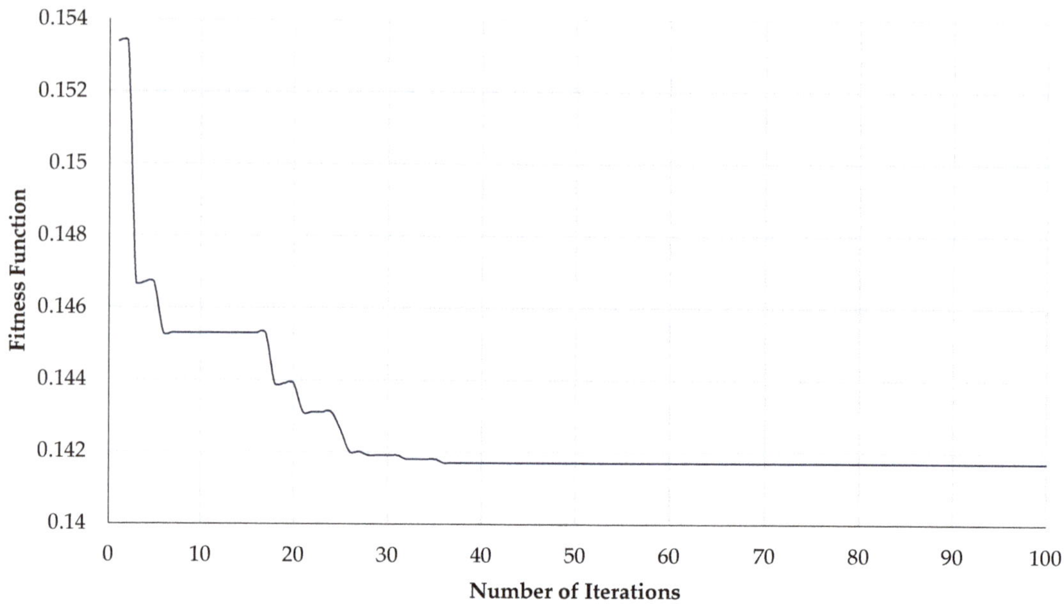

Figure 12. Convergence characteristics of fitness function pertaining to case 3 (with three PV DGs).

5. Practical PV DG Size Analysis

The optimum allocation of PV DGs using HHO indicates (refer to Table 7) that PV DGs need to be installed at bus number 13, 24, and 30 in case 2 and at bus number 17, 61 and 62 in case 3 (refer to Table 9) for optimum results. The analysis based on calculation from Section 2.2 suggests that in case 2, power to be injected at bus number 13, 24, and 30 are 831, 950, and 950 kW, respectively. In case 3, power to be injected at bus number 17, 61, and 62 are 532.9, 950, and 822 kW, respectively. Practically, to inject the targeted power as calculated through HHO, DC overload needs to be considered as described in Section 2.2. The actual field size of the PV DG plant and the number of PV modules required for case 2 and case 3 are tabulated in Table 10.

Table 10. Size of practical PV DG corresponding to case 2 and case 3 (with three PV DGs).

Case No.	Bus No.	Targeted Power to Be Injected (kW)	Actual Size of PV DG (kW)	DC Overload (kW)	Number of PV Modules
	13	831.0	1655.00	824.00	4728
Case 2	24	950.0	1891.00	941.00	5404
	30	950.0	1891.00	941.00	5404
	17	532.9	1061.08	528.18	3032
Case 3	61	950.0	1891.59	941.59	5405
	62	822.0	1636.73	814.73	4676

In case 2, the number of PV modules used in bus number 13 is 4728 to form a PV array of capacity 1.655 MW_P. The PV array of capacity 1.891 MW_P is proposed to be installed at bus number 24 and 30 using 5404 numbers of PV modules at each bus. In case 3, the practical size of the PV plant to be installed at bus number 17, 61, and 62 are of size 1.061, 1.892, and 1.637 MW_P, respectively. The analysis shows that in practical conditions, approximately 50% DC overload exists on PV DGs installation. The allocation of PV DGs in-network without considering the practical size of the PV DG may lead to underperformance.

6. Conclusions

The proposed HHO-based approach is significantly effective in finding the optimum number, optimal locations, and optimal sizes of DGs. On conventional IEEE 33 and IEEE 69 bus test systems, the efficacy of the suggested technique is evaluated. With the employment of optimally sized DGs at their optimum location, voltage profiles of load buses are improved and the losses reduced noticeably. When more PV DGs are installed, the system's performance improves. The comparison with the TLBO, GA, PSO, QOTLBO, CTLBO, CTLBO ε-method, IMOEHO, I-DBEA, SA, IWO, BFOA, MFO, and BA methods shows that the proposed method performs comparatively better among all. The active power loss is reduced by 64.42% and 68.07% using this approach with the installation of three PV DGs in IEEE 33 and IEEE 69 bus RDS, respectively. The collected findings demonstrate that the proposed technique reduces power loss by a greater proportion with a smaller DG size when compared to other algorithms, and it offers better convergence properties. According to the analysis, there is roughly a 50% DC overload on PV DGs installations under real-world situations. The in-network allocation of PV DGs without consideration of the PV DG's realistic size may result in underperformance.

The HHO is established as a reliable optimization technique for tackling the congestion problem of power systems with the application of DGs. The proposed method provides an alternative way for both system operators and energy producers to tackle complex problems such as voltage instability, transmission congestion, and huge system losses in an impressive way. In addition, current work suggests the practical size of PV DGs that may be employed for producing effective outcomes.

Author Contributions: Writing—original draft, formal analysis, and validation, S.C.; writing—original draft, conceptualization, investigation, methodology, and software, S.V.; writing—original draft, resources, supervision, and visualization, A.S.; data curation, validation and investigation, R.M.E.; writing—review and editing, D.E.; writing—review and editing, L.M.-P.; All authors have read and agreed to the published version of the manuscript.

Funding: This research received no external funding.

Institutional Review Board Statement: Not applicable.

Informed Consent Statement: Not applicable.

Data Availability Statement: Not applicable.

Conflicts of Interest: The authors declare no conflict of interest.

References

1. Sandhya, K.; Chatterjee, K. A review on the state of the art of proliferating abilities of distributed generation deployment for achieving resilient distribution system. *J. Clean. Prod.* **2021**, *287*, 125023. [CrossRef]
2. Wang, H.; Wang, J.; Piao, Z.; Meng, X.; Sun, C.; Yuan, G.; Zhu, S. The optimal allocation and operation of an energy storage system with high penetration grid-connected photovoltaic systems. *Sustainability* **2020**, *12*, 6154. [CrossRef]
3. Sahib, J.T.; Ghani, A.R.M.; Jano, Z.; Mohamed, H.I. Optimum allocation of distributed generation using PSO: IEEE test case studies evaluation. *Int. J. Appl. Eng. Res.* **2017**, *12*, 2900–2906.
4. Saad, N.M.; Sujod, M.Z.; Hui Ming, L.; Abas, M.F.; Jadin, M.S.; Ishak, M.R.; Abdullah, N.R.H. Impacts of photovoltaic distributed generation location and size on distribution power system network. *Int. J. Power Electron. Drive Syst.* **2018**, *9*, 905. [CrossRef]
5. Jabr, R.A. Linear decision rules for control of reactive power by distributed photovoltaic generators. *IEEE Trans. Power Syst.* **2018**, *33*, 2165–2174. [CrossRef]
6. Al-Ammar, E.A.; Farzana, K.; Waqar, A.; Aamir, M.; Saifullah; Ul Haq, A.; Zahid, M.; Batool, M. ABC algorithm based optimal sizing and placement of DGs in distribution networks considering multiple objectives. *Ain Shams Eng. J.* **2021**, *12*, 697–708. [CrossRef]
7. Kumawat, A.; Singh, P. Optimal placement of capacitor and DG for minimization of power loss using genetic algorithm and artificial bee colony algorithm. *Int. Res. J. Eng. Technol.* **2016**, *3*, 2482–2488.
8. Zakaria, Y.Y.; Swief, R.A.; El-Amary, N.H.; Ibrahim, A.M. Optimal distributed generation allocation and sizing using genetic and ant colony algorithms. *J. Phys. Conf. Ser.* **2020**, *1447*, 012023. [CrossRef]
9. Sambaiah, K.S.; Jayabarathi, T. Loss minimization techniques for optimal operation and planning of distribution systems: A review of different methodologies. *Int. Trans. Electr. Energy Syst.* **2020**, *30*, e12230. [CrossRef]

10. Hassan, A.A.; Fahmy, F.H.; Nafeh, A.E.-S.A.; Abu-elmagd, M.A. Hybrid genetic multi objective/fuzzy algorithm for optimal sizing and allocation of renewable DG systems: Genetic/Fuzzy Optimization of Renewable DGS. *Int. Trans. Electr. Energy Syst.* **2016**, *26*, 2588–2617. [CrossRef]
11. Patel, D.K.; Singh, D.; Singh, B. Genetic algorithm-based multi-objective optimization for distributed generations planning in distribution systems with constant impedance, constant current, constant power load models. *Int. Trans. Electr. Energy Syst.* **2020**, *30*, e12576. [CrossRef]
12. Almabsout, E.A.; El-Sehiemy, R.A.; An, O.N.U.; Bayat, O. A hybrid local search-genetic algorithm for simultaneous placement of DG units and shunt capacitors in radial distribution systems. *IEEE Access* **2020**, *8*, 54465–54481. [CrossRef]
13. Vatani, M.; Solati Alkaran, D.; Sanjari, M.J.; Gharehpetian, G.B. Multiple distributed generation units allocation in distribution network for loss reduction based on a combination of analytical and genetic algorithm methods. *IET Gener. Transm. Distrib.* **2016**, *10*, 66–72. [CrossRef]
14. Madhusudhan, M.; Kumar, N.; Pradeepa, H. Optimal location and capacity of DG systems in distribution network using genetic algorithm. *Int. J. Inf. Technol.* **2021**, *13*, 155–162.
15. Ayodele, T.R.; Ogunjuyigbe, A.S.O.; Akinola, O.O. Optimal location, sizing, and appropriate technology selection of distributed generators for minimizing power loss using genetic algorithm. *J. Renew. Energy* **2015**, *2015*, 83291. [CrossRef]
16. Sattianadan, D.; Sudhakaran, M.; Dash, S.S.; Vijayakumar, K.; Ravindran, P. Optimal placement of DG in distribution system using genetic algorithm. In *Swarm, Evolutionary, and Memetic Computing*, 1st ed.; Panigrahi, B.K., Suganthan, P.N., Das, S., Dash, S.S., Eds.; Springer International Publishing: Cham, Switzerland, 2013; Volume 8298, pp. 639–647.
17. Liu, L.; Xie, F.; Huang, Z.; Wang, M. Multi-objective coordinated optimal allocation of DG and EVCSs based on the V2G mode. *Processes* **2020**, *9*, 18. [CrossRef]
18. Hassan, A.S.; Sun, Y.; Wang, Z. Multi-objective for optimal placement and sizing DG units in reducing loss of power and enhancing voltage profile using BPSO-SLFA. *Energy Rep.* **2020**, *6*, 1581–1589. [CrossRef]
19. Fan, Z.; Yi, H.; Xu, J.; Liu, P.; Hou, H.; Cui, R.; Xie, C. Multi-objective planning of DGs considering ES and EV based on source-load spatiotemporal scenarios. *IEEE Access* **2020**, *8*, 216835–216843. [CrossRef]
20. Liu, W.; Xu, H.; Niu, S.; Xie, J. Optimal distributed generator allocation method considering voltage control cost. *Sustainability* **2016**, *8*, 193. [CrossRef]
21. Azam Muhammad, M.; Mokhlis, H.; Naidu, K.; Amin, A.; Fredy Franco, J.; Othman, M. Distribution network planning enhancement via network reconfiguration and DG integration using dataset approach and water cycle algorithm. *J. Mod. Power Syst. Clean Energy* **2020**, *8*, 86–93. [CrossRef]
22. Phuangpornpitak, W.; Bhumkittipich, K. Principle optimal placement and sizing of single distributed generation for power loss reduction using particle swarm optimization. *Res. J. Appl. Sci. Eng. Technol.* **2014**, *7*, 1211–1216. [CrossRef]
23. Tolba, M.A.; Tulsky, V.N.; Zaki Diab, A.A. Optimal allocation and sizing of multiple distributed generators in distribution networks using a novel hybrid particle swarm optimization algorithm. In Proceedings of the 2017 IEEE Conference of Russian Young Researchers in Electrical and Electronic Engineering (EIConRus), St. Petersburg/Moscow, Russia, 1–3 February 2017.
24. Raj, V.; Kumar, B.K. An improved affine arithmetic-based optimal DG sizing and placement algorithm using PSO for radial distribution networks with uncertainty. In Proceedings of the 2020 21st National Power Systems Conference (NPSC), Gandhinagar, India, 17–19 December 2020.
25. Katyara, S.; Shaikh, M.F.; Shaikh, S.; Khand, Z.H.; Staszewski, L.; Bhan, V.; Majeed, A.; Shah, M.A.; Zbigniew, L. Leveraging a Genetic Algorithm for the optimal placement of distributed generation and the need for energy management strategies using a fuzzy inference system. *Electronics* **2021**, *10*, 172. [CrossRef]
26. Barik, S.; Das, D.; Bansal, R.C. Zero bus load flow method for the integration of renewable DGs by mixed-discrete particle swarm optimisation-based fuzzy max–min approach. *IET Renew. Power Gener.* **2020**, *14*, 4029–4042. [CrossRef]
27. Bohre, A.K.; Agnihotri, G.; Dubey, M. Optimal sizing and sitting of DG with load models using soft computing techniques in practical distribution system. *IET Gener. Transm. Distrib.* **2016**, *10*, 2606–2621. [CrossRef]
28. Yahaya, A.A.; AlMuhaini, M.; Heydt, G.T. Optimal design of hybrid DG systems for microgrid reliability enhancement. *IET Gener. Transm. Distrib.* **2020**, *14*, 816–823. [CrossRef]
29. Cheng, R.; Jin, Y. A competitive swarm optimizer for large scale optimization. *IEEE Trans. Cybern.* **2015**, *45*, 191–204. [CrossRef] [PubMed]
30. Ganguly, S. Multi-objective planning for reactive power compensation of radial distribution networks with unified power quality conditioner allocation using particle swarm optimization. *IEEE Trans. Power Syst.* **2014**, *29*, 1801–1810. [CrossRef]
31. Tolabi, H.B.; Ali, M.H.; Rizwan, M. Simultaneous Reconfiguration, Optimal Placement of DSTATCOM, and Photovoltaic Array in a Distribution System Based on Fuzzy-ACO Approach. *IEEE Trans. Sustain. Energy.* **2015**, *6*, 210–218. [CrossRef]
32. Oloulade, A.; Imano Moukengue, A.; Agbokpanzo, R.; Vianou, A.; Tamadaho, H.; Badarou, R. New multi objective approach for optimal network reconfiguration in electrical distribution systems using modified ant colony algorithm. *Am. J. Electr. Power Energy Syst.* **2019**, *8*, 120. [CrossRef]
33. Das, C.K.; Bass, O.; Kothapalli, G.; Mahmoud, T.S.; Habibi, D. Optimal placement of distributed energy storage systems in distribution networks using artificial bee colony algorithm. *Appl. Energy* **2018**, *232*, 212–228. [CrossRef]

34. Seker, A.A.; Hocaoglu, M.H. Artificial Bee Colony algorithm for optimal placement and sizing of distributed generation. In Proceedings of the 2013 8th International Conference on Electrical and Electronics Engineering (ELECO), Bursa, Turkey, 28–30 November 2013.
35. Yuvaraj, T.; Ravi, K. Multi-objective simultaneous DG and DSTATCOM allocation in radial distribution networks using cuckoo searching algorithm. *Alex. Eng. J.* **2018**, *57*, 2729–2742. [CrossRef]
36. Arya, L.D.; Koshti, A. Modified shuffled frog leaping optimization algorithm based distributed generation rescheduling for loss minimization. *J. Inst. Eng. (India) Ser. B* **2018**, *99*, 397–405. [CrossRef]
37. Rajaram, R.; Sathish Kumar, K.; Rajasekar, N. Power system reconfiguration in a radial distribution network for reducing losses and to improve voltage profile using modified plant growth simulation algorithm with Distributed Generation (DG). *Energy Rep.* **2015**, *1*, 116–122. [CrossRef]
38. Othman, M.M.; El-Khattam, W.; Hegazy, Y.G.; Abdelaziz, A.Y. Optimal placement and sizing of distributed generators in unbalanced distribution systems using supervised big bang-big crunch method. *IEEE Trans. Power Syst.* **2015**, *30*, 911–919. [CrossRef]
39. Yuvaraj, T.; Ravi, K.; Devabalaji, K.R. DSTATCOM allocation in distribution networks considering load variations using bat algorithm. *Ain Shams Eng. J.* **2017**, *8*, 391–403. [CrossRef]
40. Duong, M.; Pham, T.; Nguyen, T.; Doan, A.; Tran, H. Determination of optimal location and sizing of solar photovoltaic distribution generation units in radial distribution systems. *Energies* **2019**, *12*, 174. [CrossRef]
41. Al-Bazoon, M. Harris Hawks Optimization for optimum design of truss structures with discrete variables. *Int. J. Math. Eng. Manag. Sci.* **2021**, *6*, 1157–1173.
42. Moayedi, H.; Osouli, A.; Nguyen, H.; Rashid, A.S.A. A novel Harris hawks' optimization and k-fold cross-validation predicting slope stability. *Eng. Comput.* **2021**, *37*, 369–379. [CrossRef]
43. Paital, S.R.; Ray, P.K.; Mohanty, S.R. A robust dual interval type-2 fuzzy lead-lag based UPFC for stability enhancement using Harris Hawks Optimization. *ISA Trans.* **2021**. [CrossRef]
44. Parsa, P.; Naderpour, H. Shear strength estimation of reinforced concrete walls using support vector regression improved by Teaching–learning-based optimization, Particle Swarm optimization, and Harris Hawks Optimization algorithms. *J. Build. Eng.* **2021**, *44*, 102593. [CrossRef]
45. Bandyopadhyay, R.; Basu, A.; Cuevas, E.; Sarkar, R. Harris Hawks optimisation with Simulated Annealing as a deep feature selection method for screening of COVID-19 CT-scans. *Appl. Soft Comput.* **2021**, *111*, 107698. [CrossRef]
46. Abd Elaziz, M.; Yousri, D. Automatic selection of heavy-tailed distributions-based synergy Henry gas solubility and Harris hawk optimizer for feature selection: Case study drug design and discovery. *Artif. Intell. Rev.* **2021**, *54*, 4685–4730. [CrossRef]
47. Malik, A.; Tikhamarine, Y.; Sammen, S.S.; Abba, S.I.; Shahid, S. Prediction of meteorological drought by using hybrid support vector regression optimized with HHO versus PSO algorithms. *Environ. Sci. Pollut. Res. Int.* **2021**, *28*, 39139–39158. [CrossRef]
48. Sharma, R.; Prakash, S. HHO-LPWSN: Harris hawks optimization algorithm for sensor nodes localization problem in wireless sensor networks. *ICST Trans. Scalable Inf. Syst.* **2018**, 168807. [CrossRef]
49. Gerey, A.; Sarraf, A.; Ahmadi, H. Groundwater single- and multiobjective optimization using Harris Hawks and Multiobjective Billiards-inspired algorithm. *Shock Vib.* **2021**, *2021*, 4531212.
50. Setiawan, I.N.; Kurniawan, R.; Yuniarto, B.; Caraka, R.E.; Pardamean, B. Parameter optimization of support vector regression using Harris hawks optimization. *Procedia Comput. Sci.* **2021**, *179*, 17–24. [CrossRef]
51. Mansoor, M.; Mirza, A.F.; Ling, Q. Harris hawk optimization-based MPPT control for PV systems under partial shading conditions. *J. Clean. Prod.* **2020**, *274*, 122857. [CrossRef]
52. Seyfollahi, A.; Ghaffari, A. Reliable data dissemination for the Internet of Things using Harris hawks optimization. *Peer Peer Netw. Appl.* **2020**, *13*, 1886–1902. [CrossRef]
53. Rodríguez-Esparza, E.; Zanella-Calzada, L.A.; Oliva, D.; Heidari, A.A.; Zaldivar, D.; Pérez-Cisneros, M.; Foong, L.K. An efficient Harris hawks-inspired image segmentation method. *Expert Syst. Appl.* **2020**, *155*, 113428. [CrossRef]
54. Tikhamarine, Y.; Souag-Gamane, D.; Ahmed, A.N.; Sammen, S.S.; Kisi, O.; Huang, Y.F.; El-Shafie, A. Rainfall-runoff modelling using improved machine learning methods: Harris hawks optimizer vs. particle swarm optimization. *J. Hydrol.* **2020**, *589*, 125133. [CrossRef]
55. Jia, H.; Peng, X.; Kang, L.; Li, Y.; Jiang, Z.; Sun, K. Pulse coupled neural network based on Harris hawks optimization algorithm for image segmentation. *Multimed. Tools Appl.* **2020**, *79*, 28369–28392. [CrossRef]
56. Sammen, S.S.; Ghorbani, M.A.; Malik, A.; Tikhamarine, Y.; AmirRahmani, M.; Al-Ansari, N.; Chau, K.-W. Enhanced artificial neural network with Harris hawks optimization for predicting scour depth downstream of ski-jump spillway. *Appl. Sci.* **2020**, *10*, 5160. [CrossRef]
57. Islam, M.Z.; Wahab, N.I.A.; Veerasamy, V.; Hizam, H.; Mailah, N.F.; Guerrero, J.M.; Mohd Nasir, M.N. A Harris Hawks Optimization based single- and multi-objective Optimal Power Flow considering environmental emission. *Sustainability* **2020**, *12*, 5248. [CrossRef]
58. Khalifeh, S.; Akbarifard, S.; Khalifeh, V.; Zallaghi, E. Optimization of water distribution of network systems using the Harris Hawks optimization algorithm (Case study: Homashahr city). *MethodsX* **2020**, *7*, 100948. [CrossRef]
59. Yousri, D.; Babu, T.S.; Fathy, A. Recent methodology based Harris Hawks optimizer for designing load frequency control incorporated in multi-interconnected renewable energy plants. *Sustain. Energy Grids Netw.* **2020**, *22*, 100352. [CrossRef]

60. Abbasi, A.; Firouzi, B.; Sendur, P. On the application of Harris hawks optimization (HHO) algorithm to the design of microchannel heat sinks. *Eng. Comput.* **2021**, *37*, 1409–1428. [CrossRef]
61. Teng, J.-H.; Luan, S.-W.; Lee, D.-J.; Huang, Y.-Q. Optimal charging/discharging scheduling of battery storage systems for distribution systems interconnected with sizeable PV generation systems. *IEEE Trans. Power Syst.* **2013**, *28*, 1425–1433. [CrossRef]
62. Chakraborty, S.; Kumar, R. Comparative analysis of NOCT values for mono and multi C-Si PV modules in Indian climatic condition. *World J. Eng.* **2015**, *12*, 19–22. [CrossRef]
63. Chakraborty, S. Reliable energy prediction method for grid connected photovoltaic power plants situated in hot and dry climatic condition. *SN Appl. Sci.* **2020**, *2*, 317. [CrossRef]
64. Hassan, A.S.; Othman, E.A.; Bendary, F.M.; Ebrahim, M.A. Distribution systems techno-economic performance optimization through renewable energy resources integration. *Array* **2021**, *9*, 100050. [CrossRef]
65. Mechanical Characteristics Electrical Characteristics PERC 350 Wp SPV MODULE. Available online: https://www.waaree.com/documents/WSMP-350_4BB_40mm_datasheet.pdf (accessed on 7 July 2021).
66. Heidari, A.A.; Mirjalili, S.; Faris, H.; Aljarah, I.; Mafarja, M.; Chen, H. Harris hawks optimization: Algorithm and applications. *Future Gener. Comput. Syst.* **2019**, *97*, 849–872. [CrossRef]
67. Zimmerman, R.D.; Murillo-Sánchez, C.E.; Thomas, R.J. MATPOWER: Steady state operations, planning, and analysis tools for power systems research and education. *IEEE Tran. Power Systems* **2011**, *26*, 12–19. [CrossRef]
68. Sultana, S.; Roy, P.K. Multiobjective quasi-oppositional teaching learning based optimization for optimal location of distributed genera- tor in radial distribution systems. *Int. J. Electr. Power Energy Syst.* **2014**, *63*, 534–545. [CrossRef]
69. Moradi, M.H.; Abedini, M. A combination of genetic algorithm and particle swarm optimization for optimal DG location and sizing in distribution systems. *Int. J. Electr. Power Energy Syst.* **2012**, *34*, 66–74. [CrossRef]
70. Quadri, I.A.; Bhowmick, S.; Joshi, D. A comprehensive technique for optimal allocation of distributed energy resources in radial distribution systems. *Appl. Energy* **2018**, *211*, 1245–1260. [CrossRef]
71. Meena, N.K.; Parashar, S.; Swarnkar, A.; Gupta, N.; Niazi, K.R. Improved elephant herding optimization for multiobjective DER accommodation in distri- bution systems. *IEEE Trans. Ind. Inform.* **2018**, *14*, 1029–1039. [CrossRef]
72. Ali, A.; Keerio, M.U.; Laghari, J.A. Optimal site and size of distributed generation allocation in radial distribution network using multiobjective optimization. *J. Mod. Power Syst. Clean Energy* **2021**, *9*, 404–415. [CrossRef]
73. Zimmerman, R.D.; Murillo-Sánchez, C.E. Matpower [Software]. 2020. Available online: https://matpower.org (accessed on 9 July 2021). [CrossRef]
74. Injeti, S.K.; Prema Kumar, N. A novel approach to identity optimal access point and capacity of multiple DGs in a small, medium, and large scale radial distribution systems. *Electr. Power Energy Syst.* **2013**, *45*, 142–151. [CrossRef]
75. Rama Prabha, D.; Jayabarathi, T. Optimal placement and sizing of multiple distributed generating units in distribution networks by invasive weed optimization algorithm. *Ain Shams Eng. J.* **2016**, *7*, 683–694. [CrossRef]
76. Mohamed Imran, A.; Kowsalya, M. Optimal size and siting of multiple distributed generators in distribution system using bacterial foraging optimization. *Swarm Evol. Comput.* **2014**, *15*, 58–65. [CrossRef]
77. Saleh, A.A.; Mohamed, A.-A.A.; Hemeida, A.M.; Ibrahim, A.A. Comparison of different optimization techniques for optimal allocation of multiple distribution generation. In Proceedings of the 2018 International Conference on Innovative Trends in Computer Engineering (ITCE), Aswan, Egypt, 19–21 February 2018.

energies

MDPI

Article

An Optimized Framework for Energy Management of Multi-Microgrid Systems

Komal Naz [1], Fasiha Zainab [1], Khawaja Khalid Mehmood [2], Syed Basit Ali Bukhari [2], Hassan Abdullah Khalid [1] and Chul-Hwan Kim [3,*]

1 U.S.-Pakistan Center for Advanced Studies in Energy (USPCAS-E), Department of Electrical Engineering (Power), National University of Sciences and Technology (NUST), Islamabad 44000, Pakistan; knazeep18.ces@student.nust.edu.pk (K.N.); fzainabeep18.ces@student.nust.edu.pk (F.Z.); hakhalid@uspcase.nust.edu.pk (H.A.K.)

2 Department of Electrical Engineering, The University of Azad Jammu & Kashmir, Muzaffarabad 13100, Pakistan; khalid.mehmood@ajku.edu.pk (K.K.M.); basit.bukhari@ajku.edu.pk (S.B.A.B.)

3 College of Electrical and Computer Engineering, Sungkyunkwan University, Suwon 440-746, Korea

* Correspondence: chkim@skku.edu

Citation: Naz, K.; Zainab, F.; Mehmood, K.K.; Bukhari, S.B.A.; Khalid, H.A.; Kim, C.-H. An Optimized Framework for Energy Management of Multi-Microgrid Systems. *Energies* **2021**, *14*, 6012. https://doi.org/10.3390/en14196012

Academic Editors: Abu-Siada Ahmed and Victor Becerra

Received: 20 July 2021
Accepted: 17 September 2021
Published: 22 September 2021

Abstract: Regarding different challenges, such as integration of green energy and autonomy of microgrid (MG) in the multi-microgrid (MMG) system, this paper presents an optimized and coordinated strategy for energy management of MMG systems that consider multiple scenarios of MGs. The proposed strategy operates at two optimization levels: local and global. At an MG level, each energy management system satisfies its local demand by utilizing all available resources via local optimization, and only sends surplus/deficit energy data signals to MMG level, which enhances customer privacy. Thereafter, at an MMG level, a central energy management system performs global optimization and selects optimized options from the available resources, which include charging/discharging energy to/from the community battery energy storage system, selling/buying power to/from other MGs, and trading with the grid. Two types of loads are considered in this model: sensitive and non-sensitive. The algorithm tries to make the system reliable by avoiding utmost load curtailment and prefers to shed non-sensitive loads over sensitive loads in the case of load shedding. To verify the robustness of the proposed scheme, several test cases are generated by Monte Carlo Simulations and simulated on the IEEE 33-bus distribution system. The results show the effectiveness of the proposed model.

Keywords: cost optimization; differential evolution; energy management system; multi-microgrids system; renewable generation

1. Introduction

Over the past decade, energy demand has considerably increased because of energy-dependent lifestyle of humans, industries, and electric vehicles (EVs) [1]. Conventional power systems are traditionally designed and operate on key reliability principles of security and adequacy [2]; they cannot meet these enormous rising demands adequately. The up-gradation of power systems to smart grids has been considered to address the energy shortage problem, improve reliability, and facilitate the integration of renewable energy sources (RESs). One of the significant ideas in smart grid is microgrids (MGs). An MG is comprised of a segment of distribution framework, including distributed energy resources (DERs), and diverse end clients. An MG can operate in two modes: grid-tied and islanded modes. The MG generally works in grid-tied mode. In grid-tied mode, there is bidirectional flow of energy and information between an MG and different MGs or the power grid, i.e., an MG can absorb or supply power. In stand-alone mode, it only acts as a source, i.e., only supplies power to the connected load. DERs include distributed generator (DG) and distributed storage (DS) units with various limits. The DG can be controllable DG (CDG) or renewable DG (RDG): PVs and WTs. Integration of RESs (PVs and WTs) participates in cost minimization and reduction in load shedding. In an MG, energy is generated near the point of demand which reduces transmission and distribution losses, as well as grid expansion deferral. The whole operation of the MG is supervised by the MG energy management system (EMS) which incorporates power generation, energy storage, and load management programs. A fast energy management strategy is required for sound functioning of the MG with various DERs especially in autonomous mode [3–5]. The EMS performs optimal scheduling of all available energy resources and energy storage systems (ESSs) to meet load demands [6]. The optimal operation of smart grid is a great challenge due to uncertainties in the DGs' generation, load requirement, real-time prices, and penetration of EVs. The MMGs system, the combination of different MGs, can deal with these uncertainties, to a large extent. The main objectives of considering MMGs are cost optimization and the reliability of the system.

Significant research has been conducted to put forward different algorithms and optimization models for MMG-EMS. However, these studies have some limitations in comparison to the vast literature available on the energy management of a single MG. A multi-step hierarchical optimization algorithm based on a multi-agent system considering adjustable power and demand response (DR) was proposed in Reference [6]. In Reference [7], authors proposed energy management of islanded DC MG with dual active bridge converter-based power management units and control interface. The optimized and coordinated strategies for energy management in MMGs were proposed, and a 69-bus system was used for case study in Reference [8]. According to References [9,10], the MGs could transfer energy and coordinate with each other to make the entire system reliable and stable. In Reference [11], the impact of DR, which was synchronous with the MMG-based operation of active distribution networks, was investigated, and the optimization problem was solved with non-dominated genetic algorithm-II (NSGA-II) and tested on an IEEE 69-bus distribution system. A multi-period optimal dispatch model was suggested for a distribution network with clustered MGs in Reference [12]; a modified hierarchical genetic algorithm was used to solve bi-level optimization problem and tested on IEEE 14-bus distribution system. A hierarchical decentralized system of systems has been studied for coordinating multiple autonomous microgrid systems with grid connection in Reference [13]. In Reference [14], a hierarchical multi-agent EMS was proposed to increase the utilization of RES based on the DR programs (DRP) and BESSs within the MG. Interconnected MMGs (IMMGs) have numerous benefits over the single MGs. First, IMMGs have good economic characteristics both in grid-tied [15,16] and in islanded modes [1,17] because energy sharing fulfills load demands of the MGs with their own cheap RESs, thus reducing the cost of generation based on fossil fuels and the power losses occurred in distant transference by trading energy among the MGs which are close to each other. Second, the IMMGs are more reliable than the single MG in reducing stress on the main grid.

In all of the aforementioned papers, the main focus of the research was to maximize the utilization of the cheapest controllable DG (CDG) by considering its generation price. Few papers considered the price of RESs to ensure maximum utilization of renewable sources. Most of the papers have sufficiently discussed the DRP or demand-side management (DSM) schemes to meet load demands. Moreover, some researchers have considered islanded individual MG, islanded MMG system and grid-connected MMG system for their studies. However, they overlooked the different combinations of MGs. The MMG system is the most optimized network and can operate in two ways: either as a cooperative or as an autonomous system. In the cooperative MMG system, the benefits of overall system will be considered in that it is not necessary that each individual MG is getting profit, whereas, in the autonomous MMG, each MG will consider and prefer its own benefit in that it is not necessary that the overall system is in the most optimized form. Therefore, it is possible that an MG is getting more benefit individually than to be a part of the cooperative MMG system, and vice versa. Furthermore, the types of loads and the load shedding of sensitive load have not been considered so far in the literature.

In this paper, a hierarchical EMS for the optimal operation of MMGs is proposed; the MGs are in islanded or MMGs connected mode. A two-level optimization, local and global, has been formulated. In local optimization, each MG acts autonomously and maximizes the utilization of RESs to meet load demand. In global optimization, there is a cooperative operation between the MGs which optimizes the overall system. This optimization strategy maximizes the utilization of cheap renewable energy, minimizes total cost, and reduces the load shedding. The total loads have been divided into two sets: sensitive and non-sensitive loads. The penalty prices of these loads are considered much higher than all other prices to avoid load shedding, whereas the penalty prices of sensitive loads are considered higher than that of non-sensitive loads to provide uninterrupted supply to sensitive load during load shedding. The proposed model is tested on the IEEE 33-bus test system, and the presented optimization problem is solved by differential evolution (DE) algorithm. Moreover, we have studied the proposed methodology for selected cases in order to check the robustness of the proposed scheme. For this purpose, we have used Monte Carlo Simulations (MCS) to generate the random cases of all total possible cases and run the numerical simulation of the proposed algorithm. Some of the major contributions of this study are summarized as:

1. A new energy management model to formulate day-ahead energy management problem for optimal operations of MGs is proposed which allows autonomous operation mode; each MG incorporates DG units (CDG, PV, and WT), a battery ESS (BESS), and its own EMS.
2. A two-step optimization problem is proposed. In the first step, each MG-EMS considers maximum local consumption of renewable energy, whereas, in the second step, the central EMS (CEMS) monitors the power mismatch, achieves optimal energy trading among MGs, and reduces load shedding.
3. A hierarchical EMS is developed in which the algorithm makes price-based decisions and selects the optimized options from the available resources. A methodology for the assessment of the energy management strategy is illustrated, which enables marking and examining the characteristics of MMGs.
4. Different scenarios and cases have been generated by MCS and tested on modified IEEE 33-bus distribution system; the results represent the stability of proposed algorithm and advantages of energy management system.

The rest of the paper is organized as follows. Section 2 focuses on the proposed MMG model and mathematical modeling of wind turbine and photovoltaic DGs. Section 3 explains the optimization formulation of the strategy: local and global. Section 4 refers to the simulations and case study of the modified IEEE 33-bus distribution system. Section 5 discusses the results, including the proposed scenarios and the cases generated by MCS. The conclusion is summarized in Section 6.

2. System Model

2.1. Configuration of Proposed MMG System

Figure 1 shows the proposed hierarchical two-step energy management system model. Each MG includes CDG, PV, and WT units which have sufficient power generation to fulfill its MG demands, BESS to reserve surplus energy in off-peak hours and utilize it in on-peak hours to avoid load shedding, and an EMS to make all this management possible. The EMS at the MG level ensures optimized energy management by utilizing all energy and ancillary services. In this paper, two sets of loads are taken into consideration: sensitive loads (SLs) and non-sensitive loads (NSLs). If a customer, in the event of low power generations, requires to shed loads due to shortage of power, the NSL is curtailed, and the SL gets uninterrupted supply. The CEMS is connected to each MG-EMS, central BESS (CBESS), and to the grid to coordinate the overall system to make it reliable and minimize emergency load shedding, which lowers the chance of system-collapse. The CEMS at the MMG level ensures interconnection of different autonomous MGs. The EMS of every single MG is also connected to the EMS of every other neighboring MG via power lines for power flow when needed. The MMG system, presented in Figure 1, has the following benefits: (i) they share reserves in on-peak hours or in crucial condition (e.g., faults on generation side or sudden increase in the load demand); (ii) economic dispatch in the whole system, either islanded or grid-connected; and (iii) a CBESS can serve critical loads after utilizing all available options and increases the resilience of the system.

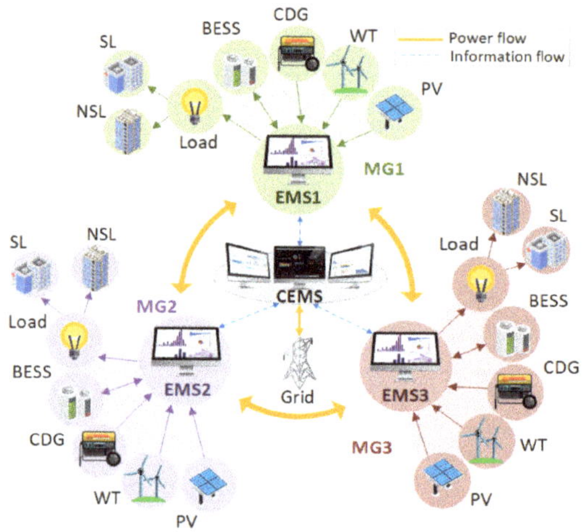

Figure 1. Proposed energy management model.

For the purpose of simulations, wind speed and solar irradiance are modeled by implementing the Weibull [18] and beta probability distributions, respectively. Each MG has the freedom to take part in the MMG system. Otherwise, it can operate as an autonomous single islanded entity.

2.2. Wind Turbine DGs

2.2.1. Wind Speed Modeling

Different probability distributions are used to model variations in the wind speed. Weibull probability distribution is often used [19] and is also used in this paper. The probability density function of this distribution is given in (1).

$$f(s_w|a,b) = \begin{cases} \frac{b}{a^b} s_w^{b-1} e^{-\left(\frac{s_w}{a}\right)^b}, \text{ for } x > 0 \\ 0, \text{ for } x \leq 0 \end{cases}, \quad (1)$$

where a and b are the scale and shape parameters of the Weibull distribution, respectively. Maximum likelihood estimation (MLE) method is used to find these parameters.

2.2.2. Wind Power Model

The wind power DG-model used for obtaining electrical energy from the generated wind speed data is given in (2) [19].

$$P_w = \begin{cases} P_r \times \frac{s_w - s_{ci}}{s_r - s_{ci}}, \text{ for } s_{ci} \leq s_w \leq s_r \\ P_r, \text{ for } s_r \leq s_w \leq s_{co} \\ 0, \text{ otherwise} \end{cases}. \quad (2)$$

where P_w and P_r are power generated from a WT and rated capacity of a WT, respectively; s_w, s_{ci}, s_{co} and s_r are current, cut-in, cut-out and rated wind speeds, respectively.

2.3. Photovoltaic DGs

2.3.1. Solar Irradiance Modeling

The beta distribution is usually used for modeling of the variations in the solar irradiance [20]. For $\alpha, \beta > 0$, the pdf of Beta distribution is given as (3).

$$f(s_{ird}|\alpha,\beta) = \begin{cases} \frac{\gamma(\alpha+\beta)}{\gamma(\alpha)\gamma(\beta)} \times {s_{ird}}^{(\alpha-1)} \\ \times(1-s_{ird})^{(\beta-1)}, \text{ for } 0 \le s_{ird} \le 1 \\ 0, \text{ for } s_{ird} < 0 \end{cases}, \quad (3)$$

where γ is the gamma function; α and β are known as the shape parameters of the beta distribution. Both parameters are estimated by the MLE method.

2.3.2. Solar Power Model

The model of a PV panel given in (4)–(8) provides electric power obtained from the generated solar irradiance samples [20].

$$T_{cell} = T_{amb} + \left(s_{ird} \times \frac{T_{not} - 20}{0.8}\right), \quad (4)$$

$$I = s_{ird} \times (I_{sc} + K_i \times (T_{cell} - 25)), \quad (5)$$

$$V = V_{oc} - K_v \times T_{cell}, \quad (6)$$

$$FF = \left(\frac{V_{maxp} \times I_{maxp}}{V_{oc} \times I_{sc}}\right), \quad (7)$$

$$P_s = N_{total} \times FF \times V \times I. \quad (8)$$

where s_{ird}, FF, N_{total}, P_s, T_{cell}, T_{amb} and T_{not} are solar irradiance, fill factor, total number of PV modules, power generated from a PV panel, temperature of a cell, ambient temperature and nominal operating temperature of a cell, respectively; I_{sc}, I_{maxp}, K_i, V_{oc}, V_{maxp} and K_v are short circuit current, current at maximum power point, temperature coefficient for current, open circuit voltage, voltage at maximum power point and temperature coefficient for voltage, respectively.

3. Optimization Formulation

In this section, a day-ahead energy management problem for proposed MMGs model is formulated based on the DE algorithm. The objective function is to minimize the total operational cost of the MG network whether it is operating as a single islanded or grid-tied MG system; the MMGs islanded or grid-connected network makes the system reliable by giving uninterrupted supply of energy and minimizes curtailment of loads. The elements of bi-level optimization model are shown in Figure 2 including decision input variables, constraints, objective function and optimal output variables. A constraints-handling technique is used in the presented DE algorithm to get the best fitness value of the decision variables, and penalty factor technique is used to attain the most feasible and optimized value of objective function. The proposed model is figured for a twenty-four-hour scheduling horizon with a time period of 't'; 't' can be any uniform interval of time. However, in the suggested model, 't' has been considered to be one hour. As mentioned earlier, it is a two-step optimization strategy: local and global. Step-wise detailed mathematical models are developed in the following subsections.

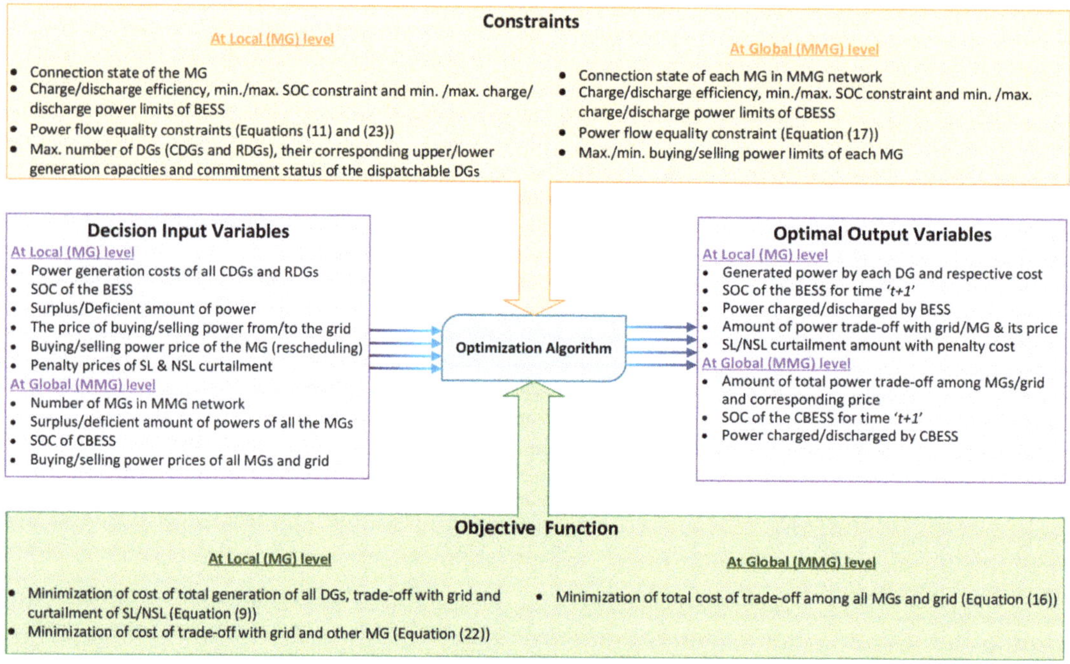

Figure 2. Elements of the optimization model.

3.1. Local Optimization

At local level, the EMS of each MG generates energy management strategy based on the proposed model, load/generation data, prices, and constraints. The objective function is given by (9) and includes CDG generation cost, RDG generation cost, prices of buying/selling power from/to the grid, and penalty prices of SL/NSL curtailments. The penalty price of curtailment of the SL is considered higher than that of the NSL. Therefore, in case of load shedding, only a part of NSL will be shed, and the SL will get continuous supply.

$$min \sum_{i=1}^{I}\sum_{t=1}^{T}(C_{i,t}^{CDG} \cdot P_{i,t}^{CDG} + C_{i,t}^{RDG} \cdot P_{i,t}^{RDG}) + \sum_{t=1}^{T}(PR_t^{Buy} \cdot P_t^{Def} + PR_t^{Sell} \cdot P_t^{Sur}) + \sum_{t=1}^{T}(C_t^{PenSL} \cdot P_t^{ShedSL} + C_t^{PenNSL} \cdot P_t^{ShedNSL}), \quad (9)$$

where P is respective power value being multiplied with corresponding cost/price (i.e., C = cost and PR = price) for time from $t = 1$ to $t = T$ and for DGs (CDGs and RDGs) from $i = 1$ to $i = I$ in an MG.

Subject to:

$$c_{i,t} \cdot P_i^{CDG_{min}} \le P_{i,t}^{CDG} \le c_{i,t} \cdot P_i^{CDG_{max}} \ \forall\, i \in \boldsymbol{I}, t \in \boldsymbol{T}, \quad (10)$$

$$P_t^{PV} + P_t^{WT} + \sum_{i=1}^{I} P_{i,t}^{CDG} + P_t^{Def} + P_t^{B-} = P_t^{L} + P_t^{Sur} + P_t^{B+} \ \forall\, t \in \boldsymbol{T}, \quad (11)$$

$$0 \le P_t^{B+} \le P_B^{Cap} \cdot \frac{1 - SOC_{t-1}^{B}}{\eta_B} \ \forall\, t \in \boldsymbol{T}, \quad (12)$$

$$0 \le P_t^{B-} \le P_B^{Cap} \cdot SOC_{t-1}^{B} \cdot \eta_B \ \forall\, t \in \boldsymbol{T}, \quad (13)$$

$$SOC_t^B = SOC_{t-1}^B - \frac{1}{P_B^{Cap}} \cdot \left(\frac{P_t^{B-}}{\eta_B} - P_t^{B+} \cdot \eta_B \right), \tag{14}$$

$$0 \leq SOC_t^B \leq 1 \;\; \forall\, t \in \boldsymbol{T}. \tag{15}$$

In (10), the inequality constraint related to the generation of CDG unit 'i' is given, which shows that the output of a CDG is between its minimum and maximum generating capacities. The power generated by all DGs units, deficit power, and power from battery must be adjusted with demand, surplus power, and power to the battery at each interval and is represented by (11). During charging, the storage system is considered as a load, and it acts as a source during the discharging of power. The constraints of the BESS of an MG, which include the charging, discharging, and SOC, are given by (12)–(15). The SOC is updated in each interval.

In single grid-tied mode, the MG adjusts surplus/deficit power by trading with the main grid, whereas, in the case of single islanded mode, the surplus power is balanced by ramping down the dispatchable generating units. Likewise, a load curtailment mechanism is employed for balancing the deficient amount of power. If the MG network is islanded or grid-tied MMGs system, after finalizing local optimization by every single MG-EMS at the first step, every MG-EMS delivers the information to the CEMS agent about the surplus and deficient amount of powers calculated from the local optimization algorithm.

3.2. Global Optimization

After being informed by each MG-EMS about surplus/deficit powers and connection/disconnection of that MG with the grid in the MMGs network, the CEMS carries out the global optimization in this step. The objective function of the MMGs network is given by (16). It contains buying-selling energy prices of all the MGs and the grid. In this paper, the CBESS is considered in subservient mode in that it will act under the command of the CEMS, instead of acting in autonomous mode. Therefore, the costs of charging and discharging of CBESS are not added in the objective function.

$$min \;\; \sum_{k=1}^{K}\sum_{t=1}^{T}(PR_{k,t}^{Rec} \cdot P_{k,t}^{Rec} + PR_{k,t}^{Send} \cdot P_{k,t}^{Send}) + \sum_{k=1}^{K}\sum_{t=1}^{T}(PR_{t}^{Buy} \cdot P_{k,t}^{Buy} + PR_{t}^{Sell} \cdot P_{k,t}^{Sell}), \tag{16}$$

where P is respective power value being multiplied with corresponding price (i.e., PR = price) for time from $t = 1$ to $t = T$ and for the MGs from $k = 1$ to $k = K$ participating in the MMG system at that time.

Subject to:

$$\sum_{k=1}^{K} P_{k,t}^{Rec} + \sum_{k=1}^{K} P_{k,t}^{Buy} + \sum_{k=1}^{K} P_{k,t}^{Sur} + P_t^{CB-} = \sum_{k=1}^{K} P_{k,t}^{Send} + \sum_{k=1}^{K} P_{k,t}^{Sell} + \sum_{k=1}^{K} P_{k,t}^{Def} + P_t^{CB+} \;\; \forall\, t \in \boldsymbol{T}, \tag{17}$$

$$0 \leq P_t^{CB+} \leq P_{CB}^{Cap} \cdot \frac{1 - SOC_{t-1}^{CB}}{\eta_{CB}} \;\; \forall\, t \in \boldsymbol{T}, \tag{18}$$

$$0 \leq P_t^{CB-} \leq P_{CB}^{Cap} \cdot SOC_{t-1}^{CB} \cdot \eta_{CB} \;\; \forall\, t \in \boldsymbol{T}, \tag{19}$$

$$SOC_t^{CB} = SOC_{t-1}^{CB} - \frac{1}{P_{CB}^{Cap}} \cdot \left(\frac{P_t^{CB-}}{\eta_{CB}} - P_t^{CB+} \cdot \eta_{CB} \right), \tag{20}$$

$$0 \leq SOC_t^{CB} \leq 1 \;\; \forall\, t \in \boldsymbol{T}. \tag{21}$$

The power bought from the neighboring MGs and the main grid, surplus power from each MG, and the CBESS discharging must be adjusted with the amount of sold power to the neighboring MG and the grid, deficit from each MG, and the CBESS charging demand as represented in (17). The limitations related to the CBESS charging, discharging, and the SOC are given by (18)–(21). In this cooperative model, selling/buying power to/from other MGs of the MMGs system is preferred. If further power is required, trading with the CBESS will be performed. If it is grid-tied MMGs system, the last option will be to trade with grid if required. Similarly, an MG with surplus energy after fulfilling its own demand and charging its BESS will sell power to other deficit MGs, charge the CBESS, and will sell power to the grid in the end. To make this strategy possible, the prices are introduced interval-wise, which will help the algorithm to make a quick, and optimized decision.

$$min \sum_{i=1}^{I}\sum_{t=1}^{T}(C_{i,t}^{CDG}\cdot P_{i,t}^{CDG}+C_{i,t}^{RDG}\cdot P_{i,t}^{RDG})+\sum_{t=1}^{T}(PR_t^{Rec}\cdot P_t^{Rec}+PR_t^{Send}\cdot P_t^{Send})+\sum_{t=1}^{T}(PR_t^{Buy}\cdot P_t^{Buy}+PR_t^{Sell}\cdot P_t^{Sell}), \quad (22)$$

subject to:

$$P_t^{PV}+P_t^{WT}+\sum_{i=1}^{I}P_{i,t}^{CDG}+P_t^{Buy}+P_t^{Rec}+P_t^{B-}+P_t^{CB-}=P_t^{L}+P_t^{Sell}+P_t^{Send}+P_t^{B+}+P_t^{CB+}\ \forall\, t\in \boldsymbol{T}. \quad (23)$$

In the proposed MMGs model, the CEMS notifies all MG-EMSs about their commitments and schedulings. Each MG-EMS reschedules its generation after global optimization via local optimization again (rescheduling). Now, the objective function of local optimization of an MG in MMGs network is (22). It includes the CDG generation cost, RDG generation cost, and prices of buying/selling power from/to other MGs and grid. The load balancing constraint is presented in (23). In addition to this constraint, objective function constraints (10)–(15) are also considered. The complete optimization procedure for the proposed MMGs network is illustrated in Figure 3.

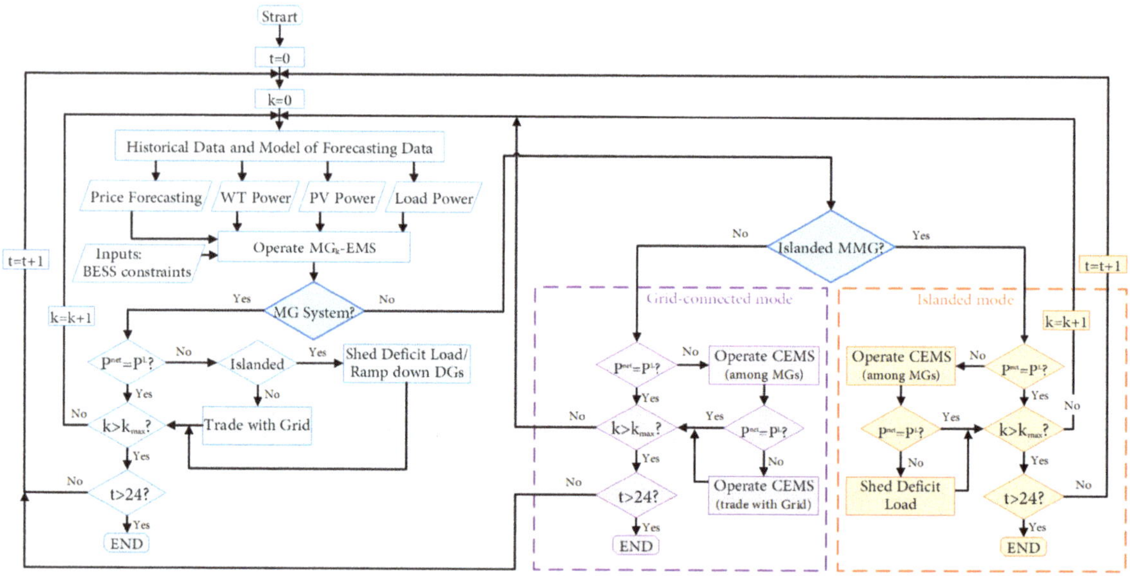

Figure 3. Flowchart of the proposed methodology.

4. Case Study

The proposed scheme has been verified on the IEEE 33-bus distribution system. The detailed specifications of the test system can be found in Reference [21]. We have modified the test system and constructed four autonomous MGs, as shown in Figure 4. For MCS, the wind speed and solar irradiance data of four seasonal data of six years (2007–2012) and seventeen years (1998–2014) have been taken [22,23].

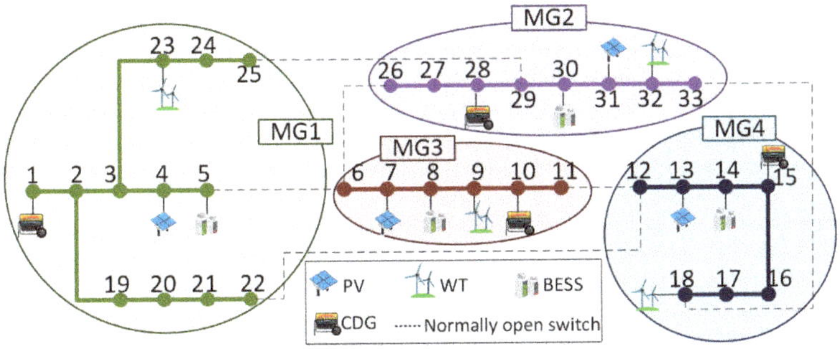

Figure 4. Modified IEEE 33-bus distribution system with four MGs.

Four different load curves, presented in Figure 5a, are used as daily load demands of MGs of the modified test system. The load curves of NYISO-CAPITAL zone, NYISO-N.Y.C. zone, ERCOT, and ISO-NE are used as demands of MG1, MG2, MG3, and MG4, respectively [24]. The parameters of a PV panel and a WT are same as considered in Reference [25]. Each MG has its own CDG, PV, WT, BESS, and EMS. Figure 5b shows the daily forecasted curves of output percentage of solar and wind DGs. The following three scenarios have been studied for the simulations. In Scenario 1, all four MGs are participating in the MMGs network; MG2 and MG3 are grid-tied. Only MG2 and MG3 can trade energy with the grid after sending or receiving power from other MGs. In Scenario 2, three MGs are participating in the MMGs grid-tied network, while the MG1 is in single islanded MG mode. In Scenario 3, three MGs are participating in MMGs grid-tied network, while MG3 is in single islanded MG mode.

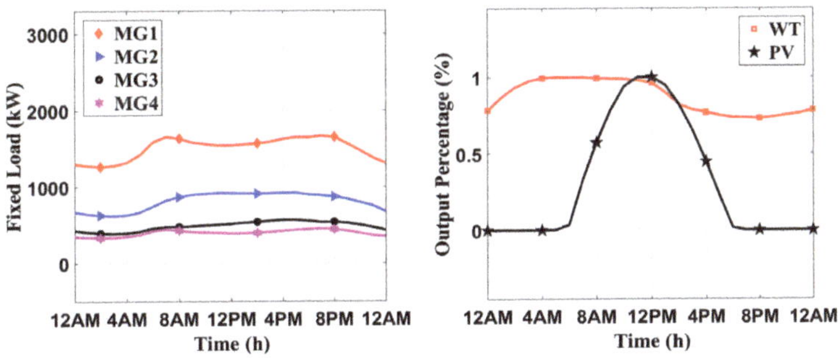

(a) (b)

Figure 5. (**a**) Fixed load profile of the different MGs and (**b**) daily forecasted curve of output percentage of solar and wind DG.

It can be observed that there can be many possible combinations of the MGs. In the proposed model, an MG can be in State 1: islanded single MG entity; State 2: grid-tied single MG; State 3: islanded MMGs system or State 4: grid-tied MMGs system, as shown in Figure 6a. Depending on these four possible states of an MG, there are total 256 cases for our test system. In order to verify the robustness of the scheme, 10% of the total cases are generated by MCS, and their details are provided in the next section. To understand the connection of MGs in different states, Cases 07 and 11 from Table 1 are illustrated in Figure 6b,c, respectively. State 1 refers to a single islanded entity, i.e., it has to fully depend on its resources, and no external source is available as MG4 in Figure 6b. In State 2, the MG is a single autonomous entity but has the grid as an external resource, as is MG1, in Figure 6c. In State 3, the MG is connected to other MGs that are also taking part in the MMG system. Other MGs may be connected with the grid depending on their respective states, but this MG is not connected to the grid, as depicted MG2 in Figure 6c, which is connected with other MGs but has no connection with the grid. In State 4, the MG is also connected to the grid besides the connection with other MGs in the MMG system. The connections of other MGs with the grid depend on their respective states. For instance, MG3 in Figure 6b is in the MMG system with MG2, and both are also grid-connected, whereas MG3 in Figure 6c is in MMG system with MG2 and MG4; MG3 and MG4 are connected to the grid, but MG2 is not depending on their states.

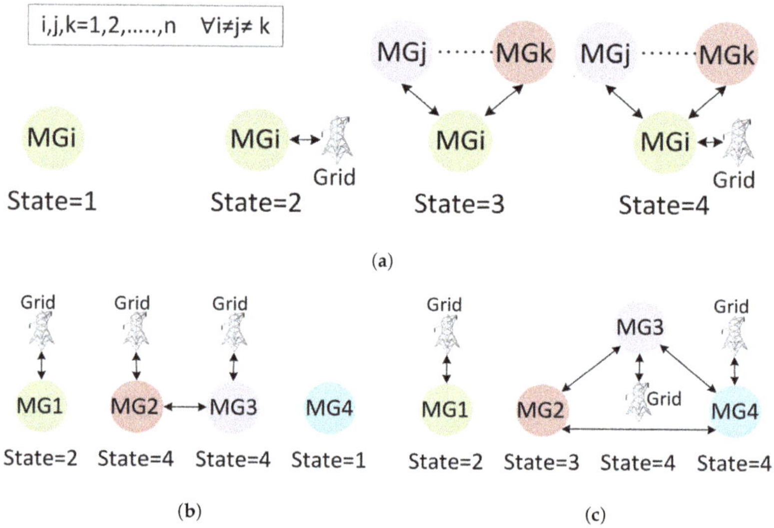

Figure 6. (**a**) States of an MG, (**b**) States in case 07, and (**c**) States in case 11.

Table 1. Optimal cost and exchange powers of the MGs.

	State of Microgrids				MG1				MG2				MG3				MG4			
Cases	MG1	MG2	MG3	MG4	PS(kW)	PB(kW)	LS(kW)	C(Rs.)	PS(kW)	PB(kW)	LS(kW)	C(Rs.)	PS(kW)	PB(kW)	LS(kW)	C(Rs.)	PS(kW)	PB(kW)	LS(kW)	C(Rs.)
Case 01	4	4	4	4	7749.82	0	0	46,189	5743.81	0	0	22,804	3965	0	0	11,874	374.44	408.59	0	13,511
Case 02	1	1	3	4	0	0	0	51,469	0	0	0	25,675	340.23	0	0	13,304	374.44	340.23	0	13,375
Case 03	1	1	3	3	0	0	0	51,469	0	0	0	25,675	340.23	0	0	13,304	0	340.23	0	13,647
Case 04	2	1	3	2	7749.82	0	0	47,594	0	0	0	25,675	0	0	0	14,352	374.44	1533	0	13,400
Case 05	4	2	1	4	7749.82	0	0	46,189	5743.81	0	0	22,804	0	0	55.73	14,562	374.44	408.59	0	13,511
Case 06	4	4	4	2	7749.82	0	0	47,594	5743.81	0	0	22,804	3965	0	0	11,874	374.44	1533	0	13,400
Case 07	2	4	4	1	7749.82	0	0	47,594	5743.81	0	0	22,804	3965	0	0	11,874	0	0	1533	18,759
Case 08	2	4	1	4	7749.82	0	0	47,594	5743.81	0	0	21,396	0	0	55.73	14,562	374.44	408.59	0	13,511
Case 09	2	4	4	3	7749.82	0	0	47,594	5743.81	0	0	21,396	3965	0	0	11,874	0	408.59	0	13,783
Case 10	1	4	2	4	0	0	0	51,469	5743.81	0	0	21,396	3965	55.73	0	11,902	374.44	408.59	0	13,511
Case 11	2	3	4	4	7749.82	0	0	47,594	408.59	0	0	24,064	3965	0	0	11,874	374.44	408.59	0	13,511
Case 12	4	4	1	4	7749.82	0	0	46,189	5743.81	0	0	22,804	0	0	55.73	14,562	374.44	408.59	0	13,511
Case 13	2	1	4	2	7749.82	0	0	47,594	0	0	0	25,676	3965	0	0	11,874	374.44	1533	0	13,400
Case 14	4	4	3	2	7749.82	0	0	47,594	5743.81	0	0	22,804	0	0	0	14,352	374.44	1533	0	13,400
Case 15	4	3	2	2	7749.82	0	0	47,594	0	0	0	25,676	3965	55.73	0	11,902	374.44	1533	0	13,400
Case 16	4	2	4	1	7749.82	0	0	47,594	5743.81	0	0	22,804	3965	0	0	11,874	0	0	1533	18,759
Case 17	4	3	1	1	7749.82	0	0	47,594	0	0	0	25,676	0	0	55.73	14,562	0	0	1533	18,759
Case 18	1	3	2	2	0	0	0	51,469	0	0	0	25,676	3965	55.73	0	11,902	374.44	1533	0	13,400
Case 19	3	2	2	3	408.59	0	0	49,859	5743.81	0	0	22,804	3965	55.73	0	11,902	0	408.59	0	13,783
Case 20	4	3	3	3	7749.82	0	0	46,189	0	0	0	25,676	0	0	0	14,352	0	408.59	0	13,783
Case 21	1	3	3	2	0	0	0	51,469	0	0	0	25,676	0	0	0	14,352	374.44	1533	0	13,400
Case 22	2	4	1	1	7749.82	0	0	47,594	5743.81	0	0	22,804	0	0	55.73	14,562	0	0	1533	18,759
Case 23	2	4	4	4	7749.82	0	0	47,594	5743.81	0	0	21,396	3965	0	0	11,874	374.44	408.59	0	13,511
Case 24	2	3	3	2	7749.82	0	0	47,594	0	0	0	25,676	0	0	0	14,352	374.44	1533	0	13,400
Case 25	1	1	1	1	0	0	0	51,469	0	0	0	25,676	0	0	55.73	14,352	0	0	1533	18,759

Note: PS = Power sold; PB = Power bought; LS = Load shed; C = Total optimal cost.

5. Results Analysis

5.1. Proposed Scenarios

Figures 7–9 show total generation, total demand, charged/discharged powers of the BESS, trading with grid and other MGs of the MMG system, and load curtailment of each MG. In the first scenario, as it can be seen in Figure 7a, MG1 has sold the surplus power from 5:00 a.m. to 7:00 a.m. to others deficit MGs. From 5:00 p.m. to 9:00 p.m., MG1 needed to discharge the BESS after utilizing the maximum capacity of its resources. In Figure 7b,c,

MG2 and MG3 were selling their surplus energy to the grid from 12:00 a.m. to 4:00 p.m. and utilizing the stored-BESS-energy from 5:00 p.m. to 9:00 p.m. MG2 was charging the BESS from 10:00 p.m. to 11:00 p.m., but MG3 was still deficient in energy, and its BESS was completely discharged already, and no neighboring MG had energy to sell the power. Therefore, it was receiving the discharged power from the CBESS for these two hours. MG4 was considered with the demand more than the generation to study the behavior and impact of the proposed scheme for worst-case scenario of any practical MG. In Figure 7d, from 12:00 a.m. to 4:00 a.m., the demand was being fulfilled by the storage system, which was completely discharged at 5:00 a.m. Therefore, it had to buy the energy from other MGs from 5:00 a.m. to 7:00 a.m. Later, it started charging the storage system from 8:00 a.m. to 12:00 p.m.; at 12:00 p.m., it was completely charged. From 4:00 p.m. to 7:00 p.m., the local energy storage was completely utilized, and no other MG had extra energy to sell. As a result, from 7:00 p.m. to 11:00 p.m., MG4 was using the discharged energy of the CBESS, and there was no load shedding in the whole day. This is the benefit of the proposed MMG system.

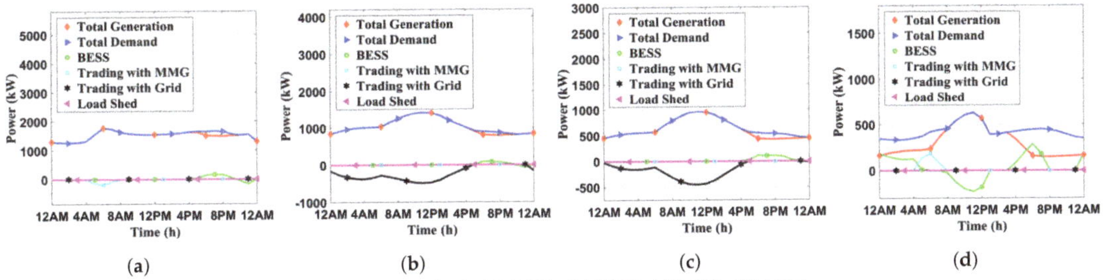

Figure 7. Scenario 1: (**a**) MG1, (**b**) MG2, (**c**) MG3, (**d**) MG4.

In the Scenario 2, MG1 was the MG which had capacity to meet the demand by its own without buying any energy. This behavior can be seen in Figure 8a. Even though MG1 was in single MG islanded mode, there was no load shedding during the whole day, and it was in the most optimized form. The behavior of MG2 was same as in Scenario 1; the only difference was: it was selling the energy to MG4 as in Figure 8b. The behavior of MG3 was same in Figure 8c as in the previous scenario. In this scenario, MG4 was grid-connected, Therefore, its behavior was slightly different from Scenario 1, i.e., it was selling surplus energy to the grid from 12:00 p.m. to 3:00 p.m. in Figure 8d.

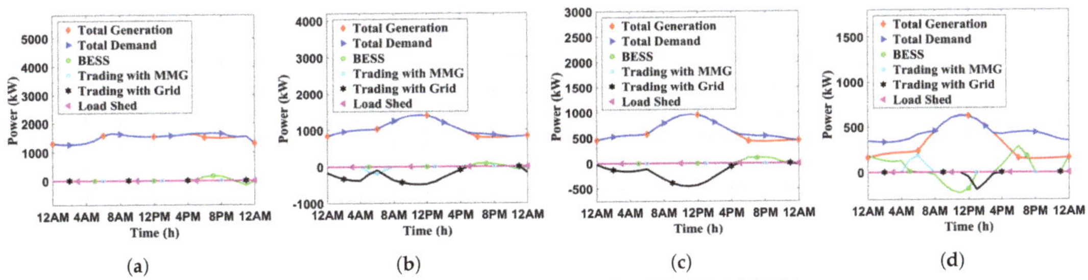

Figure 8. Scenario 2: (**a**) MG1, (**b**) MG2, (**c**) MG3, (**d**) MG4.

In Scenario 3, MG1 was grid-tied similar to MG2 and MG4, it was selling the surplus power to the grid in Figure 9a. Figure 9b,d show behaviors of MG2 and MG4, which were identical to Scenarios 1 and 2, respectively. MG3 was successfully meeting its demand from 12:00 a.m. to 9:00 p.m. by its own resources in Figure 9c, and there was a little amount of load curtailment from 10:00 p.m. to 11:00 p.m. of the NSL, which shows that the MG is reliable and providing uninterruptible energy to the SL.

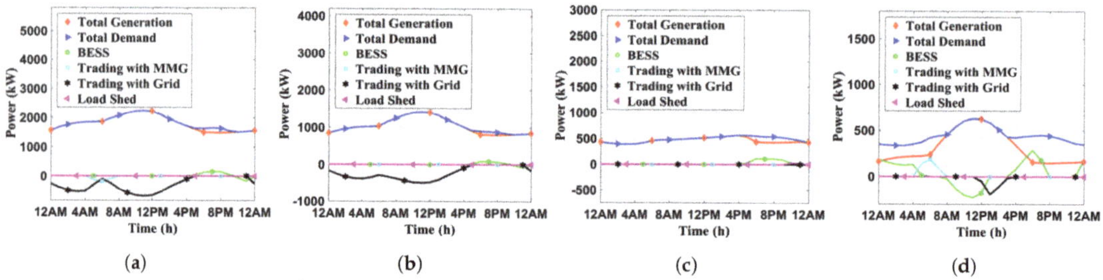

Figure 9. Scenario 3: (**a**) MG1, (**b**) MG2, (**c**) MG3, (**d**) MG4.

Figure 10a compares these three scenarios for the sum of load shedding in four MGs hour-wise. In Scenarios 1 and 2, there was no load shedding, whereas, in Scenario 3, there was a small load curtailment from 10:00 p.m. to 11:00 p.m. Figure 10b compares all scenarios for the sum of total optimal cost of all MGs hour-wise. It can be seen that the costs in all scenarios were nearly the same because, in each scenario, the proposed algorithm took the decision considering minimum cost. Therefore, all three scenarios were having the most optimized costs at each hour.

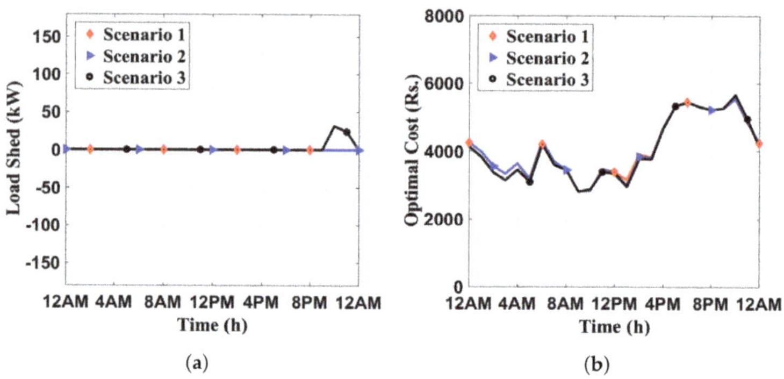

Figure 10. (**a**) Total Load shedding in all MGs and (**b**) total optimal cost of all MGs.

5.2. Cases Generated by MCS

The 25 cases were generated by MCS, and their respective results per day for each MG are presented in Table 1.

MG1 and MG2 were modeled in such a way that, most of the time, both MGs had surplus energy and never depended on external resources, whereas MG3 and MG4 were modeled in such a way that they were frequently independent of external resources, but, for a few hours, they were deficient in energy. This type of modeling revealed the effectiveness of the proposed strategy and the stability of the presented system model. In Table 1, it can be seen that the MGs in State 4, the MMG grid-connected state, were the most optimized and reliable with the minimum total cost and maximum profit. After meeting their demands, the MGs were selling/buying energy to/from other MGs, charging or discharging the CBESS, and later selling/buying energy to/from the grid. As MG1 and MG2 were microgrids with surplus energy most of the time, they were earning maximum profit in this state by selling the energy to other deficit MGs, charging the CBESS, and selling the energy to the grid. Their buying energy and load shedding were zero for all cases. In State 4, MG3 and MG4 were buying the deficit energy from MG1 and MG2 for few hours if they were in the MMG system, discharging the CBESS, and buying from the grid. In State 4, MG3 and MG4 were selling power to the MGs, charging the CBESS, or selling energy to the grid for rest of the hours, thus also earning the profit, and there was again no load shedding in these two MGs for State 4. In State 3, all MGs could only sell/buy energy to/from other MGs and charge/discharge the CBESS, and there was no trading with the grid. Again, the MG with maximum sold energy earned maximum profit, which is ultimately making its total cost minimum. The MG with buying energy is also in profit because there is no curtailment of the load, which would increase total cost with its penalty price. Thus, it proved that the proposed MMGs scheme is beneficial for each MG, either it has more or less generation capacity than the total demand, making it reliable and curtailment-free MG. In State 2, the MGs were only connected to the grid. They could neither sell/buy energy to/from other MGs nor charge/discharge the CBESS. The MGs could sell power to the grid in surplus energy hours and buy in deficit energy hours. We can see in Table 1 that, for respective states, the total cost of the MGs was nearly equal to or slightly more than that in previous states. In State 1, the MGs had only to depend on their own resources, i.e., they were single islanded entities. MG1 and MG2 in this state were fulfilling their respective demands, whereas MG3 and MG4 needed energy for few hours. No option was available to buy the power, thus having to shed the load. Now, the algorithm preferred to shed the NSL and had provided continuous supply to the SL. In this state, the MGs had more total cost than that in all three previous states. They were still reliable because there was no curtailment in the SL which would maximize the total cost, and there would be no use of energy management scheme. It can be clearly observed from Table 1 that the participation of a large number of MGs in the MMG system will be profitable for each MG, and there will be minimum local and global total costs as in Case 1. If the lesser number of MGs will participate in the MMG system, the lesser the profit will be, and the local and global total cost will increase, as in Case 25.

6. Conclusions

This paper proposed a novel energy management strategy for the day-ahead scheduling of the MMG systems. Four MGs were considered and optimized locally and globally. Each MG was fully connected with all other MGs in the MMG system. The MG-EMS performed local optimization and, after satisfying its local demand, updated the CEMS about its surplus and deficient amount of energy, thus preserving its privacy from other MGs. Only the CEMS took decisions about the power exchange between the MGs or with the grid by considering the state of the MG, generation and trading prices. As it was a two-level management strategy, thus, it increased privacy and resilience of the network. Various test cases were generated by MCS for all possible connections of the MGs. The study was performed on the IEEE 33-bus distribution system, and it has been proven that the proposed strategy and the algorithm are able to satisfy the demands of all MGs optimally. The numerical outcomes confirmed that the proposed methodology is compatible and can be executed effectively for practical microgrid applications. In summary, the proposed method can be utilized to make the management of the MMGs easy for real-world power systems, which will eventually transform the conventional power systems.

Author Contributions: Conceptualization, software, methodology, literature analysis, modeling, simulations, and writing—original draft, K.N.; writing—editing and literature analysis, F.Z.; supervision, editing and resources, K.K.M.; review, S.B.A.B.; supervision and review, H.A.K.; funding acquisition and review, C.-H.K. All authors have read and agreed to the published version of the manuscript.

Funding: This research was financed by the National Research Foundation of Korea (NRF) grant funded by the Korean Government (MSIP), Grant No. 2021R1A2B5B03086257.

Institutional Review Board Statement: Not applicable.

Informed Consent Statement: Not applicable.

Data Availability Statement: Data sharing not applicable.

Conflicts of Interest: The authors declare no conflict of interest.

Nomenclature

T	Index of time intervals.
I	Index of distributed generators.
K	Index of microgrids.
$C_{i,t}^{CDG}$	Production cost of a dispatchable unit 'i'.
$C_{i,t}^{RDG}$	Production cost of a renewable unit 'i'.
C_{t}^{PenSL}	Penalty cost for shedding of a sensitive load (l) at time 't'.
C_{t}^{PenNSL}	Penalty cost for shedding of a non-sensitive load (l') at time 't'.
$PR_{k,t}^{Rec}$	Power buying price from an MG 'k' at time 't'.
$PR_{k,t}^{Send}$	Power selling price to an MG 'k' at time 't'.
P_{CB}^{Cap}	Capacity of a central battery energy storage system (CBESS).
$P_{i,t}^{CDG_{min}}$	Minimum generation limit of a dispatchable unit 'i' at time 't'.
$P_{i,t}^{CDG_{max}}$	Maximum generation limit of a dispatchable unit 'i' at time 't'.
$P_{k,t}^{L}$	Load of an MG 'k' at time 't'.
$P_{k,t}^{Buy}$	Power of an MG 'k' bought from the grid.
$P_{k,t}^{Sell}$	Power of an MG 'k' sold to the grid.
SOC_{t}^{B}	State-of-charge (SOC) of a BESS at time 't'.
SOC_{t}^{CB}	SOC of a CBESS at time 't'.
P^{PV}/P^{WT}	Forecasted output of a photovoltaic (PV) panel/wind turbine (WT).
PR_{t}^{Buy}	Power buying price from the grid at time 't'.
PR_{t}^{Sell}	Power selling price to the grid at time 't'.
P_{B}^{Cap}	Capacity of a battery energy storage system (BESS).
η_B/η_{CB}	Efficiency of a BESS/CBESS.
$P_{i,t}^{CDG}$	Production amount of a dispatchable unit 'i'.
$P_{i,t}^{RDG}$	Production amount of a renewable unit 'i'.
$P_{k,t}^{Sur}$	Surplus amount of power in an MG 'k'.
$P_{k,t}^{Def}$	Deficient amount of power in an MG 'k'.
$P_{k,t}^{B+}$	Power required to charge a BESS in an MG 'k'.
$P_{k,t}^{B-}$	Power discharged from a BESS in an MG 'k'.
$P_{k,t}^{ShedSL}$	Amount of sensitive loads shed from an MG 'k'.
$P_{k,t}^{ShedNSL}$	Amount of non-sensitive loads shed from an MG 'k'.
$P_{k,t}^{Rec}$	Power of an MG 'k' bought from another MG.
$P_{k,t}^{Send}$	Power of an MG 'k' sold to another MG.
$c_{k,i,t}$	Commitment status of a dispatchable unit 'i' of an MG 'k'.

References

1. Gregoratti, D.; Matamoros, J. Distributed energy trading: The multiple-microgrid case. *IEEE Trans. Ind. Electron.* **2014**, *62*, 2551–2559. [CrossRef]
2. Panteli, M.; Mancarella, P. The grid: Stronger, bigger, smarter?: Presenting a conceptual framework of power system resilience. *IEEE Power Energy Mag.* **2015**, *13*, 58–66. [CrossRef]
3. Hatziargyriou, N.; Asano, H.; Iravani, R.; Marnay, C. Microgrids. *IEEE Power Energy Mag.* **2007**, *5*, 78–94. [CrossRef]
4. Katiraei, F.; Iravani, R.; Hatziargyriou, N.; Dimeas, A. Microgrids management. *IEEE Power Energy Mag.* **2008**, *6*, 54–65. [CrossRef]
5. Kanchev, H.; Lu, D.; Colas, F.; Lazarov, V.; Francois, B. Energy management and operational planning of a microgrid with a PV-based active generator for smart grid applications. *IEEE Trans. Ind. Electron.* **2011**, *58*, 4583–4592. [CrossRef]
6. Bui, V.H.; Hussain, A.; Kim, H.M. A multiagent-based hierarchical energy management strategy for multi-microgrids considering adjustable power and demand response. *IEEE Trans. Smart Grid* **2016**, *9*, 1323–1333. [CrossRef]
7. Chowdhury, D.; Hasan, A.S.M.K.; Khan, M.Z.R. Islanded DC Microgrid Architecture with Dual Active Bridge Converter-Based Power Management Units and Time Slot-Based Control Interface. *IEEJ Trans. Electr. Electron. Eng.* **2020**, *15*, 863–871. [CrossRef]
8. Arefifar, S.A.; Ordonez, M.; Mohamed, Y.A.R.I. Energy management in multi-microgrid systems—Development and assessment. *IEEE Trans. Power Syst.* **2016**, *32*, 910–922. [CrossRef]
9. Hossain, M.J.; Mahmud, M.A.; Milano, F.; Bacha, S.; Hably, A. Design of robust distributed control for interconnected microgrids. *IEEE Trans. Smart Grid* **2015**, *7*, 2724–2735. [CrossRef]
10. Che, L.; Zhang, X.; Shahidehpour, M.; Alabdulwahab, A.; Abusorrah, A. Optimal interconnection planning of community microgrids with renewable energy sources. *IEEE Trans. Smart Grid* **2015**, *8*, 1054–1063. [CrossRef]

11. Haddadian, H.; Noroozian, R. Multi-Microgrid-Based Operation of Active Distribution Networks Considering Demand Response Programs. *IEEE Trans. Sustain. Energy* **2019**, *10*, 1804–1812. [CrossRef]
12. Lu, T.; Wang, Z.; Ai, Q.; Lee, W. Interactive Model for Energy Management of Clustered Microgrids. *IEEE Trans. Ind. Appl.* **2017**, *53*, 1739–1750. [CrossRef]
13. Zhao, B.; Wang, X.; Lin, D.; Calvin, M.M.; Morgan, J.C.; Qin, R.; Wang, C. Energy Management of Multiple Microgrids Based on a System of Systems Architecture. *IEEE Trans. Power Syst.* **2018**, *33*, 6410–6421. [CrossRef]
14. Jiang, W.; Yang, K.; Yang, J.; Mao, R.; Xue, N.; Zhuo, Z. A Multiagent-Based Hierarchical Energy Management Strategy for Maximization of Renewable Energy Consumption in Interconnected Multi- Microgrids. *IEEE Access* **2019**, *7*, 169931–169945. [CrossRef]
15. Liu, Y.; Li, Y.; Gooi, H.B.; Jian, Y.; Xin, H.; Jiang, X.; Pan, J. Distributed robust energy management of a multimicrogrid system in the real-time energy market. *IEEE Trans. Sustain. Energy* **2017**, *10*, 396–406. [CrossRef]
16. Wang, S.; Gangammanavar, H.; Ekşioğlu, S.D.; Mason, S.J. Stochastic optimization for energy management in power systems with multiple microgrids. *IEEE Trans. Smart Grid* **2017**, *10*, 1068–1079. [CrossRef]
17. Hussain, A.; Bui, V.H.; Kim, H.M. A resilient and privacy-preserving energy management strategy for networked microgrids. *IEEE Trans. Smart Grid* **2016**, *9*, 2127–2139. [CrossRef]
18. Mehmood, K.K.; Kim, C.; Khan, S.U.; Haider, Z.M. Unified Planning of Wind Generators and Switched Capacitor Banks: A Multiagent Clustering-Based Distributed Approach. *IEEE Trans. Power Syst.* **2018**, *33*, 6978–6988. [CrossRef]
19. Hetzer, J.; Yu, D.C.; Bhattarai, K. An Economic Dispatch Model Incorporating Wind Power. *IEEE Trans. Energy Convers.* **2008**, *23*, 603–611. [CrossRef]
20. Atwa, Y.M.; El-Saadany, E.F.; Salama, M.M.A.; Seethapathy, R. Optimal Renewable Resources Mix for Distribution System Energy Loss Minimization. *IEEE Trans. Power Syst.* **2010**, *25*, 360–370. [CrossRef]
21. Selvan, M.P.; Swarup, K.S. Distribution system load flow using object-oriented methodology. In Proceedings of the 2004 International Conference on Power System Technology (PowerCon 2004), Singapore, 21–24 November 2004; Volume 2, pp. 1168–1173.
22. Draxl, C.; Clifton, A.; Hodge, B.M.; McCaa, J. The Wind Integration National Dataset (WIND) Toolkit. *Appl. Energy* **2015**, *151*, 355–366. [CrossRef]
23. National Solar Radiation Data Base. Available online: http://rredc.nrel.gov/solar/olddata/nsrdb/ (accessed on 23 September 2019).
24. LCG Consulting: Energy Online. Available online: http://http://www.energyonline.com/Data/GenericData.aspx?DataId=4 (accessed on 19 February 2020).
25. Khawaja, K.; Khan, S.; Lee, S.J.; Haider, Z.; Rafique, M.; Kim, C.H. A real-time optimal coordination scheme for the voltage regulation of a distribution network including an OLTC, capacitor banks, and multiple distributed energy resources. *Int. J. Electr. Power Energy Syst.* **2018**, *94*, 1–14. [CrossRef]

energies

MDPI

Article

A Novel Mean Field Game-Based Strategy for Charging Electric Vehicles in Solar Powered Parking Lots

Samuel M. Muhindo [1,2,3,*], Roland P. Malhamé [1,2,3] and Geza Joos [3,4]

1 Department of Electrical Engineering, Polytechnique Montréal, Montreal, QC H3T 1J4, Canada; roland.malhame@polymtl.ca
2 Groupe d'Études et de Recherche en Analyse des Décisions (GERAD), Montreal, QC H3T 2A7, Canada
3 Réseau Québécois sur l'Énergie Intelligente (RQEI), Trois-Rivières, QC G9A 5H7, Canada; geza.joos@mcgill.ca
4 Department of Electrical and Computer Engineering, McGill University, Montreal, QC H3A 0E9, Canada
* Correspondence: samuel.muhindo-mugisho@polymtl.ca

Abstract: We develop a strategy, with concepts from Mean Field Games (MFG), to coordinate the charging of a large population of battery electric vehicles (BEVs) in a parking lot powered by solar energy and managed by an aggregator. A yearly parking fee is charged for each BEV irrespective of the amount of energy extracted. The goal is to share the energy available so as to minimize the standard deviation (STD) of the state of charge (SOC) of batteries when the BEVs are leaving the parking lot, while maintaining some fairness and decentralization criteria. The MFG charging laws correspond to the Nash equilibrium induced by quadratic cost functions based on an inverse Nash equilibrium concept and designed to favor the batteries with the lower SOCs upon arrival. While the MFG charging laws are strictly decentralized, they guarantee that a mean of instantaneous charging powers to the BEVs follows a trajectory based on the solar energy forecast for the day. That day ahead forecast is broadcasted to the BEVs which then gauge the necessary SOC upon leaving their home. We illustrate the advantages of the MFG strategy for the case of a typical sunny day and a typical cloudy day when compared to more straightforward strategies: first come first full/serve and equal sharing. The behavior of the charging strategies is contrasted under conditions of random arrivals and random departures of the BEVs in the parking lot.

Keywords: battery electric vehicle; mean field games; nash equilibrium; parking lot; solar energy

Citation: Muhindo, S.M.; Malhamé R.P.; Joos, G. A Novel Mean Field Game-Based Strategy for Charging Electric Vehicles in Solar Powered Parking Lots. *Energies* **2021**, *14*, 8517. https://doi.org/10.3390/en14248517

Academic Editors: Victor Becerra and Ahmed Rachid

Received: 19 November 2021
Accepted: 11 December 2021
Published: 17 December 2021

1. Introduction

The massive introduction of battery electric vehicles (BEVs) [1–3] in modern power systems is bound to have important impacts, positive or negative, depending on the way this novel situation is managed [4]. There will be a great pressure to introduce numerous charging stations where the need is anticipated, but if too many high speed charging BEVs are connected at any one time (for example upon departure from work towards residence place), that may create both local transformer and eventually system wide overloads [5]. Many works in the literature [6–10] present algorithms for an optimal scheduling of vehicle-to-grid (V2G). The authors in [6] propose a centralized algorithm based on reinforcement learning which reduces the total power grid load variance by 65% in a test scenario of 300 consecutive days by charging 50 homogeneous BEVs in each hourly time slot in a neighborhood of 250 households. The authors in [7] propose a centralized weighted fair queuing (WFQ) algorithm with a 5 min time slot control switch in each smart charger to charge 300 homogeneous BEVs, by favoring those arriving with less charge. The algorithm selects a subset of BEVs to charge in each interval during peak demand when there is not enough energy. They compare the results with a first come first serve (FCFS) algorithm. They show that when the supply demand ratio (SDR) is equal to 1, there is 5% of BEVs which cannot leave their homes on time while it is 7% for FCFS. Recently, the authors in [10] study the case where the maximum charging power depends on the state of charge of the BEV's battery. They propose a centralized mixed integer linear programming (MILP) algorithm to charge a fluctuating heterogeneous population of BEVs at a single station considering availability of each BEV in order to minimize time-dependent electricity costs.

On the other hand, adequate management of the battery storage associated with an aggregate of BEVs can turn such an aggregate into a virtual power plant. Thus, in a context of integration with clean sources of energy, such as photovoltaics, despite other energy harvesting and storage techniques in the literature [11,12], BEVs' batteries could be storing the solar energy available during the day when BEVs are parked [13–16]. As is well known in the photovoltaics rich state of California for example, a so-called power demand *duck curve* [17] is observed: the peak demand occurs at the end of the day, upon return of working people to their homes. At that point, available solar radiation has all but disappeared and while solar energy may have been used by consumers during the day, there is a need for a high electric power ramp at dusk followed by several hours of sustained high power consumption. The latter power demand will be most likely met by fossil based energy sources, unless some other mitigating actions are taken. In [18,19], the authors show in their geographic context, that if the electric energy storage contained in a large number of BEVs is properly utilized, this could help significantly reduce the power needed from fossil sources during the evening peak.

The authors in [20], whose objective is close to ours in this paper, propose a centralized linear programming (LP) algorithm, in a solar powered parking lot of a car-share service to fairly distribute the available solar energy amongst 97 heterogeneous BEVs by favoring those arriving with less charge. They study the case where the SDR is strictly inferior to 1, and that all BEVs are available during the daily charging session of 5 h in the parking lot. They demonstrate, by charging a subset of 5 BEVs during each time slot, a reduction of 60% of yearly average standard deviation in the battery charge levels at the end of recharging compared to the equal sharing (ES) approach. The authors in [6,7,10,20] do not propose a decentralized algorithm. A decentralized algorithm scheme allows individual BEVs to determine their own charging pattern. Their decisions could, for example, be made on the basis of time-of-day, electricity price or battery state of health [21,22].

Table 1 below places our work in the BEVs charging optimization when the aggregators are parking lot operators (PLOs) or distribution system operators (DSOs).

Table 1. Classification of related work.

Aggregators	Potential Goal User Satisfaction	Monetary Benefits	Grid Impact
PLOs	our work, [20]	[18,21]	[9]
DSOs	[7]	[10,22]	[6]

Our objective in this paper is to propose an algorithm for sharing solar photovoltaic (PV) power amongst homogeneous BEVs parked in a parking lot, or a collection of federated parking lots. The BEVs belong to commuters working in the neighborhoods of these parking lots and could recharge at least partially depending on sunshine availability, their batteries at the parking lot charging stations. One particular business model is that the parking lots aggregator would charge a yearly fee for use of a parking space and the associated charging station. In a potential extension of the business model, if the BEVs' owners wish to recuperate part of their parking costs, they could choose to participate in a financially compensated grid support operation coordinated by the aggregator of the parking lots. In that case, *it would be in the interest of the aggregator to equalize charging of BEVs to maximize the probability of vehicle-to-grid charging participation.*

We suggest relying on an adequately tailored variation of a Mean Field Game-based algorithm scheme [23,24] which, while it fills all BEVs simultaneously, tends to provide more instantaneous charging to the BEVs with the lowest current fill levels. The problem is formulated as large population game on a finite charging interval. In [21], the authors study the existence, uniqueness and optimality of the Nash equilibrium of the charging problems to minimize local electricity costs and to fully charge. In a decentralized computational mechanism, they show in a deterministic case that the large population charging games will converge to a unique Nash equilibrium which is globally optimal for a homogeneous population.

In what follows we shall present a novel Mean Field Game-based charging algorithm to calculate the operator broadcasted decentralized algorithm laws according to the potential solar energy available. Subsequently, the performance of these laws will be compared to that of two common algorithms used in the literature. The first algorithm, *first come first full* (which, in a dynamic context, will be upgraded to *first come first serve*), consists of recharging *maximally* (or *up to an adequately updated SOC in real time*) the BEVs in order of arrivals at the parking lot. The second algorithm, *equal sharing*, consists of sharing equally at all times the available solar power amongst battery BEVs still not fully charged. All algorithms make full use of the available daily energy. Furthermore, for the purpose of meaningfully comparing the performance of the different algorithms in our case studies, we assume that the SDR is less than one.

In order to implement the charging algorithms, we make the following assumptions:

- There exists a communication infrastructure to coordinate BEVs charging in the parking lot.
- The BEVs are equipped with microprocessors in the chargers allowing them to locally compute and implement a local feedback-based charging algorithm.

The rest of this paper is organized as follows. In Section 2, we present the theoretical underpinnings and details of the MFG-based algorithm in the case of *homogeneous* BEVs. In Section 3, we present the numerical results *assuming a fixed population of BEVs which are charged in the parking lot simultaneously* with common characteristics of batteries. In Section 4, we present the algorithmic modifications and the numerical results in a more realistic situation *where BEVs arrive and depart randomly in the parking lot*. Finally, in Section 5, we conclude and give an outlook on future research.

2. Mean Field Game-Based Control of a Large Population of BEVs

2.1. Battery Model

We consider a population of n homogeneous BEVs in a parking lot. The assumption of a *large* population is needed only if, as we do in (1) below, we assume randomness in the dynamics of battery charging and later on in our analysis we will assimilate the empirical mean of SOCs with its mathematical expectation (a predictable deterministic quantity) by virtue of *the law of large numbers*. Because of the linearity of the model, the analysis will be perfectly exact for arbitrarily small numbers of BEVs if the battery charging processes remain deterministic. Each BEV i, $i = 1, \dots, n$, has a SOC $x_{i,arr}$ upon arrival which results of a daily traffic pattern from home to parking lot. We can then write the SOC stochastic dynamics for BEV i as follows [21] :

$$dx_{i,t} = bu_{i,t}dt + \nu d\omega_i, \tag{1}$$

where $t \in [t_0, T]$ is time in hours (h), $x_{i,t}$ is the SOC in per unit (pu) of capacity, $b = \frac{\alpha}{\beta}$, $\alpha \in (0, 1]$ is the charger efficiency in pu$/h$, β is the battery capacity in kWh, $u_{i,t} \in \mathbb{R}^+$ is the charging rate in kW, ω_i is a normalized Brownian process, ν is the intensity of that Brownian noise and ω_i is assumed independent of ω_j for $i \neq j$. The term $\nu d\omega_i$ defines the *stochasticity* of the SOC which can result physically from fluctuations in the charging and losses of the battery. We first present the algorithm in the simple case where all BEVs are started charging at the same time and depart at the end of control horizon, i.e., $t_0 = t_{begin}$ and $T = t_{end}$. Then the algorithm will be updated in a dynamic context, where $[t_0, T]$ will be made up of successive fixed short control horizons representing random arrivals and random departures of BEVs between t_{begin} and t_{end}.

2.2. Considerations

The charging algorithms tested in this paper will be compared with respect to the following requirements:

2.2.1. Fairness

We develop here our notion of fairness. While it is very hard to realize in practice, we wish that BEVs have near identical SOCs at the time they leave the parking lot irrespective of the time they arrived or their SOC at arrival. Furthermore, for BEVs that are simultaneously present in the parking lot, at no time should the BEV that arrived with a lower SOC be allowed to have its SOC exceed that of the other BEV. Such a criterion will lead us to the definition of a fairness coefficient by means of which we shall contrast the performance of various charging algorithms. We aim at:

(a) BEVs having relatively close SOCs at the time of their departure (under an assumption that they will spend statistically equivalent amounts of time in the parking lot).
(b) Pairs of BEVs present in the parking lot at the same time must maintain a fixed relative ordering of their SOC values (SOC of BEV i > SOC of BEV j) throughout their simultaneous stay, and until one of them leaves.

This leads us to the definition of a fairness coefficient (FC) that will help compare the fairness of the various charging strategies.

- Given (a) above, we consider that FC should be proportional to the the reduction in standard deviation of the SOCs' BEVs at departure time relative to the SOC standard deviation at arrival time.
- Given (b) above, we should count for all BEV pairs that qualify, the number of violations of the fixed SOC relative value ordering criterion during their simultaneous stay. Let N_O be the number of such violations, and let $\binom{N}{2}$ be the number of BEV pairs in a population of N BEVs. Let η be the ratio of the two. We consider that FC must decrease exponentially with the value of η.

We suggest using the following expression:

$$FC = \frac{\sigma_{x_{i,arr}} - \sigma_{x_{i,dep}}}{\sigma_{x_{i,arr}}} e^{-\eta}, \text{ where } \eta = \frac{N_O}{\binom{N}{2}} = \frac{2\ N_O\ (N-2)!}{N!}. \tag{2}$$

- The degree of fairness of the charging strategy is measured by FC if $FC > 0$.
- The charging strategy is considered *neutral* if $FC = 0$.
- The charging strategy is *rejected* if $FC < 0$.

2.2.2. Decentralization

From the point of view of the parking lot operator, decentralized charging algorithm laws are quite desirable because they minimize the need to observe the state of charge of individual batteries, a process which is both complex and invasive. Furthermore, a local algorithm allows a user to interrupt the process at any time, particularly if the parking operator has designed a charging scheme based on a poor model of the battery.

We wish to address the decentralized control of battery recharging of a set of BEVs as part of a so-called Mean Field Game (MFG). The control will be of linear-quadratic (LQ) type [25,26]. The parking lot operator broadcasts an average SOC target trajectory ($\overline{x}_t^{target}$) based on the solar energy forecast for the current day. The goal is that the BEVs store up as much of the solar energy available as possible and yet share the energy in a manner which will be deemed as fair. The proposed algorithm requires that the parking lot operator know the average SOC of the BEVs upon arrival ($\overline{x}_0$). This can be achieved by recording initial SOCs as BEVs enter the parking lot.

The key point of the approach is the prescription of a daily quadratic cost J_i for each BEV i to ensure that the BEVs that are initially fuller recharge less quickly than those that are less full, so that final SOC standard deviation of departing BEVs is reduced; while still maintaining the goal of using up the available solar power at all instants. More precisely the cost functions are designed so that by optimizing the individual BEV costs J_i, one achieves the aggregator's goals (fairness and decentralization) while using all available solar energy in the parking lot.

2.3. Establishment of Individual Battery Cost Function

The battery cost function is a mathematical expectation. It is designed by the aggregator and defined for a BEV as follows:

$$J_i(x_{i,0}, u_{i,t}) = \mathbb{E}\Big[\int_{t_0}^{T} e^{-\delta t}\big[\frac{q_t^y}{2}(x_{i,t} - y)^2 + \frac{r}{2}u_{i,t}^2 + \frac{q_{x_0}}{2}(x_{i,t} - x_{i,0})^2\big]dt \mid x_{i,0}\Big], \tag{3}$$

where $\mathbb{E}$ is the expectation operator, t_0 is the charging starting time, T is the charging stopping time, δ is a discount coefficient to ensure convergence of the cost, y is the collective direction of the BEVs' SOCs which is equal to 1 in our case (it serves as a direction signal to all BEVs, such that all BEVs should move toward y but not beyond), r is a coefficient which penalizes the level for charging rate, q_{x_0} is a pressure coefficient aimed at limiting the distance from the SOC $x_{i,0}$ (for the state of health of the user's battery and fairness to others) and q_t^y is the pressure field trajectory. The latter is common to all BEVs and is numerically obtained as the solution of a system of differential equations. It is the key quantity which will drive all SOCs towards a full state of charge while sharing instantaneously available solar energy in a way that reduces the standard deviation of SOCs. Its computation is further detailed below and it is at the heart of our inverse Nash equilibrium procedure. Note that the class of quadratic cost functions has been frequently used in the MFG literature [23,25,27].

The BEVs will collectively recharge their batteries with a time dependent coefficient (q_t^y) penalizing the gap between the current SOC ($x_{i,t}$) and the ultimate destination direction of SOCs defined by the value of y. Once the target for steady-state mean SOC of BEVs ($\overline{x}_T^{target}$) is reached, q_t^y will settle to a constant value, allowing the reaching of a mean SOC steady-state that meets the constraints set by the parking lot operator.

2.4. Optimal Control Problem and Solutions

We begin by making the following variable changes to simplify the expression of cost J_i in Equation (3):

$$\begin{aligned} X_{i,t} &= (x_{i,t} - y)e^{-\delta t/2}, U_{i,t} = u_{i,t}e^{-\delta t/2}, \\ V_t &= ve^{-\delta t/2} \text{ and } Z_{i,t} = (x_{i,t} - x_{i,0})e^{-\delta t/2}. \end{aligned} \tag{4}$$

Then:

$$\begin{aligned} dX_{i,t} &= dx_{i,t}e^{-\delta t/2} - \frac{\delta}{2}x_{i,t}e^{-\delta t/2}dt + \frac{\delta}{2}ye^{-\delta t/2}dt \\ &= (bu_{i,t}dt + vd\omega_i)e^{-\delta t/2} - \frac{\delta}{2}(x_{i,t} - y)e^{-\delta t/2}dt \\ &= -\frac{\delta}{2}X_{i,t}dt + bU_{i,t}dt + V_t d\omega_i. \end{aligned} \tag{5}$$

The solution approach is based on assuming a quadratic form of the optimal cost function [26] (with coefficients π, s and γ):

$$J_i^*(X_{i,t}) = \frac{1}{2}\pi_{i,t}X_{i,t}^2 + s_{i,t}X_{i,t}e^{-\delta t/2} + \gamma_{i,t}. \tag{6}$$

We then write the dynamic programming equation corresponding to this guess [28].

$$\frac{\partial J_i^*}{\partial t} + \min_{U} \left\{ \frac{q_t^y}{2} X_{i,t}^2 + \frac{q_{x_0}}{2} Z_{i,t}^2 + \frac{r}{2} U_{i,t}^2 + \frac{\partial J_i^*}{\partial X} \left(-\frac{\delta}{2} X_{i,t} + bU_{i,t} \right) + \frac{\partial^2 J_i^*}{\partial X^2} \frac{V_t^2}{2} \right\} = 0. \quad (7)$$

Differentiating with respect to $U_{i,t}$ yields:

$$\begin{aligned} &\nabla_U \left\{ \frac{q_t^y}{2} X_{i,t}^2 + \frac{q_{x_0}}{2} Z_{i,t}^2 + \frac{r}{2} U_{i,t}^2 \right\} + \nabla_U \left\{ \frac{\partial J_i^*}{\partial X} \left(-\frac{\delta}{2} X_{i,t} + bU_{i,t} \right) + \frac{\partial^2 J_i^*}{\partial X^2} \frac{V_t^2}{2} \right\} = 0 \\ &\longrightarrow U_{i,t}^* = -\frac{b}{r} \frac{\partial J_i^*}{\partial X}. \end{aligned} \quad (8)$$

The second derivative with respect to $U_{i,t}$ is $r > 0$, so the value found will indeed correspond to a minimum:

$$U_{i,t}^* = -\frac{b}{r} \left[\pi_{i,t} X_{i,t} + s_{i,t} e^{-\delta t/2} \right]. \quad (9)$$

The optimal control therefore depends on the values $\pi_{i,t}$ and $s_{i,t}$. The expressions of $\pi_{i,t}$ and $s_{i,t}$ are then determined by identification:

$$\begin{aligned} &\frac{1}{2} X_{i,t}^2 \frac{d\pi_{i,t}}{dt} + X_{i,t} e^{-\delta t/2} \left(\frac{ds_{i,t}}{dt} - \frac{\delta}{2} s_{i,t} \right) + \frac{d\gamma_{i,t}}{dt} \\ &= -\frac{q_t^y}{2} X_{i,t}^2 - \frac{q_{x_0}}{2} Z_{i,t}^2 + \frac{b^2}{2r} \left(\pi_{i,t}^2 X_{i,t}^2 + 2 X_{i,t} \pi_{i,t} s_{i,t} e^{-\delta t/2} + s_{i,t}^2 e^{-\delta t} \right) \\ &+ \frac{\delta}{2} X_{i,t} \left(\pi_{i,t} X_{i,t} + s_{i,t} e^{-\delta t/2} \right) - \frac{V_t^2}{2} \pi_{i,t} \\ &= X_{i,t}^2 \left(\frac{b^2}{2r} \pi_{i,t}^2 + \frac{\delta}{2} \pi_{i,t} - \frac{q_t^y}{2} - \frac{q_{x_0}}{2} \right) + X_{i,t} e^{-\delta t/2} \left(\frac{b^2}{r} \pi_{i,t} s_{i,t} + \frac{\delta}{2} s_{i,t} - y q_{x_0} + x_{i,0} q_{x_0} \right) + \ldots \end{aligned} \quad (10)$$

The analysis results in the following system of differential equations that must be solved backwards:

$$\begin{aligned} \frac{d\pi_{i,t}}{dt} &= \frac{b^2}{r} \pi_{i,t}^2 + \delta \pi_{i,t} - q_t^y - q_{x_0}, \\ \frac{ds_{i,t}}{dt} &= \left(\delta + \frac{b^2}{r} \pi_{i,t} \right) s_{i,t} + q_{x_0} (x_{i,0} - y), \end{aligned} \quad (11)$$

and the optimal control law is given by:

$$u_{i,t}^* = -\frac{b}{r} \left[\pi_{i,t} (x_{i,t} - y) + s_{i,t} \right]. \quad (12)$$

The coefficient q_t^y appearing in the differential equation of π is unknown at this stage. Nonetheless, it must respect the fact that when the BEVs use the optimal control $u_{i,t}^*$, their empirical average trajectory ($\overline{x}_t$), assimilated thanks to the law of large numbers to the mathematical expectation of the SOC of a generic BEV ($\mathbb{E}[x_{i,t}]$), will correspond to the average target trajectory ($\overline{x}_t^{target}$) imposed by the parking lot operator. The charging strategy that is developed, relies on knowing the anticipated solar energy during the day. Based on the dynamics of the SOC $x_{i,t}$ in Equation (1), the average target $\overline{x}_t^{target}$ needed is determined by integrating the curve of the forecast solar power (u_{W_t}) over the time $[t_0, T]$ of interest and dividing, for a case of homogeneous battery capacities, by the total number of BEVs present in the parking lot for recharging. The total forecast solar energy W available in the parking lot is assumed less than the total energy that all BEVs would need to fully recharge their batteries.

2.5. Calculation of q_t^y by Nash Equilibrium Inversion

We calculate the pressure field q_t^y directly by numerical resolution of differential equations. The system of differential equations to be solved is obtained by imposing that under the action of q_t^y, and the associated optimal control law in Equation (12), the average trajectory $\overline{x}_t$ of the BEVs follows the average target trajectory $\overline{x}_t^{target}$ broadcasted by the parking lot operator. This restriction allows us to write the Mean Field equations based on taking the mathematical expectation of the SOC of a generic battery in the population subject to decentralized control law (Equation (12)).

$$\begin{aligned}\frac{d\overline{x}_t^{target}}{dt} &= b\overline{u}_t^* = -\frac{b^2}{r}[\pi_t(\overline{x}_t^{target} - y) + \overline{s}_t],\\ \frac{d\pi_t}{dt} &= \frac{b^2}{r}\pi_t^2 + \delta\pi_t - q_t^y - q_{x_0},\\ \frac{d\overline{s}_t}{dt} &= (\delta + \frac{b^2}{r}\pi_t)\overline{s}_t + q_{x_0}(\overline{x}_0 - y),\end{aligned} \tag{13}$$

where $\pi_t = \pi_{i,t}$ and $\overline{s}_t = \frac{\sum_{i=1}^n s_{i,t}}{n}$. With this approach, the goal is to obtain a mathematical relationship between $\frac{d\overline{x}_t^{target}}{dt}$ and q_t^y. This is the so-called *inverse Nash algorithm*, its first steps were developed in the control of electric space heaters [27].

With the first equation of (13), we can write the relation between π_t and $\overline{s}_t$:

$$\pi_t = -\frac{\overline{s}_t}{\overline{x}_t^{target} - y} - \frac{r\frac{d\overline{x}_t^{target}}{dt}}{b^2(\overline{x}_t^{target} - y)}. \tag{14}$$

The differential equation governing the dynamics of $\overline{s}_t$ is then written as follows:

$$\frac{d\overline{s}_t}{dt} = -\frac{\overline{s}_t^2 b^2 + r\overline{s}_t\frac{d\overline{x}_t^{target}}{dt}}{r(\overline{x}_t^{target} - y)} + (\overline{x}_0 - y)\left[q_{x_0} + \frac{\frac{d\overline{x}_t^{target}}{dt}}{b^2(\overline{x}_t^{target} - y)}\right]. \tag{15}$$

To solve this differential equation numerically in the interval of time $[t_0, T]$, we need to specify a terminal condition at T. Since this equation is solved backwards in time, this is equivalent to determining $\overline{s}_T$. By choosing a time horizon T such that the solar power curve has already fallen to zero at T, $\overline{x}_t^{target}$ will settle at $\overline{x}_T^{target}$. Thus, $\frac{d\overline{x}_t^{target}}{dt} = 0,\ t \geq T$, and this is consistent with imposing $\frac{d\pi_t}{dt} = \frac{d\overline{s}_t}{dt} = 0,\ t \geq T$. We will then assimilate $q_T^y, \pi_T, \overline{s}_T$ to their steady-state values. This allows us to write:

$$\begin{aligned}0 &= -\frac{b^2}{r}[\pi_T(\overline{x}_T^{target} - y) + \overline{s}_T],\\ 0 &= \frac{b^2}{r}\pi_T^2 - q_T^y - q_{x_0},\\ 0 &= \frac{b^2}{r}\pi_T\overline{s}_T + q_{x_0}(\overline{x}_0 - y),\end{aligned} \tag{16}$$

and yields:

$$\begin{aligned}q_T^y &= q_{x_0}\left(\frac{\overline{x}_0 - \overline{x}_T^{target}}{\overline{x}_T^{target} - y}\right),\\ \pi_T &= \sqrt{\frac{r}{b^2}(q_{x_0} + q_T^y)},\\ \overline{s}_T &= \pi_T(y - \overline{x}_T^{target}) \text{ or } \overline{s}_T = \frac{rq_{x_0}(y - \overline{x}_0)}{\pi_T b^2}.\end{aligned} \tag{17}$$

Thereafter, we solve numerically and backwards the differential equation of $\frac{d\overline{s}_t}{dt}$, which yields the trajectory of $\overline{s}_t$. The latter is re-injected into the equation of π_t. Finally, we have all the necessary ingredients to calculate q_t^y from:

$$q_t^y = \frac{b^2}{r}\pi_t^2 + \delta\pi_t - \frac{d\pi_t}{dt} - q_{x_0} \tag{18}$$

3. Numerical Results in the Case of a Fixed Population of BEVs

3.1. Required Data

- Homogeneous population of BEVs: we consider an average BEV with $\alpha = 0.85$ and $\beta = 23\ kWh$ representing the average values of a realistic population of BEVs.
- Simulation parameters: we consider installing 100 solar panels in the parking lot for recharging 400 BEVs between $t_{begin} = 6\ h$ and $t_{end} = 18\ h$ in the sunny day case or the cloudy day case with a random normal distribution of SOCs upon arrival (with a mean of 0.15 and a standard deviation of 0.10). Here the charging of all BEVs starts at t_{begin} and stops at t_{end} (i.e., $t_0 = 6\ h$ and $T = 18\ h$), $dt = 0.01\ h$, $\nu = 0.01$, $\delta = 0$ (we set δ to zero to work on a *finite* control horizon), $q_{x_0} = 1000$, $r = 0.001$, $y = 1$.

3.2. MFG Inverse Nash Algorithm of Charging BEVs

Figure 1 shows the outline of the operation of the algorithm.

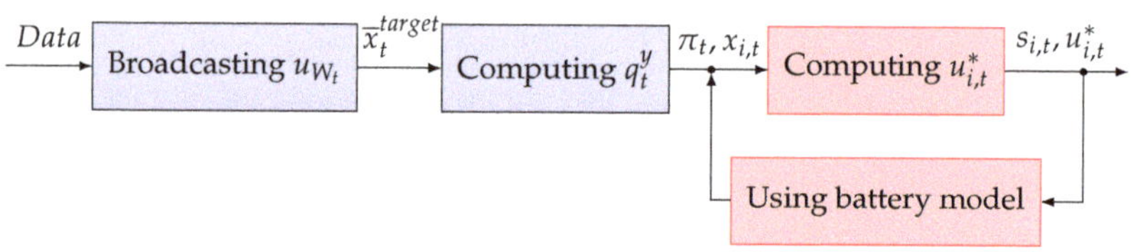

Figure 1. Block diagram of the MFG inverse Nash algorithm of BEV i at time t.

Below is the detailed Algorithm 1.

Algorithm 1 MFG inverse Nash

Require: t_0 (charging *starting* time), T (charging *stopping* time), u_{W_t} (forecast solar power at time t), n (number of BEVs present in the parking lot, i.e., at time t_0), $x_{i,0}$ (BEVs' SOCs at time t_0), $\overline{x}_0 = \frac{\sum_{i=1}^{n} x_{i,0}}{n}$, α, β, $b = \frac{\alpha}{\beta}$, dt, ν, q_{x_0}, r, y, δ.

Ensure: *The parking lot operator computes the pressure field* (q_t^y) *of n BEVs using the steps:*

1. Solve $\frac{d\overline{x}_t^{target}}{dt} = \frac{1}{n}bu_{W_t}$ and note the mean target SOC ($\overline{x}_T^{target}$) of n BEVs.
2. Calculate $q_T^y = q_{x_0}\left(\frac{\overline{x}_0 - \overline{x}_T^{target}}{\overline{x}_T^{target} - y}\right)$, $\pi_T = \sqrt{\frac{r}{b^2}(q_{x_0} + q_T^y)}$ and $\overline{s}_T = \pi_T(y - \overline{x}_T^{target})$.
3. Solve $\frac{d\overline{s}_t}{dt} = -\frac{\overline{s}_t^2 b^2 + r\overline{s}_t\frac{d\overline{x}_t^{target}}{dt}}{r(\overline{x}_t^{target} - y)} + (\overline{x}_0 - y)\left[q_{x_0} + \frac{\frac{d\overline{x}_t^{target}}{dt}}{b^2(\overline{x}_t^{target} - y)}\right]$ backwards.
4. Calculate $\pi_t = -\frac{\overline{s}_t}{\overline{x}_t^{target} - y} - \frac{r\frac{d\overline{x}_t^{target}}{dt}}{b^2(\overline{x}_t^{target} - y)}$ and determine $\frac{d\pi_t}{dt}$ (by *mean value theorem*).
5. Calculate $q_t^y = \frac{b^2}{r}\pi_t^2 + \delta\pi_t - \frac{d\pi_t}{dt} - q_{x_0}$.

Ensure: *Each BEV* $i, i = 1, 2, 3, \cdots, n$, *computes its local feedback strategy using the steps:*

1. Solve $\frac{ds_{i,t}}{dt} = (\delta + \frac{b^2}{r}\pi_t)s_{i,t} + q_{x_0}(x_{i,0} - y)$ backwards with $s_{i,T} = \frac{rq_{x_0}(y - x_{i,0})}{\pi_T b^2}$.
2. Solve $dx_{i,t} = bu_{i,t}^* dt + \nu d\omega_i = -\frac{b^2}{r}[\pi_t(x_{i,t} - y) + s_{i,t}]dt + \nu d\omega_i$.

3.3. Obtaining the Average Target SOC by Using Daily Solar Energy in the Parking Lot

Realistic generation curves based on historical meteorological data are used (Figure 2), assuming that similar generation curves can be predicted using for example, a machine learning based model. The meteorological data is obtained from photovoltaic geographical information System (PVGIS) made available by the European Commission. A typical meteorological year in the city of Montreal (45.50 North, 73.58 West) is used with a data resolution of one hour. The very same data can also be found in Canadian Weather Energy and Engineering Datasets (CWEEDS). The solar photovoltaic (PV) power output is then modeled with the simulation software TRNSYS using type 103 appropriate for modeling the electrical performance of mono and polycrystaline PV panels.

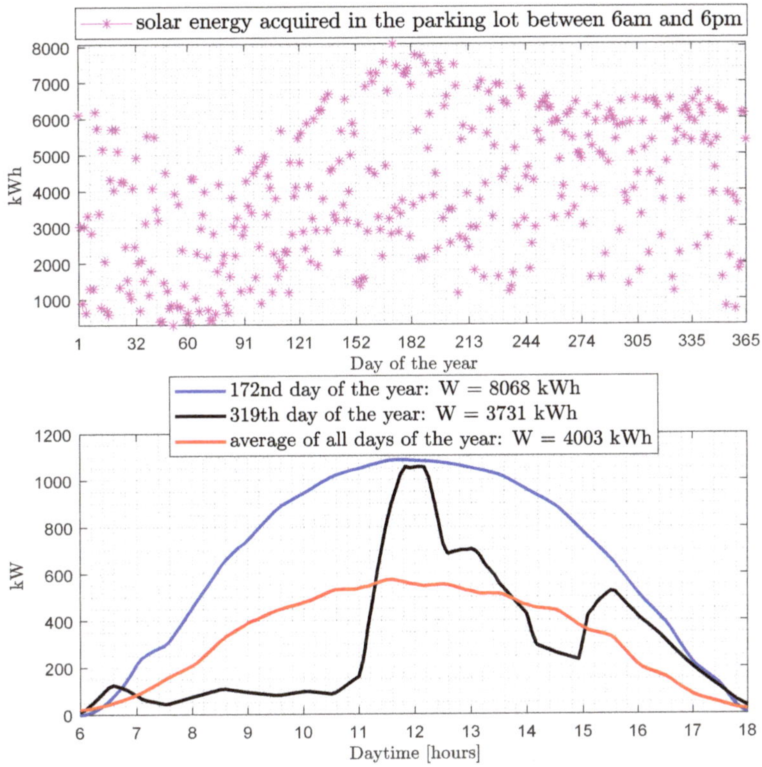

Figure 2. Solar energy production in the parking lot with 100 PV panels in Montreal (Canada).

We determine two different real solar power curves (sunny day and cloudy day cases) in order to compare the influence of the difference in generation on the behaviour of the BEVs charging. Starting with the sunny day case, and 400 BEVs, we get an average target curve that saturates at the end of the horizon ($\overline{x}_T^{target} = 0.90$). Looking at the cloudy day case and as a result of that a lower energy output at the end of the horizon, charging the same number of BEVs with the same distribution SOCs upon arrival ($\overline{x}_0 = 0.15$) would result in an average target that is low ($\overline{x}_T^{target} = 0.50$). The parking lot operator would then announce the situation the day before, so that the BEVs arrive next day more full in the parking lot. However here, for comparison purposes, we shall work with the same distribution SOCs upon arrival as for the sunny day case.

3.4. Pressure Field, Empirical per BEV Average SOC and Individual SOCs of BEVs Using MFG Inverse Nash Algorithm

In Figures 3–5, first we see that the trajectories of the empirical per BEV average SOC and the mean target SOC broadcasted by the parking lot operator are quite the same in both days, *thanks to the law of large numbers*. And as expected in steady-state we have constant values, the proposed MFG algorithm makes full use of the available daily energy between 6 a.m. and 6 p.m. The results of individual SOCs of BEVs in the sunny day case, in Figure 6, show a *strong reduction in standard deviation* $\sigma_{x_{i,T}}$ while in the cloudy day case, in Figure 7, we have a *slight reduction in standard deviation*. Additionally, all the curves' behaviour in the cloudy day case well reflects the characteristic of solar fluctuations, and more importantly the results are *less desirable* in the cloudy day case relative to the sunny day case because the charging rate needed to achieve full solar utilization is lower in the cloudy day case. Furthermore, we confirm the main features of the proposed MFG algorithm, that of *filling more batteries that were emptier to start with while bringing all batteries close to a predefined mean target*.

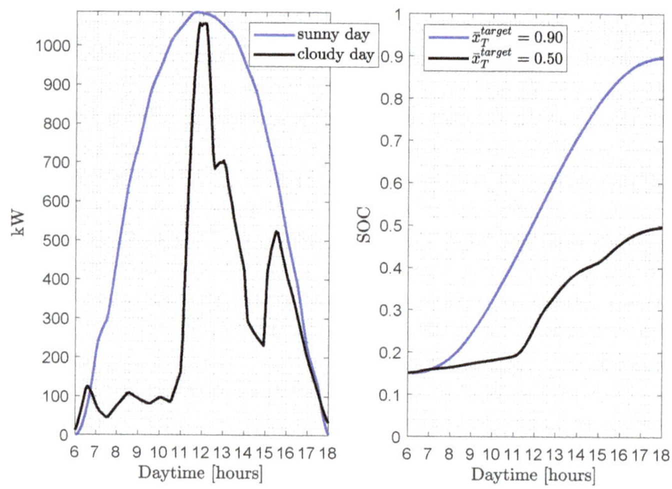

Figure 3. Daily solar power curves (u_W) for charging 400 BEVs and mean target SOC trajectories ($\bar{x}_t^{target}$) broadcasted by the parking lot operator.

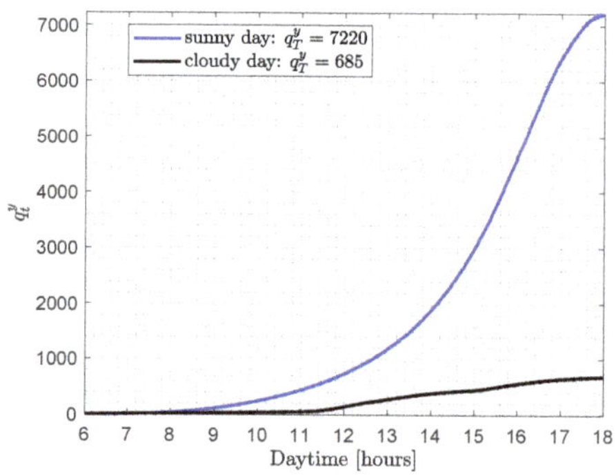

Figure 4. Pressure fields (q_t^y) of a fixed population of 400 BEVs.

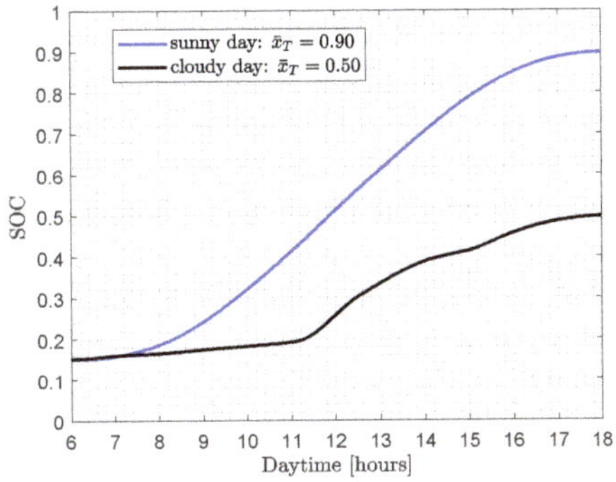

Figure 5. Empirical per BEV average SOCs ($\overline{x}_t$) of a fixed population of 400 BEVs.

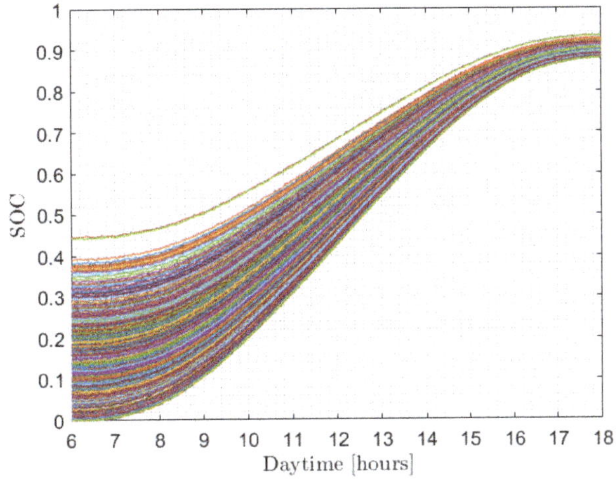

Figure 6. Individual SOCs of a fixed population of 400 BEVs in the sunny day.

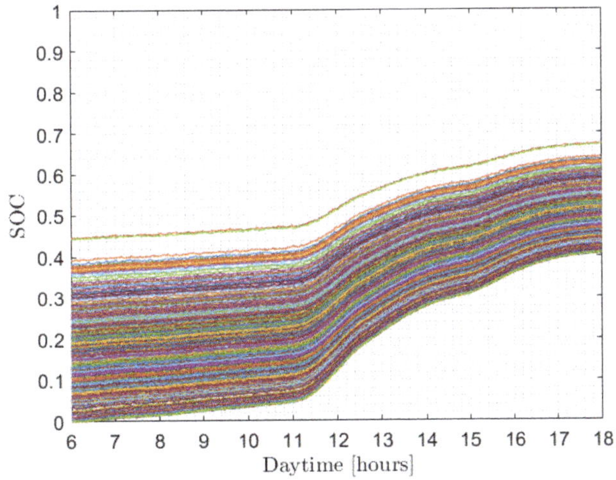

Figure 7. Individual SOCs of a fixed population of 400 BEVs in the cloudy day.

3.5. Different Charging Strategies

- *First come first full* (FCFF), which fills up to maximum capacity each BEV in the order of their arrival in the parking lot.
- *Equal sharing* (ES), which fills all the BEVs in the parking lot at equal rates until BEVs quit charging when they are full.
- *Mean Field Game* (MFG), which fills all the BEVs according to the *MFG inverse Nash algorithm* (see Algorithm 1 in Section 3.2).

3.6. Comparison of Charging Strategies

In the following, we present the results for three charging strategies in Figures 8 and 9.

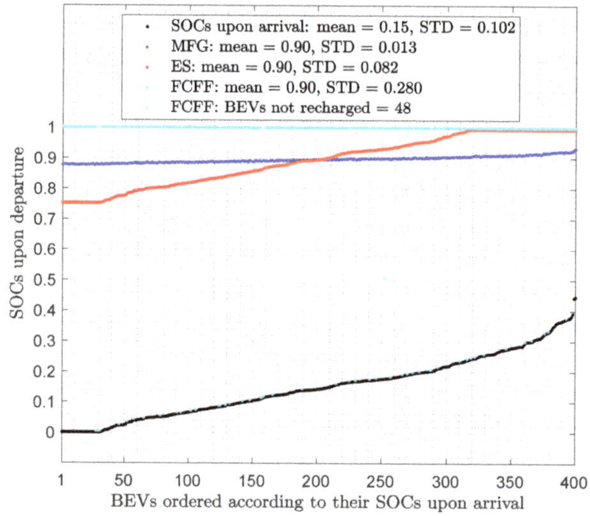

Figure 8. SOC levels at time of departure for a fixed population of 400 BEVs in the sunny day.

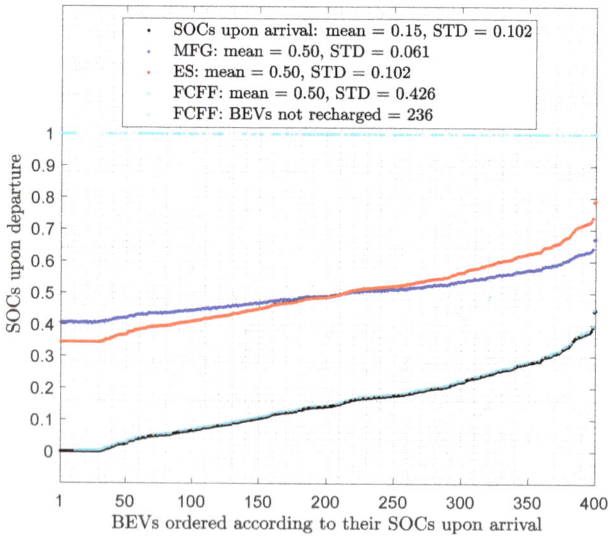

Figure 9. SOC levels at time of departure for a fixed population of 400 BEVs in the cloudy day.

3.6.1. FCFF Strategy

It is the easiest strategy to implement. A centralized signal is sent to fill maximally the BEVs in the parking lot regardless of their SOCs upon arrival, and only depending on the ordering of their arrival times. This approach is not suitable at all for our charging objective. It does not meet any of the requirements. First, it is a centralized charging strategy since the parking lot operator must not only note the order of arrivals of BEVs in the parking lot, but also when they finish filling up. Secondly, at the end of the day, there will generally be users who have not recharged their batteries at all (12% of BEVs in the sunny day case and 59% of BEVs in the cloudy day case). Thirdly, we have the highest standard deviations of SOCs upon departure: $\sigma_{x_{i,T}} = 0.280$ (an increase of 175%) in the sunny day case and $\sigma_{x_{i,T}} = 0.426$ (an increase of 318%) in the cloudy day case.

3.6.2. ES Strategy

This is a straightforward charging scheme. Using the forecast solar power available throughout the day, the parking lot operator estimates at every instant an average level per BEV available for charging and thus, the average charging rate. As BEVs get charged, the number of BEVs still in demand must be updated. Thus, this scheme, although superior to the previous one, is not decentralized. By giving the same amount to everyone, one does not significantly reduce $\sigma_{x_{i,T}}$ relative to $\sigma_{x_{i,0}}$. We have $\sigma_{x_{i,T}} = 0.082$ (a reduction of 20%) in the sunny day case and $\sigma_{x_{i,T}} = 0.102$ in the cloudy day case. Indeed, the SOCs' standard deviation remains close to what it was at the beginning, except for a slight reduction due to some BEVs completely filling up in the sunny day case.

3.6.3. MFG Strategy

This is our novel charging strategy scheme based on the idea of inverse Nash control. The parking lot operator, after broadcasting $\overline{x}_T^{target}$, prescribes a decentralized strategy via an inverse Nash algorithm in the smart charger in the parking lot. Each user applies their optimal control locally so that, the average trajectory ($\overline{x}_t$) of all users corresponds to the average target trajectory ($\overline{x}_t^{target}$) broadcasted by the parking lot operator. The standard deviation at the end of recharging has been significantly reduced. We have $\sigma_{x_{i,T}} = 0.013$ (a reduction of 87%) in the sunny day case and $\sigma_{x_{i,T}} = 0.061$ (a reduction of 40%) in the cloudy day case. Furthermore, except for the broadcasting of some initialization data by the parking lot operator, the charging scheme is decentralized. This strategy meets all the requirements of our charging objective.

In Table 2 the fairness coefficient (FC), as defined previously in Section 2.2.1, is computed for each charging strategy.

Table 2. Comparison of fairness coefficient and departing SOCs ($\min x_{i,0} = 0$ and $\max x_{i,0} = 0.45$) for a fixed population of 400 BEVs.

	Sunny Day			Cloudy Day		
	$\min x_{i,T}$	$\max x_{i,T}$	FC	$\min x_{i,T}$	$\max x_{i,T}$	FC
ES	0.76	1	0.19	0.35	0.79	0
FCFF	0	1	−0.72	0	1	−2.13
MFG	0.88	0.93	0.86	0.40	0.67	0.40

The results in both days show that the MFG strategy is the fairest strategy while the FCFF strategy is rejected.

4. Numerical Results in the Case of a fluctuating Population of BEVs

4.1. Considerations

- In the parking lot, the BEVs start arriving before 6 a.m., stop arriving at 12 p.m., start departing after 3 p.m. until after 6 p.m. (with t_{begin} = 6 h, t_{end} = 18 h).

- We consider a Poisson distribution for the arrivals (Figure 10). A Poisson distribution allows us to realistically fill parking spaces in the parking lot, where the number of arrivals at each time interval decreases exponentially until some fixed stopping time. Each 15 min arriving time interval φ_a, $a = 1, 2, 3, \cdots, 24$, a subpopulation of n_a BEVs is connected to the charging stations at random times $t_j^{(a)} < 6 + 0.25(a-1)$ a.m., $j = 1, \cdots, n_a$. We then charge, between $6 + 0.25(a-1)$ a.m. and $6 + 0.25a$ a.m., all $\sum_{k=1}^{a} n_k$ BEVs present in the parking lot. For example, the first n_1 BEVs connected to their charging stations before 6 a.m. will start charging at 6 a.m., the next n_2 BEVs connected between 6 a.m. and 6:15 a.m. will start charging at 6:15 a.m., and so on. The last n_{25} BEVs, arrive between 11:45 a.m. and 12 p.m., are connected at random times $t_j^{(25)} < 12$ p.m., $j = 1, \cdots, n_{25}$. We then charge, between 12 p.m. and 3 p.m., the total expected number of $\sum_{k=1}^{25} n_k = 400$ BEVs in the parking lot.
- We also consider a Poisson distribution for the departures (Figure 10). Here, a Poisson distribution is used for the same reason as in the case of arrivals but will be defined to empty the parking lot so that on average the departing BEVs spend the same parking time. Each 15 min departing time interval φ_d, $d = 26, 27, 28, \cdots, 38$, a subpopulation of n_d BEVs is disconnected from the charging stations at random times $t_j^{(d)} > 3 + 0.25(d-26)$ p.m., $j = 1, \cdots, n_d$. We then charge, between $3 + 0.25(d-26)$ p.m. and $3 + 0.25(d-25)$ p.m., the remaining $\left(400 - \sum_{k=26}^{d} n_k\right)$ BEVs in the parking lot. Here, we cannot delay recharging the BEVs within a given departing time interval φ_d as the charging remains continuous despite the random departures of the BEVs. It is only at the end of φ_d that we can record the exact number of vehicles that have disconnected from their charging stations. We then consider a small-size battery in the parking lot to store the unused energy $W_{\varphi_d}^*$ in a given departing time interval φ_d, i.e., not really distributed to the BEVs already left the parking lot. $W_{\varphi_d}^*$ is therefore added to the forecast solar energy $W_{\varphi_{d+1}}$ acquired in the next departing time interval φ_{d+1}.
- For all other parameters, we use the same values as in the case of a fixed population of BEVs (both in the sunny day case and in the cloudy day case).
- To summarize here, BEVs recharging is updated in 15-min cycles between 6 a.m. and 6 p.m. except between 12 p.m. and 3 p.m. when the cycle lasts 3 h.

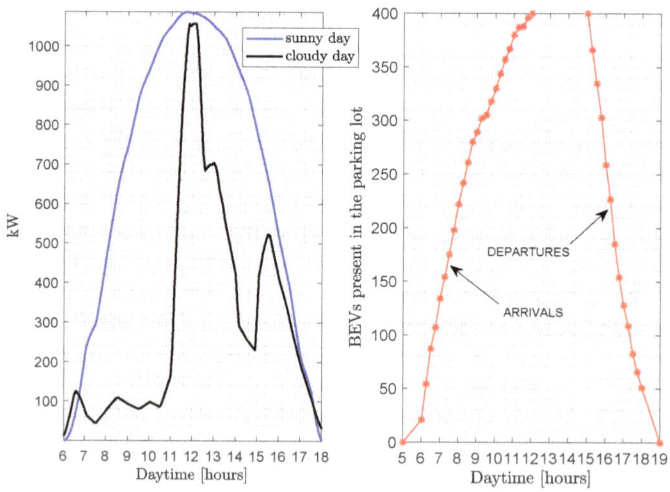

Figure 10. Daily solar power curves (u_W) for charging 400 BEVs and arrivals/departures of a fluctuating population of 400 BEVs in the parking lot.

4.2. Updated MFG Inverse Nash Algorithm

Each charging time interval φ_c (considering both arrival time interval φ_a and departing time interval φ_d), i.e., $c = 1, 2, \cdots, 25, 26, 27, \cdots, 38$ (where $\varphi_1 = [6,\ 6.25[\ h$, $\varphi_2 = [6.25,\ 6.5[\ h, \cdots, \varphi_{25} = [12,\ 15]\ h$, $\varphi_{26} =]15,\ 15.25]\ h$, $\varphi_{27} =]15.25,\ 15.5]\ h, \cdots, \varphi_{38} =]17.75,\ 18]\ h$), the parking lot operator calculates the mean target SOC trajectory of all BEVs present in the parking lot. Unlike in the case of a fixed population of BEVs in the parking lot where we worked with a solar curve over the entire control horizon between t_{begin} and t_{end}, here we work with successive fixed short control horizons at the start of which the number of BEVs present in the parking lot is recorded. In order to enforce the Riccati steady-state conditions (illustrated in Section 2), in each charging time interval φ_c we add to the solar power curve $u_{W_{\varphi_c}}$ a fictitious extension $\tilde{u}_{W_{\varphi_c}}$ into φ_{c+1} with the solar power falling smoothly to zero at the end of φ_{c+1}. Below, in Figure 11, is the example of a piece of solar power curve between 8 a.m. and 10 a.m. in the sunny day case with fictitious added parts.

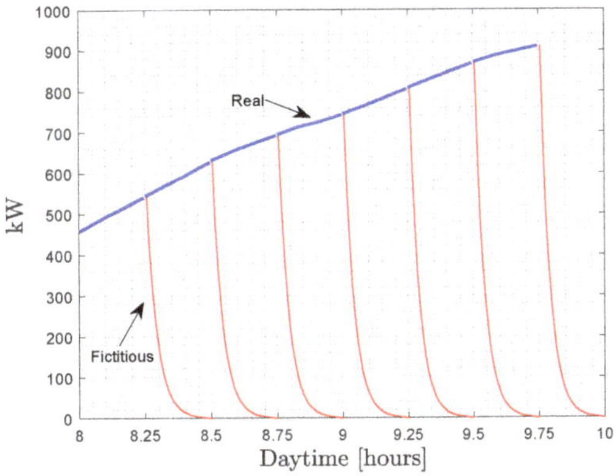

Figure 11. Section of solar power curve in the sunny day case with *fictitious added parts*.

The MFG inverse Nash algorithm (see Algorithm 1 in Section 3.2) will be applied here in each charging time interval φ_c on short control horizons (i.e., *here t_0 is the beginning of φ_c and T is the end of φ_c*) with the resulting mean target SOC trajectory $\overline{x}_t^{target}$, $t \in \{\varphi_c \cup \varphi_{c+1}\}$, knowing that resulting solar power curve is equal to $\{u_{W_{\varphi_c}} \cup \tilde{u}_{W_{\varphi_c}}\}$ (Figure 11).

4.3. Adjustment of Different Charging Strategies

- *First come first full* (FCFF), which fills up to maximum capacity each BEV among the n BEVs present in the parking lot in the order of their arrival in the parking lot in each charging time interval φ_c. We propagate this order of arrivals to the next charging time interval, φ_{c+1}.
- *First come first serve* (FCFS), is a refined version of FCFF, i.e., has the same characteristics as the FCFF except that instead of filling up to maximum capacity the n BEVs present in the parking lot, they are filled in each charging time interval φ_c up to

$$\frac{\sum_{i=1}^{n} x_{i,0}}{n} + \frac{1}{n}\int_{t_0}^{T} u_{W_t} dt.$$

- *Equal sharing* (ES), which fills the n BEVs present in the parking lot at equal rates (until BEVs quit charging when they are full) over each charging time interval φ_c.
- *Mean Field Game* (MFG), which fills the n BEVs present in the parking lot by using the inverse Nash algorithm in each charging time interval φ_c (see Algorithm 1 in Section 3.2).

4.4. Comparison of Charging Strategies Considering SOCs' BEVs at Departure Times

First, we present in Figure 12 the results of the SOCs' averages and standard deviations for the remaining BEVs in the parking lot knowing that the SoCs' averages before the departures (i.e., before 3 p.m.) are all the same for all charging strategies in both days, as they use all solar energy in the parking lot. As expected, the FCFS strategy comes first with the highest SOCs' averages and the lowest SOCs' standard deviations for the remaining BEVs as its perfectly equalizes the BEVs' SOCs regardless of their SOCs at the beginning of recharging, while the MFG strategy comes second as its also equalizes the BEVs' SOCs by maintaining some fairness criterion (with regard to their SOCs at the beginning of recharging). The ES strategy, considered as the base case, comes third, while the FCFF is clearly the worst strategy with the lowest SOCs' averages and the highest SOCs' standard deviations as its fills up to maximum capacity the BEVs' SOCs.

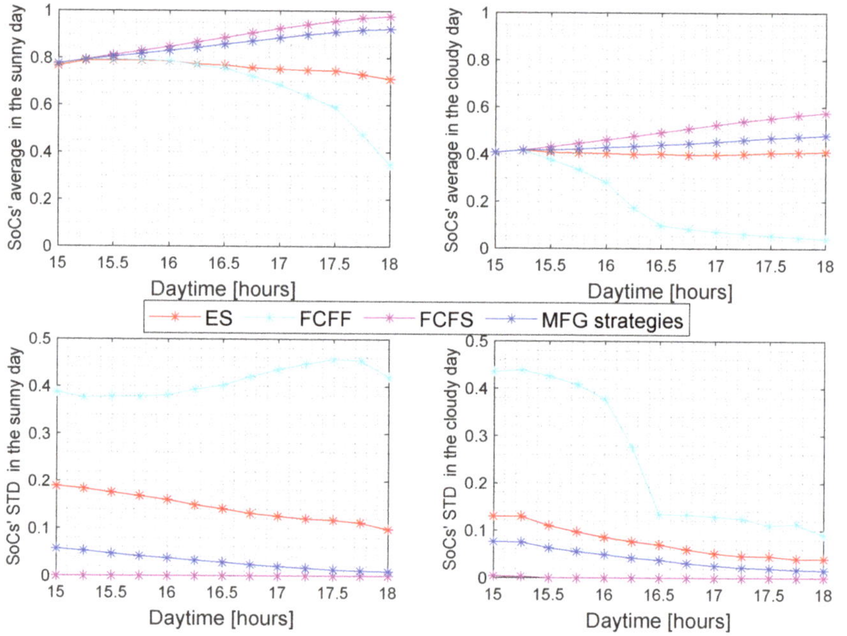

Figure 12. SOCs' averages and standard deviations of the remaining BEVs in the parking lot.

Then, we present in Figures 13 and 14 the results of *SOCs for the departing BEVs* knowing that the average parking time is 8 h.

In both days, we note that the MFG strategy remains the best between the four options insofar as the resulting *standard deviation* for the departed BEVs are concerned.

In the sunny day case, the MFG strategy reduces significantly the standard deviation ($\sigma_{x_{i,dep}} = 0.026$, a reduction of 75%). The FCFF strategy gives the worst standard deviation ($\sigma_{x_{i,dep}} = 0.296$, an increase of 190%) with 12% of users who have not recharged their batteries at all. The FCFS strategy is the improved version of the FCFF, as we can see that the BEVs are recharged with a better standard deviation ($\sigma_{x_{i,dep}} = 0.062$, a reduction of 39%). The ES strategy results in a slight increase of 16% of the standard deviation ($\sigma_{x_{i,dep}} = 0.118$) due to the fluctuating population of BEVs and to some BEVs completely filling up.

In the cloudy day case, the MFG strategy also reduces significantly the standard deviation ($\sigma_{x_{i,dep}} = 0.024$, a reduction of 76%) due to the fluctuating population of BEVs. The FCFF strategy gives the worst standard deviation ($\sigma_{x_{i,dep}} = 0.457$, an increase of 348%) with 54% of users who have not recharged their batteries at all. The FCFS strategy is the improved version of the FCFF, as we can see that the BEVs are recharged with a better standard deviation ($\sigma_{x_{i,dep}} = 0.053$, a reduction of 48%). The ES strategy results in a slight reduction of 21% of the standard deviation ($\sigma_{x_{i,dep}} = 0.081$) due to the fluctuating population of BEVs.

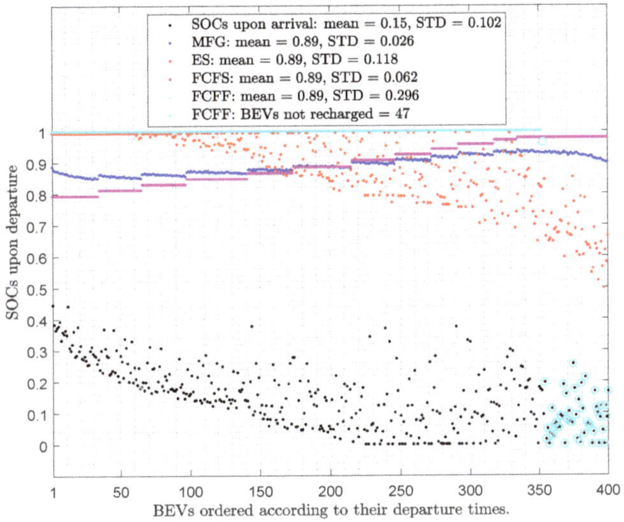

Figure 13. SOC levels at time of departure for a fluctuating population of 400 BEVs in the sunny day.

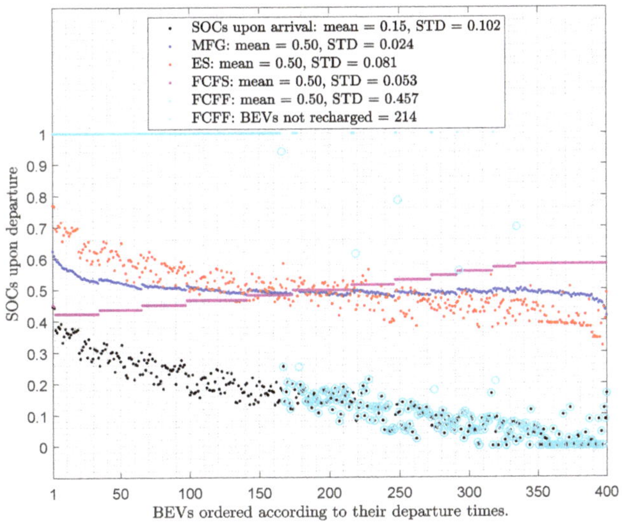

Figure 14. SOC levels at time of departure for of a fluctuating population of 400 BEVs in the cloudy day.

In Table 3 the fairness coefficient (FC), as defined previously in Section 2.2.1, is evaluated for each charging strategy.

Table 3. Comparison of fairness coefficient and departing SOCs (min $x_{i,arr} = 0$, max $x_{i,arr} = 0.45$) for a fluctuating population of 400 BEVs.

	Sunny Day			Cloudy Day		
	min $x_{i,dep}$	max $x_{i,dep}$	FC	min $x_{i,dep}$	max $x_{i,dep}$	FC
ES	0.50	1	−0.12	0.32	0.77	0.19
FCFF	0	1	−0.83	0	1	−2.78
FCFS	0.79	0.98	0.18	0.42	0.58	0.19
MFG	0.85	0.94	0.37	0.41	0.62	0.61

The results show that the MFG strategy remains the fairest strategy in both days while the FCFF strategy is rejected in both days and the ES strategy is rejected in the sunny day case.

5. Conclusions and Future Research

We have considered the situation of a large daytime work parking lot with homogeneous battery electric vehicles (BEVs) for simplicity, and solar sources based electricity charging. We have used realistic data to implement deterministic daily solar power curves with photovoltaic panels in a parking lot for a typical sunny day and a typical cloudy day. One should note that a large heterogeneous population of BEVs can be analyzed by assuming that it is possible to group the BEVs into classes considered homogeneous. Thus, all the BEVs of a class share the same physical parameters and, in order to better redistribute energy according to individual BEV needs, the forecast solar energy can be distributed by favoring a class with more BEVs, a larger-size battery and a lower charger efficiency.

In Section 3, a fair, and decentralized MFG strategy, for recharging a large fixed population of BEVs, has been developed. The goal was to reduce significantly the SOCs' standard deviation while elevating the SOCs of BEVs to a satisfactory level regardless of their SOCs upon arrival. A comparison was carried out with an equal sharing (ES) strategy and a first come first full (FCFF) strategy which we saw could result in some unsatisfied individual users with little SOCs at the end of recharging. In Section 4, we considered a large fluctuating population of homogeneous BEVs. This new situation allowed us to improve the FCFF strategy into first come first serve (FCFS) strategy. The results showed that the MFG strategy remains the most desirable charging strategy with regards to the standard deviation of SOCs upon departure and fairness criterion. Finally, we did much better than the literature [20] (60% as maximum reduction of SOCs' standard deviation in the case of a fluctuating population of BEVs) as we illustrated in the summary Table 4 below when we compared our results to the base case which is here the ES strategy.

Table 4. Standard deviation increase/reduction of SOCs' BEVs.

	400 Fixed Sunny Day	BEVs in a Cloudy Day	400 Fluctuating Sunny Day	BEVs in a Cloudy Day
FCFF	↑ 241%	↑ 318%	↑ 151%	↑ 464%
FCFS	↑ 0%	↑ 0%	↓ 47%	↓ 35%
MFG	↓ 84%	↓ 40%	↓ 78%	↓ 70%

In future research, we shall extend this work by considering stochastic solar acquisition in the parking lot and explore MFG based algorithms this time for potential partial restitution of battery energy from solar charged BEVs to the grid during evening peak hours.

Author Contributions: Conceptualization, S.M.M., R.P.M. and G.J.; Formal Analysis, S.M.M. and R.P.M.; Modeling and Simulation, S.M.M.; Experimental Results, S.M.M.; Investigation, S.M.M., R.P.M. and G.J.; Writing—Original Draft Preparation, S.M.M.; Writing—Review and Editing, S.M.M., R.P.M. and G.J.; Supervision, R.P.M. and G.J.; Funding Acquisition, R.P.M. and G.J. All authors have read and agreed to the published version of the manuscript.

Funding: This research was funded by the Réseau Québécois sur l'Énergie Intelligente and the NSERC/Hydro-Québec Industrial Research Chair on the Integration of Renewable Energies and Distributed Generation into the Electric Distribution Grid.

Acknowledgments: The authors would like to thank the Réseau Québécois sur l'Énergie Intelligente and the NSERC/Hydro-Québec Industrial Research Chair on the Integration of Renewable Energies and Distributed Generation into the Electric Distribution Grid for the financial support of this research. The first author would also like to thank Kato Vanroy from the department of Mechanical Engineering at Polytechnique Montreal for practical utilization of the TRNSYS software and Quentin Lénet for the first steps in inverse Nash algorithm.

Conflicts of Interest: The authors declare that they have no known competing financial interests or personal relationships that could have appeared to influence the work reported in this paper.

Nomenclature

$\mathbb{E}$	mathematical expectation symbol
∇	vector differential operator
$\partial\square/\partial\triangle$	partial derivative of $\square$ with respect to $\triangle$
$d\triangle$	differential of $\triangle$
t	time in h
t_{begin}, t_{end}	*beginning* of the horizon, *end* of the horizon
t_0, T	charging *starting* time, charging *stopping* time
$\varphi = [t_0, T]$	charging time interval
$t_j^{(a)}$, $t_j^{(d)}$	random *arrival* times, random *departure* times
i	a user of BEV
α	charger efficiency in pu/h
β	capacity of the battery in kWh
$b = \alpha/\beta$	characteristic of the battery in pu/kW
J, J^*	cost to minimize, *optimal* cost
$u_{i,t}$	charging rate of BEV user i at time t in kW
$u^*_{i,t}$	optimal charging rate of BEV user i at time t
u_{W_t}	forecast solar power in the parking lot at time t
$W = \int_{t_{begin}}^{t_{end}} u_{W_t} dt$	total forecast solar energy in the parking lot in kWh
$W_\varphi = \int_\varphi u_{W_t} dt$	forecast solar energy in the parking lot between t_0 and T
$x_{i,t}$	state of charge (SOC) of BEV user i at time t in pu of capacity
$x_{i,0}$, $x_{i,T}$	SOC of BEV user i at t_0, SOC of BEV user i at T
$\overline{x}_0$, $\overline{x}_T$	average SOCs of BEVs at t_0, average SOCs of BEVs at T
$\overline{x}_t$	mathematical expectation of SOCs of BEVs at time t
$\overline{x}_t^{target}$	target for mean SOC of BEVs at time t
$\overline{x}_T^{target}$	target for steady-state mean SOC of BEVs
$\sigma_{x_{i,0}}$, $\sigma_{x_{i,T}}$	standard deviation of SOCs at t_0, standard deviation of SOCs at T
q_t^y	pressure field trajectory of BEVs at time t
q_T^y	steady-state pressure field of BEVs
q_{x_0}	comfort coefficient of BEVs
r	charging rate penalty coefficient of BEVs
y	collective direction of BEVs' SOCs
ν	Brownian noise intensity
ω	Brownian motion
δ	coefficient to ensure convergence of the cost J
π, s, γ	coefficients of quadratic form of optimal cost J^*

References

1. *Global EV Outlook 2021: Accelerating Ambitions Despite the Pandemic*; Technical Report; International Energy Agency: Paris, France 2021.
2. *Canada's Energy Future 2020: Towards Net-Zero*; Technical Report; Canada Energy Regulator: Calgary, AB, Canada, 2020.
3. *EV30@30: A Campaign Launched under the Electric Vehicle Initiative*; Technical Report; Clean Energy Ministerial: Helsinki, Finland, 2019.
4. Su, W.; Wang, J.; Hu, Z. Planning, Control, and Management Strategies for Parking Lots for PEVs. In *Plug in Electric Vehicles in Smart Grids: Integration Techniques*; Rajakaruna, S., Shahnia, F., Gosh, A., Eds.; Springer: Berlin/Heidelberg, Germany, 2015; Chapter 3, pp. 61–98. [CrossRef]
5. Olivella-Rosell, P.; Villafafila-Robles, R.; Sumper, A. Impact evaluation of plug-in electric vehicle on power systems. In *Plug in Electric Vehicles in Smart Grids: Integration Techniques*; Rajakaruna, S., Shahnia, F., Gosh, A., Eds.; Springer: Berlin/Heidelberg, Germany, 2015; Chapter 6, pp. 149–178.
6. Tuchnitz, F.; Ebell, N.; Schlund, J.; Pruckner, M. Development and Evaluation of a Smart Charging Strategy for an Electric Vehicle Fleet Based on Reinforcement Learning. *Appl. Energy* **2021**, *285*, 116382. [CrossRef]
7. Zhou, Y.; Maxemchuk, N.; Qian, X.; Mohammed, Y. A weighted fair queuing algorithm for charging electric vehicles on a smart grid. In Proceedings of the 2013 IEEE Online Conference on Green Communications (OnlineGreenComm), Piscataway, NJ, USA, 29–31 October 2013. [CrossRef]
8. Tan, K.M.; Ramachandaramurthy, V.K.; Yong, J.Y.; Padmanaban, S.; Mihet-Popa, L.; Blaabjerg, F. Minimization of Load Variance in Power Grids—Investigation on Optimal Vehicle-to-Grid Scheduling. *Energies* **2017**, *10*, 1880. [CrossRef]
9. Sortomme, E.; Hindi, M.M.; MacPherson, S.D.J.; Venkata, S.S. Coordinated Charging of Plug-In Hybrid Electric Vehicles to Minimize Distribution System Losses. *IEEE Trans. Smart Grid* **2010**, *2*, 198–205. [CrossRef]
10. Schaden, B.; Jatschka, T.; Limmer, S.; Raidl, G.R. Smart Charging of Electric Vehicles Considering SOC-Dependent Maximum Charging Powers. *Energies* **2021**, *14*, 7755. [CrossRef]
11. Mariello, M.; Blad, T.; Mastronardi, V.; Madaro, F.; Guido, F.; Staufer, U.; Tolou, N.; De Vittorio, M. Flexible piezoelectric AlN transducers buckled through package-induced preloading for mechanical energy harvesting. *Nano Energy* **2021**, *85*, 105986. doi: 10.1016/j.nanoen.2021.105986. [CrossRef]
12. Kim, J.; Yamanaka, S.; Murayama, I.; Katou, T.; Sakamoto, T.; Kawasaki, T.; Fukuda, T.; Sekino, T.; Nakayama, T.; Takeda, M.; et al. Pyroelectric power generation from the waste heat of automotive exhaust gas. *Sustain. Energy Fuels* **2020**, *4*, 1143–1149. [CrossRef]
13. Ghotge, R.; Snow, Y.; Farahani, S.; Lukszo, Z.; van Wijk, A. Optimized Scheduling of EV Charging in Solar Parking Lots for Local Peak Reduction under EV Demand Uncertainty. *Energies* **2020**, *13*, 1275. [CrossRef]
14. Chandra Mouli, G.; Bauer, P.; Zeman, M. System design for a solar powered electric vehicle charging station for workplaces. *Appl. Energy* **2016**, *168*, 434–443. [CrossRef]
15. Figueiredo, R.; Nunes, P.; Brito, M.C. The feasibility of solar parking lots for electric vehicles. *Energy* **2017**, *140*, 1182–1197. [CrossRef]
16. Deshmukh, S.S.; Pearce, J.M. Electric vehicle charging potential from retail parking lot solar photovoltaic awnings. *Renew. Energy* **2021**, *169*, 608–617. [CrossRef]
17. Denholm, P.; O'Connell, M.; Brinkman, G.; Jorgenson, J. *Overgeneration from Solar Energy in California: A Field Guide to the Duck Chart*; Technical Report; National Renewable Energy Laboratory: Washington, DC, USA, 2015.
18. Drude, L.; Pereira Junior, L.C.; Rüther, R. Photovoltaics (PV) and electric vehicle-to-grid (V2G) strategies for peak demand reduction in urban regions in Brazil in a smart grid environment. *Renew. Energy* **2014**, *68*, 443–451. [CrossRef]
19. Rüther, R.; Pereira Junior, L.C.; Bittencourt, A.H.; Drude, L.; dos Santos, I.P. Strategies for plug-in electric vehicles to grid (V2G) and photovoltaics (PV) for peak demand reduction in urban regions in a smart grid environment. In *Plug in Electric Vehicles in Smart Grids: Integration Techniques*; Rajakaruna, S., Shahnia, F., Gosh, A., Eds.; Springer: Berlin/Heidelberg, Germany, 2015; Chapter 7, pp. 179–219. [CrossRef]
20. Lee, S.; Iyengar, S.; Irwin, D.; Shenoy, P. Shared solar-powered EV charging stations: Feasibility and benefits. In Proceedings of the 2016 Seventh International Green and Sustainable Computing Conference (IGSC), Hangzhou, China, 7–9 November 2016. [CrossRef]
21. Ma, Z.; Callaway, D.S.; Hiskens, I.A. Decentralized charging control of large populations of plug-in electric vehicles. *IEEE Trans. Control. Syst. Technol.* **2013**, *21*, 67–78. [CrossRef]
22. Xydas, E.; Marmaras, C.; Cipcigan, L.M. A multi-agent based scheduling algorithm for adaptive electric vehicles charging. *Appl. Energy* **2016**, *177*, 354–365. [CrossRef]
23. Huang, M.; Caines, P.E.; Malhamé, R.P. Large-Population Cost-Coupled LQG Problems with Nonuniform Agents: Individual-Mass Behavior and Decentralized ε-Nash Equilibria. *IEEE Trans. Autom. Control.* **2007**, *52*, 1560–1571. [CrossRef]
24. Cardaliaguet, P. *Notes on Mean Field Games (from Pierre-Louis Lions' Lectures at Collège de France)*; Université Paris-Dauphine: Paris, France, 2013.
25. Kizilkale, A.C.; Salhab, R.; Malhamé, R.P. An integral control formulation of Mean Field Game based large scale coordination of loads in smart grids. *Automatica* **2019**, *100*, 312–322. [CrossRef]
26. Anderson, B.D.; Moore, J.B. *Optimal Control, Linear Quadratic Methods*, 12th ed.; Dover Publications: Mineola, NY, USA, 2007.

27. Lénet, Q. Contrôle Décentralisé d'un Ensemble de Dispositifs de Chauffage électrique. Master's Thesis, Polytechnique Montreal, Montreal, QC, Canada, 2020.
28. Bellman, R.E.; Dreyfus, S.E. *Applied Dynamic Programming*, 2nd ed.; Princeton University Press: Princeton, NJ, USA, 2015.

energies

Article

Optimal Reactive Power Dispatch under Transmission and Distribution Coordination Based on an Accelerated Augmented Lagrangian Algorithm

Bin Zhou [1], Xiaodong Shen [1,*], Caimei Pan [2], Yuanbao Bai [2] and Tian Wu [2]

[1] College of Electrical Engineering, Sichuan University, Chengdu 610065, China; 2019223030002@stu.scu.edu.cn
[2] Sichuan Energy Investment Integrated Energy Co., Ltd., Chengdu 611130, China; 015740@ntsd789.com (C.P.); baiyuanbao0827@yeah.net (Y.B.); oversky2012@163.com (T.W.)
* Correspondence: shengxd@scu.edu.cn; Tel.: +86-166-0282-7868

Abstract: As many distributed power sources flood into the distribution network, the relationship between transmission and distribution grids in reactive power and voltage is becoming closer and closer. The traditional way of independent reactive power optimization in transmission and distribution grids is no longer appropriate. In this study, a collaborative and distributed reactive power optimization method for transmission and distribution grids based on the accelerated augmented Lagrangian (AAL) algorithm is proposed to adapt to the independence of the transmission and distribution grids in operation and management. The global reactive power optimization problem is decomposed into the transmission network subproblem and several distribution network subproblems. According to AAL, subproblems are solved in a distributed manner until the optimal global solution is finally reached after several iterations, and coordination between transmission and distribution grids is achieved only through the interaction of information on coordinating variables. For better convergence, a linearized and convergence-guaranteed optimal power flow model (OPF) with reactive power and voltage magnitude was applied to model the transmission grid optimization subproblem, while the second-order cone programming (SOCP) technique is used in the distribution network subproblems. The simulation results confirm that the method in this paper can effectively reduce network losses and achieve better economic performance, and converges better when compared to other algorithms.

Keywords: transmission and distribution coordination; linear approximation; reactive power optimization; accelerated augmented Lagrangian algorithm; distributed generator

Citation: Zhou, B.; Shen, X.; Pan, C.; Bai, Y.; Wu, T. Optimal Reactive Power Dispatch under Transmission and Distribution Coordination Based on an Accelerated Augmented Lagrangian Algorithm. *Energies* **2022**, *15*, 3867. https://doi.org/10.3390/en15113867

Academic Editor: Abu-Siada Ahmed

Received: 28 April 2022
Accepted: 23 May 2022
Published: 24 May 2022

1. Introduction

With the rapid development of distributed energy, distribution networks are gradually evolving into active distribution networks, and the reactive power characteristic has also changed [1]. Traditionally, the transmission and distribution network's optimal reactive power dispatch (ORPD) is carried out separately. The lower distribution networks remain unchanged when optimizing the transmission network, and the voltage of the boundary buses are used as the reference when optimizing the distribution networks, making it prone to overvoltage problems and cascading blackouts of distributed generators when there is a large-scale influx of distributed generators into distribution networks [2]. Therefore, it is necessary to give full play to the voltage support role of the transmission network and make full use of the reactive power regulation capability of the distribution network, which helps active distribution networks to eliminate overvoltage problems and reduce network losses [3,4].

Theoretically, the centralized optimization method could be used to establish a global optimal dispatch model to realize the coordinated optimal dispatch of transmission and distribution networks [5]. However, for the existing power system, transmission and distribution networks are under the jurisdiction of different dispatching agencies, which makes it hard for centralized optimization to deal with the confidentiality of information among various stakeholders and the independence of transmission and distribution networks in the process of operation and management. On the other hand, the centralized optimization method must concentrate the data of the global system to the dispatch center for calculation. However, aggregating data from multiple distribution networks will dramatically increase the complexity and computational cost of optimization problems. Therefore, it is reasonable to use a distributed optimization method when carrying out the optimal dispatch under transmission and distribution network coordination [6].

Recently, several researches have been conducted on distributed optimization methods for coordinated transmission and distribution networks. In [7], a reactive power optimization model considering coordinated transmission and distribution networks was established and solved by the generalized Benders decomposition method. However, this approach requires the subproblems to be convex to obtain the efficient Benders cut. However, shunt capacitors (SCs) and on-load transformers (OLTCs) in distribution networks are discrete control variables, which poses a challenge to the Benders decomposition method. In [8,9], the heterogeneous decomposition (HGD) was applied to the economic scheduling problem of coordinated transmission and distribution networks. However, the HGD may lead to local optima and is unsuitable for topologically complex networks, especially when there are connections between different distribution networks [10]. As a commonly used distributed optimization algorithm in optimization problems, the alternating direction multiplier (ADMM) has been widely used in multi-region network operation problems, including distributed optimal power flow and fully distributed reactive power optimization [11,12]. In [13], ADMM was applied to the reactive power optimization problem of multiple partition coordination in active distribution networks, and the results proved that the ADMM is an effective method for dealing with collaborative optimization problems. By improving the ADMM algorithm, the accelerated augmented Lagrangian algorithm (AAL) can obtain a faster convergence speed when dealing with collaborative optimization problems [14]. In [15], the AAL was applied to the distributed restoration problem of an integrated power transmission and distribution system with distributed energy sources. In [16], a framework was established to deal with the reactive power optimization of transmission and distribution networks coordination through the method of curve fitting, but the method cannot take into account the role of DGs. In addition, methods such as multi-parameter planning and the principle of auxiliary problems were also used to optimize transmission and distribution networks coordination problems [17,18].

The optimization problem of transmission and distribution networks can be formulated by nonlinear AC-OPF, DC-OPF, or other forms, such as mixed-integer linear programming, second-order cone programming (SOCP), and semidefinite programming. In [19], the DC-OPF was used for the transmission network optimization problem, and the nonlinear AC-OPF was used for the distribution network optimization problem, thereby improving the calculation speed. However, since the variables related to reactive power are not modeled in the traditional DC-OPF, it cannot be used to deal with the optimal reactive power dispatch problem. For that issue, a novel linearized OPF model with reactive power (Q) and voltage magnitude (V) was established to achieve the better performance of the DC-OPF models [20]. By applying the novel linear approximation method to the transmission network subproblem model, the complexity of the transmission network subproblem and the global optimization problem can be significantly reduced. In addition, the effectiveness of SOCP for radial networks has been demonstrated in the related literature, and it is often used in the optimization of distribution networks [21].

To give full play to the synergy between the transmission network and the distribution network, a coordinated transmission and distribution optimal reactive power dispatch (CTD-ORPD) framework is established in this paper. The main contributions are listed as follows:

(1) A fully distributed framework based on the AAL algorism is constructed for CTD-ORPD, through coordination between transmission and distribution grids to achieve the common goals of minimizing network losses and maximizing economic benefits, which performs better in terms of economy and safety when compared to the traditional independent optimization method of distribution grids [22].
(2) In the proposed CTD-ORPD framework, the global reactive power optimization problem is decomposed into the transmission network subproblem and several distribution network subproblems. Based on AAL, the subproblems are solved in a distributed manner until the optimal global solution is finally reached after several iterations, and coordination between transmission and distribution grids is achieved only through the interaction of information on coordinating variables. By adopting a distributed approach, the issues of data privacy, cybersecurity, and the computing and communication of independent system operators are reasonably addressed. Additionally, the AAL-based CTD-ORPD framework proposed in this paper is proved to have better convergence performance and solving efficiency when compared to those co-optimization methods using HGD [9] or ADMM [13].
(3) For a better convergence performance and solving efficiency of the established distributed optimization model, a novel linearized OPF and the SOCP technique are applied to the modeling of transmission network subproblem and distribution network subproblems, respectively.

The rest of this paper is organized as follows: In Section 2, the framework of the CTD-ORPD is introduced, including the separate modeling of transmission network and distribution network subproblems, as well as the novel linearized OPF and SOCP applied to the modeling of the transmission network and distribution networks, respectively. In Section 3, the distributed solving algorithm of the CTD-ORPD is detailed, and the algorithm flowchart is given. Section 3 establishes three test cases to verify the validity of the proposed CTD-ORPD and compares the performance of the proposed method in this paper with other methods. Finally, Section 4 presents our conclusions.

2. Framework of the CTD-ORPD

In this section, the CTD-ORPD framework based on AAL is introduced. In the proposed CTD-ORPD framework, the global reactive power optimization problem is decomposed into the transmission network subproblem and several distribution network subproblems. Based on AAL, the subproblems are solved in a distributed manner until the optimal global solution is finally reached after several iterations, and coordination between transmission and distribution grids is achieved through the interaction of information on coordinating variables. Coordinating variables are the variables of the points of common coupling (PCCs) between the transmission network and the distribution networks, including bus voltage magnitude and phase angle, active power output and reactive power output.

In the transmission network subproblem, control variables such as SCs, OLTCs, and reactive power output of generators need to be optimized. A novel linearized approximation method is applied to handle the nonlinearity of the power flow equations of the transmission network for better convergence [20]. In the distribution network subproblem, control variables such as SCs, OLTCs, and reactive power output of various distributed power sources need to be optimized. Additionally, SOCP method is applied to the convex modeling of the distribution network subproblems. The framework of the established CTD-ORPD is illustrated in Figure 1.

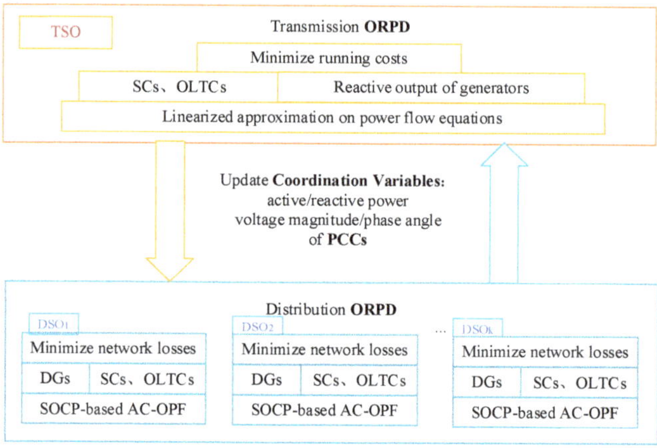

Figure 1. The framework of the proposed CTD-ORPD.

2.1. Linear Approximation of Power Flow Equations

A novel linear approximation of power flow equations to construct the linearized OPF model of the transmission subproblem is detailed in this part, which is the foundation for formulating the proposed CTD-ORPD model.

Taking advantage of the power flow equations in the polar coordination, the expression for power flows on branch *i-j* are as follows:

$$P_{ij} = \left(v_i^2 - v_i v_j \cos\theta_{ij}\right) g_{ij} - v_i v_j b_{ij} \sin\theta_{ij} \tag{1}$$

$$Q_{ij} = -\left(v_i^2 - v_i v_j \cos\theta_{ij}\right) b_{ij} - v_i v_j g_{ij} \sin\theta_{ij} \tag{2}$$

where P_{ij} and Q_{ij} are the active and reactive power passing on branch *i-j*, respectively; v_i and v_j are the voltage magnitudes of bus *i* and bus *j*, respectively; and g_{ij} and b_{ij} are conductance and susceptance of branch *i-j*. Given the initial values of voltage magnitudes and phase angles $(\boldsymbol{v}_0, \boldsymbol{\theta}_0)$, the trigonometric function in the above formula can be linearized and approximated by the first-order Taylor series as follows:

$$\sin\theta_{ij} \approx s_{ij}^1 \theta_{ij} + s_{ij}^0 \tag{3}$$

$$\cos\theta_{ij} \approx c_{ij}^1 \theta_{ij} + c_{ij}^0 \tag{4}$$

$$s_{ij}^1 = \cos\theta_{ij,0}, s_{ij}^0 = \sin\theta_{ij,0} - \theta_{ij,0}\cos\theta_{ij,0} \tag{5}$$

$$c_{ij}^1 = -\sin\theta_{ij,0}, c_{ij}^0 = \cos\theta_{ij,0} + \theta_{ij,0}\sin\theta_{ij,0} \tag{6}$$

The initial value of the voltage phase angle is very close to the final value after iteration due to the quasi-linear relationship between the active power and the voltage phase angle in the power system [23]. Therefore, Equations (3) and (4) have a good accuracy. Substituting Equations (3) and (4) into Equations (1) and (2), we can obtain:

$$P_{ij} = v_i^2 g_{ij} - v_i v_j \left(g_{ij} c_{ij}^0 + b_{ij} s_{ij}^0\right) - v_i v_j \theta_{ij} \left(g_{ij} c_{ij}^1 + b_{ij} s_{ij}^1\right) \tag{7}$$

$$Q_{ij} = -v_i^2 b_{ij} + v_i v_j \left(-g_{ij} s_{ij}^0 + b_{ij} c_{ij}^0\right) - v_i v_j \theta_{ij} \left(g_{ij} s_{ij}^1 - b_{ij} c_{ij}^1\right) \tag{8}$$

Considering $v_i v_j$ as a whole, use the first-order Taylor series expansion to decouple the variable $v_i v_j$ from the variable θ_{ij} in the expression $v_i v_j \theta_{ij}$:

$$v_i v_j \theta_{ij} \approx v_{i,0} v_{j,0} \theta_{ij} + \left(v_i v_j - v_{i,0} v_{j,0}\right) \theta_{ij,0} \tag{9}$$

By substituting Equation (9) into Equations (7) and (8), the following expressions can be obtained:

$$P_{ij} = g_{ij} v_i^2 - g_{ij}^{\mathrm{P}} v_i v_j - b_{ij}^{\mathrm{P}} \left(\theta_{ij} - \theta_{ij,0}\right) \tag{10}$$

$$Q_{ij} = -b_{ij} v_i^2 + b_{ij}^{\mathrm{Q}} v_i v_j - g_{ij}^{\mathrm{Q}} \left(\theta_{ij} - \theta_{ij,0}\right) \tag{11}$$

where g_{ij}^{P}, b_{ij}^{P}, g_{ij}^{Q}, and b_{ij}^{Q} are the active equivalent conductance, active equivalent susceptance, reactive equivalent conductance, and reactive equivalent susceptance of line *i-j*, respectively, and their expressions are as follows:

$$g_{ij}^{\mathrm{P}} = \left(g_{ij} c_{ij}^0 + b_{ij} s_{ij}^0\right) + \left(g_{ij} c_{ij}^1 + b_{ij} s_{ij}^1\right) \theta_{ij,0} \tag{12}$$

$$b_{ij}^{\mathrm{P}} = \left(g_{ij} c_{ij}^1 + b_{ij} s_{ij}^1\right) v_{i,0} v_{j,0} \tag{13}$$

$$g_{ij}^{\mathrm{Q}} = \left(g_{ij} s_{ij}^1 - b_{ij} c_{ij}^1\right) v_{i,0} v_{j,0} \tag{14}$$

$$b_{ij}^{\mathrm{Q}} = \left(-g_{ij} s_{ij}^0 + b_{ij} c_{ij}^0\right) - \left(g_{ij} s_{ij}^1 - b_{ij} c_{ij}^1\right) \theta_{ij,0} \tag{15}$$

There is still a nonlinear variable $v_i v_j$ in Equations (10) and (11), respectively. In order to eliminate the nonlinearity of the model, the following mathematical transformations are performed on $v_i v_j$:

$$v_i v_j = \frac{1}{2}\left[v_i^2 + v_j^2 - \left(v_i - v_j\right)^2\right] = \frac{v_i^2 + v_j^2}{2} - \frac{v_{ij}^2}{2} \tag{16}$$

The expression $\left(v_i^2 + v_j^2\right)/2$ is linear when considering v^2 as an independent variable. In practical power systems, the values of v_{ij}^2 are very small and often negligible. Thus, $\left(v_i^2 + v_j^2\right)/2$ can be used as an approximation of $v_i v_j$, the approximation error would be small and not affected by the initial value of the voltage magnitude [24]. For keeping the error as low as possible, the following approximation is made to v_{ij}^2 in Equation (16) based on a Taylor series expansion:

$$v_{ij}^2 \approx 2 v_{ij,0} v_{ij} - v_{ij,0}^2 \approx 2 v_{ij,0} v_{ij} \frac{v_i + v_j}{v_{i,0} + v_{j,0}} - v_{ij,0}^2 = 2 \frac{v_{i,0} - v_{j,0}}{v_{i,0} + v_{j,0}} \left(v_i^2 - v_j^2\right) - v_{ij,0}^2 = v_{ij,\mathrm{L}}^{\mathrm{s}} \tag{17}$$

where $v_{ij,\mathrm{L}}^{\mathrm{s}}$ is an approximation of v_{ij}^2. Equation (17) are linear when considered v^2 as an independent variable. Since v_{ij}^2 is non-negativity, the following non-negativity constraint should be applied:

$$v_{ij,L}^{\mathrm{s}} + \varepsilon_{ij} \geq 0, \varepsilon_{ij} \geq 0 \tag{18}$$

By substituting Equations (16) and (17) into Equations (10) and (11), the linearized approximation of the power flow equation is obtained:

$$P_{ij}^{\mathrm{L}} = g_{ij} v_i^2 - g_{ij}^{\mathrm{P}} \frac{v_i^2 + v_j^2}{2} - b_{ij}^{\mathrm{P}} \left(\theta_{ij} - \theta_{ij,0}\right) + g_{ij}^{\mathrm{P}} \frac{v_{ij,\mathrm{L}}^{\mathrm{s}}}{2} \tag{19}$$

$$Q_{ij}^{\mathrm{L}} = -b_{ij} v_i^2 + b_{ij}^{\mathrm{Q}} \frac{v_i^2 + v_j^2}{2} - g_{ij}^{\mathrm{Q}} \left(\theta_{ij} - \theta_{ij,0}\right) - b_{ij}^{\mathrm{Q}} \frac{v_{ij,\mathrm{L}}^{\mathrm{s}}}{2} \tag{20}$$

In Equations (19) and (20), the active and reactive components of the power flow are both linear functions when considering v^2 as an independent variable.

The following are the expressions of the nodal injection power equations:

$$P_i = \sum_{j=1}^{N} (v_i v_j G_{ij} \cos\theta_{ij} + v_i v_j B_{ij} \sin\theta_{ij}) \tag{21}$$

$$Q_i = -\sum_{j=1}^{N} (v_i v_j B_{ij} \cos\theta_{ij} - v_i v_j G_{ij} \sin\theta_{ij}) \tag{22}$$

where G_{ij} and B_{ij} are the real and imaginary parts of the node admittance matrix, respectively. Equations (21) and (22) can be converted into the following forms:

$$P_i = \sum_{(i,j)} P_{ij} + \left(\sum_{j=1}^{N} G_{ij}\right) v_i^2 \tag{23}$$

$$Q_i = \sum_{(i,j)} Q_{ij} + \left(\sum_{j=1}^{N} -B_{ij}\right) v_i^2 \tag{24}$$

Finally, the linearized power flow equation constraints are obtained, which are composed of Equations (19), (20), (23), and (24). They are all linear functions when v^2 is considered as an independent variable.

2.2. Transmission Network Subproblem Model

In this part, the linearized OPF model of the transmission network subproblem is established based on the above linear approximation of power flow equations.

In traditional optimal reactive power dispatch, minimizing network losses is often applied as the objective function, where network losses are functions of the voltage magnitude and the voltage phase angle [25]. However, minimizing network losses without considering generator costs may be in conflict with economic principles. Thus, minimizing active power injections at the root bus is also usually used as the objective function of the optimal reactive power dispatch problem [26]. In this paper, minimizing the operating cost was adopted as the objective function of ORPD of the transmission network subproblem, and its expression is as follows:

$$\min F_T = \sum_{g \in G} C(P_g) \tag{25}$$

2.2.1. The Cost of Coal Consumption

According to the characteristics of coal consumption, a convex quadratic function is used as the generation cost of generator g [27]. By default, the unit commitment is already achieved, so the binary variables indicating the on/off status of the unit are not included in the equation.

$$C(P_g) = \left(c_{g,2}^{P} P_g^2 + c_{g,1}^{P} P_g + c_{g,0}^{P}\right) \tag{26}$$

where $c_{g,2}^{P}$, $c_{g,1}^{P}$, and $c_{g,0}^{P}$ are the quadratic term coefficient, primary term coefficient, and constant term of the quadratic cost curve of the generator g, respectively. P_g denotes the active power output of the generator g.

For better convergence, a segmented linearization technique is applied to the linearization of the generator cost function. Figure 2 shows the linearized cost curve of generator g. We equally divided $[P_g^{\min}, P_g^{\max}]$, the power output range of generator g, into m_g segments, and the length of each segment is denoted as $PI_{g,1}, PI_{g,2}, \ldots, PI_{g,m_g}$; the slope is denoted as $KI_{g,1}, KI_{g,2}, \ldots, KI_{g,m_g}$, respectively. Since $C(P_g)$ is a convex function, then we have $KI_{g,1} < KI_{g,2} < \ldots < KI_{g,m_g}$ and the power of the later segments can only be used after the power of the previous segments has been used to the upper limit. Denoting the power at the segmented points as $P_g^{\min}$, $P_{g,1}$, $P_{g,2}$, ..., P_{g,m_g-1}, $P_g^{\max}$, then the linearized operating cost and power output can be expressed as follows:

$$C(P_g) = C(P_g^{\min}) + \sum_{k=1}^{m_g} PI_{g,k} \times KI_{g,k} \tag{27}$$

$$P_g = P_g^{\min} + \sum_{k=1}^{m_g} PI_{g,k} \tag{28}$$

$$0 \leqslant PI_{g,1} \leqslant P_{g,1} - P_g^{\min}, 0 \leqslant PI_{g,2} \leqslant P_{g,2} - P_{g,1}, \ldots, 0 \leqslant PI_{g,m_g} \leqslant P_g^{\max} - P_{g,m_g-1} \tag{29}$$

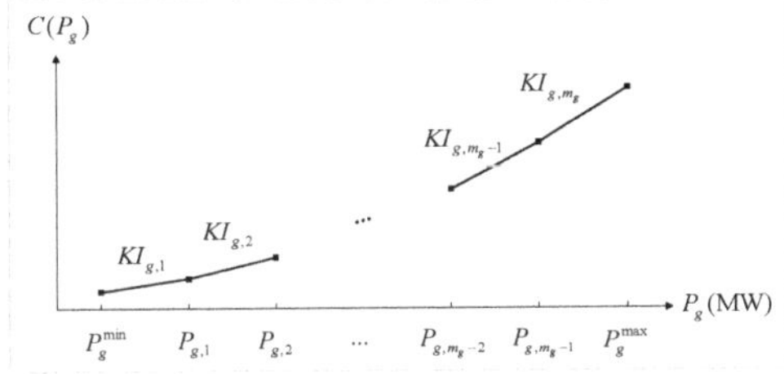

Figure 2. The linearized cost curve of generator g.

When the initial values of the voltage magnitude and phase angle (v_0, θ_0) are given, according to the description in Section 2.1, the constraints of the transmission network subproblem are constructed as follows:

2.2.2. Nodal Injection Power Balancing Constraints

For all the nodes, the following constraints need to be satisfied:

$$P_i^k = \sum_{g \in i} P_g - P_{i,d} = \sum_{(i,j)} P_{ij}^k + \left(\sum_{j=1}^{N} G_{ij}\right) V_i \tag{30}$$

$$Q_i^k = \sum_{g \in i} Q_g - Q_{i,d} = \sum_{(i,j)} Q_{ij}^k + \left(\sum_{j=1}^{N} -B_{ij}\right) V_i + Q_{i,\text{sc}} \tag{31}$$

where $P_{i,d}$ and $Q_{i,d}$ are the active and reactive loads at bus i; G_{ij} and B_{ij} are the real and imaginary parts of the node admittance matrix, respectively; $Q_{i,\text{sc}}$ is the reactive power output of SCs at bus i.; $V_i = v_i^2$ represents the square of voltage magnitude at node i, and it is used as an independent variable in this paper.

2.2.3. Branch Power Balancing Constraints

Equations (19) and (20) are the branch power balancing constraints for those branches without OLTCs. For those branches with OLTCs, the branch power balancing constraints are as follows:

$$P_{ij}^k = g_{ij} V_{i,\text{OLTC}} - g_{ij}^{\text{P}} \frac{V_{i,\text{OLTC}} + V_j}{2} - b_{ij}^{\text{P}} (\theta_{ij} - \theta_{ij,0}) + g_{ij}^{\text{P}} \frac{V_{ij,\text{L}}^{\text{OLTC}}}{2} \tag{32}$$

$$Q_{ij}^k = -b_{ij} V_{i,\text{OLTC}} + b_{ij}^{\text{Q}} \frac{V_{i,\text{OLTC}} + V_j}{2} - g_{ij}^{\text{Q}} (\theta_{ij} - \theta_{ij,0}) - b_{ij}^{\text{Q}} \frac{V_{ij,\text{L}}^{\text{OLTC}}}{2} \tag{33}$$

$$V_{ij,\text{L}}^{\text{OLTC}} = 2 \frac{v_{i,\text{OLTC},0} - v_{j,0}}{v_{i,\text{OLTC},0} + v_{j,0}} \left(V_{i,\text{OLTC}} - V_j^2\right) - \left(V_{i,\text{OLTC},0} - V_{j,0}\right)^2 \tag{34}$$

2.2.4. Constraints of OLTCs

Denote $K \in \{K_0, K_1, K_2, \ldots, K_n\}$ as the optional status of OLTCs. When the operation status of OLTCs is K_m, the following constraint should be satisfied:

$$V_i = K^2 V_{i,\text{OLTC}} = K_0^2 V_{i,\text{OLTC}} + \sum_{m=1}^{t} (K_m^2 - K_{m-1}^2) V_{i,\text{OLTC}} \tag{35}$$

However, $K_t^2 V_{i,\text{OLTC}}$ in the above equation contains a nonlinear term, which is detrimental to the solution of the model. A set of 0–1 variables $\{z_1, z_2, \ldots, z_\text{n}\}$ is introduced to linearize it.

$$V_i = K_0^2 V_{i,\text{OLTC}} + \sum_{m=1}^{\text{n}} \Delta V_m \tag{36}$$

$$0 \leq \Delta V_m \leq z_m V_{\max}\left(K_m^2 - K_{m-1}^2\right) \tag{37}$$

$$\Delta V_m \leq \left(K_m^2 - K_{m-1}^2\right) V_{i,\text{OLTC}} \tag{38}$$

$$\Delta V_m \geq \left(K_m^2 - K_{m-1}^2\right) V_{i,\text{OLTC}} - (1 - z_m) V_{\max}\left(K_m^2 - K_{m-1}^2\right) \tag{39}$$

$$z_m \geq z_{m+1}, m = 1, 2, \ldots, \text{n} - 1 \tag{40}$$

where ΔV_m stands for the term $(K_m^2 - K_{m-1}^2) V_{i,\text{OLTC}}$, replacing K as the decision variable about OLTCS in the optimization process. When $z_m = 0$, then $\Delta V_m = 0$ according to Equation (37); when $z_m = 1$, then $\Delta V_m = (K_m^2 - K_{m-1}^2) V_{i,\text{OLTC}}$ according to Equations (38) and (39). According to Equation (40), when $z_t - z_{t-1} = 1$, then the values of the 1st to the tth term of the set $\{z_1, z_2, \ldots, z_\text{n}\}$, then the values of the 1st to the tth term of the set are 1, while all the remaining values are 0. Substituting all the values into Equation (36), we can see that the set of constraints Equations (36)–(40) are equivalent to the original constraints Equation (35).

2.2.5. Constraints of SCs

The set $\{B_{i,1}, B_{i,2}, \ldots, B_{i,\text{n}}\}$ is used to represent the conductance of SCs available at bus i. Assuming that the capacitors are put in sequentially, i.e., $B_{i,t}$ is put in before $B_{i,t+1}$, then the reactive power output of SCs can be expressed as follows:

$$Q_{i,\text{sc}} = V_i \sum_{m=1}^{t} B_{i,m} = \sum_{m=1}^{t} (V_i B_{i,m}) \tag{41}$$

Similar to the modeling process of OLTCs, a set of 0–1 variables $\{x_1, x_2, \ldots, x_\text{n}\}$ is introduced to represent the operational state of SCs, then the constraint Equation (41) can be equivalently represented by the following set of linear constraints:

$$Q_{i,\text{sc}} = \sum_{m=1}^{\text{n}} Q_{i,m}^{\text{sc}} \tag{42}$$

$$0 \leq Q_{i,m}^{\text{sc}} \leq x_m V_{\max} B_{i,m} \tag{43}$$

$$Q_{i,m}^{\text{sc}} \leq V_i B_{i,m} \tag{44}$$

$$Q_{i,m}^{\text{sc}} \geq V_i B_{i,m} - (1 - x_m) V_{i,\max} B_{i,m} \tag{45}$$

$$x_t \geq x_{t+1}, t = 1, 2, \ldots, \text{n} - 1 \tag{46}$$

where $Q_{i,m}^{\text{sc}}$ stands for the term $V_i B_{i,m}$.

2.2.6. Operational Constraints

For the maintenance of the system in a normal working condition, the following constraints need to be satisfied:

$$P_g^{\min} \le P_g \le P_g^{\max} \tag{47}$$

$$Q_g^{\min} \le Q_g \le Q_g^{\max} \tag{48}$$

$$v_{i,\min}^2 \le V_i \le v_{i,\max}^2 \tag{49}$$

In addition, the linearized form of line capacity constraint can be referred to [20] and will not be repeated in this paper.

In the above transmission network ORPD subproblem, all constraints are linear when considering $V = v^2$ as an independent variable. Since discrete variables are included, the above-mentioned ORPD subproblem is a mixed-integer programming problem with linear constraints.

2.3. Distribution Network Subproblem Model

In this part, the distribution network subproblem model is established based on the SOCP method. In the ORPD of distribution network subproblems, in addition to the traditional OLTCs and SCs, the variables that need to be optimized also include Static Var Compensators (SVCs) and various controllable DGs. Since there are generally no conventional generators in the distribution network, using the minimum network loss as the objective function of the distribution network subproblems is in line with the economic principle. The objective function of the ORPD of the distribution network DSO_k is as follows:

$$\min F_{\mathrm{D},k} = \sum_{(i,j)} r_{ij,k} L_{ij,k} \tag{50}$$

where $L_{\mathrm{ij,k}}$ represents the square of the current on branch *i*-*j*. The constraints of the DSO_k subproblem are as follows:

2.4. Power Flow Equation Constraints

For a typical radial distribution network, the power flow constraint can be expressed as follows:

$$\left(P_{ij,k}\right)^2 + \left(Q_{ij,k}\right)^2 = L_{ij,k} u_{i,k} \tag{51}$$

$$\sum_{i \in u(j)} \left(P_{ij,k} - L_{ij,k} r_{ij,k}\right) + P_{\mathrm{G}j,k} = \sum_{t \in v(j)} \left(P_{jt,k}\right) + P_{\mathrm{D}j,k} \tag{52}$$

$$\sum_{i \in u(j)} \left(Q_{ij,k} - L_{ij,k} x_{ij,k}\right) + Q_{\mathrm{G}j,k} = \sum_{t \in v(j)} \left(Q_{jt,k}\right) + Q_{\mathrm{D}j,k} \tag{53}$$

$$u_{j,k} = u_{i,k} - 2\left(r_{ij,k} P_{ij,k} + x_{ij,k} Q_{ij,k}\right) + \left(\left(r_{ij,k}\right)^2 + \left(x_{ij,k}\right)^2\right) L_{ij,k} \tag{54}$$

In the above distribution network subproblem, Equation (51) contains quadratic terms, and SOCP is usually applied to transform it into the following form:

$$\left\| \begin{array}{c} 2P_{ij,k} \\ 2Q_{ij,k} \\ L_{ij,k} - u_{i,k} \end{array} \right\|_2 \le L_{ij,k} + u_{i,k} \tag{55}$$

2.5. Security Constraints

For a typical radial distribution network, the power flow constraint can be expressed as follows:

$$L_{ij,k,\min} \le L_{ij,k} \le L_{ij,k,\max} \tag{56}$$

$$u_{i,k,\min} \leq u_{i,k} \leq u_{i,k,\max} \quad (57)$$

2.6. Constraints of DGs

There are a large number of wind turbines, photovoltaics (PVs), and micro gas turbines (GTs) in distribution networks. They are generally connected to the grid through inverters, have good reactive power potentials, and participate in the distribution network's reactive power optimization process. Depending on the types of DGs and their inverter control modes, they have different reactive power regulation characteristics.

The inverter of the micro GTs generally adopts the PQ decoupling control method, and the reactive power regulation capacity is determined by the active power output and the total capacity of the inverter:

$$Q_{\mathrm{DG,max}} = \sqrt{S_{\max}^2 - P_{\mathrm{DG}}^2} \quad (58)$$

$$Q_{\mathrm{DG,min}} = -Q_{\mathrm{DG,max}} \quad (59)$$

where $Q_{\mathrm{DG,max}}$ and $Q_{\mathrm{DG,min}}$ represent the upper and lower limits of the reactive power regulation capability of the micro gas turbine; $S_{\max}$ is the maximum apparent power of the inverter; and P_{DG} is the active power output of the micro gas turbine.

For Doubly fed Induction Generators (DFIGs), the reactive power regulation capability is mainly determined by the reactive power capacity on the stator side and the total capacity of the inverter:

$$Q_{\mathrm{DG,max}} = Q_{\mathrm{s,max}} - Q_{\mathrm{c,min}} \quad (60)$$

$$Q_{\mathrm{DG,min}} = Q_{\mathrm{s,min}} - Q_{\mathrm{c,max}} \quad (61)$$

For PVs, the grid-connected inverters mainly have two control modes: voltage control and current control. Due to the limitations of the voltage control mood, the photovoltaic inverter usually adopts the current control mood. In the reactive power optimization process, the reactive power output of photovoltaics is jointly determined by the nodal voltage, active power output, and current constraints:

$$Q_{\mathrm{DG,max}} = \sqrt{V_{\mathrm{DG}}^2 I_{\mathrm{DG,max}}^2 - P_{\mathrm{DG}}^2} \quad (62)$$

$$Q_{\mathrm{DG,min}} = 0 \quad (63)$$

2.7. Other Constraints

The modeling of the OLTCs and SCs in the distribution networks is similar to that in the transmission network and will not be repeated here.

3. Solving Algorithm of the CTD-ORPD

This section describes the details of the algorithm AAL for solving the proposed CTD-ORPD framework.

The AAL algorithm is a novel distributed algorithm that combines the advantages of the Diagonal Quadratic Approximation (DQA) method and the ADMM method. It uses a local augmented Lagrangian function similar to that of DQA, while discarding its inner loop step. Compared to ADMM, the updated rules for original variables and dual variables are improved. The original variables are updated one more time compared to ADMM on the basis of the solution of the local augmented Lagrangian function, which is then used for the update of the dual variables. In addition, a fully distributed rule is used for the update of dual variables in the AAL algorithm [14].

Based on AAL, the global ORPD problem is decomposed into the TSO subproblem and several DSO$_k$ subproblems, and the subproblems are solved in a distributed manner until the optimal global solution is finally reached after several iterations. Coordination between the TSO subproblem and DSO$_k$ subproblems is achieved only through the interaction of information on coordinating variables, which consists of the variables of PCCs, including bus voltage magnitude and phase angle, active power output and reactive power output, and denoted as x_{CT} and x_{CD} in this paper. The following is the definition of the coordinating variables:

$$x_{CT} = \bigcup_{k=1}^{d} x_{CT,k},\ x_{CT,k} = \begin{bmatrix} P_{CT,k} & Q_{CT,k} & V_{CT,k} & \delta_{CT,k} \end{bmatrix} \tag{64}$$

$$x_{CD} = \bigcup_{k=1}^{d} x_{CD,k},\ x_{CD,k} = \begin{bmatrix} P_{CD,k} & Q_{CD,k} & V_{CD,k} & \delta_{CD,k} \end{bmatrix} \tag{65}$$

The CTD-ORPD problem is an optimization problem defined by the objective functions and related constraints of the TSO and DSO$_k$ subproblems, which can be expressed as follows:

$$\min_{x} F_T(x_T, x_{CT}) + \sum_{k=1}^{d} F_{D,k}(x_{D,k}, x_{CD,k}) \tag{66}$$

$$s.t.\ g_T(x_T, x_{CT}) \leq 0, h_T(x_T, x_{CT}) = 0 \tag{67}$$

$$g_{D,k}\left(x_{CD,k}, x_{D,k}\right) \leq 0, h_{D,k}\left(x_{CD,k}, x_{D,k}\right) = 0\ k = 1, 2, \ldots \mathrm{d} \tag{68}$$

$$x_{CT} - x_{CD} = 0 \tag{69}$$

In Equation (66), F_T and $F_{D,k}$ are the objective functions of the TSO and DSO$_k$ subproblems established in the previous section, respectively. Equations (67) and (68) represent the inequality constraints and equality constraints of TSO and DSO$_k$ (k = 1,2, ... ,d), respectively. x_T and $x_{D,k}$ are local variables in the respective systems of TSO and DSO$_k$ (k = 1,2, ... ,d). Equation (69) demonstrates the consistency constraint of the coordinating variables.

In the proposed distributed method, the CTD-ORPD problem Equation (66) is decomposed into (d + 1) subproblems and coordinated optimization is performed. According to AAL, the decomposed subproblems have the following expressions:

$$\min \Lambda_{TSO}(x_T, x_{CT}, \lambda) = \min F_T(x_T, x_{CT}) + \sum_{k=1}^{d} \left[\lambda_k \left(x_{CT,k} - \tilde{x}_{CD,k} \right) + \frac{\rho}{2} \| x_{CT,k} - \tilde{x}_{CD,k} \|^2 \right] \tag{70}$$

$$\begin{gathered} \min \Lambda_{DSO,k}\left(x_{D,k}, x_{CD,k}, \lambda_k\right) = \min F_{D,k}\left(x_{D,k}, x_{CD,k}\right) + \\ \left[\lambda_k \left(\tilde{x}_{CT,k} - x_{CD,k} \right) + \frac{\rho}{2} \| \tilde{x}_{CT,k} - x_{CD,k} \|^2 \right],\ k = 1, 2, \ldots, \mathrm{d} \end{gathered} \tag{71}$$

$\Lambda_{TSO}/\Lambda_{DSO,k}$ are the local augmented Lagrangian functions of the TSO/DSO$_k$ subproblems, respectively. They are, respectively, composed of the corresponding objective function, the penalty term of the consistency constraint, and a quadratic term that improves the convergence speed of the Lagrangian function. λ is the Lagrange multiplier associated with coordinating variables, and $\rho > 0$ is the penalty term coefficient. The expression of λ is as follows:

$$\lambda = \bigcup_{k=1}^{d} \lambda_k,\ \lambda_k = \begin{bmatrix} \lambda_{P,k} & \lambda_{Q,k} & \lambda_{V,k} & \lambda_{\delta,k} \end{bmatrix} \tag{72}$$

where $\tilde{x}_{CT,k}$ and $\tilde{x}_{CD,k}$ denote the reference values of the DSO$_k$ and TSO coordination variables, respectively, which are derived from the last optimization results of the TSO and DSO$_k$ subproblems. When optimizing the TSO subproblem, the coordinating variables of the DSO$_k$ subproblem $\tilde{x}_{CD,k}$ serve as the benchmark values of $x_{CT,k}$ in the TSO subproblem. Similarly, $\tilde{x}_{CT,k}$ would be served as the benchmark values of $x_{CD,k}$ when optimizing the DSO$_k$ subproblem. Figure 3 illustrates the overall pictorial view of the interactions between TSO and DSOs in the proposed distributed framework, where the solid line connections between TSOs and DSOs represent connections through transformers and the dotted line connections indicate information interactions.

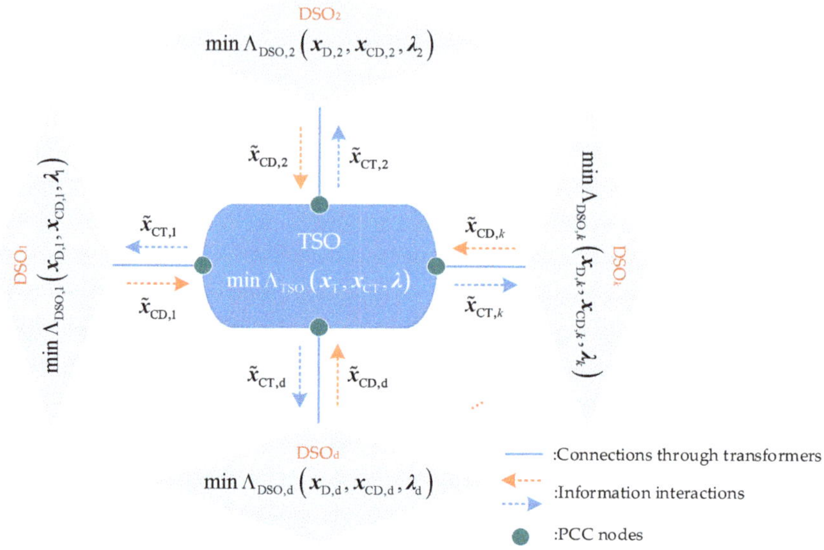

Figure 3. Interactions between TSO and DSOs in the proposed distributed framework.

In each iteration, the transmission network subproblem is solved first, and then the coordinating variables are transferred to each distribution network subproblem. After the distribution network receives the coordinating variables, the subproblems are solved in parallel, and then the updated coordinating variables are transmitted to the transmission network. After the transmission grid receives data from all distribution grids, feasibility constraints and optimality constraints are checked. The optimal solution is obtained, and the optimization process ends if these two constraints are satisfied. Otherwise, the next iteration begins. Due to the parallelism of the subproblems, the time consumed by each iteration process depends on the transmission network optimization time and the distribution network with the longest optimization time. The detailed iterative steps of the algorithm are as follows.

1. Set $t = 1$, initialize λ^t, $\rho > 0$, $\tau \in (0,0.5)$, initial local variables $x_T^0/x_{D,k}^0$, initial coordinating variables $x_{CT}^0/x_{CD,k}^0$, and $\tilde{x}_{CD,k} = x_{CD,k}^0$.
2. Optimization of the TSO subproblem: Optimize the problem Equation (70) at fixed λ and $\tilde{x}_{CD,k}$, and obtain the local/coordinating variables $\hat{x}_{CT}^t/\hat{x}_T^t$. After the following update steps are completed, transfer $\tilde{x}_{CT,k}$ to DSO_k (k = 1,2, ... ,d).

$$x_T^t = x_T^{t-1} + \tau(\hat{x}_T^t - x_T^{t-1}) \tag{73}$$

$$\tilde{x}_{CT} = x_{CT}^t = x_{CT}^{t-1} + \tau(\hat{x}_{CT}^t - x_{CT}^{t-1}) \tag{74}$$

3. Optimization of the DSO_k subproblem: Optimize the problem Equation (71) at fixed λ and $\tilde{x}_{CT,k}$, and obtain the local/coordinating variables $\hat{x}_{CD}^t/\hat{x}_D^t$. After the following update steps are completed, then transfer $\tilde{x}_{CD,k}$ to TSO.

$$x_{D,k}^t = x_{D,k}^{t-1} + \tau(\hat{x}_D^t - x_{D,k}^{t-1}) \tag{75}$$

$$\tilde{x}_{CD,k} = x_{CD,k}^t = x_{CD,k}^{t-1} + \tau(\hat{x}_{CD}^t - x_{CD,k}^{t-1}) \tag{76}$$

4. When the TSO receives all the data from DSO_k, iterations enter the next step, or else hold and wait.
5. Use Equation (77) to check whether the feasibility constraint and optimality constraint are satisfied. If both are satisfied, then the optimization process ends; otherwise, update the dual variables according to Equation (78), set $t = t + 1$, and return to step 1.

$$x_{\mathrm{CT}}^t - \tilde{x}_{\mathrm{CD}} < \psi,\ x_e^t - \hat{x}_e^t < \gamma_e,\ F_e^t - F_e^{t-1} < \xi_e,\ e \in \{\mathrm{TSO}, \mathrm{DSO}_1, \ldots, \mathrm{DSO}_{\mathrm{d}}\} \quad (77)$$

$$\lambda^{t+1} = \lambda^t + \rho\tau(x_{\mathrm{CT}}^t - \tilde{x}_{\mathrm{CD}}) \quad (78)$$

In Equation (77), ψ, γ_e, and ξ_e together determine the accuracy of the algorithm, all of which took the value of 10^{-2} in this paper.

Based on the concept of independent system operators, the program only needs to exchange a small amount of boundary information, which protects the information privacy and independent decision-making of system operators at all levels. Each distribution network subproblem can be optimized in parallel, which is beneficial to improving computational efficiency.

4. Analysis of the Simulation Case

4.1. Introduction of the Simulation Case

In order to prove the effectiveness of the proposed distributed CTD-ORPD framework, three test cases were constructed based on the IEEE-30 node system (transmission network) and the IEEE-33 node system (distribution network) for simulation calculation.

(1) Case 1: Connect an IEEE-33 node test system at node No.26 of the IEEE-30 node test system, which is denoted by TSO and D26 in the following text.
(2) Case 2: Connect an IEEE-33 node test system at the 7th, 19th, and 26th nodes of the IEEE-30 node test system, respectively, which are represented by TSO, D7, D19, and D26 in the following text.
(3) Case 3: Connect an IEEE-33 node test system at the 7th, 17th, 19th, 26th, and 29th nodes of the IEEE-30 node test system, respectively, which are represented by TSO, D7, D17, D19, D26, and D29 in the following text.

In the three test cases, the transmission network and the distribution network were connected through 132/12.66 kv transformers. The high voltage side bus of the transformer was treated as the PCC. The active and reactive power exchanged between transmission and distribution networks through PCC, as well as the magnitude and phase angle of PCC, were used as the coordinating variables in the process of transmission and distribution co-optimization. The power reference value of both the distribution network and the transmission network was set to 100 MVA.

In the IEEE-30 node system, the OLTCs were set to be adjustable within 0.95~1.05, and the adjustment step was 0.01.

In the IEEE-33 node systems, nodes 2, 4, 6, 10, 13, 24, and 28 were connected to DGs, and the detailed parameters of the DGs are shown in Table 1.

Table 1. Buses and parameters of the DGs.

DGs	Buses	Control Modes	Parameters/p.u.
GT	4	PQ decoupling	$P = 0.002$, $S_{max} = 0.005$
DFIG	2, 28	Constant voltage control	$P = 0.002$, $Q_{s,max} = 0.003$, $Q_{s,min} = -0.003$, $Q_{c,max} = 0.005$, $Q_{c,min} = -0.005$
PV	6, 13	Current control	$P = 0.002$, $I_{max} = 0.005$

Since the accuracy of the linearization of the power flow equation in the transmission network is related to the quality of the initial value $(\boldsymbol{v}_0, \boldsymbol{\theta}_0)$, the latest result of the integrated power flow calculation was used as the value of $(\boldsymbol{v}_0, \boldsymbol{\theta}_0)$.

Other parameters in the transmission and distribution grids, such as load data, impedance data, and unit cost coefficients, were derived from MATPOWER (v.7.1, Cornell University's Charles H. Dyson School, U.S.).

The program in this paper was established in MATLAB R2019b (v.9.7, MathWorks, U.S.) and ran on a computer with Intel(R) Core(TM) i7-10850H CPU @ 2.70GHz CPU and 32 GB memory.

4.2. Comparison with Other Methods

To verify the superiority of the transmission and distribution coordination method proposed in the paper over the traditional independent method and centralized method, the centralized method, the independent method and the method proposed in the paper were used for simulation calculations for three test cases. "Centralized method" means that the distribution network is regarded as a new branch connected to the PCC node in the transmission network, and the centralized optimization calculation is performed by the control center. "Independent method" means that there is no information interaction between the transmission network and the distribution network, and independent optimization is carried out according to the boundary power flow information.

Tables 2–4 shows the results of solving the three test cases by different methods, and the boundary voltage magnitudes are in the form of per-unit values. It can be seen from the tables that in the three calculation cases, the network losses and system operating costs of the independent optimization method are significantly higher than those of the centralized optimization method. However, the results obtained by the method proposed in this paper are very close to that of the centralized optimization method. This is due to the fact that the initial value $(\boldsymbol{v}_0, \boldsymbol{\theta}_0)$ used in the linearization of the power flow equations of the transmission network adopts the result of the last integrated power flow calculation. This proves that our proposed method can obtain a solution with high precision when the initial value $(\boldsymbol{v}_0, \boldsymbol{\theta}_0)$ has high accuracy. This also demonstrates that it is feasible and effective to realize the ORPD problem of the global system through the cooperative calculation between the transmission and distribution networks.

Table 2. Comparison of the results of solving test case 1 by different methods.

		Independent Method	Proposed Method	Centralized Method
System operating costs/thousand USD		61.6128	61.3055	61.3074
Network losses/MW	D26	0.0816	0.0756	0.0769
	TSO	5.1020	5.0728	5.0756
	Total	5.1836	5.1484	5.1525
Boundary voltage magnitudes (phase angles/(°))		1.0066(0)	1.0379 (−10.38)	1.0381 (−10.38)

In test case 1, the method proposed in the paper converges after three iterations, the total duration is 4.3 s, and the total duration of centralized optimization is 1.890 s. The total duration of the algorithm in this paper is 2.3 times that of centralized optimization. In test case 2, the method proposed in this paper converges to the optimal solution after four iterations. The whole process takes 5.1 s, and the total duration of centralized optimization is 7.269 s. At this time, the total duration of the method proposed in this paper is only 0.7 times that of centralized optimization. As for the situation in test case 3, the method proposed in this paper converges to the optimal solution after four iterations. The whole process takes 5.8 s, and the total duration for centralized optimization is 21.326 s. Eventually, the method proposed in this paper is obviously less time-consuming than centralized optimization, and the total duration of the algorithm in this paper is only 0.23 times that for centralized optimization. It can be inferred that, when the number of distribution networks connected to the transmission grid is further increased, the time consumption of centralized optimization increases sharply, while the total time consumption of the method proposed in this paper would be basically stable. This is because the distribution network optimization subproblems are solved in parallel in the proposed distributed computing method. In this case, the total duration is only related to the number of iterations, the time consumption of the transmission network subproblem, and the most time-consuming distribution network subproblem, but has little to do with the number of connected distribution networks. At the same time, the linearization approximation on the transmission subproblem greatly reduces the complexity and time consumption of the transmission subproblem, so the total time consumption of the method proposed in this paper is relatively low.

Table 3. Comparison of the results of solving test case 2 by different methods.

		Independent Method	Proposed Method	Centralized Method
System operating costs/thousand USD		61.6134	61.2250	61.2265
Network losses /MW	D7	0.0812	0.0769	0.0774
	D19	0.0796	0.0781	0.0794
	D26	0.0826	0.0759	0.0765
	TSO	5.0836	5.0307	5.0324
	Total	5.3270	5.2642	5.2657
Boundary voltage magnitudes (phase angles/(°))	D7	1.0031(0)	1.0192 (−6.52)	1.0190 (−6.52)
	D19	1.0155(0)	1.0287 (−9.83)	1.0284 (−9.83)
	D26	0.9985(0)	1.0305 (−10.46)	1.0300 (−10.46)

Table 4. Comparison of the results of solving test case 3 by different methods.

		Independent Method	Proposed Method	Centralized Method
System operating costs/thousand USD		61.6150	61.1229	61.1265
Network losses /MW	D7	0.0818	0.0779	0.0801
	D17	0.0750	0.0718	0.0729
	D19	0.0796	0.0791	0.0795
	D26	0.0903	0.0821	0.0826
	D29	0.0801	0.077	0.0776
	TSO	5.0801	5.0109	5.0134
	Total	5.4869	5.3988	5.4061
Boundary voltage magnitudes (phase angles/(°))	D7	1.0033(0)	1.0193 (−6.54)	1.0190 (−6.52)
	D17	1.0276 (0)	1.0433 (−8.64)	1.0429 (−8.64)
	D19	1.0144(0)	1.0296 (−9.82)	1.0284 (−9.83)
	D26	0.9901(0)	1.0309 (−10.50)	1.0300 (−10.46)
	D29	1.0066 (0)	1.0234 (−6.10)	1.0230 (−6.10)

Figures 4 and 5 show the voltage distribution of TSO and D26 using different methods to solve test case 1, respectively. (The results of test case 2 and test case 3 are similar.) Tables 5 and 6 present the optimization results of controllable resources in TSO and D26, respectively, while Table 7 shows the power exchange between TSO and D26. The result indicates that the proposed method makes fuller use of the reactive power regulation potential of the DGs in the distribution network. On the one hand, it improves the voltage distribution and reduces the network losses in D26. On the other hand, the demand of D26 for remote reactive resources in TSO is reduced, thus further improving the voltage distribution and reducing the network losses in TSO. As can be seen from the figures, the nodal voltage distributions obtained by the proposed method are very close to that of the centralized optimization method. For the transmission grid, the local voltage distribution is improved by making full use of the reactive power regulation capability of the distribution grid in the transmission and distribution coordination optimization process. In either the proposed method or the centralized optimization method, the voltage magnitudes of the PCC node of the transmission network, i.e., node 26 and its nearby nodes, are significantly improved when compared to that of the independent optimization case. In addition, Appendix A shows the detailed branch power flow of TSO, which demonstrates the influence of coordinated transmission and distribution ORPD on the branch power flow of TSO.

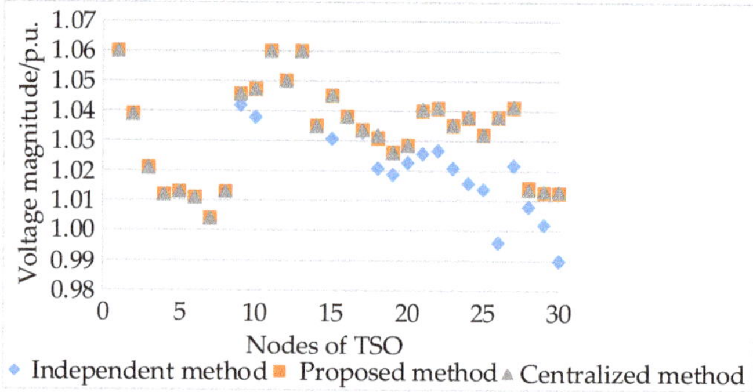

Figure 4. Voltage distribution of TSO with different methods on test case 1.

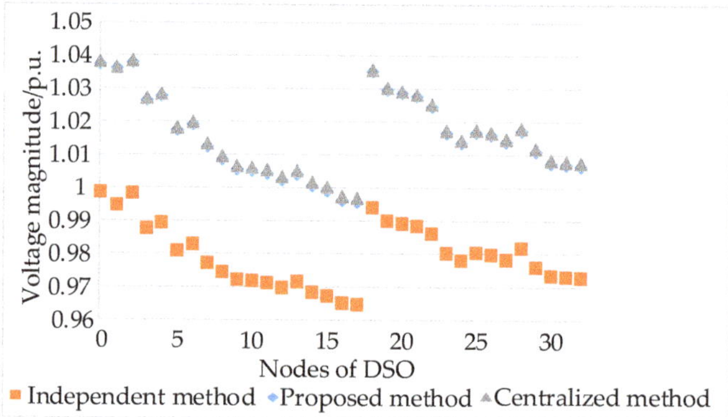

Figure 5. Voltage distribution of D26 with different methods on test case 1.

Table 5. Power output of generators in TSO with different methods on test case 1.

	Independent Method		Proposed Method		Centralized Method	
	P (MW)	Q (MVAr)	P (MW)	Q (MVAr)	P (MW)	Q (MVAr)
G1	41.52	−5.13	41.31	−5.14	41.31	−5.14
G2	55.37	2.53	55.09	2.49	55.09	2.49
G3	16.13	35.21	16.05	35.22	16.05	35.22
G4	22.7	33.59	22.58	33.35	22.58	33.35
G5	16.13	6.87	16.05	6.63	16.05	6.63
G6	39.62	31.95	39.44	30.85	39.44	30.85
Total	191.47	105.02	190.52	103.4	190.52	103.4

Table 6. Reactive power output of DGs in D26 with different methods on test case 1.

	Independent Method	Proposed Method	Centralized Method
DFIG-2/MVar	0.1264	0.3712	0.3717
GT-4/MVar	0.2012	0.3873	0.3869
PV-6/MVar	0.2367	0.3426	0.3430
PV-13/MVar	0.3246	0.3651	0.3654
DFIG-28/MVar	0.4000	0.4000	0.4000

Table 7. Power transmitted from TSO to D26 with different methods on test case 1.

	Independent Method	Proposed Method	Centralized Method
Active Power/MW	2.9310	2.9266	2.9253
Reactive Power/MVar	1.4407	0.8632	0.8628

4.3. Comparison of Different Distributed Algorithms

In order to test the superiority of the proposed method in relation to the other distributed methods, ADMM [13], G-MSS [9], ATC [28], AAL (without the linearized approximation method to model the TSO subproblem), and the proposed method in this paper were applied to test case 1, and the results are shown in Table 8.

Table 8. Comparison of the results of different distributed algorithms to solve test case 1.

Distributed Methods	System Operating Costs/thousand USD	Iterations	Total Duration/s
ADMM	61.3034	5	9.3
G-MSS	61.3022	6	11.6
ATC	61.3046	5	10.2
AAL	61.3058	5	7.1
The proposed method	61.3055	3	4.3

It can be seen from Table 8 that the operating costs of the systems optimized by various methods are not much different, and all the results are fairly close to those of the centralized optimization method, which demonstrates the effectiveness of various distributed methods for solving transmission and distribution synergy problems. However, compared with the other methods, the iterations and the total duration of the proposed method are significantly reduced. This means that the accuracies of various distributed methods are almost the same, but the method proposed in this paper has a higher efficiency. On the one hand, the AAL algorithm, which is the basis of the proposed method in the paper, combines the advantages of other algorithms to obtain improved convergence performance and solution efficiency. On the other hand, the linear approximation is applied to the power flow equations of the TSO subproblem, which greatly reduces the complexity of the optimization model, thereby improving the calculation speed and convergence performance. With the simultaneous application of the novel linearized approximation method and AAL algorithms, the proposed method achieves a significant improvement in convergence performance and solution efficiency compared to other methods.

The above results are obtained with a convergence accuracy of 10^{-2}. Figures 6 and 7 illustrate the system operating costs and iterations of the various methods for different convergence accuracies. As can be seen, the proposed method consistently achieves better optimization results as well as convergence performance compared to other distributed algorithms when the convergence accuracy varies. At the same time, as the convergence accuracy increases, the optimization results of various methods gradually converge and approach the results of centralized optimization, but the number of required iterations also increases. As the convergence accuracy increases from 10^{-2} to 10^{-3}, the quality of the optimization results of various algorithms improves significantly, while the number of iterations increases only slightly. When the convergence accuracy is increased from 10^{-4} to 10^{-5}, the quality of the optimization results of various methods improves just a little, while the number of required iterations increases a lot. We can thus see that the distributed methods can obtain the same results as the centralized optimization when the convergence accuracy is high enough, however, at the expense of the convergence performance. In practice, a reasonable compromise between optimization accuracy and convergence performance should be made before the computation process. It can also be seen that, when compared to the case of using only the AAL algorithm for distributed computation without using the novel linear approximation method to model the TSO subproblem, the proposed method in the paper has a better convergence performance at all convergence accuracies, which indicates that using both methods can effectively improve the convergence performance of the transmission and distribution collaborative framework. Furthermore, the inconsistency in the total operating cost between the two is caused when the nonlinear network loss term is linearized according to the linearized approximation method.

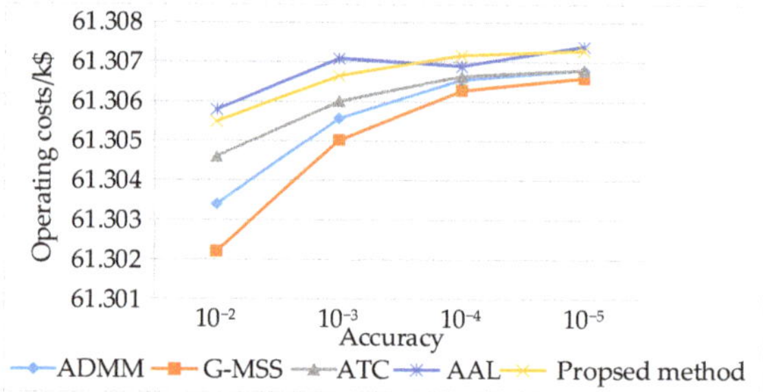

Figure 6. System operating cost with various algorithms under different accuracy.

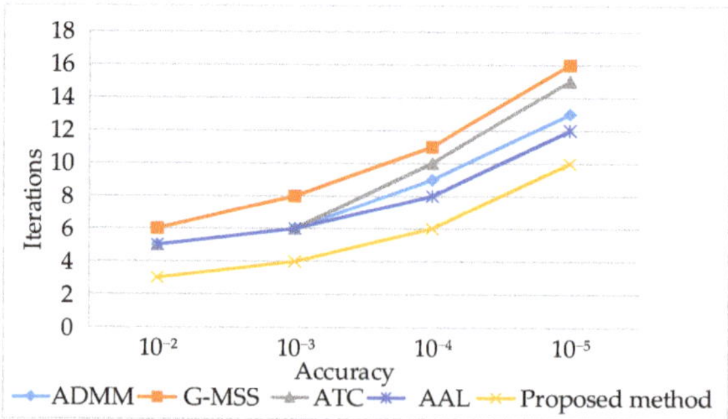

Figure 7. Iterations of various algorithms under different accuracy.

4.4. Effect of the Linearized Approximation on the TSO Subproblem

As can be seen from the analysis of the above arithmetic cases, the nonlinear optimization solver used by MATPOWER converges in all arithmetic scenarios. However, the linearized approximation method used in the paper for the TSO subproblem has the following unique advantages:

1. All formulas are convex functions, which theoretically guarantee convergence;
2. The local optimum corresponds to the global optimum;
3. The linearized OPF model of the transmission grid subproblem leads to a significant reduction in the complexity of the global optimization model, especially considering the weight of the TSO time consumption in the global optimization process.

However, the accuracy of the linearized approximation of the flow equations depends on the accuracy of the initial value $(\boldsymbol{v}_0, \boldsymbol{\theta}_0)$, and in the previous analysis, the results of the latest integrated flow calculation are used as the initial value $(\boldsymbol{v}_0, \boldsymbol{\theta}_0)$. The above analysis shows that this approach causes little error in the optimization process when the loads do not vary much. To further verify the feasibility of the linearized approximation method adopted in the paper, load variation coefficients α_P and α_Q are introduced to analyze the errors caused by the linearization of the power flow equations when the loads vary. The node loads after changes are as follows:

$$P'_{i,\mathrm{d}} = P_{i,\mathrm{d}} \times \left(1 + \alpha_P \times \frac{2i - \mathrm{N}}{\mathrm{N}}\right),\ Q'_{i,\mathrm{d}} = Q_{i,\mathrm{d}} \times \left(1 + \alpha_Q \times \frac{2i - \mathrm{N}}{\mathrm{N}}\right) \tag{79}$$

where $P'_{i,\mathrm{d}}/Q'_{i,\mathrm{d}}$ represents the changed load at bus i, and N is the number of transmission grid buses. When calculating the errors, the calculated results after the load change are used as the base values, and the errors at different load variation coefficients are shown in Table 9. As can be seen from the table, the errors in operating costs as the objective function of the optimization problem increase when the absolute values of α_P and α_Q increase, because the errors due to the linearization of the network losses become larger in that case. It can also be seen that the operating costs errors of the system and the modeling accuracy of Q are satisfactory for different load variation coefficients. This shows that the method in the paper of directly adopting the results of the latest integrated power flow calculation as the initial values $(\boldsymbol{v}_0, \boldsymbol{\theta}_0)$ and linearizing the transmission grid power flow equations is feasible and effective when the accuracy requirement is not particularly high. Alternatively, when a higher accuracy is required, the accuracy of the linearized approximation method can be further promoted by using the warm-start iterative method described in [20], where only one more iteration is required to obtain a solution significantly close to the exact modeling case.

Table 9. Errors of the linearized approximation method at different load variation coefficients.

α_P	α_Q	Errors in Total Costs	Maximum Error of Q_{ij} (p.u.)
20%	20%	0.078%	0.27
30%	20%	0.30%	0.22
40%	20%	0.47%	0.26
20%	40%	0.10%	0.27
30%	40%	0.31%	0.26
40%	40%	0.57%	0.27

5. Conclusions

This paper established a distributed framework of the coordinated transmission and distribution networks optimal reactive power dispatch (CTD-ORPD) based on the AAL method. The global reactive power optimization problem of transmission and distribution grids was decomposed into a TSO subproblem and several DSO$_k$ subproblems. Each of the subproblems ran in a distributed manner and iterated until the optimal global solution is finally reached. The transmission grid and distribution grid subproblems achieved fully distributed cooperative optimal reactive power control only through the information interaction of the coordinating variables. The main conclusions are as follows:

(1) Under the background of a large influx of new energy in the distribution networks, the coordinated reactive power optimization of transmission and distribution networks can better allocate the reactive power resources of the system than the independent optimization method. On the one hand, it gives full play to the voltage supportability of the transmission network to the distribution networks, and also makes full use of the reactive power regulation ability of the distribution networks. Thereby minimizing system network losses and maximizing economic benefits;
(2) Under the proposed framework, each subproblem is solved in a distributed manner, which protects the privacy of each independent subject and greatly reduces the complexity of the problem. The distribution network subproblems run in parallel; when there are many distribution networks connected to the transmission network, the complexity of the problem only depends on the distribution network subproblem with the longest solution time, and has little to do with the number of distribution networks. Thus, compared with the centralized optimization method, the solution speed of the optimization problem is greatly improved. Compared with the ADMM method, the AAL method updates the variables twice in each iteration, and has advantages in convergence performance and solving efficiency;
(3) The transmission subproblem plays a decisive role in the complexity of the entire model and the total time consumption. By applying a linear approximation to the power flow equations of the transmission subproblem, the complexity of the optimization problem and the total time consumption are both significantly reduced.

The AAL algorithm and the novel linearized approximation method improve the convergence performance and solving efficiency of the cooperative transmission and distribution reactive power optimization problem in terms of the distributed computation method and the model building method, respectively, to achieve more stable convergence and faster solution. Considering the demand for regulating the voltage problems in the distribution networks in the background of the influx of DERs, the proposed CTD-ORPD framework in this paper is expected to play a good role in the short-term or real-time cooperative reactive power optimal control process of transmission and distribution, thus promoting further DER consumption.

Author Contributions: Conceptualization, B.Z. and X.S.; Data curation, B.Z.; Formal analysis, B.Z. and C.P.; Funding acquisition, X.S. and C.P.; Investigation, B.Z. and Y.B.; Methodology, B.Z.; Project administration, X.S.; Resources, T.W.; Software, B.Z. and T.W.; Supervision, X.S.; Validation, B.Z. and X.S.; Visualization, B.Z.; Writing—original draft preparation, B.Z.; Writing—review and editing, X.S. All authors have read and agreed to the published version of the manuscript.

Funding: This research was funded by Natural Science Foundation of China (U2066209), by technology project of China Southern Power Grid Co., Ltd. (060400KK52190017), by Deep mining of the demand characteristics of the sales side and application and industrialization of load data template technology (2020ZHCG0022).

Institutional Review Board Statement: Not applicable.

Informed Consent Statement: Not applicable.

Data Availability Statement: Not applicable.

Conflicts of Interest: The authors declare no conflict of interest.

Appendix A

Table A1. Branch power flow in TSO with different methods evaluated on test case 1.

Branch #	From Bus	To	Independent Method						Proposed Method					
			From Bus Injection		To Bus Injection		Loss		From Bus Injection		To Bus Injection		Loss	
		Bus	P(MW)	Q(MVAr)	P(MW)	Q(MVAr)	P(MW)	Q(MVAr)	P(MW)	Q(MVAr)	P(MW)	Q(MVAr)	P(MW)	Q(MVAr)
1	1	2	21.05	−2.31	−20.88	2.81	0.17	0.50	21.04	−2.31	−20.87	2.81	0.17	0.50
2	1	3	20.47	−2.82	−20.08	4.31	0.39	1.49	20.47	−2.83	−20.08	4.32	0.39	1.49
3	2	4	18.63	−5.49	−18.22	6.66	0.41	1.17	18.62	−5.5	−18.21	6.67	0.41	1.17
4	3	4	17.85	−2.93	−17.79	3.18	0.06	0.25	17.85	−2.94	−17.79	3.19	0.06	0.25
5	2	5	14.36	−0.53	−14.17	1.30	0.19	0.77	14.35	−0.53	−14.16	1.30	0.19	0.77
6	2	6	21.64	−3.85	−21.11	5.45	0.53	1.60	21.64	−3.87	−21.10	5.47	0.54	1.60
7	4	6	17.55	5.77	−17.49	−5.52	0.06	0.25	17.55	5.76	−17.49	−5.51	0.06	0.25
8	5	7	14.25	1.13	−14.05	−0.66	0.20	0.47	14.25	1.12	−14.05	−0.65	0.20	0.47
9	6	7	8.71	8.29	−8.62	−8.06	0.09	0.23	8.71	8.3	−8.62	−8.07	0.09	0.23
10	6	8	23.81	21.37	−23.62	−20.60	0.19	0.77	23.82	21.37	−23.63	−20.60	0.19	0.77
11	6	9	7.26	−7.9	−7.26	8.37	0.00	0.47	7.25	−7.91	−7.25	8.38	0.00	0.47
12	6	10	4.15	−4.51	−4.15	4.90	0.00	0.39	4.14	−4.52	−4.14	4.92	0.00	0.40
13	9	11	0	0	0.00	0.00	0.00	0.00	0	0	0.00	0.00	0.00	0.00
14	9	10	7.26	−8.15	−7.26	8.38	0.00	0.23	7.25	−8.17	−7.25	8.40	0.00	0.23
15	4	12	11.07	−14.66	−11.07	16.31	0.00	1.65	11.06	−14.67	−11.06	16.32	0.00	1.65
16	12	13	−16.13	−33.35	16.13	36.69	0.00	3.34	−16.13	−33.36	16.13	36.70	0.00	3.34
17	12	14	4.68	2.08	−4.63	−1.95	0.05	0.13	4.68	2.08	−4.63	−1.95	0.05	0.13
18	12	15	6.05	3.19	−5.99	−3.08	0.06	0.11	6.04	3.19	−5.98	−3.08	0.06	0.11
19	12	16	5.27	5	−5.19	−4.82	0.08	0.18	5.27	4.99	−5.19	−4.81	0.08	0.18
20	14	15	−1.55	0.42	1.56	−0.40	0.01	0.02	−1.55	0.42	1.56	−0.40	0.01	0.02
21	16	17	1.73	3.1	−1.71	−3.06	0.02	0.04	1.72	3.09	−1.70	−3.05	0.02	0.04
22	15	18	7.18	3.71	−7.05	−3.46	0.13	0.25	7.18	3.71	−7.05	−3.46	0.13	0.25
23	18	19	3.91	2.67	−3.88	−2.62	0.03	0.05	3.91	2.67	−3.88	−2.62	0.03	0.05
24	19	20	−5.6	−0.76	5.62	0.80	0.02	0.04	−5.61	−0.76	5.63	0.80	0.02	0.04
25	10	20	7.87	1.62	−7.77	−1.37	0.10	0.25	7.88	1.62	−7.78	−1.37	0.10	0.25
26	10	17	7.3	2.78	−7.27	−2.69	0.03	0.09	7.31	2.78	−7.28	−2.69	0.03	0.09
27	10	21	−4.48	−11.23	4.56	11.41	0.08	0.18	−4.49	−11.25	4.57	11.43	0.08	0.18
28	10	22	−5.09	−8.19	5.21	8.44	0.12	0.25	−5.1	−8.21	5.22	8.46	0.12	0.25
29	21	22	−22.02	−22.53	22.20	22.89	0.18	0.36	−22.04	−22.55	22.22	22.91	0.18	0.36
30	15	23	−10.92	−2.67	11.14	3.12	0.22	0.45	−10.92	−2.66	11.14	3.11	0.22	0.45
31	22	24	−4.58	2.53	4.64	−2.44	0.06	0.09	−4.61	2.26	4.67	−2.17	0.06	0.09
32	23	24	1.89	2.34	−1.87	−2.30	0.02	0.04	1.88	2.11	−1.86	−2.07	0.02	0.04
33	24	25	−11.43	−1.86	11.87	2.63	0.44	0.77	−11.47	−2.35	11.92	3.14	0.45	0.79
34	25	26	2.96	2.49	−2.90	−2.40	0.06	0.09	2.95	0.9	−2.91	−0.85	0.04	0.05
35	25	27	−14.64	−4.78	15.07	5.61	0.43	0.83	−14.67	−3.69	15.08	4.48	0.41	0.79
36	28	27	−11.48	−21.05	11.48	25.32	0.00	4.27	−11.48	−21.06	11.48	25.34	0.00	4.28
37	27	29	6.16	1.65	−6.02	−1.38	0.14	0.27	6.16	1.65	−6.02	−1.38	0.14	0.27
38	27	30	7.1	1.63	−6.83	−1.13	0.27	0.50	7.1	1.63	−6.83	−1.13	0.27	0.50
39	29	30	3.68	0.6	−3.63	−0.49	0.05	0.11	3.68	0.6	−3.63	−0.49	0.05	0.11
40	8	28	−6.29	−9.06	6.41	9.47	0.12	0.41	−6.29	−9.06	6.41	9.47	0.12	0.41
41	6	28	−5.07	−14.46	5.15	14.71	0.08	0.25	−5.07	−14.47	5.15	14.72	0.08	0.25

References

1. Martins, V.F.; Borges, C.L.T. Active Distribution Network Integrated Planning Incorporating Distributed Generation and Load Response Uncertainties. *IEEE Trans. Power Syst.* **2011**, *26*, 2164–2172. [CrossRef]
2. Chen, P.-C.; Salcedo, R.; Zhu, Q.; de Leon, F.; Czarkowski, D.; Jiang, Z.-P.; Spitsa, V.; Zabar, Z.; Uosef, R.E. Analysis of Voltage Profile Problems Due to the Penetration of Distributed Generation in Low-Voltage Secondary Distribution Networks. *IEEE Trans. Power Deliv.* **2012**, *27*, 2020–2028. [CrossRef]
3. Zhao, J.; Zhang, Z.; Yao, J.; Yang, S.; Wang, K. A distributed optimal reactive power flow for global transmission and distribution network. *Int. J. Electr. Power Energy Syst.* **2019**, *104*, 524–536. [CrossRef]
4. Li, P.; Ji, H.R.; Wang, C.S.; Zhao, J.L.; Song, G.Y.; Ding, F.; Wu, J.Z. Coordinated Control Method of Voltage and Reactive Power for Active Distribution Networks Based on Soft Open Point. *IEEE Trans. Sustain. Energy* **2017**, *8*, 1430–1442. [CrossRef]
5. Mohseni, B.S.M.; Kamwa, I.; Moeini, A.; Rabiee, A. Voltage Security Constrained Stochastic Programming Model for Day-Ahead BESS Schedule in Co-Optimization of T&D Systems. *IEEE Trans. Sustain. Energy* **2020**, *11*, 391–404.
6. Sun, H.; Guo, Q.; Zhang, B.; Guo, Y.; Li, Z.; Wang, J. Master-Slave-Splitting Based Distributed Global Power Flow Method for Integrated Transmission and Distribution Analysis. *IEEE Trans. Smart. Grid.* **2015**, *6*, 1484–1492. [CrossRef]

7. Lin, C.; Wu, W.; Zhang, B.; Wang, B.; Zheng, W.; Li, Z. Decentralized Reactive Power Optimization Method for Transmission and Distribution Networks Accommodating Large-Scale DG Integration. *IEEE Trans. Sustain. Energy* **2017**, *8*, 363–373. [CrossRef]
8. Li, Z.; Guo, Q.; Sun, H.; Wang, J. Coordinated Economic Dispatch of Coupled Transmission and Distribution Systems Using Heterogeneous Decomposition. *IEEE Trans. Power Syst.* **2016**, *31*, 4817–4830. [CrossRef]
9. Li, Z.; Guo, Q.; Sun, H.; Wang, J. A New LMP-Sensitivity-Based Heterogeneous Decomposition for Transmission and Distribution Coordinated Economic Dispatch. *IEEE Trans. Smart Grid* **2018**, *9*, 931–941. [CrossRef]
10. Li, Z.; Guo, Q.; Sun, H.; Wang, J. Coordinated Transmission and Distribution AC Optimal Power Flow. *IEEE Trans. Smart Grid* **2018**, *9*, 1228–1240. [CrossRef]
11. Erseghe, T. Distributed Optimal Power Flow Using ADMM. *IEEE Trans. Power Syst.* **2014**, *29*, 2370–2380. [CrossRef]
12. Zheng, W.; Wu, W.; Zhang, B.; Sun, H.; Liu, Y. A Fully Distributed Reactive Power Optimization and Control Method for Active Distribution Networks. *IEEE Trans. Smart Grid* **2016**, *7*, 1021–1033. [CrossRef]
13. Ruan, H.; Gao, H.; Liu, Y.; Wang, L.; Liu, J. Distributed Voltage Control in Active Distribution Network Considering Renewable Energy: A Novel Network Partitioning Method. *IEEE Trans. Power Syst.* **2020**, *35*, 4220–4231. [CrossRef]
14. Chatzipanagiotis, N.; Zavlanos, M.M. On the Convergence of a Distributed Augmented Lagrangian Method for Nonconvex Optimization. *IEEE Trans. Autom. Control.* **2017**, *62*, 4405–4420. [CrossRef]
15. Nejad, R.R.; Sun, W.; Golshani, A. Distributed Restoration for Integrated Transmission and Distribution Systems with DERs. *IEEE Trans. Power Syst.* **2019**, *34*, 4964–4973. [CrossRef]
16. Ding, T.; Li, C.; Huang, C.; Yang, Y.; Li, F.; Blaabjerg, F. A Hierarchical Modeling for Reactive Power Optimization With Joint Transmission and Distribution Networks by Curve Fitting. *IEEE Syst. J.* **2018**, *12*, 2739–2748. [CrossRef]
17. Lin, C.; Wu, W.; Chen, X.; Zheng, W. Decentralized Dynamic Economic Dispatch for Integrated Transmission and Active Distribution Networks Using Multi-Parametric Programming. *IEEE Trans. Smart Grid* **2018**, *9*, 4983–4993. [CrossRef]
18. Liu, B.-Y.; Yang, R.-G. Multi-subarea parallel reactive power optimization based on APP. *Proc. CSEE* **2009**, *29*, 47–51.
19. Mohammadi, A.; Mehrtash, M.; Kargarian, A. Diagonal Quadratic Approximation for Decentralized Collaborative TSO plus DSO Optimal Power Flow. *IEEE Trans. Smart Grid* **2019**, *10*, 2358–2370. [CrossRef]
20. Yang, Z.; Zhong, H.; Bose, A.; Zheng, T.; Xia, Q.; Kang, C. A Linearized OPF Model With Reactive Power and Voltage Magnitude: A Pathway to Improve the MW-Only DC OPF. *IEEE Trans. Power Syst.* **2018**, *33*, 1734–1745. [CrossRef]
21. Chen, Y.B.; Yao, Y.; Zhang, Y. A Robust State Estimation Method Based on SOCP for Integrated Electricity-Heat System. *IEEE Trans. Smart Grid* **2021**, *12*, 810–820. [CrossRef]
22. Rostampour, V.; Keviczky, T. Probabilistic Energy Management for Building Climate Comfort in Smart Thermal Grids with Seasonal Storage Systems. *IEEE Trans. Smart Grid* **2019**, *10*, 3687–3697. [CrossRef]
23. Yang, Z.; Bose, A.; Zhong, H.; Zhang, N.; Xia, Q.; Kang, C. Optimal Reactive Power Dispatch With Accurately Modeled Discrete Control Devices: A Successive Linear Approximation Approach. *IEEE Trans. Power Syst.* **2017**, *32*, 2435–2444. [CrossRef]
24. Yang, Z.; Zhong, H.; Xia, Q.; Bose, A.; Kang, C. Optimal power flow based on successive linear approximation of power flow equations. *IET Gener. Transm. Distrib.* **2016**, *10*, 3654–3662. [CrossRef]
25. Zhao, B.; Guo, C.X.; Cao, Y.J. A multiagent-based particle swarm optimization approach for optimal reactive power dispatch. *IEEE Trans. Power Syst.* **2005**, *20*, 1070–1078. [CrossRef]
26. ElKeib, A.A.; Ma, X. Calculating short-run marginal costs of active and reactive power production. *IEEE Trans. Power Syst.* **1997**, *12*, 559–564. [CrossRef]
27. Huang, Q.; Han, S.; Rong, N.; Luo, J.; Hu, X. Stochastic Economic Dispatch of Hydro-Thermal-Wind-Photovoltaic Power System Considering Mixed Coal-Blending Combustion. *IEEE Access* **2020**, *8*, 218542–218553. [CrossRef]
28. Li, P.; Wu, Q.W.; Yang, M.; Li, Z.S.; Hatziargyriou, N.D. Distributed Distributionally Robust Dispatch for Integrated Transmission-Distribution Systems. *IEEE Trans. Power Syst.* **2021**, *36*, 1193–1205. [CrossRef]

energies

MDPI

Article

Optimal Power Sharing in Microgrids Using the Artificial Bee Colony Algorithm

Kalim Ullah, Quanyuan Jiang *, Guangchao Geng, Sahar Rahim and Rehan Ali Khan

College of Electrical Engineering, Zhejiang University, Hangzhou 310027, China; kalim125@zju.edu.cn (K.U.); ggc@zju.edu.cn (G.G.); sahar@ciit.edu.pk (S.R.); rehan@zju.edu.cn (R.A.K.)
* Correspondence: jqy@zju.edu.cn

Abstract: In smart grids, a hybrid renewable energy system that combines multiple renewable energy sources (RESs) with storage and backup systems can provide the most cost-effective and stable energy supply. However, one of the most pressing issues addressed by recent research is how best to design the components of hybrid renewable energy systems to meet all load requirements at the lowest possible cost and with the best level of reliability. Due to the difficulty of optimizing hybrid renewable energy systems, it is critical to find an efficient optimization method that provides a reliable solution. Therefore, in this study, power transmission between microgrids is optimized to minimize the cost for the overall system and for each microgrid. For this purpose, artificial bee colony (ABC) is used as an optimization algorithm that aims to minimize the cost and power transmission from outside the microgrid. The ABC algorithm outperforms other population-based algorithms, with the added advantage of requiring fewer control parameters. The ABC algorithm also features good resilience, fast convergence, and great versatility. In this study, several experiments were conducted to show the productivity of the proposed ABC-based approach. The simulation results show that the proposed method is an effective optimization approach because it can achieve the global optimum in a very simple and computationally efficient way.

Keywords: microgrid; ABC; power-sharing; cost optimization; renewable energy

Citation: Ullah, K.; Jiang, Q.; Geng, G.; Rahim, S.; Khan, R.A. Optimal Power Sharing in Microgrids Using the Artificial Bee Colony Algorithm. *Energies* **2022**, *15*, 1067. https://doi.org/10.3390/en15031067

Academic Editor: Adel Merabet

Received: 20 December 2021
Accepted: 19 January 2022
Published: 31 January 2022

1. Introduction

Today, the world's greatest challenges are the rapidly growing demand for electrical energy [1], rising electricity prices, the increasing use of non-renewable energy, the limits of conventional energy for power generation, global warming, global climate change, and related environmental problems [2]. All these challenges have resulted in the world being in a catastrophic economic and political crisis. Because of this, every government in the world is eager to expand renewable energy generation [3]. Renewable energy generation can help nations achieve their long-term development goal of providing safe, cheap, clean, environmentally friendly, and sustainable energy [4]. Despite the many advantages offered by renewable energy compared to conventional energy, they all have weaknesses in common, such as high weather vulnerability, low stability, and high unpredictability [5], all of which lead to low reliability and efficiency of energy generation [6]. Consequently, the hybrid renewable energy system can solve important problems and constraints in efficiency, reliability, and economy [7], which makes it an effective choice to meet the load requirements and support and improve the system [8]. The use of a central control unit to control the integrated power management of microgrids improves the efficiency, flexibility, and response time to fluctuations. The optimal power flow in the grid, the voltage level in each bus, and the transmission losses are all essential variables to be considered in efficient power systems [9]. The actual problem is an MILP problem, and the constraints on the optimal power flow are linear. For this purpose, a stochastic two-stage model of the interactions between the distribution system operator and many microgrids on the distribution line is used in [10].

The Stackelberg game concept is used in [11] for the interconnections between generators and microgrids. This model considers all electrical power flow constraints, voltage constraints, and line losses. The main objective of each is to maximize the profit.

MGs are classified in terms of power, control, mode, phase, and applications, as shown in Figure 1.

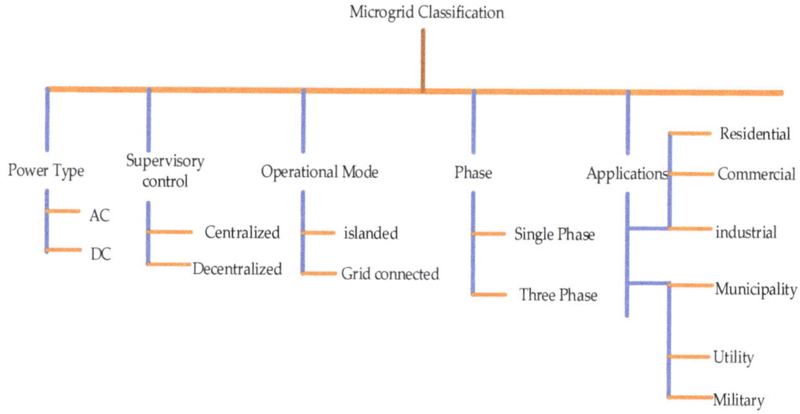

Figure 1. Classification of microgrids.

A microgrid (MG), as shown in Figure 2, consists of various distributed energy sources (DERs), responsive loads, and critical loads. A common connection point (PCC) connects the MG to the main grid [12]. Each DER is connected to the power electronic interface (PEI) in both grid-connected and islanded modes to perform control, measurement, protection, and plug-and-play functions. An MG in the grid-connected mode benefits from sharing power with the main grid. However, in the event of a fault or failure of the main grid, MG switches to the islanded mode to ensure system stability. In this mode, critical loads are continuously supplied with power by efficiently integrating DERs, demand response (DR), and load shedding (LS). The central microgrid controller (MGCC) and local controllers (LCs) manage and coordinate the entire MG operation [13]. Efficient DER management and coordination in MG lead to higher system performance and long-term development [14]. Due to increased environmental awareness, socioeconomic growth, and the need to reduce greenhouse gas emissions, MGs are mainly composed of sustainable energy systems, such as renewable energy sources and energy-efficient systems that use local heat waste [15]. Various energy storage technologies that have potential for high penetration and integration in microgrids are mentioned in [16]. Different energy trading systems are examined for interactive energy trading, multienergy management, and resilient operations in [17]. The DSM technique is used in [18] to reduce the operating cost of the grid-connected microgrid. An optimization-based energy management system is used to reduce the generation and curtailment costs [19]. In [20], an optimization controller is developed to control the energy management system for distributed energy sources in microgrids.

Table 1 shows a few examples of sustainable energy being used to run the energy management system from the literature review.

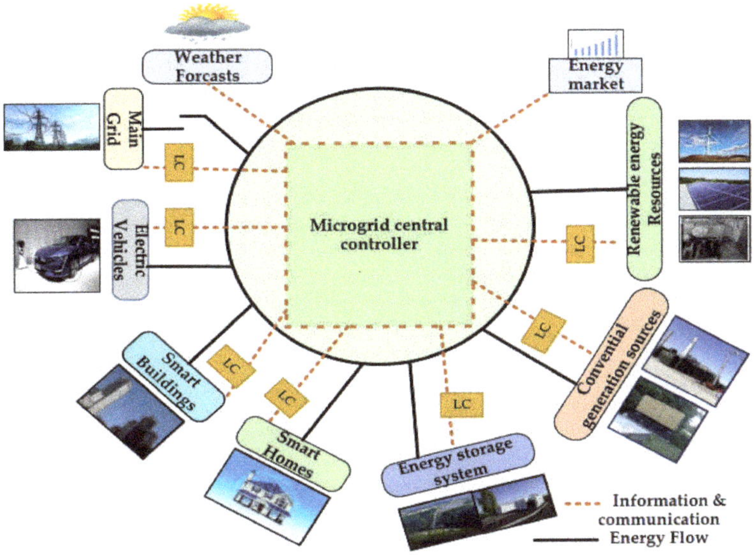

Figure 2. Microgrid setup.

Table 1. Sustainable energy (SE) system in microgrids.

References	Solar	WT	FC	CHP	EES	Biomass	Hydro	Tidal
[21]	✓						✓	
[22]	✓	✓		✓				
[23]	✓		✓					✓
[24]					✓			
[25]					✓			
[26]					✓			
[27]	✓	✓				✓		
[28]	✓	✓	✓	✓				
[29–31]					✓			

The microgrid EMS is a decision-making technique. For sustainable development, these techniques improve system efficiency, boost system reliability, decrease energy consumption, reduce DER operating costs, decrease system losses, and eliminate GHG emissions.

This paper presents a model for a hybrid renewable energy system integrated with a smart grid. The hybrid system includes wind turbines (WTs), photovoltaic (PV) systems, an electricity distribution company (Disco), gas turbines (GTs), and battery storage. Each component of the generation and load side is represented by a model. The hourly data of the wind speed and solar radiation on a daily basis are used as a case study in this paper. The proposed approach is used to optimize the power-sharing in the MGs to minimize the total cost of the system and the cost of each integrated microgrid. An artificial bee colony (ABC), as the optimization algorithm, is designed here to deal with the proposed model, with the objective of minimizing the cost and power imported from outside the MG. The idea of this paper is represented in Figure 3.

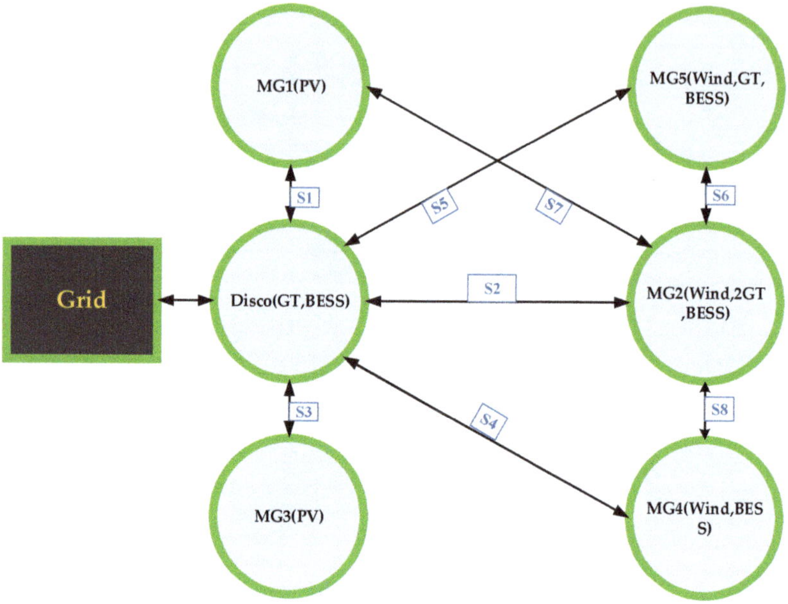

Figure 3. The proposed idea.

Figure 3 shows a microgrid (MG) connected to a distribution network. This MG contains multiple small MGs, which are connected with each other by electric lines. The electric line can be allowed to export and impart energy from small MGs. The small MG contains a mixture of components that consume and produce electricity (solar and wind plants) in addition to electricity storage units. The operational status of the small MG is divided into three sections: first, self-sufficiency in the event of equal production and consumption, so the MG will neither import nor export energy; second, import, which occurs when consumption is greater than production, as the MG needs energy; and third, export, which occurs when there is a surplus of energy within the MG and it needs to be exported.

Optimization is a mathematical problem that may be found in all engineering domains. This term's literal definition is "best possible or desirable". Because optimization problems are so broad and varied, it is a significant academic field.

Optimization algorithms are classified into two types: deterministic and stochastic. Previously, tackling optimization issues required tremendous computational effort, which frequently failed as the problem size grew larger. This is why bio-inspired stochastic optimization algorithms are being used as computationally efficient alternatives to deterministic approaches. Metaheuristics are based on the iterative improvement of either a population of solutions (evolutionary algorithms or swarm-based algorithms) or a single solution (Tabu Search) to solve a given optimization problem, and they primarily use randomization and local search.

The literature on bio-inspired algorithms (BIAs) for solving a wide variety of issues is vast, and various studies have reported on the usefulness of such tactics for handling difficult problems in the main disciplines of engineering in recent years. The two most prevalent and successful BIA classes or routes are evolutionary algorithms and swarm-based algorithms, both inspired by animals' collective behavior and natural development. In order to obtain a broader view on the subject, the algorithms were divided into regions based on where the inspiration for them came from in nature.

Swarm intelligence is a novel and rising paradigm used for creating adaptive systems in bio-inspired computing. In this sense, evolutionary computation (EC) is an extension of this. Swarm intelligence is based on organisms' collective social behaviors, whereas evolutionary algorithms are based on species' genetic adaptability. Swarm intelligence, as defined in the literature, is the use of the collective intelligence of groups of simple organisms to solve problems, based on the behavior of actual insect swarms. The term "swarm" refers to the chaotic movement of particles in the affected region. Some important SI algorithms are particle swarm optimization (PSO), the ant colony optimization algorithm (ACO), the fish swarm algorithm (FSA), the firefly algorithm, and the artificial bee colony (ABC) methods, which have all been employed as optimization methodologies. In the foraging process, these algorithms, which were inspired by animals' collective behavior, exhibit decentralized self-organized patterns. However, the artificial bee colony (ABC) approach was used in this article [32].

The ABC algorithm outperforms or is similar to other population-based algorithms, with the added advantage of requiring fewer control parameters. The ABC algorithm also features good resilience, fast convergence, and great versatility.

The ABC algorithm, developed by Karaboga and Basturk, replicates the intelligent foraging behavior of a honeybee swarm. The ABC algorithm's artificial bee colony is made up of three categories of bees: hired bees, bystanders, and scouts. An employed bee is a spectator who does not participate in the dance but instead travels to the food source being frequented by the observer. The scout bee, on the other hand, performs random searches for fresh sources. The quality (or fitness) of a solution may be measured by comparing the location of a food source to the amount of nectar it generates. Beehives are built and then released into the two-dimensional search space. Bees build social relationships with one another while foraging for nectar. Intense bee–bee interactions are essential to the discovery of a solution.

2. Problem Formulation

2.1. ABC Algorithm

In 2005, Karaboga discovered the ABC algorithm, influenced by honey bee behavior. The algorithm of a honey bee colony has the ability to find the best quality food sources in nature with ease. Therefore, the concept of ABC was derived from the clever foraging behavior of honey bees to find suitable solutions to optimization problems. Generally, bee colonies are classified into three types according to their foraging ability: employed bees, onlooker bees, and scout bees. The employed bees are responsible for collecting nectar (food). They investigate the location of the food supply in advance and alert the scout bees about the quality of the food. Based on the information relayed by the employed bees, the scout bees wait in the swarm and decide whether to take advantage of a food source. The scout bees randomly search the environment for a new nectar supply, either from internal motivation or from likely external cues [33]. The quality (fitness) of the feasible solution to the optimization questions is related to the profitability of a nectar source. The presence of a nectar source indicates a feasible solution to the optimization issues. Each nectar source is visited by only one honey bee. In other words, the number of employed or onlooker bees is proportional to the number of nectar sources [34]. Employed bees maintain an excellent solution, onlooker bees accelerate convergence, and scout bees improve the ability to eliminate local optimums [35,36].

ABC Algorithm Iteration Steps

ABC algorithm's main steps are listed as follows [37,38]:

1. Initialization. Generate N random solutions (food sources) Xi $(i = 1, 2, 3, \ldots, N)$ in a dimensional searching space D, where N represents the number of food sources, which is half the size of the colony Xi $(i = 1, 2, 3, \ldots, D)$ is a D-dimensional solution vector. For $i = 1, 2, 3, \ldots, N$, the ith food source in the original population, and the number of optimization population parameters is denoted by D.

2. During the honey collection stage, each employed bee creates a new nectar source in the food source's vicinity. When a new nectar source is compared to the previous one, the

high probability will be memorized. Each onlooker bee assesses the attractiveness of nectar sources received from all employed bees and selects a food source with a high probability. Like the employed bees, she changes the source location in her memory and maintains a higher nectar supply. In these two phases, the following formula is employed to regenerate nectar sources:

$$l_{ij} = h_{ij} + \theta \cdot \left(h_{ij} - h_{kj}\right) \tag{1}$$

where i $(k = 1,2,3,\ldots,N)$, $(j = 1,2,3,\ldots,D)$, and $\theta[0,\ 1]$ is a random number that determines the generation range of $hij's$ neighborhoods. As the search comes closer to an optimal solution, the number of neighborhoods available will decrease.

3. Food source selection. In the next step, the onlooker bees compare the probability calculated by the fitness value to select a food source. Nectar sources with a high probability are chosen with a high degree of certainty. The chance of being chosen for food sources is computed using the equation below:

$$P_i = \frac{Fit_i}{\sum_i^N Fit_i} \tag{2}$$

The fitness value of the ith solution, $Fit_i i(i = 1,\ 2,\ 3,\ldots,N)$ may be determined using following equation:

$$Fit_i = \begin{cases} \dfrac{1}{1+f_i},\ if\ f_i \geq 0, \\ 1+|f_i|,\ if\ f_i \leq 0 \end{cases} \tag{3}$$

If the quality of the new food source location is the same as or better than the previous one, the old one is updated with a new one, where f_i is the value of the objective function for $i = 1,\ 2,\ 3,\ldots,N$, which is unique to the optimization problem. Otherwise, the old one will be kept the same as the stage of employed bees.

4. Population elimination. A solution is considered to have fallen into a local optimum solution if it has not improved significantly after a specified number of trials, known as "max iteration", and the starting position is abandoned. As a consequence, the matching employed bees will become scout bees, and a new solution, which may be described as follows, will be generated at random in place of the discarded solution:

$$h\_ij = \theta * (h\_maxj - h\_minj) + h\_minj \tag{4}$$

where h_maxj and h_minj are the jth and ith individual maximum and lowest values, respectively, and i and j are the same (1).

The ABC algorithm may be expressed in the following stages based on the preceding phases [39].Flowchart of ABC algorithm is shown in Figure 4.

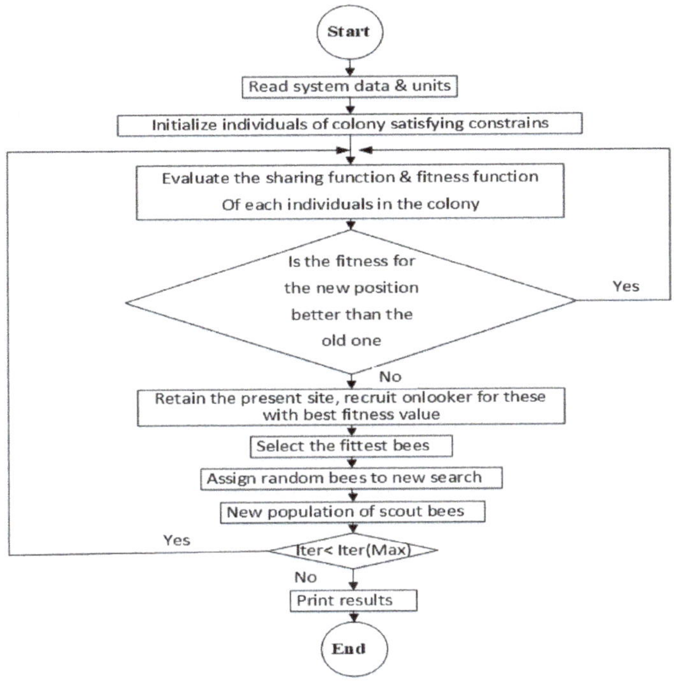

Figure 4. ABC flow chart.

3. Mathematical Modeling

The microgrid architecture studied in this paper is represented in Figure 5. It includes wind turbines (WTs), gas turbines (GTs), photovoltaic (PV), batteries (BT), additional storage components, general loads, and essential loads with different characteristics. The connection point, also known as the point of common coupling (PCC), is the interface between this architecture's utility and microgrid systems. As a result, there are two types of modes.

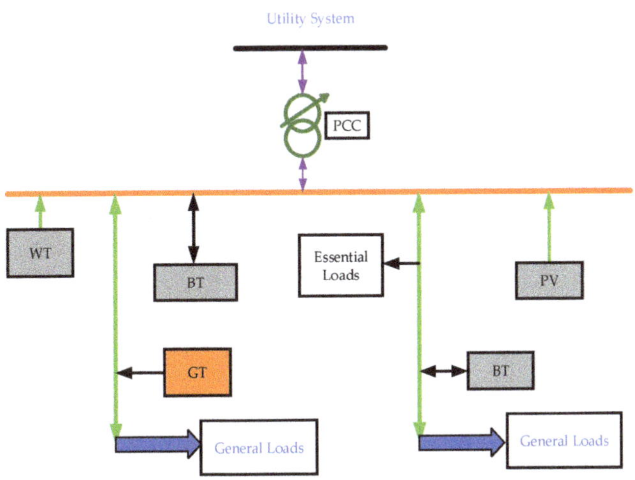

Figure 5. Typical architecture of the microgrid system.

When the MG is connected to the main grid via PCC, it is in the grid-connected mode and when it does not connect with the main grid, it is in the islanded mode.

Renewable energy sources (RESs) can be connected via a DC, AC, or a hybrid DC/AC bus. For most generators and loads, the appropriate configuration is determined by the type of output power. Therefore, DC bus coupling is preferred when both loads and generators are DC [40]. When the loads and generators are AC, then AC bus coupling is preferred [41] when the generation and load are mixed, such as AC and DC hybrid renewable energy sources (HRESs). A hybrid AC, DC bus coupling system is used [42], as shown in Figure 6.

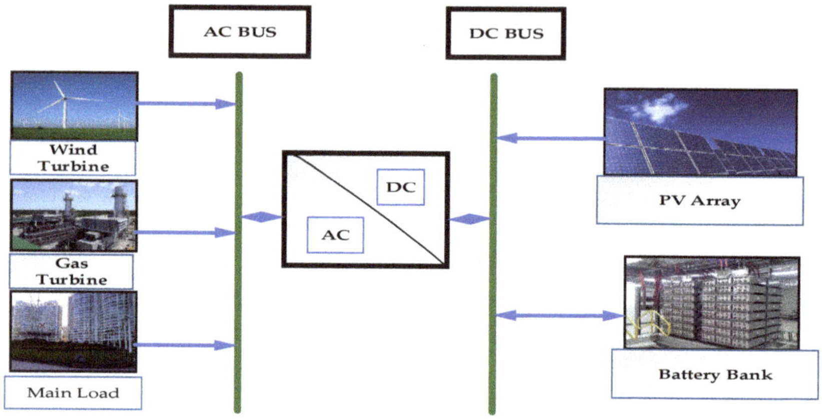

Figure 6. Schematic diagram of hybrid wind/PV/GT/battery storage.

3.1. *Hybrid Wind/PV/Battery Storage/Gas Turbine*

The configuration shown in Figure 6 consists of wind energy, a PV energy system, battery storage, a gas turbine, and the main load. The response of this configuration is simple and easy to understand. Due to the bidirectional converter, the WT and the PV system are mainly responsible for supplying the main load. The excess power generated by wind and/or PV is stored in a battery storage system until the battery is fully charged to SOC_{max}. Excess power generated above SOC_{max} is supplied to dedicated loads, i.e., dummy loads, such as loads for cooling, home appliances and heating, and charging the batteries of emergency lights when the battery storage is full. When the load power exceeds the generated power, the batteries are used to make up the difference until they reach the minimum SOC (SOC_{min}). Suppose the battery is fully discharged by SOC_{min} and the hybrid renewable energy sources cannot meet the microgrid's load demand. In this case, it imports energy from another microgrid. Moreover, when it is unable to purchase energy from another microgrid, the microgrid purchases power from the main grid to balance the load demand [43].

3.1.1. Wind Energy System

Wind generation depends on both the wind speed and the height of the hub at a given location. The power–law equation [44] is used to calculate the wind speed at the hub height of WT using data collected at the anemometer height: $u(h)$ and $u(hg)$ are the wind speeds at the hub height (h) and the anemometer height (hg), respectively, and α is the roughness factor. The value changes from location to location and over time at the same location:

$$V(h) = V(h_g)\left(\frac{h}{h_g}\right)^{\alpha} \tag{5}$$

where $V(h)$ and $V(h_g)$ are the wind speeds at hub height h and anemometer height h_g. α is a roughness factor and varies from location to location and time to time.

The output power of WT from the typical WT curve is as follows [45]:

$$p_{wt} = \begin{cases} 0, & v_{wt} < v_{ci}, v_{wt} > v_{co} \\ p_r * \frac{v_{wt}^2 - v_{co}^2}{v_r^2 - v_{ci}^2}, & v_{ci} < v_{wt} < v_r \\ p_r, & v_{ci} < v_{wt} < v_{co} \end{cases} \quad (6)$$

where p_{wt} is the output power of the wind turbine, p_r is the rated output power of the wind turbine, v_{ci} is the cut in wind speed of WT, v_r is the rated wind speed of WT, and v_{co} is the cut off wind speed of the wind turbine.

3.1.2. PV Energy System

The solar radiation on a tilted surface (Ht) can be calculated using solar insolation, the ambient temperature, data from the PV panel manufacturer, the PV panel slope, and the site latitude and longitude [46,47]. The following equation [48] is used to compute the PV system's output power:

$$p_{pv}(t) = h_t(t) * PVA * \mu_c(t) \quad (7)$$

where $\mu_c(t)$ is the PV system's hourly generating efficiency, which may be calculated in terms of the cell temperature using Equation (8) [49]:

$$\mu_c(t) = \mu_{cr}[1 - \beta_t * (T_c(t) - T_{cr})]) \quad (8)$$

where β_t is the temperature coefficient, μ_{cr} and T_{cr} are the solar cell efficiency and temperature at maximum radiation solar flux $T_c(t)$, It is an hourly solar cell temperature at ambient temperature (T_a):

$$T_c(t) = T_a + \lambda h_t(t) \quad (9)$$

where λ (Ross coefficient) is a coefficient that represents how the temperature increases above ambient as solar flux increases. The overall output of the PV array is:

$$p_s = \left(\frac{G_{pv}}{G_o}\right) * p_r \quad (10)$$

where p_s is the PV array output power, G_{pv} is the irradiation of solar, G_o is the solar irradiance under standard test conditions (1000 W/m^2). p_r is the rated power of solar [49]. Table 2 represents the PV parameter values.

Table 2. PV panel parameters. Reproduced from [49], the (Journal of Energy storage): 2021.

Parameters	Values	Unit
G_o	1000	W/m^2
μ	20	%
$T_{M,O}$	25	°C
NOCT	44	°C
Π_{PV}	0	Cent/kWh

3.1.3. Battery Storage

The state of charge (SOC) of a battery is determined based on the energy balance between the wind, PV energy systems, and the load as given by the following equations after a particular time (t):

$$E_b(t) = E_b(t-1)(1-\sigma) + P_s * \eta_{bc} \quad (11)$$

where (11) is the battery charging mode equation:

$$E_b(t) = E_b(t-1)(1-\sigma) - \frac{p_d}{\eta_{bd}} \quad (12)$$

where (12) is the battery discharging mode equation. E_b is the energy of the battery bank, and η_{bc}, η_{bd} is the charging and discharging efficiency of the battery storage system. It is considered to be 90% and 85%, respectively, in [50], where σ is the battery self-discharge rate and is assumed to be 0.2% per day for most batteries [51]. Ps is the surplus power, and p_d is the deficit power.

The battery bank should always follow the following limitations:

$$E_{b,min} \leq E_b(t) \leq E_{b,max} \quad (13)$$

$$E_b(t) = E_b(t-1)(1-\sigma) \quad (14)$$

where $E_{b,max}$ and $E_{b,min}$ are the battery bank's maximum and minimum storage capacity, respectively. The following equation can be used to calculate $E_{b,min}$:

$$E_{b,min} = DOD * E_b \quad (15)$$

where E_b is the battery nominal storage capacity, and DOD is the depth of discharge of a battery opposite to the SOC of a battery.

4. Mathematical Modeling of the Proposed Approach

The proposed approach of this article is to optimize the power transfer between microgrids to minimize the overall cost of the system and each microgrid. For this purpose, a mathematical model is designed for a net load for each microgrid first. The proposed idea is represented in Figure 3, and the proposed work is represented by the flowchart shown in Figure 7.

The proposed approach has two main tasks: storage; the other is energy sharing in the first part, which is energy storage. If the net load is greater than zero, storage is used to discharge energy to meet the load requirements. If the net load is less than zero, the extra system energy is either shared with other microgrids or stored in the storage system to reduce the total cost of electricity. Data about the load, PV, and wind in each MG is calculated. After obtaining data of the load, PV, and wind in each MG, the sharing parameters are initialized, such as the *SOC* of the battery, the size of the battery, and the minimum and maximum generation in each MG, the net load is calculated using the generation and load balance equations.

The proposed work shown in Figure 7 is about energy management in small MGs and is an attempt to reduce imports from energy distribution networks as much as possible, by linking several small microgrids together. These MGs contain different mixtures of energy production and consumption, which helps to increase the reliability of these small MGs and gradually dispense with power distribution networks. The novelty is that the difference in the energy mix between these small MGs and linking them together will help dispense with distribution networks, which will reduce energy losses and the price of electricity and provides the possibility for small consumers to benefit from the production and sale of energy.

Suppose a microgrid has a battery energy storage system. In this case, it will have two possibilities if the net load exceeds zero. The *SOC* of a battery is checked to observe whether it is above 20%, and the battery will be discharged to meet the load requirements. Still, if the net load is not greater than zero, there is some extra energy in the system, which may be used to charge the battery if the *SOC* of a battery is less than the maximum storage capacity.

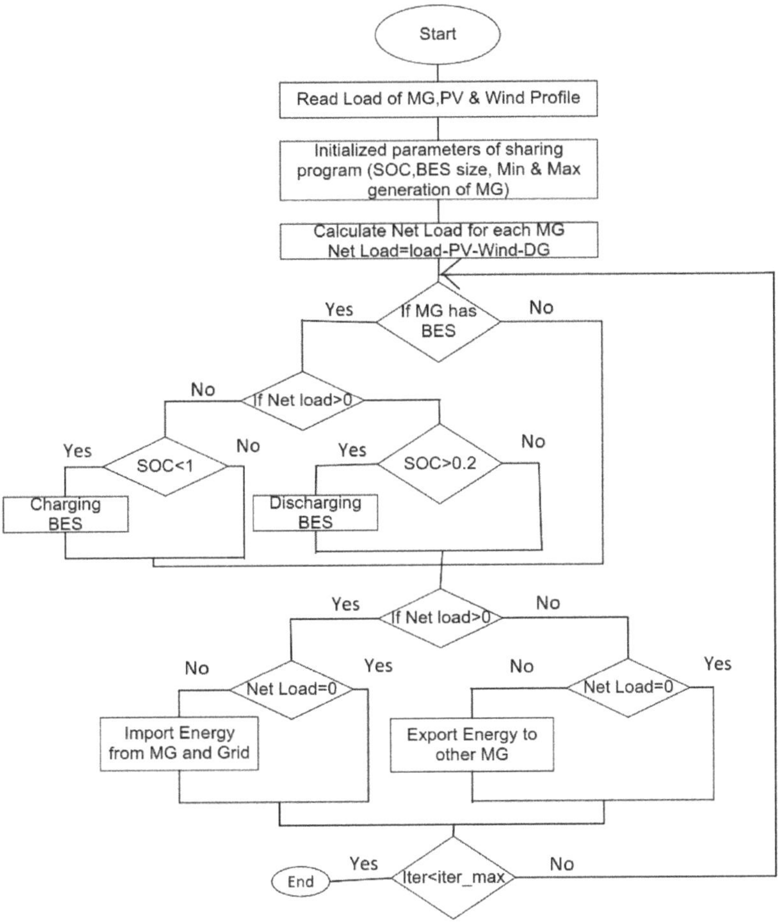

Figure 7. Flowchart of the proposed idea.

In the second task of the proposed approach, whether the net load is greater than or less than zero is checked. If it is greater than zero, the microgrid will import energy from other microgrids or the main grid. If the net load is less than zero, the microgrid will export energy to other microgrids or the main grid.

The above process is iterated until optimal power-sharing among the microgrids is achieved.

4.1. Microgrids' Net Load

The mathematical equation for *Disco* is:

$$Disco\ Net\ Load = L_D - G_D \tag{16}$$

where L_D is the load of the distribution company (*Disco*) and G_D is the conventional generation in *Disco*:

$$MG1\ Net\ Load = L_{MG1} - PV_{MG1} * PF_{PV} \tag{17}$$

where L_{MG1} is the load of $MG1$, PV_{MG1} is the rated power of PV in $MG1$, and PF_{PV} is the 24 h PV profile:

$$MG2\ Net\ Load = L_{MG2} - G_{D1} - G_{D2} - WT_{MG2} * PF_{WT} \tag{18}$$

where L_{MG2} is the load of $MG2$, G_{D1}, G_{D2} are the generation of conventional generators 1 and 2, WT_{MG2} is the wind turbine generation in $MG2$, and PF_{WT} is the wind profile of 24 h.

$$MG3\ Net\ Load = L_{MG3} - PV_{MG3} * PF_{PV} \tag{19}$$

L_{MG3} is the load of $MG3$, PV_{MG3} is the rated PV power, and PF_{PV} is the PV profile for 24 h.

$$MG4\ Net\ Load = L_{MG4} - WT_{MG4} * PF_{WT} \tag{20}$$

L_{MG4} is the load of $MG4$, WT_{MG4} is the wind turbine generation in $MG4$, PF_{WT} is the 24 h wind profile:

$$MG5\ Net\ Load = L_{MG5} - G_D - WT_{MG5} * PF_{WT} \tag{21}$$

where L_{MG5} is the load of $MG5$, G_D is the generation of a conventional generator, WT_{MG5} is the wind generation in $MG5$, and PF_{WT} is the 24 h wind profile.

The net load is the difference between the load and generation inside the MG itself. It is used to determine whether the microgrid has a shortage or excess of energy to import or export to other microgrids with a shortage of energy, store it in battery energy storage, or sell it to the main grid.

4.2. Energy Management Strategy

The proposed HRES management algorithm is described in the following strategy. If the amount of power generated by RES surpasses the amount needed to meet the load requirements, the excess power will be used to charge the batteries until they reach their maximum capacity, $E_{b,max}$. The extra power in the batteries will be used to power the dummy load, P_{dummy}. The reason behind this can be described as follows:

$$\begin{gathered} if\ P_g(t) > P_L(t)\ and\ SOC < E_{b,max}\ then; \\ P_{bc} = [P_w(t) - P_L(t) + P_{PV}(t)]\eta_{bc}\ Charging\ Process \end{gathered} \tag{22}$$

$$\begin{gathered} if\ P_g > P_L(t)\ and\ SOC \geq E_{b,max}\ then \\ P_{bc}(t) = 0 \\ P_d(t) = [P_w(t) - P_L(t) + P_{PV}(t)] \end{gathered} \tag{23}$$

where P_d is the power to the dummy load:

$$\begin{gathered} if\ P_w(t) < P_L(t)\ ,\ [P_w(t) + P_{PV}(t)] > P_L(t)\ and \\ SOC < E_{b,max}\ then \\ P_{bc}(t) = [P_{PV}(t) - (P_L(t) - P_w(t))]\eta_{bc}, \end{gathered} \tag{24}$$

Equation (24) shows the charging process:

$$\begin{gathered} if\ P_w(t) < P_L(t)\ and\ (P_w(t) + P_{PV}(t)) > P_L(t)\ and \\ SOC > E_{b,max}\ then; \\ P_{bc}(t) = 0 \\ P_d(t) = [P_w(t) - P_L(t) + P_{PV}(t)] \end{gathered} \tag{25}$$

If the load demand requires more power than RES can provide, the battery will meet the demand until it is reduced to its minimal level $E_{b,min}$. If there is still a power shortage, the DG will make up for the deficit load demand. The equations that describe this logic are as follows:

$$\begin{gathered} if\ [P_w(t) + P_{PV}(t)] < P_L(t)\ and\ SOC > E_{b,min}\ then; \\ P_{bd}(t) = \frac{[P_L(t) - P_w(t) - P_{PV}(t)]}{\eta_{bd}}, \end{gathered} \tag{26}$$

where $P_{bd}(t)$ is discharging power, and this is a discharging process:

$$\begin{gathered} if\ [P_{wt}(t) + P_{PV}(t)] < P_L(t)\ and\ SOC < E_{b,min}\ then; \\ P_{bd}(t) = 0 \end{gathered}$$

All the above equations and details of the generation, load, and storage indicate how the energy management strategy is effectively working in this case. When the generation of HRES in the microgrid is higher than the load, the surplus energy is stored in the battery energy storage system. When the generation of the HRES is less than the load of the microgrid, then the shortage of energy is met by the battery storage system. If the generation of HRES is less than the load and the battery energy is less than the minimum level of stored energy, then the microgrid will import energy from another microgrid to meet the demand. In the above equations, SOC is the state of the charge of the battery, P_{bc} is the power used to charge the battery, and P_{bd} is the discharging power of the battery. η_{bc}, η_{bd} is the battery charging and discharging efficiencies, respectively.

The power flow constraints of the HRES and battery are as follows:

$$0 \leq P_G \leq P_{max}\ Conventional\ Generation \tag{27}$$

$$0 \leq P_{PV} \leq P_{max}\ PV\ Generation \tag{28}$$

$$0 \leq P_{wt} \leq P_{max}\ WT\ Generation \tag{29}$$

$$P_{min} \leq P_b \leq P_{max}\ Battery\ power \tag{30}$$

$$SOC_{min} \leq SOC \leq SOC_{max}\ Battery\ State\ of\ Charge \tag{31}$$

4.3. Microgrid Energy Sharing Problem

The microgrid has an excess or shortage of energy. The microgrid with excess energy will export the energy to other microgrids, which need a power through-line between them $(S1, S2, S3, \ldots, S15)$. The microgrid with a shortage of energy will import the energy from other microgrids, which have an excess power through-line between them $(S1, S2, S3, \ldots, S15)$. If all microgrids do not have enough energy to cover the shortage in the system, the system will import energy from the main grid. If we consider a sequence of $1 : Disco, 2 : MG1, 3 : MG2, 4 : MG3, 5 : MG4, 6 : MG5$, if sending end $SN = 1$ and receiving end $RN = 2$, it means the line is connected from Disco to $MG1$ and so on.

5. Power Balance

The power balance in the generation and load is represented as:

$$System\ Load - P_{PV} - P_w - P_{con} - P_{Grid} = 0 \tag{32}$$

$$MG\ Load + Export\ Energy - MG\ Generation - Import\ Energy = 0 \tag{33}$$

Objective Function

The objective function is a cost reduction of the generated power, transfer power, and import and export power:

$$C_T = \sum_1^G C_G + C_{Dis} + C_{MG} + C_B + C_P \tag{34}$$

where C_G is the total conventional generation cost, C_{Dis} is the total transfer energy cost from Disco to MG, C_{MG} is the total transfer energy cost from MG to MG, C_B is the total energy cost from the battery, and C_P is the penalty cost for the unsupplied energy.

Mathematically, all of the above costs are represented as:

$$C_G = (G_{Dis} + G1_{MG2} + G2_{MG2} + G5_{MG5}) * P_{gas} \quad (35)$$

where G_{Dis} is conventional generation in Disco, $G1_{MG2}$ is conventional generation in $MG2$, $G2_{MG2}$ is conventional generation in $MG2$, $G5_{MG5}$ is conventional generation in $MG5$, and P_{gas} is the gas price:

$$C_{Dis} = (S1 + S2 + S3 + S4 + S5) * P_{Dis} \quad (36)$$

where $S1,\ S2,\ S3,\ S4,\ S5$ transfer power from Disco to $MG1,\ MG2,\ MG3,\ MG4,\ MG5$, respectively, and P_{Dis} is the energy cost of Disco:

$$C_{MG} = (S6 + S7 + S8 + \ldots + S15) * P_{MG} \quad (37)$$

where $(S6 + S7 + S8 + \cdots + S15)$ is the power transfer from MG to MG, and P_{MG} is the cost of energy from MG to MG:

$$C_B = (BES_{Disco} + BES_{MG2} + BES_{MG4} + BES_{MG5}) * P_B \quad (38)$$

In Equation (38), BES_{Disco} is the energy of the battery in Disco, BES_{MG2} is the energy of the battery in $MG2$, BES_{MG4} is the energy of the battery *in* $MG4$, BES_{MG5} is the energy of the battery in $MG5$, and P_B is the cost of energy from the battery:

$$C_P = E_{ns} * P_P \quad (39)$$

where E_{ns} is the unsupplied energy and P_P is the penalty price that is fixed [49].

Table 3 represents the cost data of different parameters (also taken from [49]).

Table 3. Cost data. Reproduced from [49], the (Journal of Energy Storage): 2021.

Parameters	Values	Unit
P_{gas}	5	Cent/kWh
P_{MG}	15.75	Cent/kWh
P_{Dis}	15.3	Cent/kWh
P_B	3	Cent/kWh
P_P	40	Cent/kWh

6. Simulations and Results Analysis

As discussed above, in all the mathematical formulations regarding the objective functions and microgrid components, an ABC optimization technique reduces the system's total cost and minimal sharing cost of all the microgrids. Microgrids consist of conventional generation and intermittent energy sources, such as PV, wind, and battery energy storage systems. Among all the sources, conventional generators produce 24 h electricity, and wind generation is wind speed dependent and PV dependent on solar irradiations. Conventional generation occurs in Disco, $MG2$, and $MG5$. If their capacity fails to meet the desired load demands, they will import energy from the main grid if there is an unavailability of energy from other microgrids.

There are different power-through lines between Disco and microgrids and from microgrids to microgrids. The lines connecting Disco and microgrids are $(S1, S2, S3, S4, S5)$, i.e., from $MG1$ *to* $MG5$. From $MG1$ *to* $(MG2,\ MG3,\ MG4,\ MG5)$, the lines are $(S6, S7, S8, S9)$; from $MG2$ *to* $(MG3,\ MG4,\ MG5)$, the lines are $(S10, S11, S12)$; from $MG3$ *to*$(MG4,\ MG5)$, *the lines are* $(S13, S14)$; and the last connection between $MG4$ *and* $MG5$ is $S15$. These lines from $S1, S2, \ldots, S15$ represent power transfer from Disco to MG and MGs to MGs. If the value of $S1$ *to* $S15$ is negative, power is transferred from the receiving to the sending end and vice versa if it is positive.

A cost convergence curve is shown below in Figure 8, which shows how the best solution deals with each iteration. We can see the best solution decreases when the iteration increases until it reaches the best one. The maximum number of iterations considered in the proposed idea is 150, upon which it converges to an optimal value.

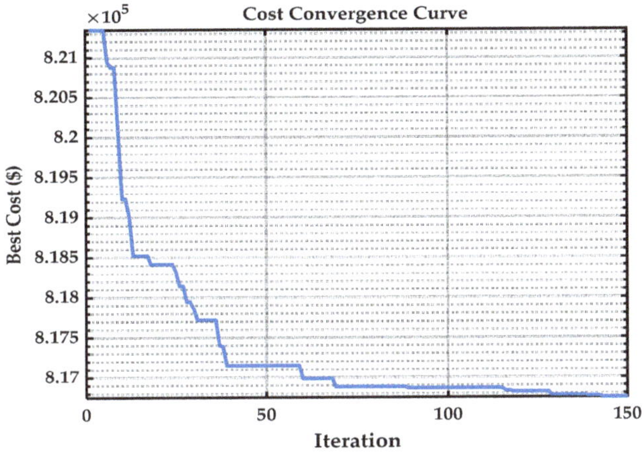

Figure 8. Cost convergence graph.

Regarding the MG with only PV generation source, Figure 9 shows that it can only generate power according to the solar irradiance data in the PVWT profile. For the rest of the hours, it imports energy from outside, and from 11 to 13 h, it sells energy as it exceeds the load demand. MG1 has only one PV generation source, and as per Equation (7), PV generation is dependent on thee tilted surface, PV array, and its efficiency. Maximum generation is achieved when maximum sunlight is available for the tilted surface, and it is generally from 7 to 19 h that power is generated. Moreover, the intensity of light is very high, from 11 to 13 h, during which the generated power reaches th maximum and is available to sell to other MGs or the main grid. In the absence of sunlight, MG1 will not generate electricity. Hence, it will purchase electricity from other MGs.

Microgrid 2 contains wind generation, battery storage, and 2 thermal units. The wind is almost always the available generation source, as shown in Figure 10. From 1–6 h, the load demand can easily be met by the wind generation source and thermal unit generation. During this time, extra power is used for charging the battery. From 7–8 h, EES is used to meet the load demand along with the wind and thermal generation sources. From 9–23 h, wind generation, thermal generation, and EES are not able to meet the load demand; hence, electricity is purchased from outside.

Microgrid 3 has one source of PV and load demand, as shown in Figure 11. To meet the load demand with just one PV source, it will purchase electricity from other microgrids in the absence of sunlight. When sunlight becomes available, it will gradually overcome the load demand and at high intensity, sunlight from 11–13 h MG3 will be able to meet the entire demand. The extra amount of electricity will be sold to other MGs.

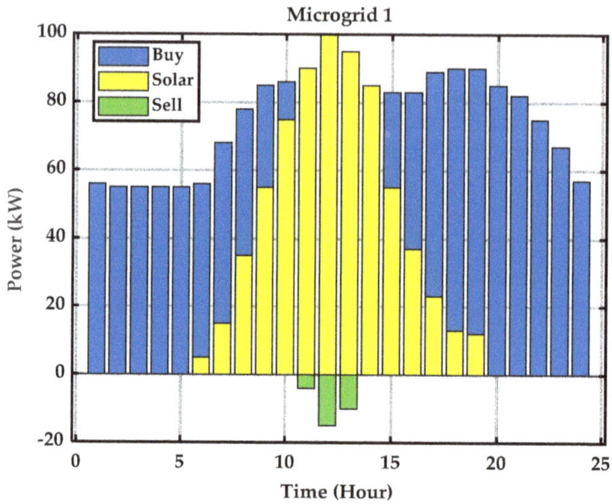

Figure 9. Microgrid 1.

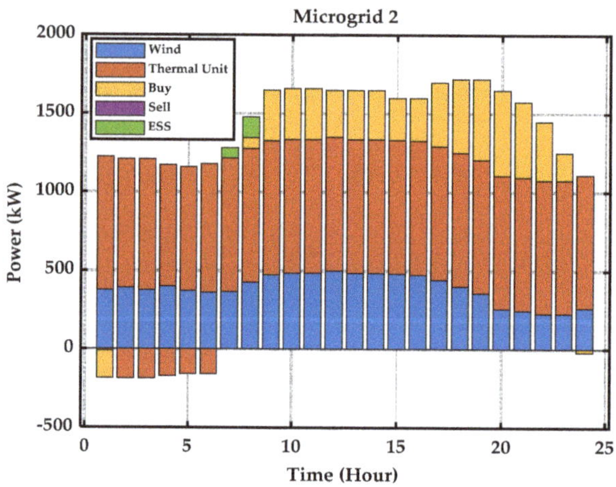

Figure 10. Microgrid 2 response.

MG4 contains wind generation and battery storage, and the load demand is shown in the Figure 12. The result in Figure 12 shows that from 1–6 h, wind generation is able to meet the required load demand and the extra amount of electricity will be used to charge the battery. Due to the intermittent nature of wind, it will purchase electricity if the wind generation is less than the load demand. As shown in Figure 12, from 7–10 h, the load demand exceeds wind generation; hence, EES is used to meet the load demand along with wind generation, and some energy is still required to meet the required demand, it will be purchased from other microgrids. From 11–24 h, wind generation cannot meet the required load demand; hence, the extra amount of electricity needed will be purchased from other microgrids.

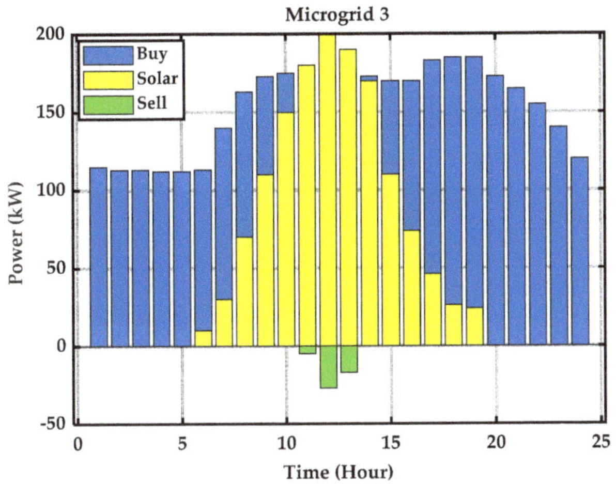

Figure 11. Microgrid 3 response.

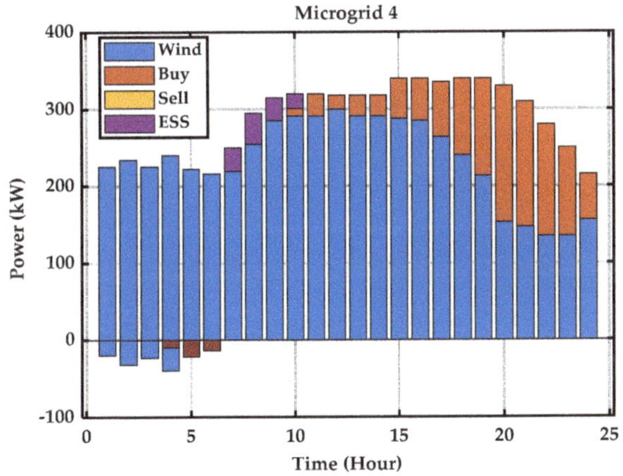

Figure 12. Load profile of MG4.

MG5 contains wind generation, battery storage, and one thermal unit. The load demand is shown in Figure 13. Wind generation is available 24 h as shown in the figure. From 1–6 h, the load demand is met by wind and thermal generation and the extra amount of electricity is used to charge the battery. From 7–8 h, the load demand exceeds the wind and thermal generation; hence, EES is used to meet the required demand. From 9–24 h, as the load demand is higher than the inside generation, the extra energy required to meet the demand is purchased from other microgrids.

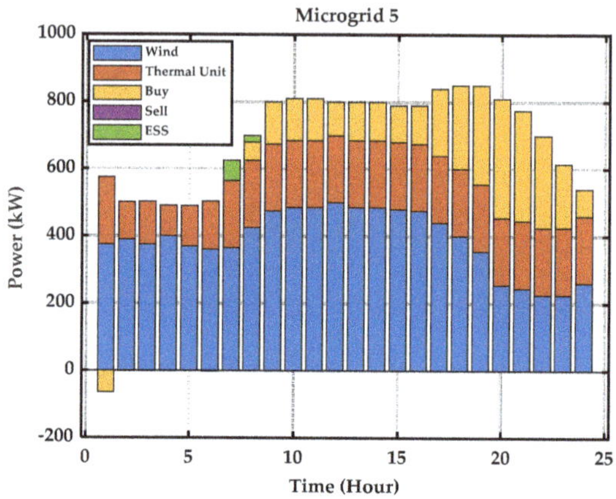

Figure 13. Load profile of MG5.

Disco contains battery storage and one thermal unit. The thermal generation is a constant source of generation as shown in Figure 14, and the load demand of Disco is shown in the figure. From 1–6 h, thermal generation alone cannot meet the load demand. Therefore, energy is purchased from other microgrids at the optimal rate and the battery is charged during this time. From 7–8 h, thermal generation and EES are both used, and the extra amount needed to meet the load demand is purchased from outside. From 9–24 h, thermal generation is used, but the generation is not sufficient to meet the required demand. Therefore, the extra amount is purchased from outside.

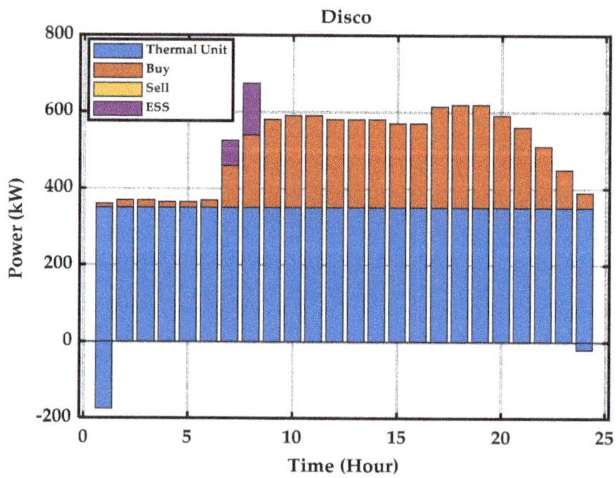

Figure 14. Load profile of Disco.

The cost optimization results of the microgrids and total systems are shown in Figure 15. It is clearly shown in the figure that the total cost of MG2, MG4, and MG5 is zero, which means that they do not import energy from other microgrids or the amount of electricity that is bought or sold is the same; hence, they compensate for their cost effects, which is why their cost is zero.

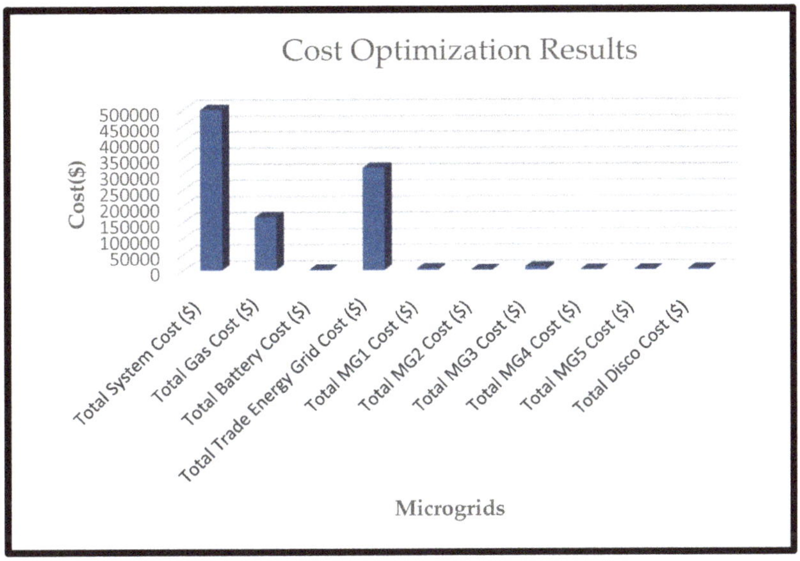

Figure 15. Cost optimization results.

7. Conclusions and Future Work

The microgrid has emerged as a new paradigm shift for global energy systems to replace remote centralized power plants with more efficient, localized, and distributed generation systems, especially in various cities and towns around the globe. It provides more stable, flexible, and energy-efficient solutions for the power grid so that an increasing number of loads can be handled without new infrastructure needing to be built. This work provides a new iterative ABC optimization method that addresses HRESs, such as solar, wind, GT, and battery. In particular, this work optimizes the power transmission between different microgrids to minimize the cost of each microgrid and the whole system while maintaining the load requirements and stability. The optimization problem was solved by the new iterative method ABC. This study also conducted several experiments with historical data to test the proposed method for the HRES model. The simulation results show that the proposed method is efficient in finding the best solution and adjusting the optimization parameters and constraints. This method can also be applied to any site with meteorological data and any WT with technical information.

Future research will focus on the use of hybrid swarm intelligence systems, such as the particle swarm algorithm and the artificial bee colony algorithm or the differential evolution algorithm, for the economic dispatch of microgrids. In addition, flexible switching techniques for real-time scheduling in microgrids that consider more constraint situations is an interesting future research direction.

Author Contributions: Conceptualization, K.U. and Q.J.; methodology, software and validation, K.U. formal analysis, investigation and resources, R.A.K.; writing—original draft preparation, K.U.; writing—review and editing, G.G.; visualization, S.R.; supervision, Q.J. All authors have read and agreed to the published version of the manuscript.

Funding: This research received no external funding.

Institutional Review Board Statement: Not available.

Informed Consent Statement: Not applicable.

Data Availability Statement: The data that support the study's findings, such as numerical simulation, model, or code generated or used during the study, are available upon request from the journal and the corresponding author.

Acknowledgments: Thanks to the China Scholarship Council, Zhejiang University, Hangzhou, China (www.zju.edu.cn, accessed on 18 January 2022) and the University of Science & Technology Bannu, Pakistan (www.ustb.edu.pk, accessed on 18 January 2022) for providing us great research environment during this work.

Conflicts of Interest: The corresponding author declares that there is no contradiction on behalf of all authors.

References

1. Mohammed, O.H.; Amirat, Y.; Benbouzid, M.E.H.; Feld, G. Optimal Design and Energy Management of a Hybrid Power Generation System Based on Wind/Tidal/PV Sources: Case Study for the Ouessant French Island. In *Smart Energy Grid Design for Island Countries: Challenges and Opportunities*; Springer: Berlin/Heidelberg, Germany, 2017; ISBN 9783319501970.
2. Child, M.; Bogdanov, D.; Breyer, C. The role of storage technologies for the transition to a 100% renewable energy system in Europe. *Energy Procedia* **2018**, *155*, 44–60. [CrossRef]
3. Aslam, S.; Herodotou, H.; Mohsin, S.M.; Javaid, N.; Ashraf, N.; Aslam, S. A survey on deep learning methods for power load and renewable energy forecasting in smart microgrids. *Renew. Sustain. Energy Rev.* **2021**, *144*, 110992. [CrossRef]
4. Zhang, D.; Mu, S.; Chan, C.C.; Zhou, G.Y. Optimization of renewable energy penetration in regional energy system. *Energy Procedia* **2018**, *152*, 922–927. [CrossRef]
5. Stroe, D.I.; Zaharof, A.; Iov, F. Power and energy management with battery storage for a hybrid residential PV-wind system—A case study for Denmark. *Energy Procedia* **2018**, *155*, 464–477. [CrossRef]
6. Zhou, Z.; Benbouzid, M.E.H.; Charpentier, J.F.; Scuiller, F. Hybrid Diesel/MCT/Battery Electricity Power Supply System for Power Management in Small Island. In *Smart Energy Grid Design for Island Countries: Challenges and Opportunities*; Springer: Berlin/Heidelberg, Germany, 2017; ISBN 9783319501970.
7. Mohammed, O.H.; Amirat, Y.; Benbouzid, M.; Elbast, A. Optimal Design of a PV/Fuel Cell Hybrid Power System for the City of Brest in France. In Proceedings of the 2014 First International Conference on Green Energy ICGE 2014, Sfax, Tunisia, 25–27 March 2014.
8. Zhang, J.; Huang, L.; Shu, J.; Wang, H.; Ding, J. Energy Management of PV-Diesel-Battery Hybrid Power System for Island Stand-Alone Micro-Grid. *Energy Procedia* **2017**, *105*, 2201–2206. [CrossRef]
9. Khursheed, A.; Aslam, S.; Haider, S.I.; Mohsin, S.M.; ul Islam, S.; Khattak, H.A.; Shah, S. Energy forecasting using multiheaded convolutional neural networks in efficient renewable energy resources equipped with energy storage system. *Trans. Emerg. Telecommun. Technol.* **2019**, e3837. [CrossRef]
10. Toutounchi, A.N.; Seyedshenava, S.; Contreras, J.; Akbarimajd, A. A Stochastic Bilevel Model to Manage Active Distribution Networks with Multi-Microgrids. *IEEE Syst. J.* **2019**, *13*, 4190–4199. [CrossRef]
11. Aboli, R.; Ramezani, M.; Falaghi, H. A hybrid robust distributed model for short-term operation of multi-microgrid distribution networks. *Electr. Power Syst. Res.* **2019**, *177*, 106011. [CrossRef]
12. Lasseter, R.; Abbas, A.; Marnay, C.; Stevens, J.; Dagle, J.; Guttromson, R.; Meliopoulos, A.S.; Yinger, R.; Eto, J. Integration of Distributed Energy Resources: The CERTS Microgrid Concept. *Consult. Rep. Prep. Calif. Energy Comm.* **2003**, 89.
13. LI, Y.; Nejabatkhah, F. Overview of control, integration and energy management of microgrids. *J. Mod. Power Syst. Clean Energy* **2014**, *2*, 212–222. [CrossRef]
14. Hatziargyriou, N.; Jenkins, N.; Strbac, G.; Pecas Lopes, J.A.; Ruela, J.; Engler, A.; Oyarzabal, J.; Kariniotakis, G.; Amorim, A. Microgrids—Large scale integration of microgeneration to low voltage grids. In Proceedings of the 41st International Conference on Large High Voltage Electric Systems 2006, CIGRE 2006, Paris, France, 27 August–1 September 2006.
15. Li, M.; Zhang, X.; Li, G.; Jiang, C. A feasibility study of microgrids for reducing energy use and GHG emissions in an industrial application. *Appl. Energy* **2016**, *176*, 138–148. [CrossRef]
16. Chaudhary, G.; Lamb, J.J.; Burheim, O.S.; Austbø, B. Review of energy storage and energy management system control strategies in microgrids. *Energies* **2021**, *14*, 4929. [CrossRef]
17. Zhou, B.; Zou, J.; Chung, C.Y.; Wang, H.; Liu, N.; Voropai, N.; Xu, D. Multi-Microgrid Energy Management Systems: Architecture, Communication, and Scheduling Strategies. *J. Mod. Power Syst. Clean Energy* **2021**, *9*, 463–476. [CrossRef]
18. Raghav, L.P.; Rangu, S.K.; Dhenuvakonda, K.R.; Singh, A.R. Optimal energy management of microgrids-integrated nonconvex distributed generating units with load dynamics. *Int. J. Energy Res.* **2021**, *45*, 18919–18934. [CrossRef]
19. Restrepo, M.; Cañizares, C.A.; Simpson-Porco, J.W.; Su, P.; Taruc, J. Optimization- and Rule-Based Energy Management Systems at the Canadian Renewable Energy Laboratory Microgrid Facility. *Appl. Energy* **2021**, *290*, 116760. [CrossRef]
20. Roslan, M.F.; Hannan, M.A.; Jern Ker, P.; Begum, R.A.; Indra Mahlia, T.M.; Dong, Z.Y. Scheduling controller for microgrids energy management system using optimization algorithm in achieving cost saving and emission reduction. *Appl. Energy* **2021**, *292*, 116883. [CrossRef]

21. Li, F.F.; Qiu, J. Multi-objective optimization for integrated hydro-photovoltaic power system. *Appl. Energy* **2016**, *167*, 377–384. [CrossRef]
22. Askarzadeh, A. A Memory-Based Genetic Algorithm for Optimization of Power Generation in a Microgrid. *IEEE Trans. Sustain. Energy* **2018**, *9*, 1081–1089. [CrossRef]
23. Obara, S.; Kawai, M.; Kawae, O.; Morizane, Y. Operational planning of an independent microgrid containing tidal power generators, SOFCs, and photovoltaics. *Appl. Energy* **2013**, *102*, 1343–1357. [CrossRef]
24. De Lannoy, G.J.M.; Reichle, R.H.; Peng, J.; Kerr, Y.; Castro, R.; Kim, E.J.; Liu, Q. Converting between SMOS and SMAP Level-1 Brightness Temperature Observations over Nonfrozen Land. *IEEE Geosci. Remote Sens. Lett.* **2015**, *12*, 1908–1912. [CrossRef]
25. Samuel, O.; Javaid, N.; Khalid, A.; Khan, W.Z.; Aalsalem, M.Y.; Afzal, M.K.; Kim, B.S. Towards Real-Time Energy Management of Multi-Microgrid Using a Deep Convolution Neural Network and Cooperative Game Approach. *IEEE Access* **2020**, *8*, 161377–161395. [CrossRef]
26. Sedighizadeh, M.; Fazlhashemi, S.S.; Javadi, H.; Taghvaei, M. Multi-objective day-ahead energy management of a microgrid considering responsive loads and uncertainty of the electric vehicles. *J. Clean. Prod.* **2020**, *267*, 121562. [CrossRef]
27. Arefifar, S.A.; Ordonez, M.; Mohamed, Y.A.R.I. Energy Management in Multi-Microgrid Systems—Development and Assessment. *IEEE Trans. Power Syst.* **2017**, *32*, 910–922. [CrossRef]
28. Baboli, P.T.; Shahparasti, M.; Moghaddam, M.P.; Haghifam, M.R.; Mohamadian, M. Energy management and operation modelling of hybrid AC-DC microgrid. *IET Gener. Transm. Distrib.* **2014**, *8*, 1700–1711. [CrossRef]
29. Hasankhani, A.; Hakimi, S.M. Stochastic energy management of smart microgrid with intermittent renewable energy resources in electricity market. *Energy* **2021**, *219*, 119668. [CrossRef]
30. Querini, P.L.; Chiotti, O.; Fernádez, E. Cooperative energy management system for networked microgrids. Sustain. *Energy Grids Netw.* **2020**, *23*, 100371. [CrossRef]
31. Lan, T.; Jermsittiparsert, K.; Alrashood, S.T.; Rezaei, M.; Al-Ghussain, L.; Mohamed, M.A. An advanced machine learning based energy management of renewable microgrids considering hybrid electric vehicles' charging demand. *Energies* **2021**, *14*, 569. [CrossRef]
32. Ghavifekr, A.A. Application of heuristic techniques and evolutionary algorithms in microgrids optimization problems. In *Microgrids*; Power Systems; Springer International Publishing: Cham, Switzerland, 2021.
33. Chen, S.M.; Sarosh, A.; Dong, Y.F. Simulated annealing based artificial bee colony algorithm for global numerical optimization. *Appl. Math. Comput.* **2012**, *219*, 3575–3589. [CrossRef]
34. Zhang, C.; Zheng, J.; Zhou, Y. Two modified Artificial Bee Colony algorithms inspired by Grenade Explosion Method. *Neurocomputing* **2015**, *151*, 1198–1207. [CrossRef]
35. Mernik, M.; Liu, S.H.; Karaboga, D.; Črepinšek, M. On clarifying misconceptions when comparing variants of the Artificial Bee Colony Algorithm by offering a new implementation. *Inf. Sci.* **2015**, *291*, 115–127. [CrossRef]
36. Li, L.; Yao, F.; Tan, L.; Niu, B.; Xu, J. A novel DE-ABC-based hybrid algorithm for global optimization. In Proceedings of the Lecture Notes in Computer Science (Including Subseries Lecture Notes in Artificial Intelligence and Lecture Notes in Bioinformatics), Zhengzhou, China, 11–14 August 2011; Volume 6840.
37. Mao, W.; Lan, H.Y.; Li, H.R. A New Modified Artificial Bee Colony Algorithm with Exponential Function Adaptive Steps. *Comput. Intell. Neurosci.* **2016**, *2016*, 9820294. [CrossRef] [PubMed]
38. Satapathy, S.C.; Naik, A. Modified Teaching-Learning-Based Optimization algorithm for global numerical optimization—A comparative study. *Swarm Evol. Comput.* **2014**, *16*, 28–37. [CrossRef]
39. Xiang, Y.; Peng, Y.; Zhong, Y.; Chen, Z.; Lu, X.; Zhong, X. A particle swarm inspired multi-elitist artificial bee colony algorithm for real-parameter optimization. *Comput. Optim. Appl.* **2014**, *57*, 493–516. [CrossRef]
40. Setiawan, A.A.; Zhao, Y.; Nayar, C.V. Design, economic analysis and environmental considerations of mini-grid hybrid power system with reverse osmosis desalination plant for remote areas. *Renew. Energy* **2009**, *34*, 374–383. [CrossRef]
41. Wang, C.; Nehrir, M.H. Power management of a stand-alone wind/photovoltaic/fuel cell energy system. *IEEE Trans. Energy Convers.* **2008**, *23*, 957–967. [CrossRef]
42. Eltamaly, A.M.; Mohamed, M.A. A novel design and optimization software for autonomous PV/wind/battery hybrid power systems. *Math. Probl. Eng.* **2014**, *2014*, 637174. [CrossRef]
43. Bernal-Agustín, J.L.; Dufo-López, R. Simulation and optimization of stand-alone hybrid renewable energy systems. *Renew. Sustain. Energy Rev.* **2009**, *13*, 2111–2118. [CrossRef]
44. Eltamaly, A.M.; Addoweesh, K.E.; Bawah, U.; Mohamed, M.A. New software for hybrid renewable energy assessment for ten locations in Saudi Arabia. *J. Renew. Sustain. Energy* **2013**, *5*, 033126. [CrossRef]
45. Sreeraj, E.S.; Chatterjee, K.; Bandyopadhyay, S. Design of isolated renewable hybrid power systems. *Sol. Energy* **2010**, *84*, 1124–1136. [CrossRef]
46. Mohamed, M.A.; Eltamaly, A.M.; Alolah, A.I. Sizing and techno-economic analysis of stand-alone hybrid photovoltaic/wind/diesel/battery power generation systems. *J. Renew. Sustain. Energy* **2015**, *7*, 063128. [CrossRef]
47. Eltamaly, A.M.; Mohamed, M.A. A novel software for design and optimization of hybrid power systems. *J. Braz. Soc. Mech. Sci. Eng.* **2016**, *38*, 1299–1315. [CrossRef]
48. Habib, M.A.; Said, S.A.M.; El-Hadidy, M.A.; Al-Zaharna, I. Optimization procedure of a hybrid photovoltaic wind energy system. *Energy* **1999**, *24*, 919–929. [CrossRef]

49. Javanmard, B.; Tabrizian, M.; Ansarian, M.; Ahmarinejad, A. Energy management of multi-microgrids based on game theory approach in the presence of demand response programs, energy storage systems and renewable energy resources. *J. Energy Storage* **2021**, *42*, 102971. [CrossRef]
50. Mohamed, M.A.; Eltamaly, A.M.; Alolah, A.I. Swarm intelligence-based optimization of grid-dependent hybrid renewable energy systems. *Renew. Sustain. Energy Rev.* **2017**, *77*, 515–524. [CrossRef]
51. Yang, H.; Zhou, W.; Lu, L.; Fang, Z. Optimal sizing method for stand-alone hybrid solar-wind system with LPSP technology by using genetic algorithm. *Sol. Energy* **2008**, *82*, 354–367. [CrossRef]

energies MDPI

Article

Multi-Objective Immune-Commensal-Evolutionary Programming for Total Production Cost and Total System Loss Minimization via Integrated Economic Dispatch and Distributed Generation Installation

Mohd Helmi Mansor [1,*], Ismail Musirin [2] and Muhammad Murtadha Othman [2]

1 Department of Electrical and Electronics Engineering, College of Engineering, Universiti Tenaga Nasional, Kajang 43000, Selangor, Malaysia
2 School of Electrical Engineering, College of Engineering, Universiti Teknologi MARA, Shah Alam 40450, Selangor, Malaysia; ismailbm@uitm.edu.my (I.M.); m_murtadha@uitm.edu.my (M.M.O.)
* Correspondence: mhelmi@uniten.edu.my

Abstract: Economic Dispatch (ED) problems have been solved using single-objective optimization for so long, as Grid System Operators (GSOs) previously only focused on minimizing the total production cost. In modern power systems, GSOs require not only optimizing the total production cost but also, at the same time, optimizing other important objectives, such as the total emissions of the greenhouse gasses, total system loss and voltage stability. This requires a suitable multi-objective optimization approach in ensuring the ED solution produced is satisfying all the objectives. This paper presents a new multi-objective optimization technique termed Multi-Objective Immune-Commensal-Evolutionary Programming (MOICEP) for minimizing the total production cost and total system loss via integrated Economic Dispatch and Distributed Generation installation (ED-DG). This involved the application of a weighted-sum multi-objective approach that combined with an optimization technique called Immune-Commensal-Evolutionary Programming (ICEP). The proposed MOICEP has been compared with other multi-objective techniques, which are Multi-Objective-Evolutionary Programming (MOEP) and Multi-Objective-Artificial Immune System (MOAIS). It was found that MOICEP performs very well in producing better optimization results for all the three types of Economic Dispatch (ED) problems compared to MOEP and MOAIS in terms of cheap total production costs and low total system loss.

Keywords: multi-objective optimization; economic dispatch; distributed generation installation; evolutionary programming; artificial immune system; symbiotic organisms search; commensalism

Citation: Mansor, M.H.; Musirin, I.; Othman, M.M. Multi-Objective Immune-Commensal-Evolutionary Programming for Total Production Cost and Total System Loss Minimization via Integrated Economic Dispatch and Distributed Generation Installation. *Energies* **2021**, *14*, 7733. https://doi.org/10.3390/en14227733

Academic Editors: Victor Becerra and Ahmed F. Zobaa

Received: 1 October 2021
Accepted: 8 November 2021
Published: 18 November 2021

1. Introduction

In the early years after the introduction of Economic Dispatch (ED), most of the ED problems were solved based on single optimization [1]. This means that only the total production cost was optimized to get the minimum value. Nowadays, researchers and engineers are interested in solving ED problems with multiple objectives [2]. Besides the total production cost, the total greenhouse gasses emissions and total system loss can also be optimized at the same time. The reason for this multi-objective optimization introduced to solve ED problems is that greenhouse gasses emissions from fossil-fueled generating units and total system loss are also grid system operators' main concerns in modern electrical power systems [3]. There are two popular approaches to solve multi-objective problems, which are the weighted-sum and Pareto optimal [4]. The weighted-sum approach has been widely used to solve multi-objective problems due to its simplicity, while Pareto optimal solution is more complicated but could give a better nondominated solution for certain optimization problems [5].

Numerous multi-objective ED problems have been solved by researchers and engineers for the past ten years. Thenmalar et al. [3] solved Economic Dispatch and Economic Emission Dispatch using multi-objective optimization. Two objective functions are optimized by them, which are minimizing the total production cost and minimizing the emission of nitrogen oxides (NO_x). Quadratic programming by Wolfe's method was chosen to solve the multi-objective ED problem. M. Musau et al. [6] proposed to solve the multi-objective ED problem by considering the presence of renewable energy. They optimized both the renewable cost and total production cost functions by combining them to be solved using one single-objective function. Roy et al. [7] optimized the total production cost and total emissions simultaneously. In this study, a Normal Boundary Intersection (NBI)-based decomposition scheme was utilized to implement multi-objective optimization. In Reference [8], Dash and Mohanty applied the weighted-sum approach to solve a multi-objective ED problem. There three objectives were solved simultaneously using a simulated annealing optimization technique. The objectives were to minimize the total production cost, to minimize the nitrous oxide (NO_x) emissions and to improve the security of the transmission lines. Therefore, there were three weights that were associated with the three fitness function equations. Meiqin et al. [9] also solved three objectives of the ED problem, which were the total production cost, consumer outage cost and total emissions. They used weighted-sum integrated with a membership degree of the fuzzy set theory to solve the multi-objective ED problem. Man-im et al. [10] applied the Pareto solution instead of the weighted-sum approach to solve a multi-objective ED problem that simultaneously minimized the total production cost and power system security risk. The Pareto optimal solution approach was associated with the optimization technique and termed Non-dominated Sorting Particle Swarm Optimization (NSPSO). In Reference [11], minimizing the total system loss was also considered as one of the ED problem objectives, together with minimizing the total production cost and minimizing the NO_x emissions. Minimizing the total production cost and total system loss are more crucial than minimizing the total emissions in a modern power system that consists of substantial amounts of renewable energy sources. The emissions and fuel costs were optimized simultaneously using the Pareto optimal approach in Reference [12]. In this study, the multi-objective results of ED were compared with single-objective results. It was found that the results of the single-objective optimization were better than the multi-objective optimization in terms of producing the lowest fuel costs. However, in the multi-objective optimization, two objectives were achieved, which were to minimize the fuel costs and emissions. This is far better compared to only one objective being achieved. In a later study, Huang et al. [13] optimized three objective functions while solving for ED. The objective functions were the total fuel consumption function, oxynitride emission function and electricity purchase cost function. The ED problem was solved using a hybrid optimization technique termed the Hybrid Intelligent Algorithm. The technique is based on Particle Swarm Optimization (PSO) and Artificial Fish Swarm Algorithm (AFSA). From this review, it was discovered that multi-objective optimization is more powerful than single-objective optimization in solving ED problems, as more objectives can be optimized using multi-objective optimization at one time.

Single-objective optimization has its limitation in terms of satisfying more than one ED problem's objectives. For instance, some GSOs prefer to minimize both the total system loss and total production cost or to minimize the total production cost and emissions at the same time. Furthermore, GSOs will have flexibility in solving the ED problem of their power systems. Instead of solving the ED problem with only one objective, they can add other important objectives to be optimized concurrently. In previous practices, a power system problem was solved independently without integrating with other power system problems. For instance, the ED problem was solved independently to find the optimal production cost, and the Distributed Generation (DG) installation problem was solved to find the optimal location and sizing of DG units in a power system so that the system loss is the minimum. This practice is not efficient for GSOs to solve various problems of their power systems, especially in modern power systems.

This paper presents the implementation of a multi-objective optimization technique termed Multi-Objective Immune-Commensal-Evolutionary Programming (MOICEP) for solving integrated ED-DG problems. Two power system problems (ED problem and DG installation problem) are solved concurrently using MOICEP. Furthermore, two important objectives in power system optimization, which are the total production cost and total system loss, are chosen to be optimized simultaneously using the weighted-sum multi-objective approach with the help of ICEP as the optimizer. The weight coefficients of the two objectives are varied to observe and analyze the multi-objective optimization solution produced by MOICEP. There are three ED problems solved in this study, which are the basic ED problem (smooth cost function), ED problem with prohibited operating zones and ED problem with valve-point loading effect. Each of these ED problems is solved with the DG installation problem of three DG units. It was decided to install three DG units in this study based on the sizes of the test systems. It was assumed that the more DG units in the system, the lower the total production cost of the power system. The solutions for these integrated problems are the optimal output of the conventional generating units and the optimal location and sizing of the DG units. The proposed MOICEP technique is validated by implementing it on the IEEE 30-Bus RTS and IEEE 57-Bus RTS. Besides that, the solution of the integrated ED-DG problems produced by MOICEP is compared with the solution produced by the existing multi-objective techniques of Multi-Objective-Evolutionary Programming (MOEP) and Multi-Objective-Artificial Immune System (MOAIS) for comparative studies purposes.

2. Proposed Weighted-Sum Multi-Objective Immune-Commensal-Evolutionary Programming for Total Production Cost and Total System Loss Minimization

In this study, the total production cost and total system loss are optimized simultaneously using the weighted-sum Multi-Objective Immune-Commensal-Evolutionary Programming (MOICEP) technique. The total production is calculated using Equation (1) for the basic ED problems and ED problems with prohibited operating zones, while Equation (2) is used to calculate the total production cost for the ED problems with a valve-point loading effect. The total system loss is calculated using Equation (3).

$$C_{total} = \sum_{i=1}^{n} C_i(P_i) = \sum_{i=1}^{n} a_i P_i^2 + b_i P_i + c_i \quad (1)$$

where

C_i is the production cost of ith generating unit, and
P_i is the real power output of the ith generating unit.
a_i, b_i and c_i are the cost coefficients of the ith generating unit and
n is the number of dispatchable generating units:

$$C_{total} = \sum_{i=1}^{n} C_i(P_i) = \sum_{i=1}^{n} a_i P_i^2 + b_i P_i + c_i + \left| e_i sin\left(f_i \left(P_i^{min} - P_i \right) \right) \right| \quad (2)$$

where e_i and f_i are the valve-point loading effect coefficients of the ith generating unit.

$$P_{loss} = \sum_{k=1}^{l} g_k [V_i^2 + V_j^2 - 2V_i V_j cos(\delta_i - \delta_j)],\ k \in \{1, 2, \ldots, l\} \quad (3)$$

where

g_k is the conductance of kth line,
V_i and δ_i are the voltage magnitude and angle of bus i, respectively,
V_j and δ_j are the voltage magnitude and angle of bus j, respectively, and
l is the number of lines in the system.

The implementation of this proposed technique is illustrated in Figure 1. The procedure of minimizing the total production costs and total system loss using MOICEP shown in Figure 1 is explained in the following steps.

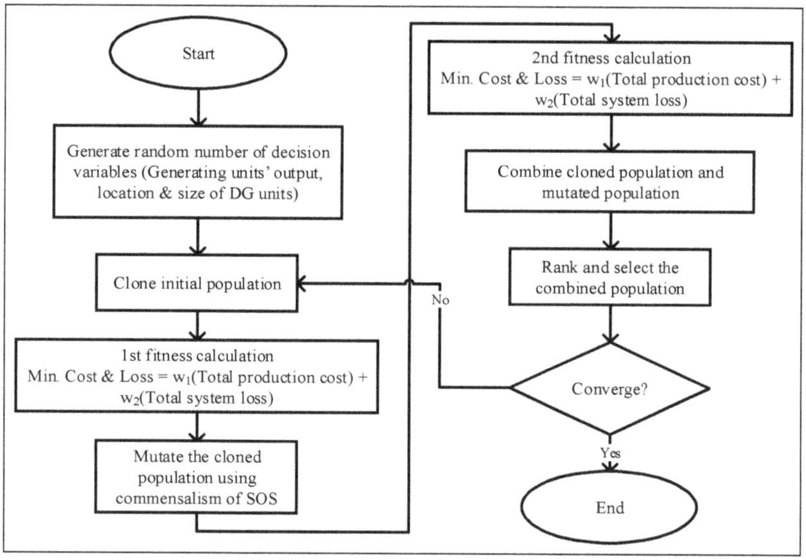

Figure 1. Overall MOICEP processes for the total production costs and total system loss minimization.

Step 1: Generate decision variable

MOICEP starts with the randomization of the decision variable values based on their constraints. There are three types of decision variables for this problem: real power output of generating units, location of DG units and size of DG units. This randomization is a part of the initialization process of MOICEP. Twenty individuals of each decision variable are stored in the initialization pool that satisfied the constraint violation test. The group of individuals in the pool is called as the initial population.

Step 2: Cloning process

Subsequently, the initial population is cloned to increase its size to two hundred. The fitness of this multi-objective is then calculated using Equation (4). The weight coefficients are varied depending on the problem's desired output. For this study, there were five settings of weight coefficients used to minimize the total production costs and total system loss, which are $w_1 = 0.9$, $w_2 = 0.1$, $w_1 = 0.5$, $w_2 = 0.5$, and $w_1 = 0.1$, $w_2 = 0.9$.

$$F(x) = w_1 f_1(x) + w_2 f_2(x) \quad (4)$$

where

$F(x)$ is weighted objective function,
w_1 is weight coefficient for the first objective function,
w_2 is weight coefficient for the second objective function
$f_1(x)$ is the first objective function (total production cost minimization) and
$f_2(x)$ is the second objective function (total system loss minimization).

Step 3: Mutation process

Then, the best two hundred cloned individuals that produce the best value of fitness are mutated using the commensalism equation of the Symbiotic Organisms Search (SOS), as shown in Equation (5). This mutation process is conducted to produce offspring of the mutated population. There are two hundred individual offspring after this process.

$$X_{i,new} = X_i + rand(-1,1) \times (X_{best} - X_j) \quad (5)$$

where

$X_{i,new}$ is the offspring of the ith individual,
X_i is the parent of the ith individual,
X_{best} is the fittest individual and
X_j is another individual besides the ith individual.

Step 4: Combination process

The cloned population and the mutated populated are then combined to become a large population with the size of four hundred.

Step 5: Ranking and selection process

From these four hundred individuals, the twenty best individuals will be selected to undergo the convergence process. However, prior to that, the combined individuals are ranked based on their fitness. Those who give the best value of fitness will be included into the best twenty group; otherwise, they will be excluded from undergoing the next process.

Step 6: Convergence test

Subsequently, the best twenty individuals will undergo the convergence process. In this convergence process, a stopping criterion is set to indicate that the algorithm has found the global optima of the multi-objective optimization problem. The stopping criterion used is as shown in Equation (6). If the stopping criterion is not met, the process will proceed to the cloning process and repeat the previous whole steps until the solution converges.

$$fitness_{max} - fitness_{min} \leq 0.0001 \quad (6)$$

where

$fitness_{max}$ is the maximum fitness value and
$fitness_{min}$ is the minimum fitness value.

3. Results and Discussion

This section presents the results and discussion on Multi-Objective Immune-Commensal-Evolutionary Programming (MOICEP) for the total production costs and total system loss minimization via integrated Economic Dispatch and Distributed Generation installation (ED-DG). As mentioned previously, the weight coefficients of the objective functions can be varied from 0 to 1.0 between the two objective functions (total production cost minimization and total system loss minimization) to find the trade-off between the two objective functions. The set of weight coefficients for the two fitness functions are as shown in Table 1. The objective functions for this multi-objective problem are the total production cost and total system loss minimization. It can be seen from the table that the summation of w_1 and w_2 is equal to one. If the weight of one of the fitness functions is set to one, then the weight for the other fitness function will be zero. This means that only the objective function with the value of one is optimized, and the other fitness function is not optimized. If the weight value is more than zero, it means that the fitness function is optimized with the amount of the weight. The difference between the fitness and objective functions needs to be understood. Fitness is the equation for the problem formulation, while the objective function is the minimization or the maximization of the fitness function. Let us say, our fitness function $f_1 = f_1(x)$; thus, the objective function will be $max\ (f_1(x))$ or $min\ (f_1(x))$.

Table 1. Weight coefficients values for two objective functions (total production cost and total system loss).

Setting	Weight Coefficients		Fitness Function	
	w_1	w_2	Total Production Cost	Total System Loss
1	1	0	✓	×
2	0.9	0.1	✓	✓
3	0.8	0.2	✓	✓
4	0.7	0.3	✓	✓
5	0.6	0.4	✓	✓
6	0.5	0.5	✓	✓
7	0.4	0.6	✓	✓
8	0.3	0.7	✓	✓
9	0.2	0.8	✓	✓
10	0.1	0.9	✓	✓
11	0	1.0	×	✓

As suggested by N. A. Rahmat [14], the best settings of the weight coefficients (trade-off) for optimizing the two objective functions were settings 2, 6 and 10, as shown in Table 1. This was because the other settings produced the same nondominated values as settings 2, 6 and 11. Therefore, in this study, the weighted-sum Multi-Objective Immune-Commensal-Evolutionary Programming (MOICEP) technique applied settings 2, 6 and 10 for minimizing the total production costs and total system loss minimization simultaneously.

The overall implementation of the MOICEP is illustrated in Figure 2. The proposed technique has been used to minimize the total production costs and total system loss for three different ED problems. The problems are a basic ED problem, ED problem with prohibited operating zones and ED problem with a valve-point loading effect. Furthermore, for each ED problem, two test systems are used for implementing the MOICEP, which are IEEE 30-Bus RTS and IEEE 57-Bus RTS.

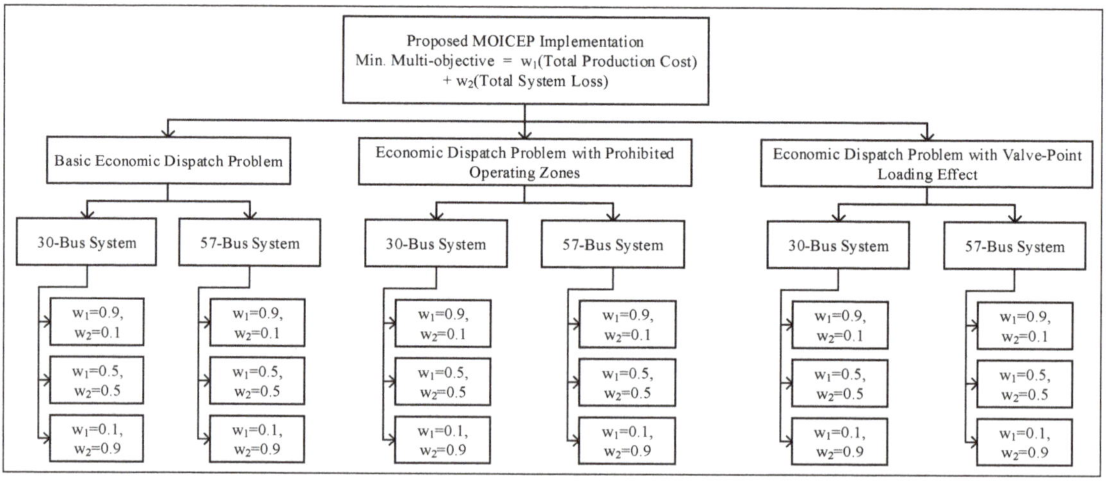

Figure 2. Overall implementation of the MOICEP for total production cost and total system loss minimization.

3.1. MOICEP-Based Technique for Basic Economic Dispatch Problem

For this basic ED problem (smooth cost function), Equations (1) and (3) are used to calculate the total production cost and total system loss, respectively. The MOICEP was implemented on the IEEE 30-Bus RTS and IEEE 57-Bus RTS. For each test system, there are five settings of weight coefficients of the two objectives (total production cost minimization and total system loss minimization). The settings are $w_1 = 0.9, w_2 = 0.1, w_1 = 0.5, w_2 = 0.5$ and $w_1 = 0.1, w_2 = 0.9$. w_1 is the weight coefficient of objective 1 (total production cost minimization), while w_2 is weight coefficient of objective 2 (total system loss minimization).

3.1.1. For $w_1 = 0.9$ and $w_2 = 0.1$

The results of the integrated Basic ED and DG installation of the IEEE 30-Bus and 57-Bus RTSs for the second setting of the weight coefficients ($w_1 = 0.9$ and $w_2 = 0.1$) are tabulated in Tables 2 and 3, respectively.

Table 2. Results of the integrated basic ED and DG installation of the IEEE 30-Bus RTS ($w_1 = 0.9$ and $w_2 = 0.1$).

Optimization Technique		MOICEP	MOEP	MOAIS
Locations of DGs (Bus no.)	Lo_1	29	19	19
	Lo_2	9	28	28
	Lo_3	5	5	25
Sizes of DGs (MW)	DG_1	5.00	4.62	4.62
	DG_2	50.00	27.70	27.70
	DG_3	100.00	75.81	101.63
Generating Unit Output (MW)	P_{G1}	16.42	58.48	32.36
	P_{G2}	20.00	48.66	48.66
	P_{G5}	15.00	23.32	23.32
	P_{G8}	26.21	16.87	16.87
	P_{G11}	23.26	17.19	17.19
	P_{G13}	28.95	13.33	13.33
Total Production Cost ($/h)		355.91	463.64	402.50
Total System Loss (MW)		1.44	2.60	2.28

Table 3. Results of the integrated basic ED and DG installation of the IEEE 57-Bus RTS ($w_1 = 0.9$ and $w_2 = 0.1$).

Optimization Technique		MOICEP	MOEP	MOAIS
Locations of DGs (Bus no.)	Lo_1	25	33	33
	Lo_2	22	10	10
	Lo_3	13	40	40
Sizes of DGs (MW)	DG_1	4.84	2.66	2.66
	DG_2	49.99	36.55	36.55
	DG_3	150.00	141.03	141.03
Generating Unit Output (MW)	P_{G1}	137.42	191.80	191.80
	P_{G2}	47.83	6.68	6.68
	P_{G5}	22.57	58.45	58.45
	P_{G6}	71.59	68.68	68.68
	P_{G8}	381.87	323.93	323.93
	P_{G9}	39.43	93.04	93.04
	P_{G12}	355.16	346.93	346.93
Total Production Cost ($/h)		33,285.97	35,369.69	35,369.69
Total System Loss (MW)		9.90	18.95	18.95

From Table 2, it can be seen that MOICEP continued to give the best solution of the basic ED and DG installation of the IEEE 30-Bus RTS by producing the lowest total production costs and total system loss of 355.91 $/h and 1.44 MW, respectively. The locations of the three DG units found via MOICEP were bus 29, bus 9 and bus 5 with the sizes of 5.00, 50.00 and 100.00 MW, respectively. While the locations and sizes for the three DG units found via MOEP were bus 19 with 4.62 MW, bus 28 with 27.70 MW and bus 5 with 75.81 MW. MOAIS identified bus 19 with 4.62 MW, bus 28 with 27.70 MW and bus 25 with 101.63 MW, respectively. MOEP and MOAIS produced total production costs and total system losses of 463.64 $/h and 2.60 MW and 402.50 $/h and 2.28 MW, respectively. These values were higher than the MOICEP. The proposed MOICEP performed much better than MOEP and MOAIS.

For the results of the IEEE 57-Bus RTS shown in Table 3, similar pattern of results as for the IEEE 30-Bus RTS could be seen. MOICEP maintained producing the lowest total production cost and total system loss of 33285.97 $/h and 9.90 MW, respectively, whereas MOEP and MOAIS found the same total production cost and total system loss of 35,369.69 $/h and 18.95 MW. The optimal location determined by the MOICEP to install the three DG units were bus 25, bus 22 and bus 13 with the sizes of 4.84 MW, 49.99 MW and 150.00 MW, respectively, while MOEP and MOAIS identified bus 33, bus 10 and bus 40 to install the three DG units (Lo_1, Lo_2 and Lo_3) with the sizes of 2.66, 36.55 and 141.03 MW, respectively. MOICEP forced generating unit 8, P_{G8}, to produce the highest power to the system, while generating unit 12, P_{G12}, produced the highest power in MOEP and MOAIS.

3.1.2. For $w_1 = 0.5$ and $w_2 = 0.5$

The results value for w_1 was set to 0.5, and w_2 was 0.5 as the third setting. For the third setting of the weight coefficients, the results of the IEEE 30-Bus RTS and IEEE 57-Bus RTS are tabulated in Tables 4 and 5, respectively.

Table 4. Results of the integrated basic ED and DG installation of the IEEE 30-Bus RTS ($w_1 = 0.5$ and $w_2 = 0.5$).

Optimization Technique		MOICEP	MOEP	MOAIS
Locations of DGs (Bus no.)	Lo_1	29	27	19
	Lo_2	21	7	28
	Lo_3	5	5	25
Sizes of DGs (MW)	DG_1	5.00	4.06	4.62
	DG_2	31.87	33.33	27.70
	DG_3	101.63	60.69	94.86
Generating Unit Output (MW)	P_{G1}	18.54	21.75	32.36
	P_{G2}	24.51	60.95	48.66
	P_{G5}	15.00	38.82	23.32
	P_{G8}	35.00	15.13	16.87
	P_{G11}	30.00	26.73	17.19
	P_{G13}	30.00	24.18	13.33
Total Production Cost ($/h)		402.50	557.01	429.30
Total System Loss (MW)		1.40	2.25	2.28

From Table 4, the MOICEP achieved 402.50 $/h as the total production cost and 1.40 MW for the total system loss. These values were the lowest compared to MOEP (557.01 $/h and 2.25 MW) and MOAIS (429.30 $/h and 2.28 MW). The best total production cost and total system loss in MOICEP were contributed through the placement of DG_1, DG_2 and DG_3 at bus 29, bus 21 and bus 5 with the sizes of 5.00 MW, 31.87 MW and 101.63 MW, respectively. Generating unit 5, P_{G5}, produced the least power in MOICEP, while generating unit 8, P_{G8}, and generating unit 13, P_{G13}, in MOEP and MOAIS, respectively. Apparently, MOICEP consistently outperformed MOEP and MOAIS.

Table 5. Results of the integrated basic ED and DG installation of the IEEE 57-Bus RTS ($w_1 = 0.5$ and $w_2 = 0.5$).

Optimization Technique		MOICEP	MOEP	MOAIS
Locations of DGs (Bus no.)	Lo_1	35	45	34
	Lo_2	38	30	23
	Lo_3	3	55	11
Sizes of DGs (MW)	DG_1	3.65	3.53	3.44
	DG_2	44.23	24.77	40.05
	DG_3	68.97	132.47	92.34
Generating Unit Output (MW)	P_{G1}	206.60	237.10	24.14
	P_{G2}	16.24	25.01	53.09
	P_{G5}	34.87	120.52	91.62
	P_{G6}	52.04	47.13	40.76
	P_{G8}	351.68	278.28	450.99
	P_{G9}	81.82	60.04	71.83
	P_{G12}	400.55	335.20	397.14
Total Production Cost ($/h)		37,513.91	38,111.41	37,752.82
Total System Loss (MW)		9.90	13.26	14.61

From the results of the IEEE 57-Bus shown in Table 5, it can be seen that MOICEP continued to produce better total production costs and total system loss as compared to MOEP and MOAIS. MOICEP produced 37,513.91 $/h and 9.90 MW, while MOEP and MOAIS produced 38,111.41 $/h and 13.26 MW and 37,752.82 $/h and 14.61 MW, respectively. Bus 35, bus 38 and bus 3 were determined as the optimal locations to install DG_1, DG_2 and DG_3. The corresponding sizes for the three DG units were 3.65 MW, 44.23 MW and 68.97 MW. Generating unit 2, P_{G2}, produced the least power output compared to the other generating units in MOICEP and MOEP, while generating unit 1, P_{G1}, in MOAIS. From the results when $w_1 = 0.5$ and $w_2 = 0.5$, it is observed that the total production cost was slightly expensive, and the total system loss was lower than the previous two settings of weight coefficients.

3.1.3. For $w_1 = 0.1$ and $w_2 = 0.9$

The fourth setting was $w_1 = 0.1$ and $w_2 = 0.9$. For this fourth setting of the weight coefficients, the results of integrated Basic ED and DG installation of the IEEE 30-Bus RTS and IEEE 57-Bus RTS are tabulated in Tables 6 and 7, respectively. From the comparison of the results between the three techniques (MOICEP, MOEP and MOAIS), it was observed that the total production cost produced by MOAIS was the lowest, while the total system loss produced by MOICEP was the lowest. The total production cost in MOAIS was lower than in MOICEP and MOEP, because the total DG size for the three DG units was the highest, worth 133.95 MW compared to MOICEP (121.26 MW) and MOEP (98.08 MW). However, MOICEP still exhibited the lowest total system loss of 1.39 MW as compared to MOEP (2.25 MW) and MOAIS (2.28 MW). Bus 29, bus 21 and bus 5 were the optimal locations to install DG_1, DG_2 and DG_3. The corresponding sizes for the three DG units were 5.00, 29.77 and 86.49 MW. Generating unit 1, P_{G1}, in MOICEP produced the lowest power output, while, in MOEP, generating unit 8, P_{G8}, and, in MOAIS, generating unit 13, P_{G13}, were the ones with the lowest power outputs.

Table 6. Results of the integrated basic ED and DG installation of the IEEE 30-Bus RTS ($w_1 = 0.1$ and $w_2 = 0.9$).

Optimization Technique		MOICEP	MOEP	MOAIS
Locations of DGs (Bus no.)	Lo_1	29	27	19
	Lo_2	21	7	28
	Lo_3	5	5	25
Sizes of DGs (MW)	DG_1	5.00	4.06	4.62
	DG_2	29.77	33.33	27.70
	DG_3	86.49	60.69	101.63
Generating Unit Output (MW)	P_{G1}	18.65	21.75	32.36
	P_{G2}	27.91	60.95	48.66
	P_{G5}	22.01	38.82	23.32
	P_{G8}	34.96	15.13	16.87
	P_{G11}	30.00	26.73	17.19
	P_{G13}	30.00	24.18	13.33
Total Production Cost ($/h)		461.64	557.01	402.50
Total System Loss (MW)		1.39	2.25	2.28

Table 7. Results of the integrated basic ED and DG installation of the IEEE 57-Bus RTS ($w_1 = 0.1$ and $w_2 = 0.9$).

Optimization Technique		MOICEP	MOEP	MOAIS
Locations of DGs (Bus no.)	Lo_1	25	45	34
	Lo_2	23	30	23
	Lo_3	13	55	11
Sizes of DGs (MW)	DG_1	4.99	3.53	3.44
	DG_2	44.63	24.77	40.05
	DG_3	123.36	132.47	92.34
Generating Unit Output (MW)	P_{G1}	125.84	237.10	24.14
	P_{G2}	26.37	25.01	53.09
	P_{G5}	87.41	120.52	91.62
	P_{G6}	95.96	47.13	40.76
	P_{G8}	238.78	278.28	450.99
	P_{G9}	100.00	60.04	71.83
	P_{G12}	409.98	335.20	397.14
Total Production Cost ($/h)		36,159.62	38,111.41	37,752.82
Total System Loss (MW)		6.52	13.26	14.61

For the results of IEEE 57-Bus RTS shown in Table 7, both the total production cost and total system loss for this setting of weight coefficients ($w_1 = 0.1$ and $w_2 = 0.9$) were found to be the lowest in MOICEP, which were 36,159.62 $/h and 6.52 MW as compared to MOEP with 38,111.41 $/h and 13.26 MW and MOAIS with 37,752.82 $/h and 14.61 MW. The optimal locations to install the three DG units identified by MOICEP were bus 25, bus 23 and bus 13 with the corresponding sizes of 4.99, 44.63 and 123.36 MW, respectively. It could be observed that MOICEP forced the DG units to operate near to their maximum capacity. This was the reason why the total production cost reduced significantly as compared to MOEP and MOAIS.

3.2. MOICEP-Based Technique for Economic Dispatch Problem with Prohibited Operating Zones

The proposed MOICEP technique was also used to solve the ED problem with prohibited operating zones. In this study, the same equations of the total production cost (Equation (1)) and total system loss (Equation (3)) in the first ED problem were reutilized. However, there was a difference in the constraint of the operating limits of the generating units. The generating units would only be allowed to operate in the allowable zones to make sure the stability of their operation. The allowable and prohibited zones for the IEEE 30-Bus RTS and IEEE 57-Bus RTS are shown in Tables 8 and 9, respectively.

Table 8. Cost coefficients and generating unit limits for the ED problem with the prohibited operating zones (IEEE 30-Bus RTS).

Gen. Unit	P_{min}	P_{max}	Prohibited Zones		Cost Coefficients		
			Zone 1	Zone 2	a_i	b_i	c_i
1	50	200	55–66	80–120	0.00375	2.00	0
2	20	80	21–24	45–55	0.01750	1.75	0
5	15	50	30–36	-	0.06250	1.00	0
8	10	35	25–30	-	0.00834	3.25	0
11	10	30	25–28	-	0.02500	3.00	0
13	10	30	24–30	-	0.02500	3.00	0

Table 9. Cost coefficients and generating unit limits for the ED problem with the prohibited operating zones (IEEE 57-Bus RTS).

Gen. Unit	P_{min}	P_{max}	Prohibited Zones		Cost Coefficients		
			Zone 1	Zone 2	a_i	b_i	c_i
1	0	575.88	10–50	480–520	0.0775795	20	0
2	0	100	5–10	75–80	0.0100000	40	0
3	0	140	10–25	100–110	0.2500000	20	0
6	0	100	5–10	-	0.0100000	40	0
8	0	550	10–30	-	0.0222222	20	0
9	0	100	5–10	-	0.0100000	40	0
12	0	410	10–35	-	0.0322581	20	0

3.2.1. For w_1 = 0.9 and w_2 = 0.1

The integrated ED problem with prohibited operating zones and DG installation continued to be solved using MOICEP, MOEP and MOAIS for the second setting of weight coefficients ($w_1 = 0.9$ and $w_2 = 0.1$). The results for this setting are shown in Table 10 (IEEE 30-Bus RTS) and Table 11 (IEEE 57-Bus RTS). Similar to the first setting, the solutions produced by MOEP and MOAIS were the same for both systems. From Table 10 of the IEEE 30-Bus RTS, the total production cost and total system loss produced by MOICEP were 380.05 $/h and 1.71 MW, respectively. While MOEP and MOAIS produced a total production cost of 503.05 $/h and total system loss of 3.81 MW. MOICEP significantly outperformed MOEP and MOAIS.

For the DG installation, it was found that bus 29 with 5.00 MW, bus 21 with 50.00 MW and bus 5 with 100.00 MW were the optimal locations and sizes for the three DG units in the IEEE 30-Bus RTS, while, for the IEEE 57-Bus RTS, as shown in Table 11, bus 25, bus 14 and bus 13 were found to be the suitable locations to install the three DG units with the sizes of 3.87, 50.00 and 150.00 MW, respectively, via MOICEP. This resulted in a total production cost and total system loss worth 33,315.31 $/h and 11.29 MW, respectively.

Generating unit 6, P_{G6}, was found to be the least generating unit supplying its real power output to the IEEE 57-Bus RTS in MOICEP, while generating unit 2, P_{G2}, in MOEP and MOAIS. Comparing the total production cost produced in this second setting with the first setting, it can be observed that the total production cost in the second setting was higher for both systems. This was due to the value of w_1 decreasing to 0.1 and the value of w_2 increasing to 0.1 in the second setting.

Table 10. Results of the integrated ED problem with prohibited operating zones and DG installation of the IEEE 30-Bus RTS ($w_1 = 0.9$ and $w_2 = 0.1$).

Optimization Technique		MOICEP	MOEP	MOAIS
Locations of DGs (Bus no.)	Lo_1	29	24	24
	Lo_2	21	15	15
	Lo_3	5	22	22
Sizes of DGs (MW)	DG_1	5.00	4.03	4.03
	DG_2	50.00	49.60	49.60
	DG_3	100.00	68.27	68.27
Generating Unit Output (MW)	P_{G1}	12.26	19.48	19.48
	P_{G2}	26.70	44.96	44.96
	P_{G5}	27.43	39.54	39.54
	P_{G8}	34.25	32.76	32.76
	P_{G11}	10.27	15.25	15.25
	P_{G13}	19.20	13.33	13.33
Total Production Cost ($/h)		380.05	503.05	503.05
Total System Loss (MW)		1.71	3.81	3.81

Table 11. Results of the integrated ED problem with prohibited operating zones and DG installation of the IEEE 57-Bus RTS ($w_1 = 0.9$ and $w_2 = 0.1$).

Optimization Technique		MOICEP	MOEP	MOAIS
Locations of DGs (Bus no.)	Lo_1	25	50	50
	Lo_2	14	33	33
	Lo_3	13	10	10
Sizes of DGs (MW)	DG_1	3.87	3.62	3.62
	DG_2	50.00	23.72	23.72
	DG_3	150.00	120.12	120.12
Generating Unit Output (MW)	P_{G1}	121.07	201.79	201.79
	P_{G2}	96.27	0.77	0.77
	P_{G5}	43.92	53.68	53.68
	P_{G6}	34.14	41.75	41.75
	P_{G8}	379.12	377.72	377.72
	P_{G9}	41.12	58.90	58.90
	P_{G12}	342.59	381.48	381.48
Total Production Cost ($/h)		33,315.31	36,146.27	36,146.27
Total System Loss (MW)		11.29	12.75	12.75

3.2.2. For $w_1 = 0.5$ and $w_2 = 0.5$

Tables 12 and 13 show the results of the integrated ED problem with the prohibited operating zones and DG installation of the IEEE 30-Bus RTS and IEEE 57-Bus RTS, respectively, with weight coefficients $w_1 = 0.5$ and $w_2 = 0.5$.

Again, MOEP and MOAIS produced the same solution for both systems. As referred to in Table 12, MOICEP produced a total production cost of 492.48 $/h and total system loss of 1.84 MW, while MOEP and MOAIS produced a similar total production cost of 559.24 $/h and a similar total system loss of 3.34 MW. MOICEP managed to save about 66.76 $/h in the total production cost. Besides that, the total system loss found via MOICEP was almost half of the total system loss produced via MOEP and MOAIS.

According to Table 13 for the IEEE 57-Bus RTS, MOICEP managed to produce the lowest total production cost of 35,926.32 $/h and total system loss of 11.83 MW. MOEP and MOAIS produced a similar total production cost of 36,146.27 $/h and a similar total system loss of 12.75 MW. MOICEP identified the optimal locations of DG_1, DG_2 and DG_3 in the IEEE 57-Bus RTS at bus 57, bus 36 and bus 43, respectively. The sizes for the three DG units were 4.52, 44.13 and 101.28 MW. MOEP and MOAIS identified bus 50, bus 33 and bus 10 as the optimal locations to install DG_1, DG_2 and DG_3 with the sizes of 3.62, 23.72 and 120.12 MW, respectively. Generating unit 8, P_{G8}, was found to be the highest generating unit to supply a real power output to the IEEE 57-Bus RTS in MOICEP. In MOEP and MOAIS, generating unit 12, P_{G12}, was the highest contributor of real power output to the system.

Table 12. Results of the integrated ED problem with prohibited operating zones and DG installation of the IEEE 30-Bus RTS ($w_1 = 0.5$ and $w_2 = 0.5$).

Optimization Technique		MOICEP	MOEP	MOAIS
Locations of DGs (Bus no.)	Lo_1	29	23	23
	Lo_2	5	10	10
	Lo_3	9	27	27
Sizes of DGs (MW)	DG_1	5.00	3.84	3.84
	DG_2	45.57	42.29	42.29
	DG_3	81.18	54.73	54.73
Generating Unit Output (MW)	P_{G1}	3.99	57.48	57.48
	P_{G2}	53.22	20.05	20.05
	P_{G5}	41.51	42.07	42.07
	P_{G8}	17.06	17.98	17.98
	P_{G11}	10.14	18.35	18.35
	P_{G13}	27.56	30.00	30.00
Total Production Cost ($/h)		492.48	559.24	559.24
Total System Loss (MW)		1.84	3.34	3.34

Table 13. Results of the integrated ED problem with prohibited operating zones and DG installation of the IEEE 57-Bus RTS ($w_1 = 0.5$ and $w_2 = 0.5$).

Optimization Technique		MOICEP	MOEP	MOAIS
Locations of DGs (Bus no.)	Lo_1	57	50	50
	Lo_2	36	33	33
	Lo_3	43	10	10
Sizes of DGs (MW)	DG_1	4.52	3.62	3.62
	DG_2	44.13	23.72	23.72
	DG_3	101.28	120.12	120.12
Generating unit output (MW)	P_{G1}	182.78	201.79	201.79
	P_{G2}	57.29	0.77	0.77
	P_{G5}	68.03	53.68	53.68
	P_{G6}	73.47	41.75	41.75
	P_{G8}	363.59	377.72	377.72
	P_{G9}	29.91	58.90	58.90
	P_{G12}	337.61	381.48	381.48
Total Production Cost ($/h)		35,926.32	36,146.27	36,146.27
Total System Loss (MW)		11.83	12.75	12.75

3.2.3. For $w_1 = 0.1$ and $w_2 = 0.9$

The results of the fourth setting of the weight coefficients ($w_1 = 0.1$ and $w_2 = 0.9$) for the IEEE 30-Bus RTS and IEEE 57-Bus RTS are tabulated in Tables 14 and 15, respectively. As referred to in the two tables, it can be seen that MOEP and MOAIS continued to produce the same solution of the integrated ED problem with prohibited operating zones and DG installation. This could be due to the prohibited operating zones of the generating units that make the searching space of the MOEP and MOAIS algorithms smaller and lead to the same global optima. MOICEP achieved a total production cost of 517.76 $/h and total system loss of 1.49 MW for the IEEE 30-Bus RTS, while, in the IEEE 57-Bus RTS, 35,184.12 $/h and 6.94 MW. As referred to in Table 14, the optimal locations to install DG_1,

DG_2 and DG_3 in the IEEE 30-Bus RTS were bus 29, bus 21 and bus 5 with the sizes of 5.00, 27.69 and 85.03 MW, respectively.

Table 14. Results of the integrated ED problem with prohibited operating zones and DG installation of the IEEE 30-Bus RTS ($w_1 = 0.1$ and $w_2 = 0.9$).

Optimization Technique		MOICEP	MOEP	MOAIS
Locations of DGs (Bus no.)	Lo_1	29	23	23
	Lo_2	21	10	10
	Lo_3	5	27	27
Sizes of DGs (MW)	DG_1	5.00	3.84	3.84
	DG_2	27.69	42.29	42.29
	DG_3	85.03	54.73	54.73
Generating Unit Output (MW)	P_{G1}	13.55	57.48	57.48
	P_{G2}	38.62	20.05	20.05
	P_{G5}	27.96	42.07	42.07
	P_{G8}	34.17	17.98	17.98
	P_{G11}	16.25	18.35	18.35
	P_{G13}	36.61	30.00	30.00
Total Production Cost ($/h)		517.76	559.24	559.24
Total System Loss (MW)		1.49	3.34	3.34

Table 15. Results of the integrated ED problem with prohibited operating zones and DG installation of the IEEE 57-Bus RTS ($w_1 = 0.1$ and $w_2 = 0.9$).

Optimization Technique		MOICEP	MOEP	MOAIS
Locations of DGs (Bus no.)	Lo_1	21	3	3
	Lo_2	26	53	53
	Lo_3	13	16	16
Sizes of DGs (MW)	DG_1	4.99	4.87	4.87
	DG_2	34.73	28.54	28.54
	DG_3	150.00	77.88	77.88
Generating Unit Output (MW)	P_{G1}	148.50	175.15	175.15
	P_{G2}	1.04	10.46	10.46
	P_{G5}	81.08	93.14	93.14
	P_{G6}	76.86	83.34	83.34
	P_{G8}	254.90	319.71	319.71
	P_{G9}	99.82	65.86	65.86
	P_{G12}	405.81	402.63	402.63
Total Production Cost ($/h)		35,184.12	38,362.28	38,362.28
Total System Loss (MW)		6.94	10.80	10.80

On the other hand, in MOEP and MOAIS, the locations and sizes were bus 23 with 3.84 MW, bus 10 with 42.29 MW and bus 27 with 54.73 MW. The six generating units (P_{G1}, P_{G2}, P_{G5}, P_{G8}, P_{G11} and P_{G13}) in MOICEP produced 13.55, 38.62, 27.96, 34.17, 16.25 and 36.61 MW, respectively. In MOEP and MOAIS, they produced 57.48, 20.05, 42.07, 17.98, 18.35 and 30.00 MW, respectively. Generating unit 1, P_{G1}, in MOICEP produced the lowest real power output compared to the other units. Generating unit 8, P_{G8}, produced the lowest real power output in MOEP and MOAIS.

According to Table 15 of the IEEE 57-Bus RTS, bus 21, bus 26 and bus 13 were found to be the optimal locations to install DG_1, DG_2 and DG_3, respectively. Their sizes were 4.99 MW, 34.73 and 150.00 MW, respectively.

Meanwhile, in MOEP and MOAIS, bus 3, bus 53 and bus 16 were the optimal locations to install DG_1, DG_2 and DG_3 with the sizes of 4.87, 28.54 and 77.88 MW, respectively. It was observed that generating unit 2, P_{G2}, in MOICEP produced a very little real power output of 1.04 MW compared to the other units. The lowest real power output produced by a generating unit in MOEP and MOAIS was 10.46 MW, which was generating unit 2, P_{G2}.

3.3. MOICEP-Based Technique for Economic Dispatch Problem with Valve-Point Loading Effect

The total production cost and total system loss for this type of ED problem were calculated using Equations (2) and (3), respectively. The cost function of this ED was not smooth because of the valve-point loading effect of the generating unit considered in the cost function. The results of the integrated ED problem with a valve-point loading effect and DG installation are observed and discussed in this section.

3.3.1. For $w_1 = 0.9$ and $w_2 = 0.1$

Tables 16 and 17 show the results of the integrated ED problem with a valve-point loading effect and DG installation of the IEEE 30-Bus RTS and 57-Bus RTS with $w_1 = 0.9$ and $w_2 = 0.1$, respectively. It can be seen from Table 16 that MOICEP maintained producing the best solution of an ED with a valve-point loading effect and DG installation for the second setting of the weight coefficients by giving the lowest total production cost of 11,817.81 $/h and the lowest total system loss of 2.04 MW. MOEP and MOAIS produced 13,331.25 $/h and 3.61 MW and 11,627.07 $/h and 13.32 MW, respectively. The best solution from MOICEP was backed by the placement of DG_1, DG_2 and DG_3 at bus 26, bus 5 and bus 9 with the sizes of 5.00, 50.00 and 100.00 MW, respectively. Generating unit 1, P_{G1}, and generating unit 8, P_{G8}, produced the lowest and the highest real power outputs in MOICEP, respectively. The most expensive total production cost was obtained in MOEP, which was 13,331.25 $/h. However, MOAIS produced the highest total system loss of 13.32 MW compared to the other two techniques. It was five times more than the total system loss produced via MOICEP.

As referred to in Table 17 of the IEEE 57-BUS RTS, it can be seen that MOICEP produced the cheapest total production cost of 32,560.84 $/h and total system loss of 9.78 MW compared to MOEP and MOAIS. MOEP and MOAIS produced the same solution, with the total production cost of 35,369.69 $/h and total system loss of 18.95 MW.

MOICEP found bus 26, bus 10 and bus 14 as the optimal locations to install DG_1, DG_2 and DG_3 with the sizes of 5.00, 49.96 and 149.98 MW, respectively. MOEP and MOAIS identified bus 33, bus 10 and bus 40 as the optimal locations to install DG_1, DG_2 and DG_3 with the sizes of 2.66, 36.55 and 141.03 MW, respectively. It was observed that the real power output of the three DG units was higher in MOICEP compared to the real power output in MOEP and MOAIS.

Table 16. Results of the integrated ED problem with a valve-point loading effect and DG installation of the IEEE 30-Bus RTS ($w_1 = 0.9$ and $w_2 = 0.1$).

Optimization Technique		MOICEP	MOEP	MOAIS
Locations of DGs (Bus no.)	Lo_1	26	25	29
	Lo_2	5	27	4
	Lo_3	9	7	18
Sizes of DGs (MW)	DG_1	5.00	1.06	2.50
	DG_2	50.00	35.98	39.45
	DG_3	100.00	93.41	1.30
Generating Unit Output (MW)	P_{G1}	8.39	3.75	13.23
	P_{G2}	20.00	51.78	22.77
	P_{G5}	22.29	31.26	18.40
	P_{G8}	34.46	33.52	30.59
	P_{G11}	15.30	11.00	23.90
	P_{G13}	30.00	25.23	16.34
Total Production Cost ($/h)		11,817.81	13,331.25	11,627.07
Total System Loss (MW)		2.04	3.61	13.32

Table 17. Results of the integrated ED problem with a valve-point loading effect and DG installation of the IEEE 57-Bus RTS ($w_1 = 0.9$ and $w_2 = 0.1$).

Optimization Technique		MOICEP	MOEP	MOAIS
Locations of DGs (Bus no.)	Lo_1	26	33	33
	Lo_2	10	10	10
	Lo_3	14	40	40
Sizes of DGs (MW)	DG_1	5.00	2.66	2.66
	DG_2	49.96	36.55	36.55
	DG_3	149.98	141.03	141.03
Generating Unit Output (MW)	P_{G1}	132.46	191.82	191.82
	P_{G2}	43.98	6.68	6.68
	P_{G5}	49.48	58.44	58.44
	P_{G6}	72.02	68.67	68.67
	P_{G8}	327.03	323.93	323.93
	P_{G9}	73.78	93.04	93.04
	P_{G12}	356.90	346.92	346.92
Total Production Cost ($/h)		32,560.84	35,369.69	35,369.69
Total System Loss (MW)		9.78	18.95	18.95

3.3.2. For $w_1 = 0.5$ and $w_2 = 0.5$

The third setting of the weight coefficients for solving the integrated ED problem with a valve-point loading effect and DG installation for the IEEE 30-Bus RTS and IEEE 57-Bus RTS are tabulated in Tables 18 and 19. Referring to Table 18, it can be seen that MOICEP managed to provide a better solution of ED with a valve-point loading effect and DG installation by producing a total production cost of 11,758.98 $/h and total system loss of 1.50 MW compared to MOEP (13,340.00 $/h and 3.31 MW) and MOAIS (13,178.04 $/h and 2.28 MW). This was caused by the placement of DG_1, DG_2 and DG_3 at bus 18, bus 9 and bus 5 with the sizes of 5.00 MW, 50.00 MW and 100.00 MW, respectively. The total DG size was the highest in MOICEP (155.00 MW), while, for MOEP, it was 125.91 MW and, for MOAIS, it was 133.95 MW. Generating unit 1, P_{G1}, produced the lowest real power output of 7.05 MW compared to the other generating units in MOICEP.

Table 18. Results of the integrated ED problem with a valve-point loading effect and DG installation of the IEEE 30-Bus RTS ($w_1 = 0.5$ and $w_2 = 0.5$).

Optimization Technique		MOICEP	MOEP	MOAIS
Locations of DGs (Bus no.)	Lo_1	18	16	19
	Lo_2	9	19	28
	Lo_3	5	9	5
Sizes of DGs (MW)	DG_1	5.00	3.35	4.62
	DG_2	50.00	43.05	27.70
	DG_3	100.00	79.51	101.63
Generating Unit Output (MW)	P_{G1}	7.05	12.77	32.36
	P_{G2}	20.00	39.05	48.66
	P_{G5}	15.00	46.04	23.32
	P_{G8}	34.65	26.30	16.87
	P_{G11}	24.78	13.01	17.18
	P_{G13}	28.42	23.63	13.33
Total Production Cost ($/h)		11,758.98	13,340.00	13,178.04
Total System Loss (MW)		1.50	3.31	2.28

Table 19. Results of the integrated ED problem with a valve-point loading effect and DG installation of the IEEE 57-Bus RTS ($w_1 = 0.5$ and $w_2 = 0.5$).

Optimization Technique		**MOICEP**	**MOEP**	**MOAIS**
Locations of DGs (Bus no.)	Lo_1	25	45	45
	Lo_2	22	30	30
	Lo_3	13	55	55
Sizes of DGs (MW)	DG_1	4.97	3.53	3.53
	DG_2	42.96	24.77	24.77
	DG_3	150.00	132.47	132.47
Generating Unit Output (MW)	P_{G1}	131.64	237.10	237.10
	P_{G2}	36.11	25.01	25.01
	P_{G5}	65.13	120.52	120.52
	P_{G6}	72.78	47.13	47.13
	P_{G8}	273.04	278.28	278.28
	P_{G9}	100.00	60.04	60.04
	P_{G12}	381.24	335.20	335.20
Total Production Cost ($/h)		33,447.51	38,013.28	38,013.28
Total System Loss (MW)		7.10	13.26	13.26

The results of the IEEE 57-Bus are shown in Table 19. It can be observed that MOICEP continued to produce a better solution of the integrated ED problem with a valve-point loading effect and DG installation. This was based on the total production cost and total system loss produced, which were the lowest at 33,447.51 $/h and 7.10 MW, respectively. MOEP and MOAIS produced the same results, with a total production cost of 38,013.28 $/h and total system loss worth 13.26 MW.

In MOICEP, bus 25, bus 22 and bus 13 were identified as the best locations to install DG_1, DG_2 and DG_3 with the sizes of 4.97, 42.96 and 150.00 MW, respectively. The generating units 1, 2, 5, 6, 8, 9 and 12 produced real power outputs of 131.64, 36.11, 65.13, 72.78, 273.04, 100.00 and 381.24 MW, respectively.

MOEP and MOAIS identified bus 45, bus 30 and bus 55 as the optimal locations to place DG_1, DG_2 and DG_3 with the sizes of 3.53, 24.77 and 132.47 MW, respectively. The total DG size was higher in MOICEP, which was 197.93 MW compared to MOEP and MOAIS, which was 160.77 MW. As a result, MOICEP achieved a lower total production cost than MOEP and MOAIS.

3.3.3. For $w_1 = 0.1$ and $w_2 = 0.9$

The breadth of this study was expanded with the next weight coefficient setting, i.e., $w_1 = 0.1$ and $w_2 = 0.9$. Tables 20 and 21 show the results of the integrated ED problem with a valve-point loading effect and DG installation of the IEEE 30-Bus RTS and IEEE 57-Bus RTS, respectively.

From Table 20, it can be seen that MOICEP achieved the lowest total system loss of 1.47 MW compared to MOEP (2.34 MW) and MOAIS (2.28 MW). However, the total production cost produced in MOAIS was the lowest compared to MOICEP and MOEP. MOAIS produced a total production cost of 13,178.04 $/h, while MOICEP and MOEP produced 13,472.72 $/h and 15,179.41 $/h, respectively. The highest total production cost was found in MOEP. The optimal locations and sizes found via MOICEP for DG_1, DG_2 and DG_3 were bus 29 with 5.00 MW, bus 22 with 28.71 MW and bus 5 with 86.75 MW. It was observed that, to make sure the total system loss was the lowest, MOICEP avoided the three DG units operating near their maximum capacity.

A similar observation was experienced by the IEEE 57-Bus RTS, as shown in Table 21; the total system loss produced via MOICEP was the lowest. However, the total production cost was the highest compared to MOEP and MOAIS. MOICEP produced a total production cost of 38,352.96 $/h and total system loss of 8.35 MW. MOEP and MOAIS produced the same results, with a total production cost of 38,013.28 $/h and total system loss of 13.26 MW.

The optimal locations to install DG_1, DG_2 and DG_3 in the IEEE 57-Bus RTS via MOICEP were bus 29, bus 22 and bus 13 with the sizes of 1.03, 44.02 and 59.53 MW, respectively. The total system loss produced via MOICEP was 37.03% lower than the total system loss produced via MOEP and MOAIS. However, in terms of the total production cost, MOICEP did not force the three DG units to operate near their maximum capacity. This resulted in a high total production cost produced via MOICEP. It must be kept in mind that the weight coefficients also play an important role that leads to the results. As the value of w_1 decreases and the value of w_2 increases, the algorithm tends to focus on optimizing objective function 2 (total system loss), as expected.

Table 20. Results of the integrated ED problem with a valve-point loading effect and DG installation of the IEEE 30-Bus RTS ($w_1 = 0.1$ and $w_2 = 0.9$).

Optimization Technique		MOICEP	MOEP	MOAIS
Locations of DGs (Bus no.)	Lo_1	29	27	19
	Lo_2	22	4	28
	Lo_3	5	5	5
Sizes of DGs (MW)	DG_1	5.00	3.12	4.62
	DG_2	28.71	35.20	27.70
	DG_3	86.75	50.38	101.63
Generating Unit Output (MW)	P_{G1}	12.52	26.03	32.36
	P_{G2}	41.12	60.73	48.66
	P_{G5}	20.46	38.78	23.32
	P_{G8}	35.00	33.37	16.87
	P_{G11}	29.18	17.67	17.18
	P_{G13}	26.12	20.39	13.33
Total Production Cost ($/h)		13,472.72	15,179.41	13,178.04
Total System Loss (MW)		1.47	2.34	2.28

Table 21. Results of the integrated ED problem with a valve-point loading effect and DG installation of the IEEE 57-Bus RTS ($w_1 = 0.1$ and $w_2 = 0.9$).

Optimization Technique		MOICEP	MOEP	MOAIS
Locations of DGs (Bus no.)	Lo_1	29	45	45
	Lo_2	22	30	30
	Lo_3	13	55	55
Sizes of DGs (MW)	DG_1	1.03	3.53	3.53
	DG_2	44.02	24.77	24.77
	DG_3	59.53	132.47	132.47
Generating Unit Output (MW)	P_{G1}	143.43	237.10	237.10
	P_{G2}	54.96	25.01	25.01
	P_{G5}	84.08	120.52	120.52
	P_{G6}	79.60	47.13	47.13
	P_{G8}	309.65	278.28	278.28
	P_{G9}	85.57	60.04	60.04
	P_{G12}	397.27	335.20	335.20
Total Production Cost ($/h)		38,352.96	38,013.28	38,013.28
Total System Loss (MW)		8.35	13.26	13.26

4. Conclusions

The new Multi-Objective Immune-Commensal-Evolutionary Programming (MOICEP) technique for the total production cost and total system loss minimization via integrated Economic Dispatch and Distributed Generation installation (ED-DG) was presented in this paper. It was employed for the total production cost and total system loss minimization in an integrated weighted-sum multi-objective optimization technique for solving the ED problem with DG installation consideration. This scheme is uniquely termed as Economic

Dispatch and Distributed Generation installation (ED-DG). The proposed weighted-sum multi-objective function that integrated both the total production cost and total system loss was highlighted in this study. The introduction of weight coefficients in the proposed fitness exhibited convincing results in addressing both components. The weight coefficients were varied between 0.1 and 0.9 to find the optimal solution that satisfied both the total production cost and total system loss. From the variations, it could be seen that, as the value of w_1 was greater than w_2, the solution produced favored total production cost minimization, while, if the value of w_2 was greater than w_1, the solution favored total system loss minimization. GSOs can decide which solution is the best for their system requirements and their desired output. The proposed MOICEP technique managed to achieve better results over benchmarked techniques, i.e., MOEP and MOAIS, for the minimization of the total production costs and total system loss for three types of ED problems (basic ED, ED problem with prohibited operating zones and an ED problem with a valve-point loading effect). This phenomenon highlights the superiority of MOICEP over MOEP and MOAIS. This phenomenon has been identified as cutting edge in ED problems as the original independent scheme, i.e., ED and DG installation. The proposed multi-objective technique could be feasible for solving other optimization problems with considerable fine-tuning. It is also beneficial for power system utilities in solving their ED problems. It can also be concluded that, while optimizing the two objectives, total production cost and total system loss, the integrated Economic Dispatch and Distributed Generation installation ED-DG problems solved using the proposed MOICEP technique simultaneously solved two power system problems, ED and DG installation, which have been previously solved separately by many researchers. The two objectives are inextricably linked to the two problems.

The number of objectives for the multi-objective ED problems can be increased in the future. This can be done by adding more objectives like voltage stability improvement, emission minimization and DG installation cost minimization into the integrated ED-DG problems.

Author Contributions: All the authors listed conceived and designed the project. Conceptualization, M.H.M. and I.M.; acquired the funding, M.H.M.; methodology, M.H.M., I.M. and M.M.O.; validation, M.H.M. and I.M.; analysis, M.H.M. and M.M.O.; writing—original draft preparation, M.H.M. and I.M. and writing—review and editing, M.M.O. All authors have read and agreed to the published version of the manuscript.

Funding: This research was funded by the Ministry of Higher Education Malaysia (MOHE), grant number FRGS/1/2020/TK0/UNITEN/03/4.

Institutional Review Board Statement: Not applicable.

Informed Consent Statement: Not applicable.

Conflicts of Interest: The authors declare no conflict of interest.

References

1. Mehigan, L.; Deane, J.P.; Gallachóir, B.P.Ó.; Bertsch, V. A review of the role of distributed generation (DG) in future electricity systems. *Energy* **2018**, *163*, 822–836. [CrossRef]
2. Zhan, J.; Wu, Q.H.; Guo, C.; Zhou, X. Economic Dispatch With Non-Smooth Objectives—Part I: Local Minimum Analysis. *IEEE Trans. Power Syst.* **2015**, *30*, 710–721. [CrossRef]
3. Thenmalar, K.; Anujak, S.; Ramesh, S. Multi-Objective economic emission load dispatch solution using wolf's method in various generation plants with wind power penetration. In Proceedings of the 2014 International Conference on Electronics and Communication Systems, ICECS 2014, Coimbatore, Tamilnadu, 13–14 February 2014; pp. 1–13.
4. Jakob, W.; Blume, C. Pareto optimization or cascaded weighted sum: A comparison of concepts. *Algorithms* **2014**, *7*, 166–185. [CrossRef]
5. Marler, R.T.; Arora, J.S. The weighted sum method for multi-objective optimization: New insights. *Struct. Multidiscip. Optim.* **2010**, *41*, 853–862. [CrossRef]
6. Musau, M. Multi Area Multi Objective Dynamic Economic Dispatch with Renewable Energy and Emissions. In Proceedings of the 2016 IEEE International Energy Conference (ENERGYCON), Leuven, Belgium, 4–8 April 2016; pp. 112–117.

7. Roy, N.; Ghosh, A.; Sanyal, K. Normal Boundary Intersection based multi-objective Harmony Search algorithm for environmental Economic Load Dispatch problem. In Proceedings of the 2016 IEEE 6th International Conference on Power Systems, ICPS 2016, New Delhi, India, 4–6 March 2016; pp. 1–6.
8. Mohanty, P.S. Multi-objective economic emission load dispatch with nonlinear fuel cost and noninferior emission level functions for IEEE-118 bus system. In Proceedings of the 2nd International Conference on Electronics and Communication Systems (ICECS), Coimbatore, India, 26–27 February 2015; pp. 1371–1376.
9. Mao, M.; Ji, M.; Dong, W.; Chang, L. Multi-objective economic dispatch model for a microgrid considering reliability. In Proceedings of the 2nd International Symposium on Power Electronics for Distributed Generation Systems, PEDG 2010, Hefei, China, 16–18 June 2010; pp. 993–998.
10. Man-Im, A.; Ongsakul, W.; Singh, J.G.; Boonchuay, C. Multi-objective economic dispatch considering wind generation uncertainty using non-dominated sorting particle swarm optimization. In Proceedings of the 2014 International Conference and Utility Exhibition on Green Energy for Sustainable Development, ICUE 2014, Pattaya City, Thailand, 19–21 March 2014; pp. 19–21.
11. Bilil, H.; Ellaia, R.; Maaroufi, M. A New Multi-objective Particle Swarm Optimization for Economic Environmental Dispatch. In Proceedings of the 2012 IEEE International Conference on Complex Systems (ICCS), Agadir, Morocco, 5–6 November 2012; Volume 3, pp. 1–6.
12. Shen, X.; Zou, D.; Duan, N.; Zhang, Q. An efficient fitness-based differential evolution algorithm and a constraint handling technique for dynamic economic emission dispatch. *Energy* **2019**, *186*, 1–28. [CrossRef]
13. Yuan, G.; Yang, W. Study on optimization of economic dispatching of electric power system based on Hybrid Intelligent Algorithms (PSO and AFSA). *Energy* **2019**, *183*, 926–935. [CrossRef]
14. Rahmat, N.A. Computational Intelligence Based Technique for Solving Economic Dispatch Problem. Ph.D. Thesis, Universiti Teknologi MARA, Shah Alam, Malaysia, 2016.

energies

MDPI

Article

DSO Strategies Proposal for the LV Grid of the Future

Bartłomiej Mroczek [1,2,*] and Paweł Pijarski [1]

[1] Department of Power Engineering, Lublin University of Technology, Nadbystrzycka St. 38D, 20-618 Lublin, Poland; p.pijarski@pollub.pl

[2] Head of Strategy Department, ENERGA SA, Aleja Grunwaldzka 472, 80-309 Gdańsk, Poland

* Correspondence: b.mroczek@pollub.pl

Abstract: A significant challenge for the DSO (Distribution System Operator) will be to choose the optimum strategy for flexibility service in the LV area with high RES (renewable energy sources) penetration. To this end, a representative LV grid operated in Poland was selected for analysis. Three research scenarios with RES generation were presented in the range of 1–8 kW for the power factor from 0.9 to 1. The grid PV capacity was determined for four load profiles. Based on this factor, optimum RES volume management service types were determined. Under the flexibility service, the proposed power conversion services and active RES operations for DSO were proposed. The research was conducted using the Matlab and PowerWorld Simulator environment. Optimum active power values were obtained for the RES generation function for single and dual operation systems of the power conversion system. In future, the knowledge in the field of grid capacity will enable the DSO to increase the operating efficiency of the LV grid. It will enable the optimum use of the RES generation maximisation function and proper strategy selection. It will improve the energy efficiency of the power input through the MV/LV node.

Keywords: grid feeding; energy storage; PV; flexibility service; optimisation model

Citation: Mroczek, B.; Pijarski, P. DSO Strategies Proposal for the LV Grid of the Future. *Energies* **2021**, *14*, 6327. https://doi.org/10.3390/en14196327

Academic Editors: Victor Becerra and Ahmed Rachid

Received: 30 August 2021
Accepted: 29 September 2021
Published: 3 October 2021

1. Introduction

The European Union strategy presented in the European Green Deal aims at reducing greenhouse gas emissions by at least 55% by 2030 [1]. The set target will be attained among other things by installing renewable energy sources, while improving the energy efficiency. The direction to de-carbonise the power generation sector has already been accepted by the member states, including Poland. One of the main premises of the transformation process is action intended to protect the planet and human health. The validity of the assumed transformation path is confirmed by section VI of the IPCC report (AR6) [2], which indicates that global warming of 1.5–2 deg C/F can be stopped by quick greenhouse gas reduction.

The commonly observed mega-trend towards the development of prosumer power generation in Poland is confirmed by the statistics in the reports presented by the President of Energy Regulatory Office [3], who point to the significant dynamics of the year-to-year increase in new installations. The dynamics of the increase in prosumer microinstallations was approximately 191% in the period of 2018/2019 and approximately 202% in the period of 2019/2020. Due to the availability of subsidies and financial support for such projects, territorial units work with the residents to de-carbonise the urban and suburban sections of the LV grid. The electricity supply and consumption balance for end consumers changes locally.

At the national level, Poland's energy policy until 2040 predicts a fivefold increase in the number of prosumers by 2030, relative to the current condition. In addition, the number of energy-sustainable areas will increase at the local level [4]. It is predicted that microgrids will be formed based on separated sections of the LV grid. The power and electrical energy settlement for end consumers will be shifted to local balance areas as the local dispersed generation capacity increases.

It should be noted that the photovoltaic panel technology changes at the same time At present, manufacturers are introducing to the market a new type of panel in bifacial technology [5–7], with higher capacity levels (500–670 kWp) than previous types, defined as ultra-high power modules. This will provide the prosumers with higher generation capacities, often exceeding their momentary power requirements.

The process of power generation sector transformation through dispersed power generation changes the operating characteristics of the LV grid [8,9], to which these sources are connected. This phenomenon has at least two new contexts. The first context is the technical adjustment of the LV grid and the second context is the current balancing of the demand and the supply according to the established quality of service parameters. The European Commission has identified the need to manage these contexts. It defined the mitigation tool as energy grid flexibility and addressed it in art. 32 of chapter IV of Directive 944/2019 [10]. This enabled the DSO to apply a freely selected strategy (demand size response, storage system, etc.), depending on the context of the LV grid operation.

Considering the above-listed factors changing the existent LV grid operation, this research was motivated by the need to develop and describe the possible strategies for the DSO in the area. At present, the authors of the paper point to two possible technical solutions. The first solution is centralised and consists in using power converter systems as energy storage systems (ESS) in the grid operation system on the DSO side [11–15], as part of the infrastructure. In this solution, there is no description of a particular location for grid connection (beginning, middle, end of the LV grid) and the number of such systems as ESS. These considerations are exclusively connected with a single specific test case. The second solution is decentralised and consists in using power conversion systems with energy storage systems at the consumer (prosumers) [16–20]. The use of dispersed ESS systems depends on the times of response to the service of the ESS user [21] and/or the availability of the ESS systems at the given adjustment moment [22]. At the given moment, each ESS features different parameters: Capacity, power, ramp, rate, cycle time [23]. The flexibility of this service is negligible, including the control/adjustment capacity on the side of the DSO. Therefore, the authors of this paper identified the research problem in the form of the selection of the strategy for the LB grid based on a measurable factor. The grid PV capacity indicator was selected [24–26] for the LV grid operating in Poland. The power flow throughput limits of the LV grid were the subject of this research and the resulting evaluation, positive or negative, is a proposal for strategy development by the DSO.

2. Materials and Methods

2.1. Research Problem

The basic assumption of the research is the absorption of the maximum output of electrical energy generation of all PV sources in the LV grid, without applying restrictions to micro-producers. The role of the DSO is to use the available grid flexibility utilities to:

1. Maintain the voltage value at the individual connection points $(0.9\div1.1)\cdot U_n$, as per EN 50160 [27,28].
2. Reduce the impact of the energy flow towards the LV grid [8,29–31].
3. Reduce the ML/LV transformer overload during PV generation [11,30–32].

The authors of the paper assumed the research hypothesis that flexibility service rating exists, connected with the prosumer and generation load profile. The rating was built on the basis of the cost index [33,34] (unit power price of the converter system, number of ESS in relation to the existing PV inverters) and effective grid regulation. The obtained results are presented in Section 4. The ultimate purpose was to provide the DSO with a proposal including effective service flexibility strategies in accordance with the process presented in Figure 1.

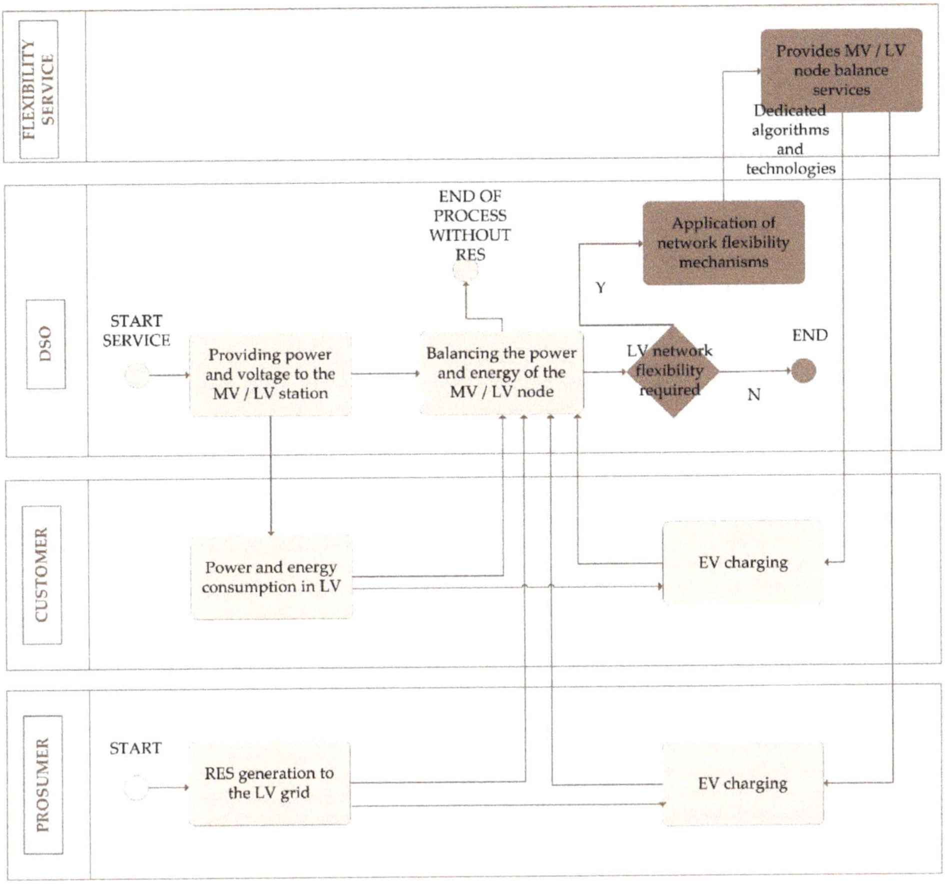

Figure 1. Launch the grid LV flexibility service.

The light beige colour was used to mark the tasks that the DSO currently performed in the distribution service provision process. The dark brown colour was used to mark the tasks that the DSO would perform in the future [10].

2.2. Research Object

The DSO infrastructure of the LV grid selected for research purposes was typical of suburban areas of agglomerations, to which new dispersed energy sources were connected systematically. The LV grid is radial and overhead, with two feeders (F1 and F2). The bus is made using a wire with a section of 70 mm^2; this results directly from the DSO standard in Poland [35]. The LV grid diagram is presented in Figure 2. The LV grid is connected with 23 consumer/prosumers with the connection power of 11 kW. Figure 2 shows the distances of the individual sections of the F1 and F2 bus (unit: Meter, e.g., −45—between the power poles Nr S1-1 and Nr S1-2).

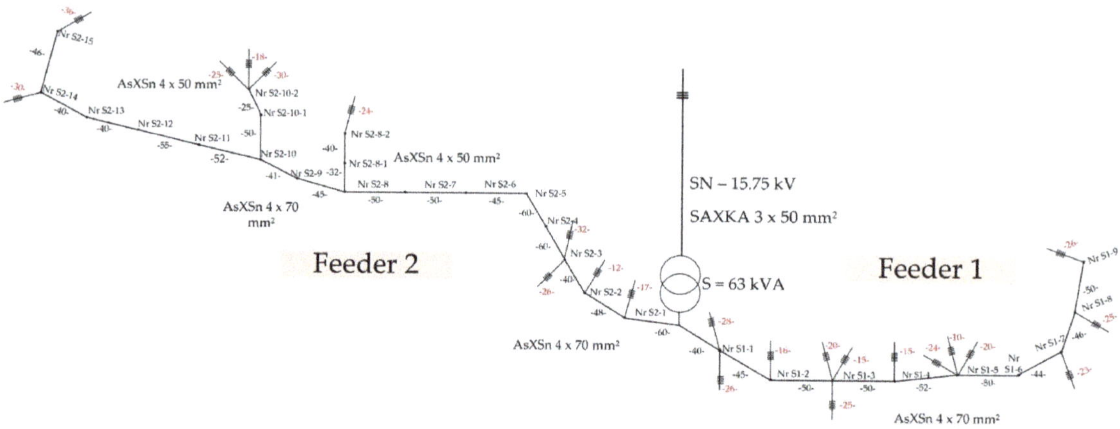

Figure 2. Diagram of the research LV grid.

The technical specifications of the analysed grid were specified in Tables 1–3. The technical specifications were extracted from the LV line design and supplemented with catalogue specifications.

Table 1. Technical specifications of the overhead line [36].

AsXSn LV	50 mm^2	70 mm^2
R [Ω/km]	0.641	0.443
X [Ω/km]	0.085	0.083
SAXKA MV	**50 mm^2**	
R [Ω/km]	0.641	-
X [Ω/km]	0.144	-

Table 2. Technical specifications of the transformer.

Power Rate	63 kVA
Voltage rate MV/LV	15.75/0.4 kV
Winding configuration	Dyn5
No load loss	0.81 kW
Load loss	1.2 kW
Impedance	4.5%

Table 3. Distances between the MV/LV transformer and the connection points (only data per consumer/connection point).

Connector Number	Feeder 1 [km] (Name)	Feeder 2 [km] (Name)
1	0.066 (1.01)	0.077 (2.01)
2	0.068 (1.02)	0.120 (2.02)
3	0.10 (1.03)	0.174 (2.03)
4	0.150 (1.04)	0.18 (2.04)
5	0.155 (1.05)	0.509 (2.05)
6	0.160 (1.06)	0.592 (2.06)
7	0.247 (1.07)	0.599 (2.07)
8	0.257 (1.08)	0.604 (2.08)

Table 3. *Cont.*

Connector Number	Feeder 1 [km] (Name)	Feeder 2 [km] (Name)
9	0.261 (1.09)	0.716 (2.09)
10	0.262 (1.10)	0.768 (2.10)
11	0.354 (1.11)	-
12	0.402 (1.12)	-
13	0.453 (1.13)	-

The total length of LV line for feeder 1 = 0.427 km, for feeder 2 = 0.732 km.

2.3. Boundary Conditions for Simulations

Research works were conducted based on specifications of the actual LV grid in Central Poland (Tables 1–3). To this end, the Matlab environment was used, integrated through dedicated scripts with the PowerWorld Simulator environment (Figure 3). Power distribution for the given condition, for each connection point, was calculated in accordance with the following formula (Equations (1) and (2)) [22,37].

$$P_k(V_k, V_j, \theta_k, \theta_j) = \sum_{j=1}^{N} |V_k||V_j|(G_{kj}\cos(\theta_{kj}) + B_{kj}\sin(\theta_{kj})) \quad (1)$$

$$Q_k(V_k, V_j, \theta_k, \theta_j) = \sum_{j=1}^{N} |V_k||V_j|(G_{kj}\sin(\theta_{kj}) - B_{kj}\cos(\theta_{kj})) \quad (2)$$

where:

P_k—active power at bus k;
Q_k—reactive power at bus k;
V_k—Voltage magnitude at bus k;
V_j—Voltage magnitude at bus j;
G_{kj}—Mutual conductance between buses k and j;
B_{kj}—Mutual susceptance between buses k and j;
θ_{kj}—Voltage angle difference between buses k and j;

One of the most important components for the development of proper evaluation of LV grid operation with a large number of PV generators is proper selection of the P_l load model by the consumer/prosumers. The consumer profiles used in simulation are selected to include the broadest possible range of cases possible under the given conditions, the most extreme for the LV grid operation [38]. In addition, P_l has a direct impact on the LV grid operation, balancing or not the current/present generation profile—P_{gc}. The four selected profiles P_l are presented in Table 4.

Table 4. Load profiles used in power distribution simulations.

Profile P_l	Name	Feeder 1	Feeder 2
1. Simultaneity factor for peak power demand in accordance with the Polish standard SEP-E-002, Figure 4a	SF	3.27 kW	3.78 kW
2. Right skewed Figure 4c,e distribution exponential function $f(x) = a^x$	RS	$\sum_{n}^{n=13} P_l = 42.51$ kW	$\sum_{n}^{n=10} P_l = 37.8$ kW
3. Left skewed Figure 4d,f distribution exponential function $f(x) = x^a$	LS	$\sum_{n}^{n=13} P_l = 42.51$ kW	$\sum_{n}^{n=10} P_l = 37.8$ kW
4. Random data, Figure 4b	RD	<0.5–2.5> kW	<0.5–2.5> kW

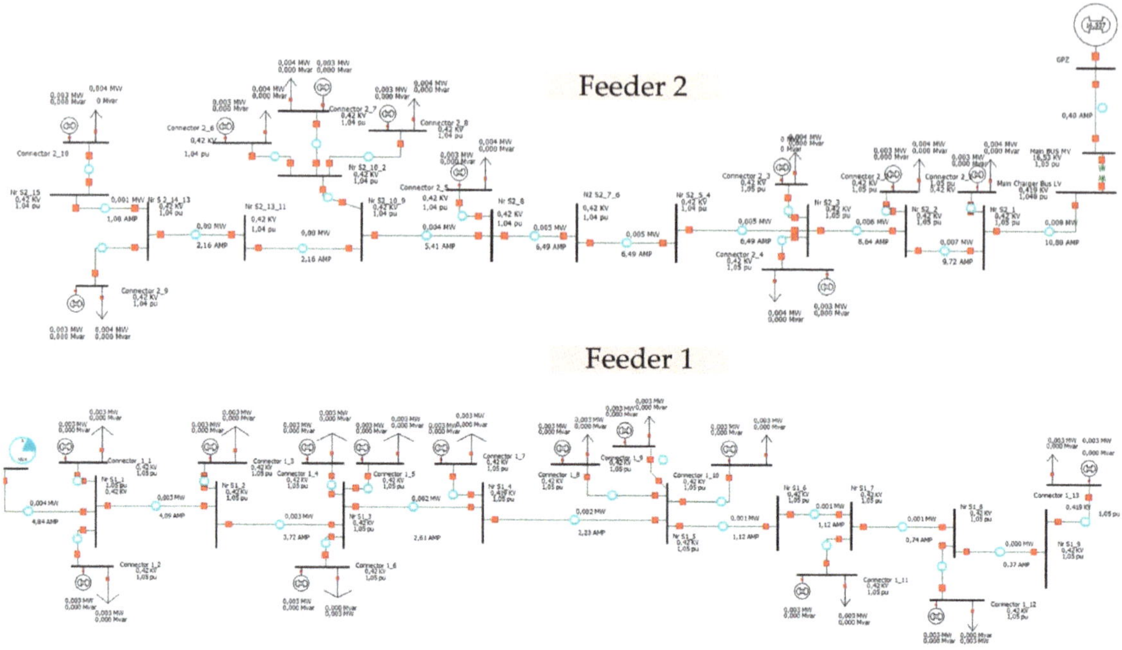

Figure 3. Section of the LV grid (Figure 1) implemented in the PowerWorld environment.

The generation PV profiles P_{gc} for Central Poland are the production hours from 9 am to 6 pm within prosumer load demand P_l, in the same time [39]. The value P_l on the side of prosumers is different for period: (A) Saturday–Sunday/holiday, (B) Monday–Friday. Generally, P_l in the LV grid is stochastic. Therefore, an attempt was made to define the boundary conditions so that they contain the largest number of solutions corresponding to the real P_l profile. The first three profiles—SF, RS, LS—correspond to period A. With this end in view, defined load with a constant value (SF) was calculated in accordance with the standard SEP-E-002 (power connection 11 kW). The other two profiles, RS and LS, are unequal loads. There are differences at the beginning and end of the circuit F1 and F2. This has been described mathematically (Table 4). The fourth profile (Table 4.4) is typical loads for period B, where the profile is developed based on small household appliances (refrigerator) and electronics in stand-by mode.

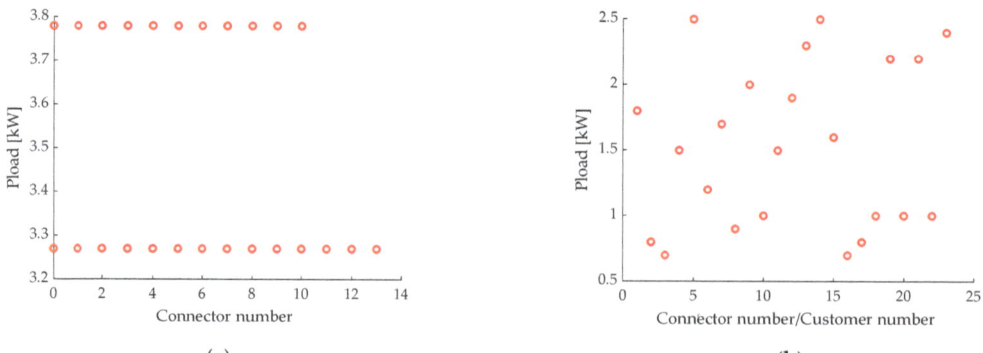

(a) (b)

Figure 4. *Cont.*

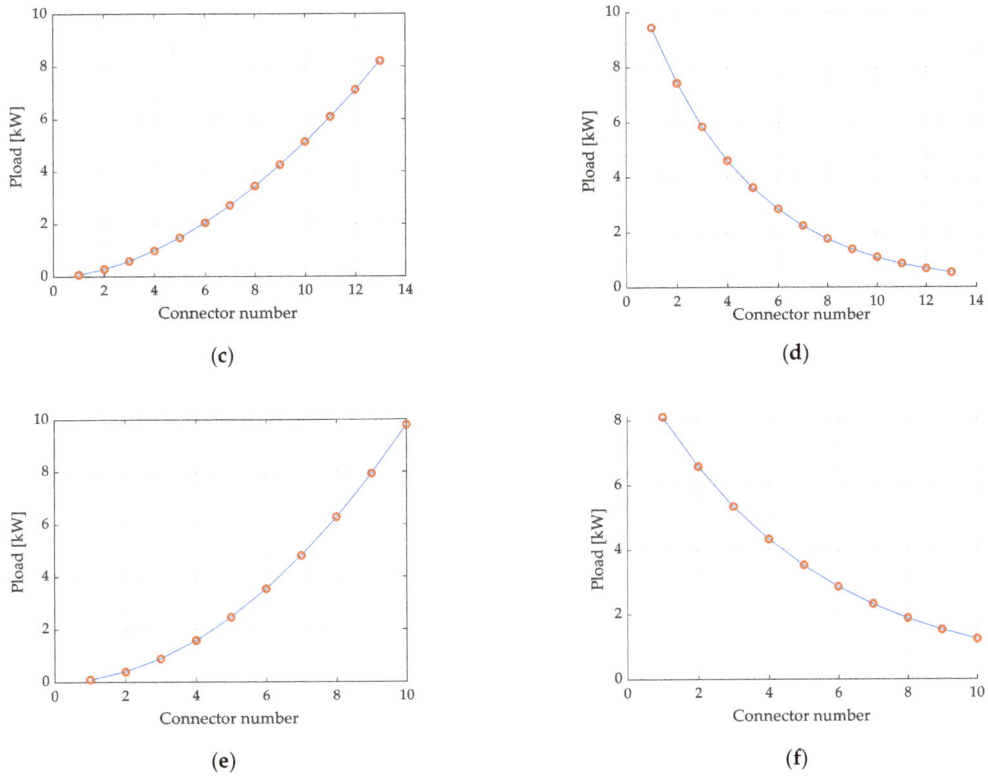

Figure 4. Distribution of load profiles: (**a**) SR; (**b**) RS; (**c**) LS F1; (**d**) RD F1; (**e**) LS F2; (**f**) RD F2.

The apparent power distribution in the LV grid for each of the described profiles P_l, were simulated for P_{gc} in the range of 1 kW to 8 kW and tracking of 0.25 kW. Since the LV grid, despite the high R/X ratio [37], does not operate in pure resistance mode (PV power electronic system operation) [40], PV generation is simulated for the power factor (p.f.) from 0.9 to 1. The encountered voltage asymmetry in the LV grid was eliminated by the four-wire power conversion system, controlled independently for each phase [41,42]. In the research, this fact was omitted as negligible; phases F1 and F2 can be considered as technically separate and controlled independently.

2.4. Scenarios

The boundary conditions described in the previous section as components of data input for the research were framed in three research scenarios. In addition, the operating system of the LV grid (Figure 1) is supplemented with ESS operating in the grid feeding mode [43] (Figure 5).

The operation of the LV grid with ESS was controlled by the inputs P (active power) and Q (reactive power) of the power converter system (Figure 5) to achieve the objectives described in Section 2.1 (Numbers 1–3).

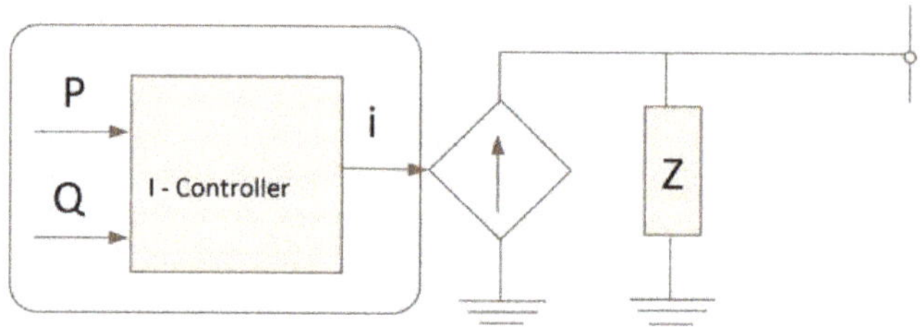

Figure 5. Control system—Grid Feeding.

The first research scenario is the evaluation of the grid PV capacity for LV, the "AS IS" condition analysis. The input data values P_{gc} and P_l were searched for the LV grid, where the voltage requirement of 1.1 *Un* is not met. In addition, power flows towards MV through the transformer were checked.

The second research scenario was implemented using ESS in the operation system, Figure 6a. The analysis covered the LV grid and its operating characteristics with the boundary conditions described in Section 2.3 to define the values of *P* and *Q* for ESS. Location at one point of the grid—the MV/LV station (Figure 6a).

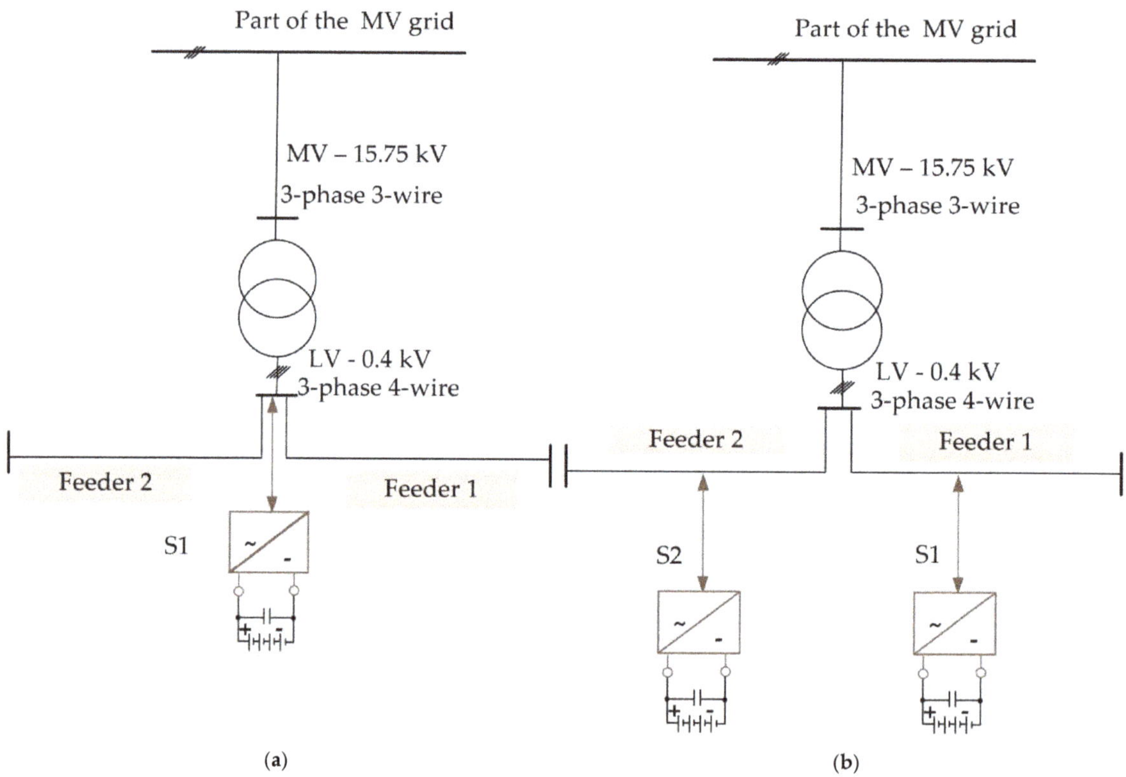

Figure 6. Scenario with grid feeding as: (**a**) 1-ESS; (**b**) 2-ESS.

The third research scenario was implemented using two ESS systems: *S1* and *S2* in the operation system, Figure 6b. The analysis covered the LV grid and its operating characteristics with the boundary conditions described in Section 2.3 to define the values of *P1*, *P2* and *Q1*, *Q2* for *S1* and *S2* (Figure 6b). The location of the ESS in the depth of the LV grid is based on the data of scenario no. 1.

3. Results

By iterative power flow analysis in the dedicated LV grid, within the scope of three scenarios and four load profiles, the answer was obtained in the field of grid PV capacity. Developed results presented in Table 5. Based on them, a non-linear function was developed, described in three dimensions *f(x, y, z)* where: *x*—distance from the MV/LV station (distance); *y*—values of PV current generation (P_{gc}); *z*—voltage at the given node (*U p.u.*). It should be emphasized that the results in Table 5 are the maximum values of PV power generation for which the *Un* = 1.1 condition is met for each node in the LV grid.

Table 5. Capacity Grid for four load profiles.

Capacity Grid	SF	RS	LS	RD
Feeder 1, cosf(φ) = 1.00 for PV	<8 kW	<8 kW	<8 kW	7.5 kW
Feeder 2, cosf(φ) = 1.00 for PV	7.75 kW	<8 kW	7.0 kW	5.75 kW
Feeder 1, cosf(φ) = 0.95 for PV	6.5 kW	7.25 kW	5.75 kW	5.5 kW
Feeder 2, cosf(φ) = 0.95 for PV	6.0 kW	6.75 kW	5.25 kW	4.25 kW
Feeder 1, cosf(φ) = 0.90 for PV	5.75 kW	6.25 kW	5.0 kW	4.5 kW
Feeder 2, cosf(φ) = 0.90 for PV	5.25 kW	6.0 kW	4.75 kW	4.0 kW

The values presented in Table 5 are the starting point for the regulation of power converter systems (power tracking) in accordance with the diagram in Figure 5. The obtained grid PV capacity results will also be the basis for the evaluation of the strategy selection on the DSO side. Green: Grid PV capacity values that do not require adjustment for a given load profile; red: Those cases of LV network operation that require a response from the DSO; yellow: Result values are shown for little regulation adjustment.

Analytical research was carried out in two ways. The first track of works was carried out in the field of searching for a grid PV capacity for one value of the maximum RES generation capacity per every node. The second course was carried out by the optimal grid PV capacity values for all nodes (different values of RES per node). Below is a mathematical model for the second track.

The purpose of the research is to maximise RES generation. Therefore, the objective function is grid PV capacity measured by the maximum generation value of *P* at cos(φ) = const., for which the voltage requirements for the electrical energy supply are met. For the purposes of optimisation of the objective function *f(x, y, z)*, the equations of Karush–Kuhn–Tucker [44,45] for non-linear programming algorithms were used, following the formulae (Equations (3)–(6)) [46]:

$$\max_{x \in \mathbb{R}^n} = f(x) \tag{3}$$

$$L(x,u,v) = f(x) + \sum_{i=1}^{m} \mu_i g_i(x) + \sum_{j=1}^{r} \lambda_j h_j(x), \tag{4}$$

$$-\nabla f(x) + \sum_{i=1}^{m} \mu_i \nabla g_i(x) + \sum_{j=1}^{r} \lambda_i \nabla h_j(x) = 0, \tag{5}$$

$$\begin{gathered} g_i(x) \leq 0,\ for\ i = 1,\ \ldots, m \\ h_i = 0,\ for\ j = 1, \ldots, r \\ \mu_i g_i(x) = 0,\ for\ i = 1,\ \ldots, m. \end{gathered} \tag{6}$$

The objective function has two limits: g_i—in the RES power range (P_{gc}); h_i—in the 1.1 *p.U.* value range for each node (*U per unit*).

3.1. The First Scenario

The first scenario is dedicated to the evaluation of grid PV capacity without flexibility service. It clearly indicates the limits, at which the given feeder (F1 and F2) achieves the maximum operating values. In addition, Figure 6 presents the results of optimisation of RES operation. However, Figure 5 presents the operation profile of the MV/LV transformer in this type of grid.

3.1.1. Grid PV Capacity

In Table 5, the max P_{gc} values obtained for each P_l profile were listed. At the same time, reactive power from the RES side was input at connection points.

3.1.2. Optimisation of the Non-Linear Function

In Figure 7, only for P_l = SR (limited due to the number of possible diagrams), optimum RES generation values were presented for each connection point as a function of the distance. These are the optimum values of grid PV capacity per node. The analysis points to the limits for the individual feeders and individual nodes. This points to the potential ESS connection point in the depth of the grid, the minimum value of the function in the given distance range.

3.1.3. Operation Balance for the MV/LV Transformer

For the load profile of P_l = SR, as in the previous sub-section, power inputs through the MV/LV station were determined as a function of RES generation increase. Figure 8 shows how the power input from the MV grid to the LV grid changes the vector to LV to MV. The process is completed regardless of the type of load profile and p.f., in connection with the increase of the RES generation value (P_{gc}).

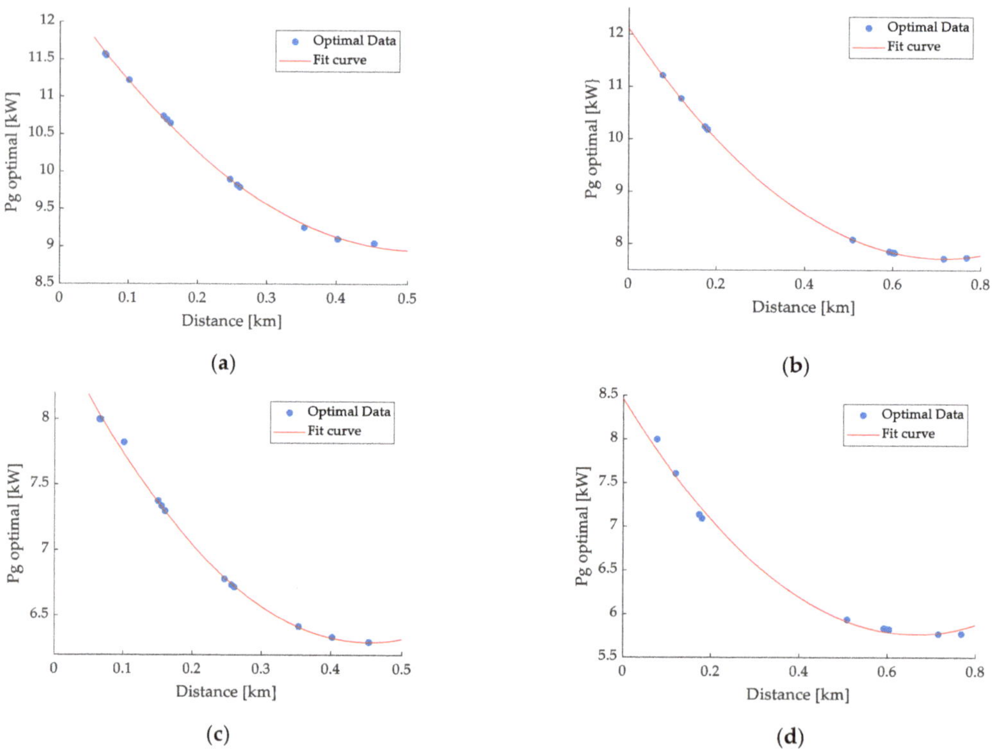

Figure 7. *Cont.*

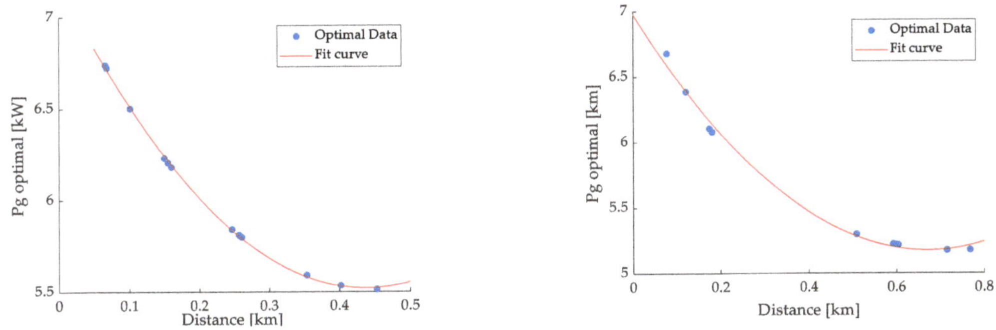

(e) (f)

Figure 7. Optimisation results for SR: (**a**) F1, p.f. = 1; (**b**) F2, p.f. = 1; (**c**) F1, p.f. = 0.95; (**d**) Figure 2. p.f. = 0.95; (**e**) F1, p.f. = 0.9; (**f**) F2, p.f. = 0.9.

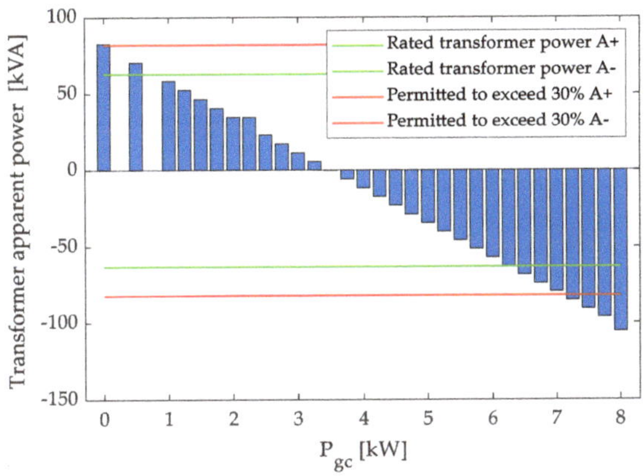

Figure 8. Operation profile of the MV/LV transformer as a function of RES generation increase (P_{gc}).

Therefore, the selection of transformer power at high RES penetration is important. Nominal service operation with PV requires re-selection of the transformer power or in addition requires the use of flexibility service.

3.1.4. Reactive Power during PV Generation

The requirements for PV inverters on the side of the DSO are p.f. = 1. Due to the quality of the power electronic systems used by the prosumers, deviations from this value exist. Notwithstanding, this type of operation in case of small P_l power values does not serve its purpose in terms of grid PV capacity (Table 5).

The second method is active collaboration with the prosumers for reactive power consumption from the LV grid. Reactive power could be managed by PV inverters in the p.f. range of 0.9–1. This will reduce the RES production, but at the same time it will enable the prosumer to provide service to the DSO. Below is presented the most demanding case: P_l = RD and p.f. = 0.9, the results of the described prosumer service. Figure 9 features the RMS value of the nodal voltage for the two feeders—blue, while red states restriction level for node 1.1 *Un*.

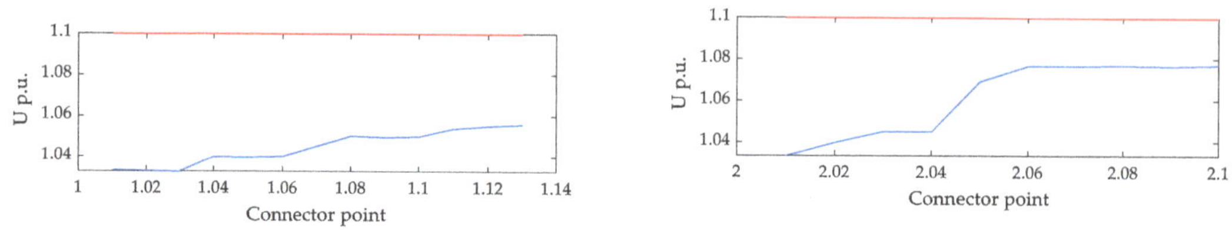

(a) (b)

Figure 9. The voltage value distribution at connection points, with reactive power consumption by RES: (**a**) RD, p.f. = 0.9 for F1; (**b**) RD, p.f. = 0.9 for F2.

The conducted research proved that in this case, regardless of the load profile, and for RES generation in the range of ⟨1 : 8⟩ kW, there is no voltage excess at node points. This is the most effective method; however, it depends on a third party.

3.2. The Second Scenario

The second scenario is dedicated to the evaluation of the required capacity of the power conversion system for voltage condition adjustment at nodes. The operation system is presented in Figure 6a. The volume of data collected for *S1* at the given P_l depends on the grid PV capacity (Table 5). Figures 10–12 present the active power values *P* for *S1*, with identification of the function $S1 = f(P_{gc})$.

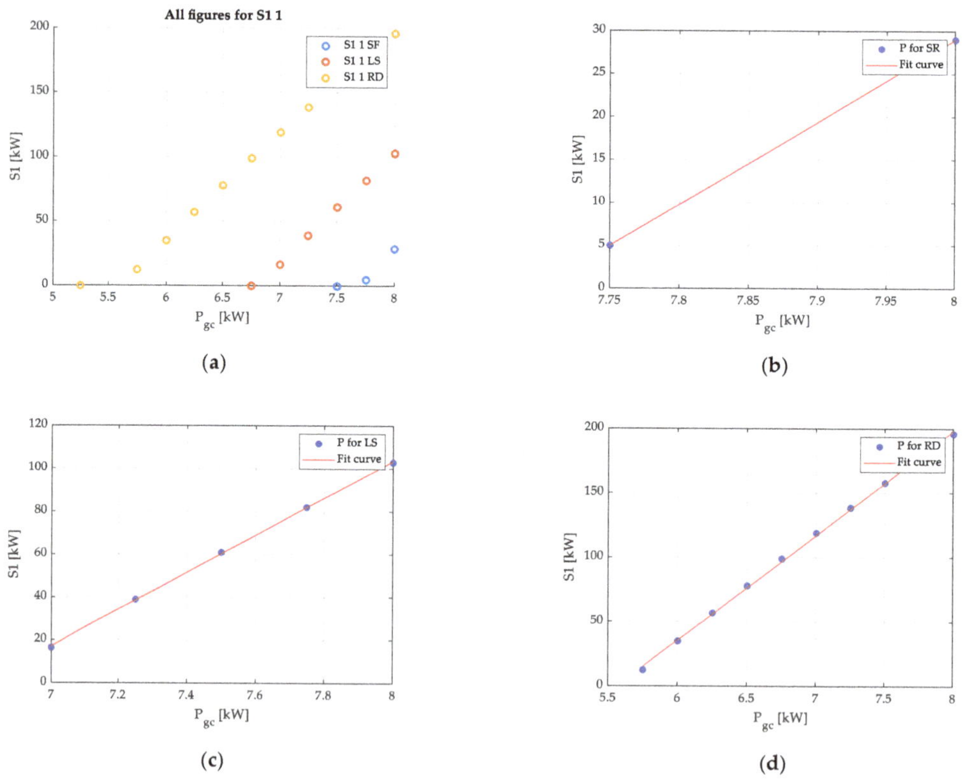

(a) (b) (c) (d)

Figure 10. *S1*, p.f. = 1: (**a**) All figures for *S1*; (**b**) SR; (**c**) LS; (**d**) RD.

In accordance with the obtained results (Table 5), only three load profiles are subject to power input adjustment. Adjusted R-square (R^2) for SR = 1 (small data volume), R^2 for LS = 0.9997, R^2 for RD = 0.999.

In accordance with the obtained results (Table 5), four load profiles are subject to power input adjustment. R^2 for SR = 0.9993, R^2 for PS = 1, R^2 for LS = 0.9998, R^2 for RD = 0.9996.

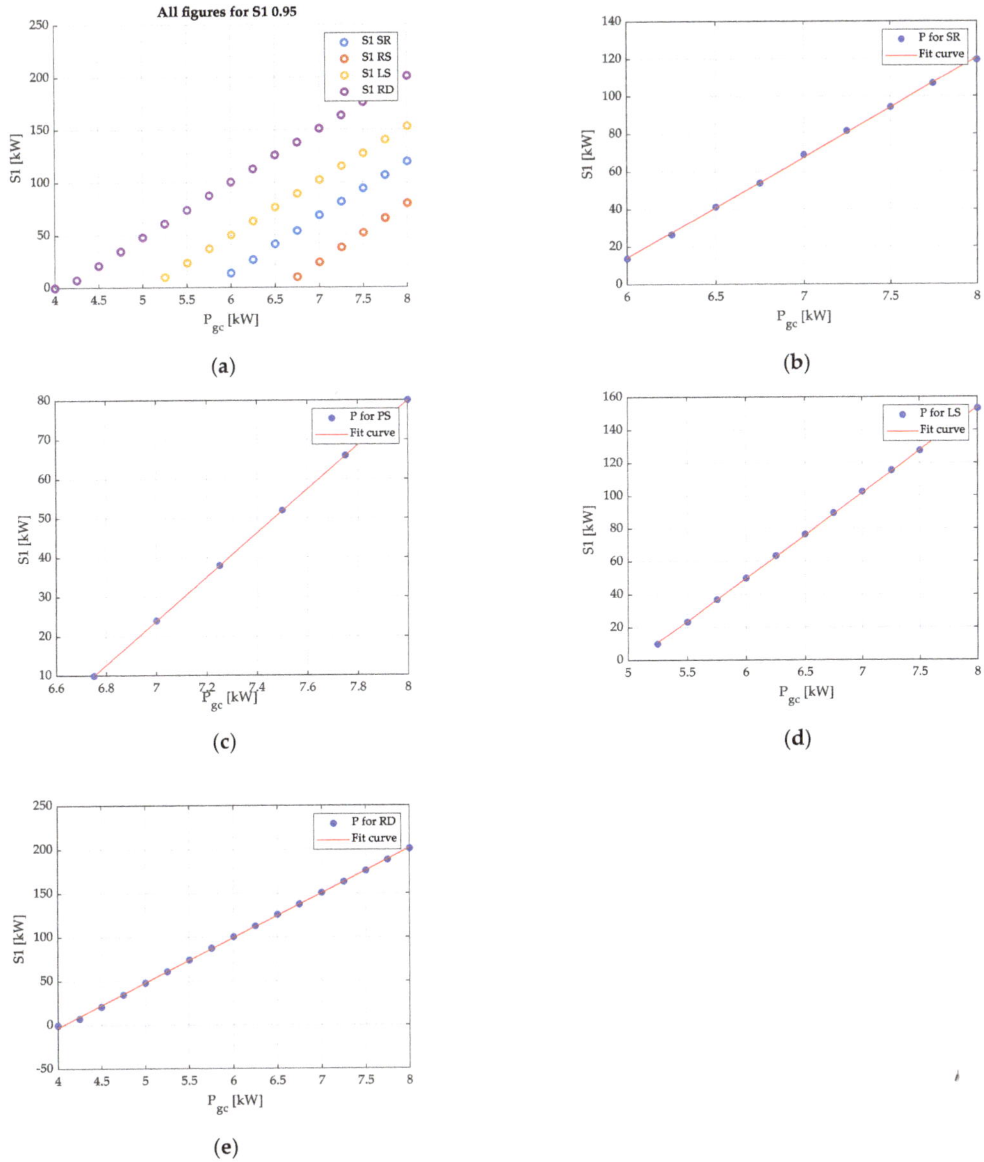

Figure 11. *S1*, p.f. = 0.95: (**a**) All figures for *S1*; (**b**) SR; (**c**) PS; (**d**) LS; (**e**) RD.

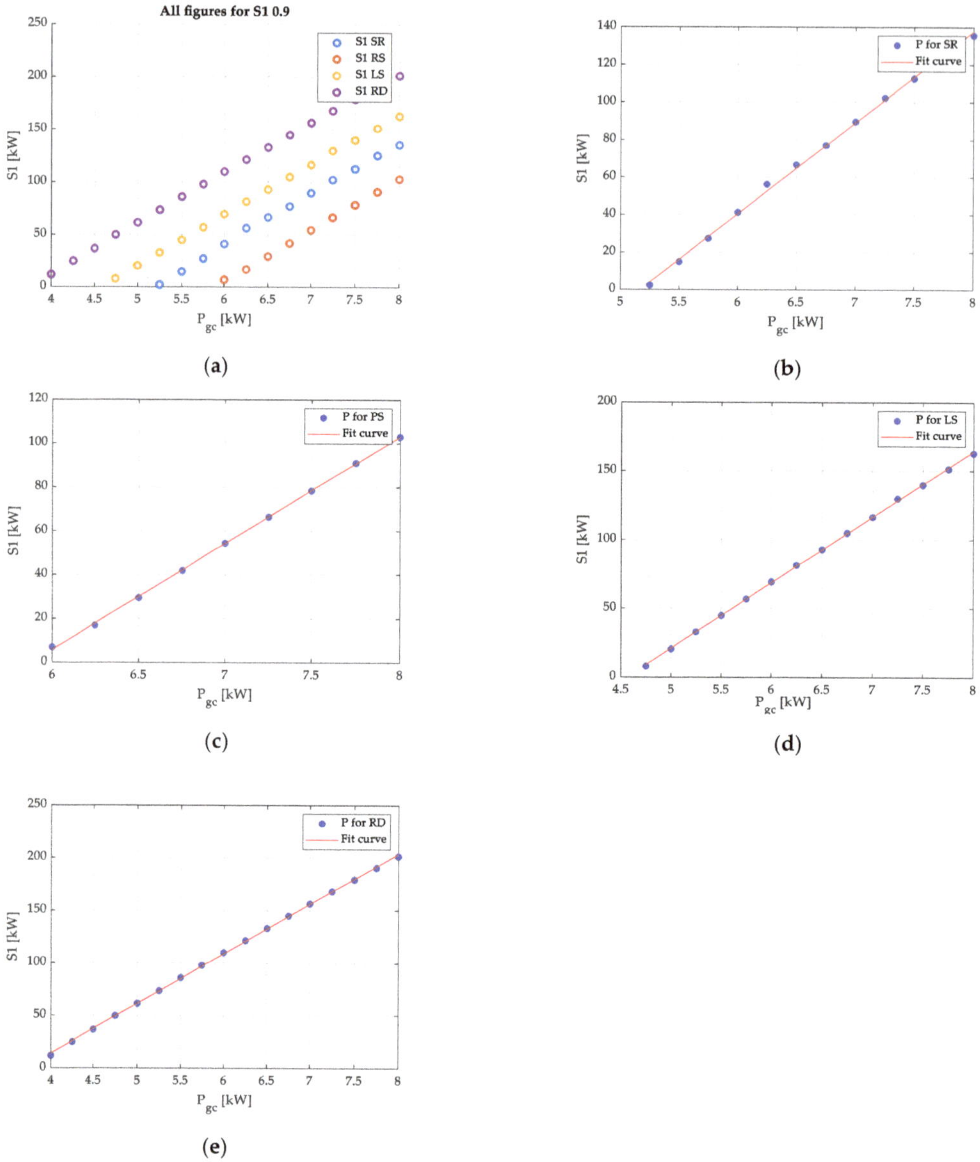

Figure 12. *S1*, p.f. = 0.9: (**a**) All figures for *S1*; (**b**) SR; (**c**) PS; (**d**) LS; (**e**) RD.

In accordance with the obtained results (Table 5), four load profiles are subject to power input adjustment. R^2 for SR = 0.998, R^2 for PS = 0.9996, R^2 for LS = 0.9997, R^2 for RD = 0.9997. The application of only *S1* for operation adjustment only is to maintain the voltage conditions with active power from 2.5 kW to 201.5 kW.

3.3. The Third Scenario

The third scenario is dedicated to the evaluation of the required capacity for two power conversion systems for voltage condition adjustment at nodes, independent of *F1* and *F2*. The operation system is presented in Figure 6b. Two ESS systems: *S1* and *S2* are connected in the depth of the LV grid. For *F1*, it is the node No. *S2_10_2* (Figure 2); for *F2* it is the node No. *S1_7* (Figure 2). The volume of data collected for *S1* at the given load profile depends on the grid PV capacity (Table 5). Due to the collected data volume, the results are presented in tables (Tables 6–8).

Table 6. Results for *S1* and *S2*, p.f. = 1.

Load Profile/Pgc (kW)	5.75	6.0	6.25	6.5	6.75	7.0	7.25	7.5	7.75	8.0
SR										
S1 (kW)	-	-	-	-	-	-	-	-	0	0
S2 (kW)	-	-	-	-	-	-	-	-	0.5	2.5
LS										
S1 (kW)	-	-	-	-	-	0	0	0	0	0
S2 (kW)	-	-	-	-	-	1.5	4.0	6.0	8.5	11.0
RD										
S1 (kW)	0	0	0	0	0	0	0	0	0	2.0
S2 (kW)	1.5	4.0	6.0	8.0	10.0	12.5	14.5	16.5	19.0	21.0

R^2 for SR S2 = 1 (small data volume), R^2 for LS S2 = 0.9982; R^2 for RD S1 = 0.9997; RD S2 = 0.9994. On average, approximately 40% smaller active power is required to control the LV grid if two, not one, ESS systems are used. In extreme cases (P_{gc} = 8 kW), even approximately 90% smaller.

Table 7. Results for S1 and S2, p.f. = 0.95.

Load Profile/Pgc (kW)	5.75	6.0	6.25	6.5	6.75	7.0	7.25	7.5	7.75	8.0
SR										
S1 (kW)	-	0	0	0.5	3.0	5.0	7.5	10.0	12.0	14.5
S2 (kW)	-	2.5	5.5	7.0	9.5	12.0	13.5	15.5	18.0	20.0
RS										
S1 (kW)	-	-	-	-	0	0	0.5	3.5	6.0	8.5
S2 (kW)	-	-	-	-	1.0	4.0	6.5	8.5	11.0	13.0
LS										
S1 (kW)	0	2.0	4.5	7.0	9.0	11.5	14.0	16.5	18.5	21
S2 (kW)	6.5	9.0	11.0	13.0	15.0	17.5	19.5	21.5	23.5	25.5
RD										
S1 (kW)	3.5	5.5	8.0	10.5	13.0	15.5	17.5	20.0	22.5	24.5
S2 (kW)	15.5	18.0	19.5	21.5	23.5	25.5	27.5	30.0	32.0	34.0

R^2 for SR S1 = 0.9991, S2 = 0.9967; R^2 for LS S1 = 0.9995, S2= 0.9994; R^2 for PS S1 = 0.9968, S2 = 0.9762; R^2 for RD S1 = 0.9995, RD S2 = 0.9972.

Table 8. Results for S1 and S2, p.f. = 0.9.

Load Profile/Pgc (kW)	5.75	6.0	6.25	6.5	6.75	7.0	7.25	7.5	7.75	8.0
SR										
S1 (kW)	1.5	3.5	6.0	8.5	10.5	13.5	16.0	18.5	20.5	23.0
S2 (kW)	6.0	8.5	10.0	12.5	14.5	16.5	18.5	20.5	23.0	25.0
RS										
S1 (kW)	-	0.0	1.0	3.5	6.0	9.0	11.5	14.5	17.5	20.5
S2 (kW)	-	1.5	4.0	6.0	8.0	10.5	12.5	14.5	17.0	19.5
LS										
S1 (kW)	7.0	9.5	11.5	14.0	16.5	19.0	21.5	24.0	26.0	28.5
S2 (kW)	11.5	13.5	15.5	17.5	19.5	22.0	24.0	26.0	28.0	30.5
RD										
S1 (kW)	10.0	12.5	15.0	17.5	20.0	22.0	24.5	27.5	30.0	33.5
S2 (kW)	19.0	21.5	23.5	25.5	27.5	30.0	31.5	33.0	36.0	37.5

R^2 for SR S1 = 0.9989, S2 = 0.986; R^2 for LS S1 = 0.9994, S2= 0.9988; R^2 for PS S1 = 0.9986, S2 = 0.9989; R^2 for RD S1 = 0.9985, RD S2 = 0.9985.

4. Discussion

The collected simulation data clearly indicate that the consumer load profile with simultaneous RES generation has a direct impact on the grid PV capacity (Table 5). Based on the completed mathematical power flow simulation in the LV grid, in different research scenarios, it should be ascertained that the control (maintenance of the voltage value of 1.1 *Un*) and efficient balance optimisation (Figure 5) of the grid are possible. The research indicates that there is no single method of flexibility service. Proper selection of the method depends on the effective grid regulation and profitability (capex and/or opex) and the day of the week (confirmed hypothesis—Section 2.1). These above-mentioned items will build a method rating for a given area of the grid.

Table 9 lists flexibility service proposals according to the load profile type (Table 4). It also represents a method rating using a colour and mark (+, −, +/−). Green is used to mark the types of service that provides technical efficiency at the lowest cost per service. Yellow is used to mark the services that do not provide the full range of LV grid adjustment and/or that may have a high cost per service. Red is used to mark the services that do not provide LV grid adjustment and/or are too expensive [47,48].

Table 9. Proposed flexibility service rating.

Type of Flexibility Service	SF	RS	LS	RD
PV inverter Inverter parameters power factor = 1, Additional requirements in Networks Codes DSO.	+	+	+/−	−
PV inverter Reactive power consumption,	+	+	+	+
One ESS system central installation, connected with main bus	+/−	+	-	−
Two ESS systems Storage systems installed in the depth of the grid	+	+	+	+/−
DSR signals calling for an increase in power consumption	−	+/−	+	+
Hybrid bundling services	+/−	+/−	+	+

In addition, the research indicates a quasi-linear (laboratory confirmation required) relationship between the P and jQ of RES generation, and the P and jQ of the power conversion system *S1* and/or *S2*. There is an error connected with non-linearity, measured by the indicator of function alignment with measurement data, adjusted R-square statistic, which is close to 1 (descriptions under Tables 6–8). The error results only from the precision of ESS control by power increase in 0.25 kW increments [49]. The simplest and cheapest service method is the maintenance of the p.f. close to 1 by RES or reactive power consumption by the inverter [50,51]. The developed results (data) will be used in the future to build machine learning for the purposes of power and voltage regulation in the LV grid [52–54].

5. Conclusions

There is still need for research works on bridging the gap between power generation based on end consumer supplied from central generation units and power generation based on local RES. The assumed research hypothesis (section research problem) was confirmed by the results obtained in dedicated scenarios (Tables 5–8, Figures 7–12). The basic task of the future DSO is to prepare the bus LV grid—the minimum section of 70 mm^2—for the mega-process that is RES growth in the grid. In addition, the accumulation of knowledge of the load profile and generation is necessary for the development of grid PV capacity, as well as future strategies of dedicated services.

Proper selection of the MV/LV transformer with ESS remains an extremely important issue. Even constant PV generation at the level of 3.5 kW in the investigated LV grid resulted in the change of the power flow direction. The services proposed in Table 9 directly contribute to the transformer load balance, but full coverage of the power flow towards MV will only be possible in connection with ESS.

Author Contributions: Oprogramowanie Naukowo-Techniczne sp. z o.o. sp. k, Software Vendor of Matlab in Poland in the scope of technical support in the optimisation of non-linear functions. Conceptualization, B.M. and P.P.; methodology, B.M.; software, B.M. and P.P.; validation, B.M.; formal analysis, B.M.; investigation, B.M. and P.P.; resources, B.M.; data curation, B.M.; writing—original draft preparation, B.M.; writing—review and editing, B.M. and P.P.; visualization, B.M.; supervision, P.P.; project administration, B.M.; funding acquisition, B.M. and P.P. All authors have read and agreed to the published version of the manuscript.

Funding: This research received no external funding.

Institutional Review Board Statement: Not applicable.

Informed Consent Statement: Not applicable.

Data Availability Statement: Not applicable.

Conflicts of Interest: The authors declare no conflict of interest.

References

1. European Commission. Available online: https://ec.europa.eu/info/strategy/priorities-2019-2024/european-green-deal/delivering-european-green-deal_en#transforming-our-economy-and-societies (accessed on 30 September 2021).
2. IPCC Home. Available online: https://www.ipcc.ch/report/ar6/wg1/ (accessed on 30 September 2021).
3. Biuletyn Informacji Publicznej Urząd Regulacji Energetyki. Available online: https://bip.ure.gov.pl/bip/o-urzedzie/zadania-prezesa-ure/raport-oze-art-6a-ustaw/3793,Raport-dotyczacy-energii-elektrycznej-wytworzonej-z-OZE-w-mikroinstalacji-i-wpro.html (accessed on 30 September 2021).
4. Ministerstwo Klimatu i Środowiska. Available online: https://www.gov.pl/web/klimat/polityka-energetyczna-polski-do-2040-r-przyjeta-przez-rade-ministrow (accessed on 30 September 2021).
5. Olczak, P.; Olek, M.; Matuszewska, D.; Dyczko, A.; Mania, T. Monofacial and Bifacial Micro PV Installation as Element of Energy Transition—The Case of Poland. *Energies* **2021**, *14*, 499. [CrossRef]
6. Kopecek, R.; Libal, J. Bifacial Photovoltaics 2021: Status, Opportunities and Challenges. *Energies* **2021**, *14*, 2076. [CrossRef]
7. Jang, J.; Pfreundt, A.; Mittag, M.; Lee, K. Performance Analysis of Bifacial PV Modules with Transparent Mesh Backsheet. *Energies* **2021**, *14*, 1399. [CrossRef]
8. Pereira, C.O.; Cunha, V.C.; Riciiradi, T.R.; Riboldi, V.B.; Tuo, J. Pre-Installation Studies of a BESS in a Real LV Network with High PV Penetration. In Proceedings of the 2019 IEEE PES Innovative Smart Grid Technologies Conference—Latin America (ISGT Latin America), Gramado City, Brazil, 15–18 September 2019.
9. Bangash, K.N.; Farrag, M.E.A.; Osman, A.H. Manage Reverse Power Flow and Fault Current Level in LV Network with High Penetration of Small-Scale Solar and Wind Power Generation. In Proceedings of the 2018 53rd International Universities Power Engineering Conference (UPEC), Glasgow, Scotland, 4–7 September 2018.
10. EUR-Lex. Available online: https://eur-lex.europa.eu/legal-content/PL/ALL/?uri=CELEX%3A32019L0944 (accessed on 30 September 2021).
11. Gao, X.; Sossan, F.; Christakou, K.; Paolone, M.; Liserre, M. Concurrent Voltage Control and Dispatch of Active Distribution Networks by means of Smart Transformer and Storage. *IEEE Trans. Ind. Electron.* **2018**, *65*, 6657–6666. [CrossRef]
12. Zarrilli, D.; Giannitrapani, A.; Paoletti, S.; Vicino, A. Energy Storage Operation for Voltage Control in Distribution Networks: A Receding Horizon Approach. *IEEE Trans. Control Syst. Technol.* **2018**, *26*, 599–609. [CrossRef]
13. Zupancic, J.; Lakic, E.; Medved, T.; Gubina, F.A. Advanced Peak Shaving Control Strategies for Battery Storage Operation in Low Voltage Distribution Network. In Proceedings of the 2017 IEEE Manchester PowerTech, Manchester, UK, 18–22 June 2017.
14. Alam, M.J.E.; Kashem, M.; Sutanto, D. Community Energy Storage for Neutral Voltage Rise Mitigation in Four-Wire Multi-grounded LV Feeders With Unbalanced Solar PV Allocation. *IEEE Trans. Smart Grid* **2015**, *6*, 2845–2855. [CrossRef]
15. Ji, H.; Wang, C.H.; Li, P.; Zhao, J.; Song, G.; Ding, F.; Wu, J. A centralized-based method to determine the local voltage control strategies of distributed generator operation in active distribution networks. *Appl. Energy Vol.* **2018**, *228*, 2024–2036. [CrossRef]
16. Okubo, R.; Yoshizawa, S.; Hayashi, Y.; Kawano, S.; Takano, T.; Itaya, N. Decentralized Charging Control of Battery Energy Storage Systems for Distribution System Asset Management. In Proceedings of the 2019 IEEE Milan PowerTech, Milan, Italy, 23–27 June 2019.
17. Nousdilis, A.I.; Kryonidis, G.C.; Kontis, E.O.; Papagiannis, G.K.; Christoforidis, G.C. Economic Viability of Residential PV Systems with Battery Energy Storage Under Different Incentive Schemes. In Proceedings of the 2018 IEEE International Conference on Environment and Electrical Engineering and 2018 IEEE Industrial and Commercial Power Systems Europe (EEEIC/I&CPS Europe), Palermo, Italy, 12–15 June 2018.
18. Ambrosio-Albalá, P.; Bale, C.S.E.; Pimm, A.J.; Taylor, P.G. What Makes Decentralised Energy Storage Schemes Successful? An Assessment Incorporating Stakeholder Perspectives. *Energies* **2020**, *13*, 6490. [CrossRef]
19. Mocci, S.; Natale, N.; Pilo, F.; Ruggeri, S. Exploiting distributed energy storage to increase network hosting capacity with a Multi-Agent control system. In Proceedings of the 2016 AEIT International Annual Conference (AEIT), Capri, Italy, 5–7 October 2016.

20. Unigwe, O.; Okekunle, D.; Kiprakis, A. Towards benefit-stacking for grid-connected battery energy storage in distribution networks with high photovoltaic penetration. In Proceedings of the 11th IET International Conference on Advances in Power System Control, Operation and Management (APSCOM 2018), Hong Kong, China, 11–15 November 2018. [CrossRef]
21. Karrari, S.; Ludwig, N.; Hagenmeyer, V.; Noe, M. A Method for Sizing Centralised Energy Storage Systems Using Standard Patterns. In Proceedings of the 2019 IEEE Milan PowerTech, Milan, Italy, 23–27 June 2019.
22. Costa, H.M.; Sumaili, J.; Madureira, A.G.; Gouveia, C. A multi-temporal optimal power flow for managing storage and demand flexibility in LV networks. In Proceedings of the 2017 IEEE Manchester PowerTech, Manchester, UK, 18–22 June 2017.
23. Nousdilis, A.I.; Kryonidis G., C.; Kontis, E.O.; Christoforidis, G.C.; Papagiannis, G.K. An Exponential Droop Control Strategy for Distributed Energy Storage Systems Integrated with Photovoltaics. *IEEE Trans. Power Syst.* **2021**, *36*, 3317–3328. [CrossRef]
24. Hartvigsson, E.; Odenberger, M.; Chen, P.; Nyholma, E. Estimating national and local low-voltage grid capacity for residential solar photovoltaic in Sweden, UK and Germany. *Renew. Energy Vol.* **2021**, *171*, 915–926. [CrossRef]
25. Priebe, J.; Wehbring, N.; Chang, H.; Moser, A.; Çakmak, H.K.; Kühnapfel, U.; Hagenmeyer, V. Exploiting Unused Capacity in the Grid. In Proceedings of the 2019 IEEE Innovative Smart Grid Technologies—Asia (ISGT Asia), Chengdu, China, 21–24 May 2019.
26. Bharath, B.R.; Stefan, M.; Schwalbe, R.; Zeilinger, F.; Schenk, A.; Frischenschlager, A.; Stetn, P. Grid Capacity Management for peer-to-peer Local Energy Communities. In Proceedings of the 2020 IEEE Power & Energy Society General Meeting (PESGM), Montreal, QC, Canada, 3–6 August 2020.
27. Bereczki, B.; Hartmann, B. LV Grid Voltage Control with Battery Energy Storage Systems. In Proceedings of the 2020 IEEE International Conference on Environment and Electrical Engineering and 2020 IEEE Industrial and Commercial Power Systems Europe (EEEIC/I&CPS Europe), Madrid, Spain, 9–12 June 2020.
28. Start, D.J. A review of the new CENELEC standard EN 50160. In *IEEE Colloquium on Issues in Power Quality*; IET: London, UK, 1995; pp. 4/1–4/7. [CrossRef]
29. Torquato, R.; Salles, D.; Pereira, C.O.; Meira, P.C.M.; Freitas, W. A Comprehensive Assessment of PV Hosting Capacity on Low-Voltage Distribution Systems. *IEEE Trans. Power Deliv.* **2018**, *33*, 1002–1012. [CrossRef]
30. Małkowski, R.; Izdebski, M.; Miller, P. Adaptive Algorithm of a Tap-Changer Controller of the Power Transformer Supplying the Radial Network Reducing the Risk of Voltage Collapse. *Energies* **2020**, *13*, 5403. [CrossRef]
31. Lamberti, F.; Calderaro, V.; Galdi, V.; Piccolo, A.; Graditi, G. Impact analysis of distributed PV and energy storage systems in unbalanced LV networks. In Proceedings of the 2015 IEEE Eindhoven PowerTech, Eindhoven, The Netherlands, 29 June–2 July 2015. [CrossRef]
32. Costa, L.F.; Carne, G.D.; Buticchi, G.; Liserre, M. The smart transformer: A solid-state transformer tailored to provide ancillary services to the distribution grid. *IEEE Power Electron. Mag.* **2017**, *4*, 56–67. [CrossRef]
33. Norbu, S.; Couraud, B.; Robu, V.; Andoni, M.; Flynn, D. Modeling Economic Sharing of Joint Assets in Community Energy Projects Under LV Network Constraints. *IEEE Access* **2021**, *9*, 112019–112042. [CrossRef]
34. Petrou, K.; Procopiou, A.T.; Gutierrez-Lagos, L.; Liu, M.Z.; Ochoa, L.F.; Langstaff, T.; Theunissen, J.M. Ensuring Distribution Network Integrity Using Dynamic Operating Limits for Prosumers. *IEEE Trans. Smart Grid* **2021**, *12*, 1–11. [CrossRef]
35. Available online: https://pgedystrybucja.pl/strefa-klienta/przydatne-dokumenty/akordeon-przydatne-dokumenty/wbse (accessed on 30 September 2021).
36. Farrokhi, M.; Fotuhi-Firuzabad, M.; Heidari, S. A novel voltage and Var control model in distribution networks considering high penetration of renewable energy sources. In Proceedings of the 2015 23rd Iranian Conference on Electrical Engineering, Tehran, Iran, 10–14 May 2015.
37. Ishaq, J.; Fawzy, Y.T.; Buelo, T.; Engel, B.; Witzmann, R. Voltage control strategies of low voltage distribution grids using photovoltaic systems. In Proceedings of the 2016 IEEE International Energy Conference (ENERGYCON), Leuven, Belgium, 4–8 April 2016.
38. Zangs, M.J.; Adams, P.B.E.; Yunusov, T.; Holderbaum, H.; Potter, B.A. Distributed Energy Storage Control for Dynamic Load Impact Mitigation. *Energies* **2016**, *9*, 647. [CrossRef]
39. Górnowicz, R.; Castro, R. Optimal design and economic analysis of a PV system operating under Net Metering or Feed-In-Tariff support mechanisms: A case study in Poland. *Sustain. Energy Technol.* **2020**, *42*, 100863. [CrossRef]
40. Dhaneria, A. Grid Connected PV System with Reactive Power Compensation for the Grid. In Proceedings of the 2020 IEEE Power & Energy Society Innovative Smart Grid Technologies Conference (ISGT), Washington, DC, USA, 17–20 February 2020.
41. Kim, S.P.A.; Song, S.G.; Park, S.J.; Kang, F.S. Imbalance Compensation of the Grid Current Using Effective and Reactive Power for Split DC-Link Capacitor 3-Leg Inverter. *IEEE Access* **2021**, *9*, 81189–81201. [CrossRef]
42. Li, Y.C.H.; Tsai, T.W.; Yang, C.J.; Chen, J.M.; Chang, Y.R. Per-Phase Control Strategy of the Three-Phase Four-Wire Inverter. In Proceedings of the 2018 International Power Electronics Conference (IPEC-Niigata 2018—ECCE Asia), Niigata, Japan, 20–24 May 2018.
43. Zhang, D.; Fletcher, J. Operation of Autonomous AC Microgrid at Constant Frequency and with Reactive Power Generation from Grid-forming, Grid-supporting and Grid-feeding Generators. In Proceedings of the TENCON 2018—2018 IEEE Region 10 Conference, Jeju Island, Korea, 28–31 October 2018.
44. Ali, Y.; Shen, Z.; Zhu, F.; Xiong, G.; Chen, S.; Xia, Y.; Wang, F.Y. Solutions Verification for Cloud-Based Networked Control System using Karush-Kuhn-Tucker Conditions. In Proceedings of the 2018 Chinese Automation Congress (CAC), Xi'an, China, 30 November–2 December 2018. [CrossRef]

45. Gibilisco, P.; Ieva, G.; Marcone, F.; Porro, G.; De Tuglie, E. Day-ahead operation planning for microgrids embedding Battery Energy Storage Systems. A case study on the PrInCE Lab microgrid. In Proceedings of the 2018 AEIT International Annual Conference, Bari, Italy, 3–5 October 2018. [CrossRef]
46. Pijarski, P.; Kacejko, P. Voltage Optimization in MV Network with Distributed Generation Using Power Consumption Control in Electrolysis Installations. *Energies* **2021**, *14*, 993. [CrossRef]
47. Karanja, J.M.; Hinga, P.K.; Ngoo, L.M.; Muriithi, C.M. Optimal Battery Location for Minimizing the Total Cost of Generation in a Power System. In Proceedings of the 2020 IEEE PES/IAS PowerAfrica, Nairobi, Kenya, 25–28 August 2020.
48. Li, X.; Wang, L.; Yan, N.; Ma, R. Economic Dispatch of Distribution Network with Distributed Energy Storage and PV Power Stations. In Proceedings of the 2020 IEEE International Conference on Applied Superconductivity and Electromagnetic Devices (ASEMD), Tianjin, China, 16–18 October 2020.
49. Rui, H.; Wellssow, W.H. Assessing distributed storage management in LV grids using the smart grid metrics framework. In Proceedings of the 2015 IEEE Eindhoven PowerTech, Eindhoven, The Netherlands, 29 June–2 July 2015. [CrossRef]
50. Hasan, S.; Luthander, R.; de Santiago, J. Reactive Power Control for LV Distribution Networks Voltage Management. In Proceedings of the 2018 IEEE PES Innovative Smart Grid Technologies Conference Europe (ISGT-Europe), Bucharest, Romania, 29 September–2 October 2018.
51. Divshali, P.H.; Söder, L. Improving PV Dynamic Hosting Capacity Using Adaptive Controller for STATCOMs. *IEEE Trans. Energy Conv.* **2019**, *34*, 415–425. [CrossRef]
52. Czerwiński, D.; Gęca, J.; Kolano, K. Machine Learning for Sensorless Temperature Estimation of a BLDC Motor. *Sensors* **2021**, *21*, 4655. [CrossRef] [PubMed]
53. Mokhtar, M.; Valentin, R.; Flynn, D.; Higgins, C.; Whyte, J.; Loughran, C.; Fulton, F. Predicting the Voltage Distribution for Low Voltage Networks using Deep Learning. In Proceedings of the 2019 IEEE PES Innovative Smart Grid Technologies Europe (ISGT-Europe), Bucharest, Romania, 29 September–2 October 2019; pp. 1–5. [CrossRef]
54. Bellizio, F.; Karagiannopoulos, S.; Aristidou, P.; Hug, G. Optimized Local Control for Active Distribution Grids using Machine Learning Techniques. In Proceedings of the 2018 IEEE Power & Energy Society General Meeting (PESGM), Portland, OR, USA, 5–9 August 2018; pp. 1–5. [CrossRef]

energies

MDPI

Article

Distributed Generation Based Virtual STATCOM Configuration and Control Method

Su-Han Pyo [1], Tae-Hun Kim [1], Byeong-Hyeon An [1], Jae-Deok Park [1], Jang-Hyun Park [1], Myoung-Jin Lee [2] and Tae-Sik Park [1,*]

[1] Department of Electrical and Control Engineering, Mokpo National University, Mokpo City 534-729, Korea; vytngks@naver.com (S.-H.P.); kaeu123@naver.com (T.-H.K.); shfmql12@naver.com (B.-H.A.); asfgd2000@naver.com (J.-D.P.); jhpark72@mokpo.ac.kr (J.-H.P.)

[2] Department of Electronics and Computer Engineering, Chonnam University, Gwangju 500-757, Korea; mjlee@jun.ac.kr

* Correspondence: tspark@mokpo.ac.kr

Abstract: Recently, because of the increase in the number of connections to Distributed Generation (DG), the problem of lowering voltage stability in the distribution system has become an issue. Reactive power compensators, such as Static Synchronous Compensators (STATCOM), may be used to solve the problem of voltage stability degradation. However, because of the complexity of the distribution system, it is very difficult to select the installation location for STATCOM. Furthermore, when installed in the wrong location, economical efficiency and availability problems may occur. This paper proposes a Virtual STATCOM Configuration and Control method that would operate like a single STATCOM based on multiple DGs connected to the system. The proposed Virtual STATCOM has the merit of being economical by using existing facilities without adding new power facilities, and it solves the problem of the difficulty of selecting the installation location because of the complexity of the distribution system. In addition, while the conventional STATCOM uses an independent control method in consideration of the power quality of the access point, the Virtual STATCOM performs the Point of Common Coupling (PCC) power quality compensation using the integrated control of multiple DGs connected to the system. In the proposed method, the Virtual STATCOM integrated control algorithm is configured by adopting linear programming, and the compensation is performed while considering the distance between DG and PCC, the inverter's rated capacity, and the power generation. The performance of the Virtual STATCOM power quality compensation was verified using MATLAB/SIMULINK and Real Time Simulator (OPAL-RT).

Keywords: STATCOM; reactive power compensation; power quality

Citation: Pyo, S.-H.; Kim, T.-H.; An, B.-H.; Park, J.-D.; Park, J.-H.; Lee, M.-J.; Park, T.-S. Distributed Generation Based Virtual STATCOM Configuration and Control Method. *Energies* **2022**, *15*, 1762. https://doi.org/10.3390/en15051762

Academic Editor: Mario Marchesoni

Received: 24 January 2022
Accepted: 24 February 2022
Published: 26 February 2022

1. Introduction

In the existing power system, electricity generated from large-scale power sources such as nuclear power generation and thermal power generation is supplied to loads tens or hundreds of kilometers away through long-distance high-voltage transmission lines. For about 50 to 60 years, this has been operating well without changing the configuration of the power system, providing inexpensive and stable electric power to industrial, commercial, and residential loads. However, existing large-scale power generation sources emit a lot of carbon, causing problems such as global warming. To solve these problems, the configuration of the power system is changing for de-carbonization worldwide [1]. As small-scale DGs such as wind power and solar power are rapidly increasing on the distribution system level, the configuration of the distribution system is changing to become very different from the existing ones [2–5]. However, there may be problems in that the power supply is not constant because of the increase in the connection of intermittent power sources to the distribution system, and the degradation of voltage stability caused by the Ferranti Effect occurs as the voltage of the distribution system increases [6].

Many studies have been conducted on the Flexible AC Transmission System (FACTS) to solve the following problems [7]. STATCOM, as one kind of FACTS, is a facility that is connected to the power system in parallel to compensate for the power quality of the connected point. STATCOM is mainly installed in transmission systems and substations with large capacities to compensate for the power quality using the reactive power compensation for the power system [8–11]. Depending on the STATCOM installation location, the power quality compensation performance has large deviations. When the optimal installation location is selected, the active power loss may be minimized, and the voltage stability of the power system may be improved [12,13]. The DG inverter is composed of a smart inverter, so it is possible to control active and reactive power. This means that the DG connected to the system can be controlled in an integrated way [14–17]. When a PV solar farm is used as a STATCOM, it has been proven to increase the grid interconnection of the wind power plants connected to the surrounding area and to improve the power transmission capacity [18–20]. Power oscillation damping (POD) can be performed using a PV power plant as a STATCOM. As soon as power oscillations caused by a system disturbance are detected, the solar farm discontinues its real power generation function very briefly and makes its entire inverter capacity available to operate as a STATCOM for POD [20–24].

STATCOM installed in the transmission system cannot compensate for the power quality considering the distribution system. When installing STATCOM to compensate for the power quality of the distribution system, it is difficult to select the installation location because of the complexity of the distribution system. Furthermore, when installing a large number of STATCOMs, the investment cost is high, resulting in economic degradation. The installation location can be selected using an optimization algorithm [25]. However, as the number of DGs increases, the complexity of the distribution system increases and the optimal installation location may be changed. To solve the above-mentioned problems, this paper proposes a solution to the problem of selecting the optimal installation location by configuring a Virtual STATCOM that operates like a single STATCOM based on DGs.

2. Overview of STATCOM Power Quality Compensation

The active power of the power system affects frequency fluctuations, and the reactive power affects the voltage fluctuations [26]. To improve the power quality, voltage, frequency, and waveform should be kept constant. It is very important to keep the voltage constant. When the supply and demand balance of reactive power is not achieved, voltage fluctuations occur. A generator is used to supply the active power to the power system. However, facilities such as a synchronous generator, a passive filter, and an active filter may be used to supply the reactive power.

The frequency and voltage of the power system are kept constant using the active power—frequency $(P - f(\delta))$ control by the generator and the reactive power—voltage $(Q - V)$ control by the reactive power compensator. When the load voltage V_L and the line reactance component X are constant, the voltage fluctuation ΔV has a proportional relationship with the reactive power Q [27]. Where V_s = source voltage:

$$\Delta V = V_s - V_L = \frac{QX}{V_L}\ (R \ll X) \tag{1}$$

Equation (1) shows the relationship between the voltage fluctuation ΔV and the reactive power Q. It is found that the voltage fluctuation can be controlled by controlling the reactive power from STATCOM.

Principle of STATCOM Voltage Stability Improvement

Figure 1 is an equivalent circuit diagram of a system in which STATCOM is installed to explain the principle of voltage stability improvement. V_S is the equivalent of the bus voltage, Y_L is the equivalent of the admittance, and Z_S is the equivalent of the line impedance generated between the bus voltage V_S and the load. STATCOM is connected to the grid in parallel to improve the voltage stability by the amount of the reactive power compensated for when the load-end voltage stability is improved. The voltage drop between the load terminal and the bus voltage is caused by the line impedance Z_S. When the voltage drops because of load admittance Y_L and the line impedance Z_S is ΔV, it can be expressed by the following equation [28]:

$$\Delta V = V_S - V_L = Z_S I_L \tag{2}$$

$$\begin{aligned}\Delta V &= (R_S + jX_S)\frac{(P_L - jQ_L)}{V_L} \\ &= \frac{(R_S P_L + X_S Q_L)}{V_L} + j\frac{(X_S P_L - R_S Q_L)}{V_L} \\ &= \Delta V_R + \Delta V_X = I_S R_S + jI_S X_S\end{aligned} \tag{3}$$

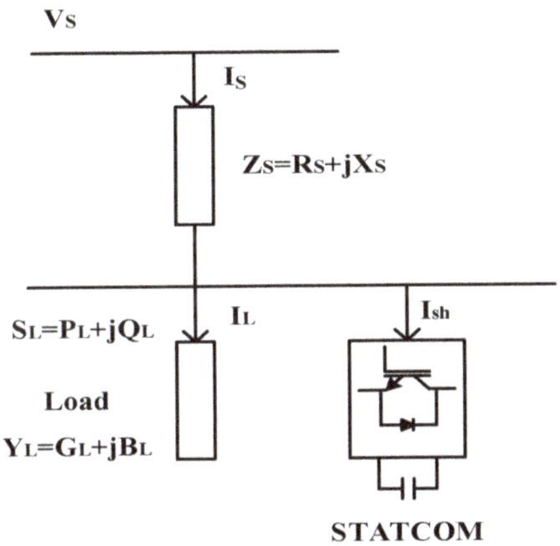

Figure 1. Principle of STATCOM voltage stability improvement.

Figure 2 is a vector diagram of voltage drop ΔV caused by the line impedance Z_S. Depending on the power system situation, STATCOM operates in the form of supplying reactive power (Capacitive Reactive Power) or absorbing the reactive power (Inductive Reactive Power), or in the form of supplying the active power or absorbing the active power. When a voltage drop occurs at the receiving end because of the inductance of the line impedance Z_S, the reactive power compensation current I_{sh} is supplied for compensating the voltage drop of the system. A voltage drop is generated by resistance component $I_s R_s$ and reactance component $jI_s X_s$, and before compensation the load current angle is θ_L. V_S and V_L can be kept the same through I_{sh} current compensation, and the load current angle is compensated to θ_S. The reactive power compensation current I_{sh} can be expressed as Equation (4) [29]:

$$I_S = I_L + I_{sh} \tag{4}$$

When the loads is an inductive load, STATCOM can make a capacitive current and scale it to change the phase angle of I_S. By adjusting ΔV according to the compensation current of STATCOM, the voltage of the load stage and the magnitude of the bus voltage can be kept constant.

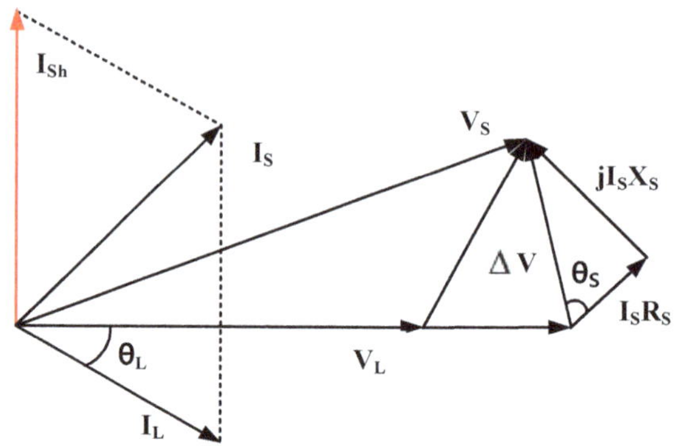

Figure 2. Vector Diagram of STATCOM Power Quality Compensation.

3. Proposed Virtual STATCOM System

DGs, such as solar power and wind power ESS, are connected to the power system via inverters, and a number of inverters connected to the DG are capable of outputting active power and reactive power. However, distributed power sources currently connected to the grid operate as the power sources and supply only active power. The reactive power generated in the power system is not compensated for. Since each inverter is individually and independently controlled, it is difficult to effectively compensate for the reactive power of the power system. The proposed Virtual STATCOM system can effectively control the reactive power of the power system with an integrated control rather than the independent control of multiple inverters of distributed power sources connected to the power system. Since no new facilities are installed for compensating for the reactive power, the installation cost can be economically feasible. Furthermore, compensating for the reactive power of the PCC stage to 0 can be expected to improve voltage stability, to remove harmonics, to increase transmission capacity, and to enhance the lifespan of facilities. Figure 3 shows the configuration schematic of Virtual STATCOM.

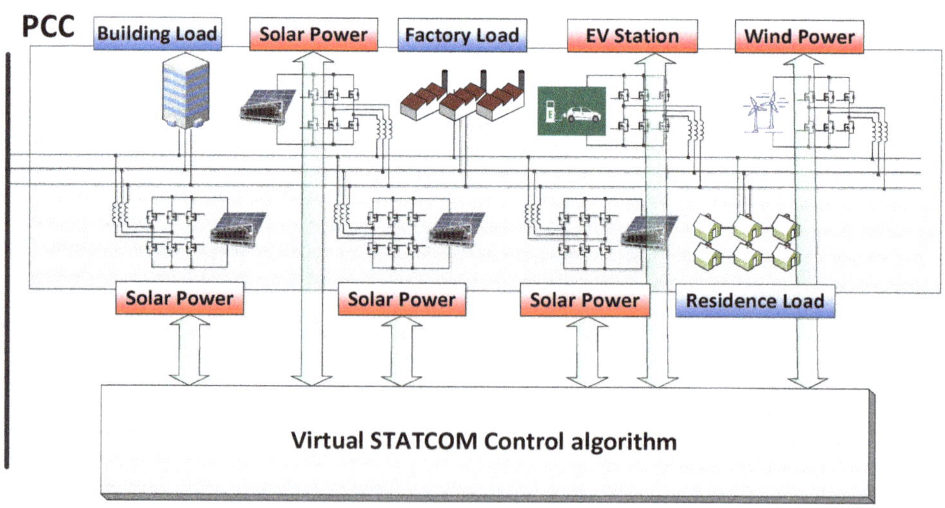

Figure 3. Virtual STATCOM configuration schematic.

3.1. Configuration of Virtual STATCOM

In general, STATCOM is a facility that improves the stability of the power system by controlling the reactive power of the connection point. When multiple DG inverters connected to the system independently compensate for reactive power at the connection point, like in the case of a general STATCOM, the system stability may deteriorate because of the hunting that occurs during the control operation. The purpose of this paper is to compensate for or control the reactive power at a designated location in the system by controlling the inverters of DGs connected to multiple connection points. That is, it is expected that the proposed Virtual STATCOM can improve the stability of the entire power system by controlling reactive power at a desired point, such as a weak or important location of the connection or distribution end of the transmission and distribution system, like the PCC of the power system. In this paper, PCC was designated as the control reference point that is the connection point of transmission and distribution, and the Virtual STATCOM integrated control algorithm was configured by the linear programming. To improve the voltage stability of the PCC, the reactive power of the PCC should be controlled to 0. The reactive power of the PCC stage is divided into a component generated by the line impedance and a component generated by the load. To compensate for the reactive power of the PCC stage to 0, it is necessary to compensate for the reactive power while considering both causes of the reactive power.

Figure 4 shows the configuration of the Virtual STATCOM system. Three solar power plants are connected between the voltage source and load. It is assumed that each solar power plant has a different distance from the bus of the PCC stage, and the amounts of connected load, solar power plant capacities, and power generation amounts are different from each other. The inverter of the solar power plant can output the active power and the reactive power as much as the rated capacity S. When there is room in the inverter output rated capacity after supplying the active power P generated by the PV panel to the grid, the reactive power can be supplied. The inverter rated capacity S of a solar power plant can be expressed as Equation (5):

$$S_{inv\_n} = \sqrt{P_{inv\_n}^2 + Q_{inv\_n}^2} \tag{5}$$

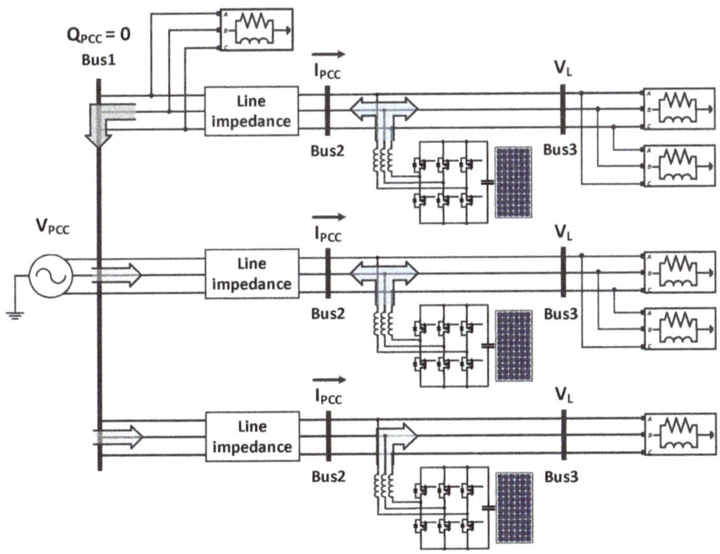

Figure 4. Configuration Scheme of the Virtual STATCOM System.

The Virtual STATCOM integrated control algorithm selects the inverter closest to the PCC to give it priority, and when the amount of the reactive power that can be output is insufficient, the compensation will be performed from the inverter as the next priority.

Figure 5 shows the configuration of the integrated control algorithm of Virtual STATCOM. By combining Solar Station, PMU, and Distance Information, the Virtual STATCOM Control algorithm is configured based on LP, and the control reference of the inverter is derived. A plurality of inverters improves the voltage stability by the reactive power compensation of the PCC stage.

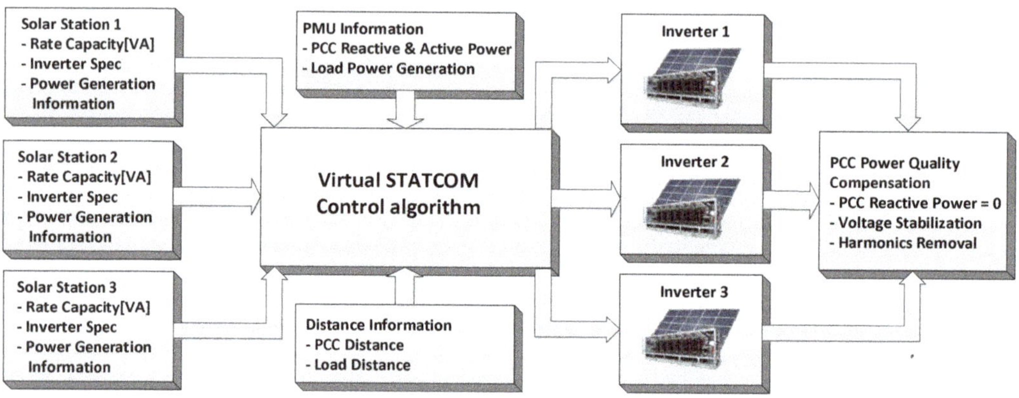

Figure 5. Configuration of Virtual STATCOM Integrated Control Algorithm.

3.2. *Scheme of Virtual STATCOM Control*

To configure the integrated control algorithm of Virtual STATCOM based on LP, the decision variables, the objective functions, and the constraints must be selected according to the control purpose [30].

3.2.1. Decision Variable

The decision variable is for setting the variable that is the control criterion of the LP. Since the purpose of the Virtual STATCOM is to output the inverter reactive power compensation reference, the inverter reactive power output amount was selected as a decision variable.

$$\text{Decision Variable} = Q_{inv}\text{: Inverter Reactive Power Output}$$

3.2.2. Objective Function

The objective function is for setting the goal of the decision variable. As the distance from the PCC terminal increases, the loss caused by the line impedance increases, so it is designated as the goal to use an inverter with a shorter distance. To select a minimum distance, a point where the product of the decision variable and the weight w_n is minimized is set as the optimal point.

$$\text{Objective Function} = \min(w_1 * Q_{inv1} + w_2 * Q_{inv2} + \cdots + w_n * Q_{inv_n})$$

where, w_n refers to the distance weight of inverter, and l_n refers to the line distance of inverter from the PCC end point. w_n is decided by the following equation as a consideration of the distance between the inverter and the PCC end point.

$$w_n = \frac{l_n}{l_1 + l_2 + l_3 + \cdots + l_n} \tag{6}$$

3.2.3. Constraints

The constraints determine the minimum compensation value and the maximum compensation value for the decision variable of the inverter to determine the reactive power that is to be compensated for by the inverter. The constrains for the minimum and maximum amounts of the inverter reactive power output can be expressed as Equation (7):

$$Q_{inv\_n\_min} \leq Q_{inv\_n} \leq Q_{inv\_n\_max} \tag{7}$$

where $Q_{inv\_n\_min}$ = Minimum of Inverter Reactive Power Compensation (var), and $Q_{inv\_n\_max}$ = Maximum of Inverter Reactive Power Compensation (var).

When the output of the inverter is maximized, the life of the inverter rapidly decreases, so the maximum output of the inverter is limited to 80%. To determine the amount of the reactive power compensation of the inverter, the minimum and maximum amounts of reactive power compensation of the inverter should be set as the constraints. Constraints are determined in consideration of the effective load amount of the inverter terminal, the reactive load amount of the inverter terminal, the rated capacity, the inverter active power generation amount, and the inverter generation rated limit.

$Q_{inv\_n\_max}$ means the maximum reactive power compensation amount that can be compensated for by the inverter. The maximum amount of inverter reactive power compensation is determined by considering the rated capacity of the inverter, the amount of active power currently being generated by the inverter, and the inverter power generation rating limit.

$$(S_{inv_n} * 0.8) = \sqrt{P_{inv\_n}^2 + Q_{inv_n}^2} \tag{8}$$

$$Q_{inv\_n\_max} = \sqrt{(S_{inv_n} * 0.8)^2 - P_{inv_n}^2} \tag{9}$$

where $Q_{inv\_n\_min}$ means the minimum reactive power compensation amount that can be compensated by the inverter. The reactive power can be expressed in negative or positive number according to inductivity and capacity, so the negative number of the maximum amount of compensable reactive power is determined as the minimum compensable amount of the inverter.

$$Q_{inv\_n\_min} = -Q_{inv\_n\_max} \tag{10}$$

Considering the reactive power Q_{PCC} generated in the PCC stage, the reference value of the reactive power compensation for each inverter is calculated. The reactive power of PCC stage consists of the reactive power Q_{Line} generated by line impedance and the reactive power Q_{Load} generated by load. When the total amount of reactive power generated in the PCC stage is equal to the total amount of reactive power compensated for by the inverter, the reactive power in the PCC stage is compensated to 0.

$$Q_{PCC} = Q_{Line} + Q_{Load} \tag{11}$$

$$Q_{PCC} - (Q_{inv1} + Q_{inv2} + Q_{inv3} + \cdots + Q_{inv_n}) = 0 \tag{12}$$

Table 1 shows the parameters for configuring the Virtual STATCOM system.

Table 1. Virtual STATCOM System Parameters.

Parameters	*inv_n*
Load Active Power (W)	P_{L_n}
Load Reactive Power (var)	Q_{L_n}
Inverter Rated Capacity (VA)	S_{inv_n}
Inverter Active Power Output (W)	P_{inv_n}
Inverter Output Rating Limit (%)	80%
Inverter Compensation min (var)	$Q_{inv\_min} = -Q_{inv\_max}$
Inverter Compensation max (var)	$Q_{inv\_max} = \sqrt{(S_{inv_n} * 0.8)^2 - P_{inv\_n}^2}$
PCC Distance (km)	l_n

Figure 6 shows the Virtual STATCOM Integrated Control Algorithm Diagram. The compensation algorithm is configured through five steps. Step 1 measures the parameter values required for Virtual STATCOM configuration. Step 2 determines the reactive power compensation's possible range based on the measured parameter value. Step 3 determines the optimal reference Q_{LP} for reactive power compensation using LP. Step 4 performs reactive power compensation through an inverter and measures the reactive power of PCC. If the reactive power of PCC converges to 0, proceed to Step 5. If it does not converge, return to Step 1 and perform the algorithm again. Step 5 is the step of storing the reactive power compensation value determined by Steps 1–4. Equation (13) shows the reactive power compensation value of the algorithm diagram, where $Q_{PCC\_n\_pre}$ is the stored PCC reactive power value of Step 5.

$$Q_{PCC} = Q_{PCC\_n\_pre} + \Delta Q_{PCC} \tag{13}$$

If the reactive power compensation value is not saved and the algorithm is executed, only the value of ΔQ_{PCC} is compensated for when a load change occurs, and PCC reactive power is generated to as much as the $Q_{PCC\_n\_pre}$ value.

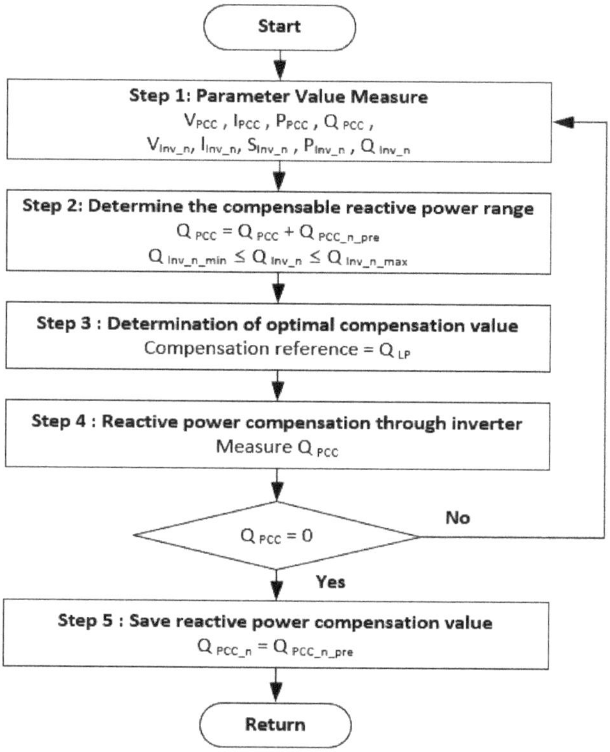

Figure 6. Virtual STATCOM Integrated Control Algorithm Diagram.

4. Virtual STATCOM Simulation

To verify the Virtual STATCOM's performance, simulation was conducted using Matlab Simulink and the Real Time Simulator (OPAL-RT) OP5700.

The Real Time Simulator is a simulation technology that proceeds by setting the simulation time to be same as the real time. By running the simulation under the same conditions as in reality, problems that do not appear in the off-line simulation can be derived, and the reliability of the simulation can be improved. By establishing a virtual controller and a virtual plant in the target PC, a Model In the Loop Simulation (MILS) was performed. The plant of the simulation model was designated as Master, and the controller was designated as Slave, and tow cores were used. The result data of the simulation were output via the target PC. Figure 7 shows the Simulation Model of Virtual STATCOM.

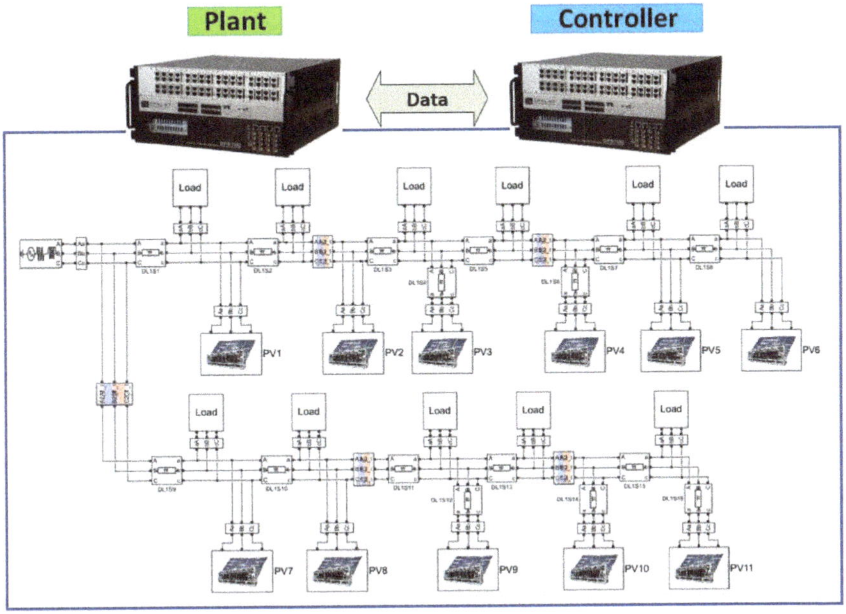

Figure 7. Virtual STATCOM Simulation Model.

The system model used in the simulation was that of Anseong, in South Korea. The distributed resources connected to the Anseong area, and the system diagram of the entire systems was made equivalent by integrating the system load into sections as shown in Figure 8. Table 2 shows the DG inverter and load data of the Anseong area system.

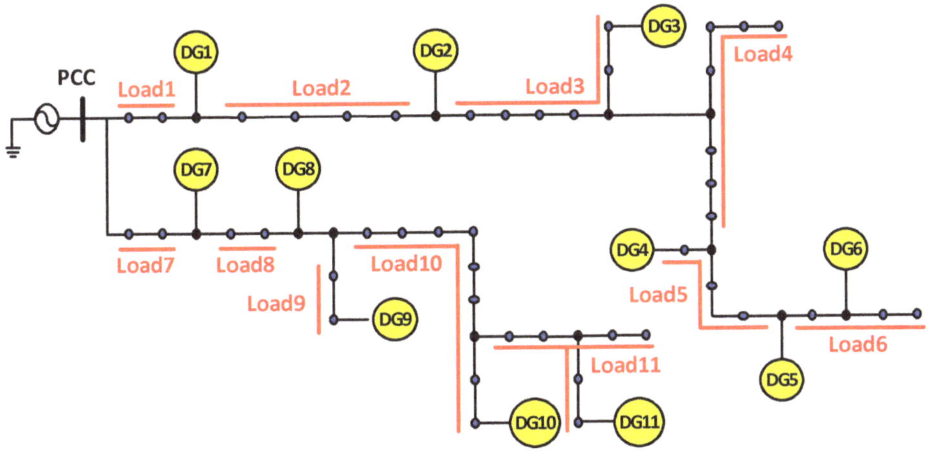

Figure 8. Virtual STATCOM Simulation Grid Diagram.

Table 2. Virtual STATCOM Simulation Model Parameters.

	Inverter Rating (MVA)	Distance (km)		Load Active Power (MW)	Load Reactive Power (Mvar)
$inverter_1$	0.5	2	$Load_1$	0.988	0.33
$inverter_2$	1.5	6	$Load_2$	1.25	0.541
$inverter_3$	3	9	$Load_3$	1.75	0.707
$inverter_4$	1	17	$Load_4$	3.12	1.3
$inverter_5$	5	18	$Load_5$	0.294	0.104
$inverter_6$	3	18	$Load_6$	0.159	76.5
$inverter_7$	1	2	$Load_7$	0.178	74.2
$inverter_8$	0.5	4	$Load_8$	1.68	0.743
$inverter_9$	1.5	5	$Load_9$	3.32	1.32
$inverter_{10}$	2	7	$Load_{10}$	1.48	0.588
$inverter_{11}$	2	10	$Load_{11}$	0.211	88.6

For the simulation system model, the simulation was performed with equivalence to a system including one voltage power source, 11 DGs, and 11 loads. The power source output AC of 22.9kV, 60Hz via the Three Phase Source. The load was simulated by dividing the section of the system using the three-phase parallel RLC load and making the load within the section equivalent to one load. The load between PCC and DG1 was set as $Load_1$, and the subsequent loads were designated as $Load_n$ by grouping the loads between DGs. All loads used in the simulation were divided into 11 sections. To perform compensation considering the distance between the PCC terminal and the DGs, the line impedance according to the distance was made equivalent using the PI Section Line.

For the simulation sample time, the simulation was conducted by dividing it into three categories: the power system, the inverter controller, and the algorithm according to the purpose. The sample time of the power system was set to 50 (μs) for simulating the instantaneous system change situation, and the sample time for the inverter controller was set to 100 (μs). The Virtual STATCOM integrated control algorithm sample time was set to 0.1 (s) in consideration of the data measurement period of the PMU.

Table 3 shows the Virtual STATCOM simulation scenario. This table indicates whether the Virtual STATCOM Control Algorithm operates according to the simulation time, and also indicates the load change time. The data for the load change are shown in Table 4. For the simulation time of 0~1 (s), a situation, in which the Virtual STATCOM Control Algorithm does not operate and the load fluctuation does not occur, is simulated.

For the simulation time of 1 (s), the load variation situation was simulated. $Load_1$ generated a variation of +0.7 (Mvar) in 1 (s), $Load_3$ generated +0.3 (Mvar), $Load_4$ generated −0.25 (Mvar), $Load_5$ generated +1.1 (Mvar), $Load_8$ generated +2.1 (Mvar), $Load_9$ generated +0.2 (Mvar), and $Load_{10}$ generated the variation of −0.15 (Mvar), so the reactive power of PCC generated the variation of +4 (Mvar). At the simulation time of 2 (s), $Load_1$ had the variation of −0.4 (Mvar), $Load_2$ had −0.2 (Mvar), $Load_3$ had +0.5 (Mvar), $Load_4$ had −0.3 (Mvar), $Load_5$ had −0.5 (Mvar), $Load_6$ had +1 (Mvar), $Load_7$ had +0.2 (Mvar), $Load_8$ had −2 (Mvar), and $Load_{10}$ had the variation of −0.3 (Mvar), so the reactive power of PCC having the variation of −2 (Mvar) was simulated.

Table 3. Virtual STATCOM Simulation Scenario.

Simulation Time (s)	Virtual STATCOM Control Algorithm	Load Variation (Mvar)
0~0.5	Off	0
0.5~1	On	0
1~2	On	+4
2~3	On	−2

Table 4. Virtual STATCOM Simulation Model Parameters (Load Variation).

	0~1 (s) Load Reactive Power (Mvar)	1 (s) Load Variation (Mvar)	1~2 (s) Load Reactive Power (Mvar)	2 (s) Load Variation (Mvar)	2~3 (s) Load Reactive Power (Mvar)
$Load_1$	0.33	+0.7	1.03	−0.4	0.63
$Load_2$	0.541	0	0.541	−0.2	0.341
$Load_3$	0.707	+0.3	1.007	+0.5	1.507
$Load_4$	1.3	−0.25	1.05	−0.3	0.75
$Load_5$	0.104	+1.1	1.204	−0.5	0.704
$Load_6$	0.0765	0	0.0765	+1	1.076
$Load_7$	0.0742	0	0.0742	+0.2	0.274
$Load_8$	0.743	+2.1	2.843	−2	0.843
$Load_9$	1.32	+0.2	1.52	0	1.52
$Load_{10}$	0.588	−0.15	0.438	−0.3	0.138
$Load_{11}$	0.0886	0	0.0886	0	0.0886

Figure 9 shows the simulation results, assuming that power quality compensation using the Virtual STATCOM control algorithm is not conducted. During the simulation run time of 0~1 (s), the distributed power supply was connected to the system and the steady state operation was in progress. The voltage of PCC has a voltage drop of 22.87 kV caused by the inductance L and resistance R. In PCC reactive power, the ground reactive power of 5.8 (Mvar) is generated by line impedance components and non-linear loads. In 1~2 (s), it is found that the reactive power of +4 (Mvar) is increased by the load variation, and the voltage of PCC is 22.85 kV, so that the voltage drop is increased. In 2~3 (s), the occurrence of voltage transients and reactive power of −2 (Mvar) were reduced because of the increase of the capacitive load, and reactive power of 7.8 (Mvar) is generated at the PCC point, and a voltage drop of 22.86 kV occurs.

Figure 10 shows the power quality compensation simulation results using the Virtual STATCOM control algorithm. Virtual STATCOM was configured using inverters of multiple DGs connected to the system, and the optimum compensation command value of inverters was controlled using LP. Each inverter was controlled for the purpose of improving voltage stability by compensating for the reactive power of PCC to 0. Figure 10a,b show the PCC voltage and reactive power, and Figure 10c shows the amount of reactive power compensation of Virtual STATCOM inverters.

As the simulation time of 0~0.5 (s) is the status in which the Virtual STATCOM algorithm is not applied, the power quality is not compensated for, so the reactive power of the PCC stage is about 5.8 (Mvar). Furthermore, the voltage at the PCC terminal is 2.287 kV due to the line and load impedance components, and a voltage drop has occurs. It was found that when the Virtual STATCOM control algorithm operated at 0.5 (s), the reactive power of PCC stage converged to 0 by the reactive power compensation, and the voltage of the PCC stage was compensated to 22.9 kV. At 1 (s), as the reactive power of PCC increased with +4 (Mvar) by Load Variation, a voltage drop of 22.88 kV occurred, but the reactive power and the voltage of the PCC were compensated for to the reference values by the Virtual STATCOM integrated algorithm. For 1~2 (s), there was no change. At 2 (s), Load Variation occurred, and the capacitance reactive power of −2 (Mvar) occurred in PCC, so the voltage rose to 22.93 kV. However, it was found that the reactive power of the PCC stage was compensated to 0 and the voltage was sustained to 22.9 kV by the integrated control algorithm.

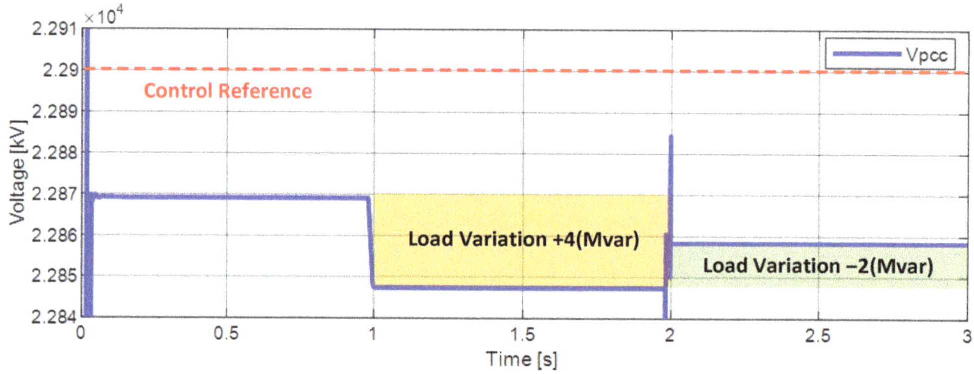

(a)

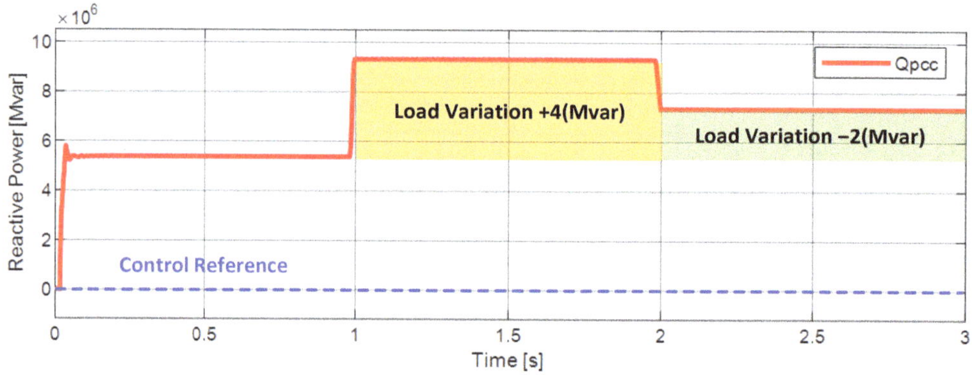

(b)

Figure 9. PCC Variation according to the Load Variation of Virtual STATCOM (Before Compensation by Algorithm) (**a**) PCC Voltage, (**b**) PCC Reactive Power.

Figure 11 shows the amount of reactive power compensation of Virtual STATCOM inverters 1–11. To control the reactive power of the PCC to 0 using the inverter integrated control, the reactive power compensation amount of the inverters was determined in consideration of the available reactive power amount of the inverter and the distance between the PCC and the inverter. Since inverters have different distances from the PCC, the loss caused by the line impedance may increase when the long-distance inverter compensates for a large amount. Therefore, compensation was carried out by selecting an inverter that is relatively close to the PCC as the compensation priority. In addition, the rated capacity is different for each connected inverter, and the amount of the reactive power that can be output is different because the amount of power currently being output is different. Therefore, by considering the rating and remaining capacity of each inverter, the optimum compensation command is generated with priority according to the maximum amount of reactive power and distance.

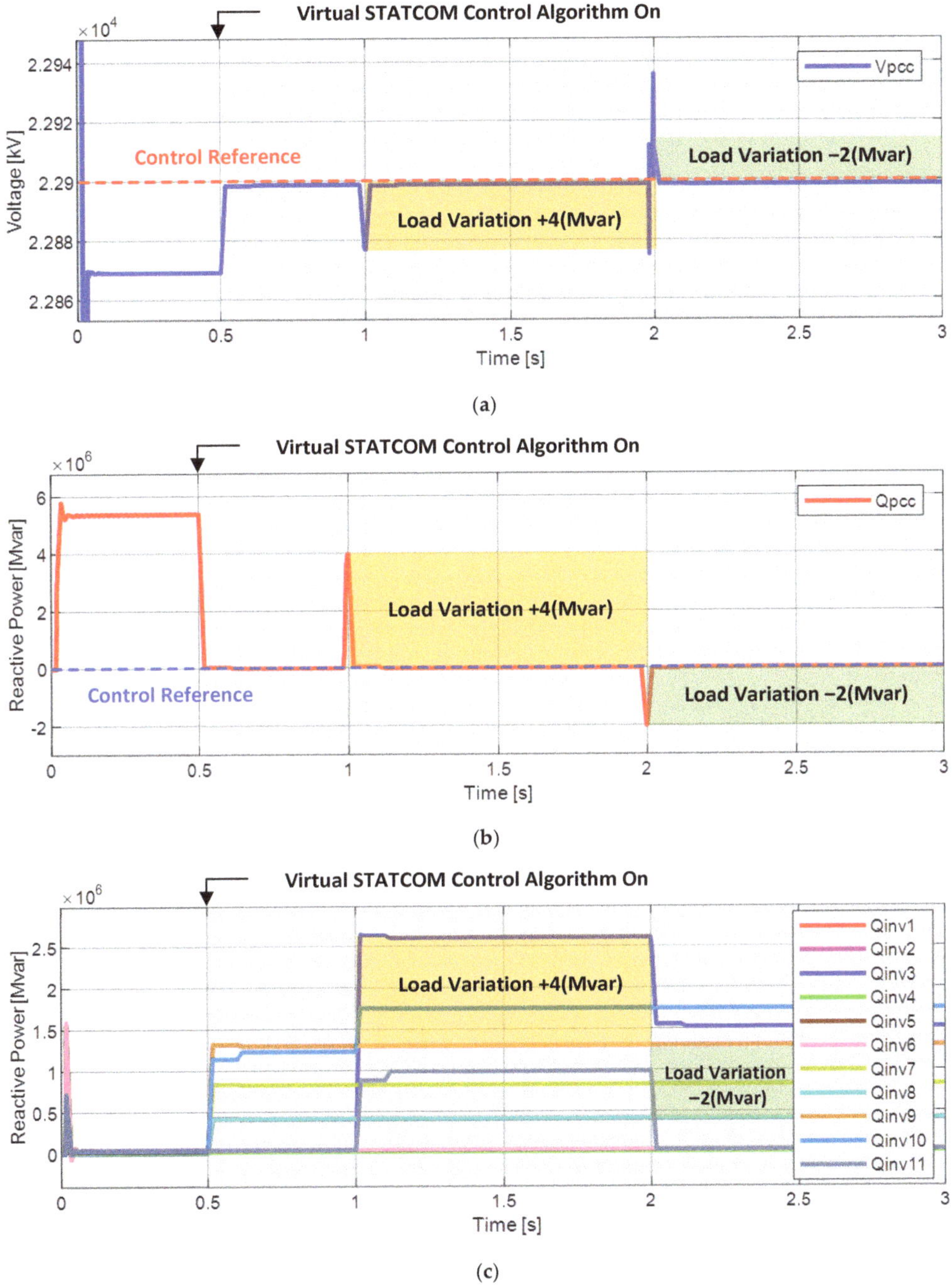

Figure 10. Virtual STATCOM Power Quality Compensation (**a**) PCC Voltage, (**b**) PCC Reactive Power, (**c**) Compensation Amount for Inverter Reactive Power.

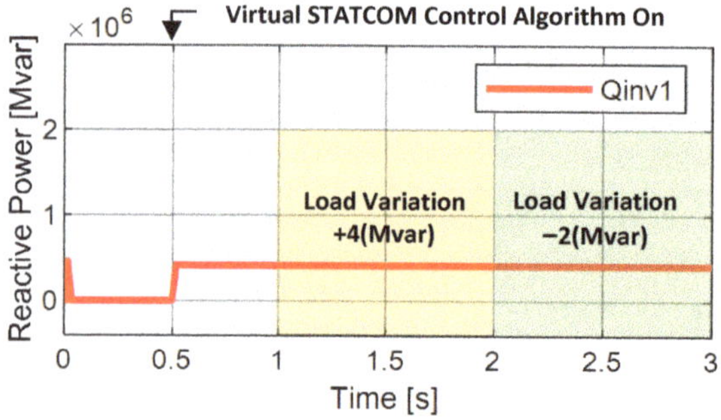

(**a**)

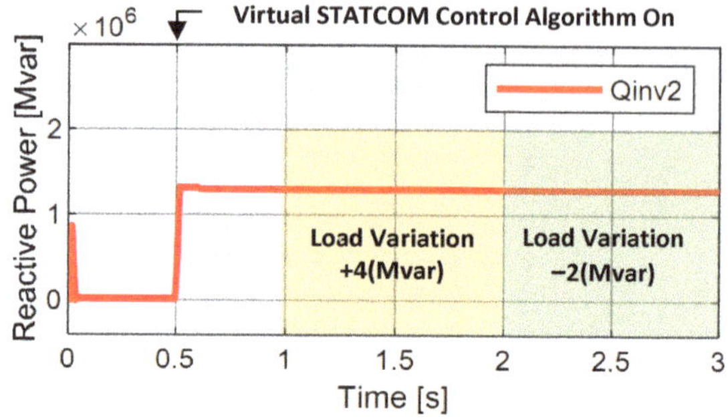

(**b**)

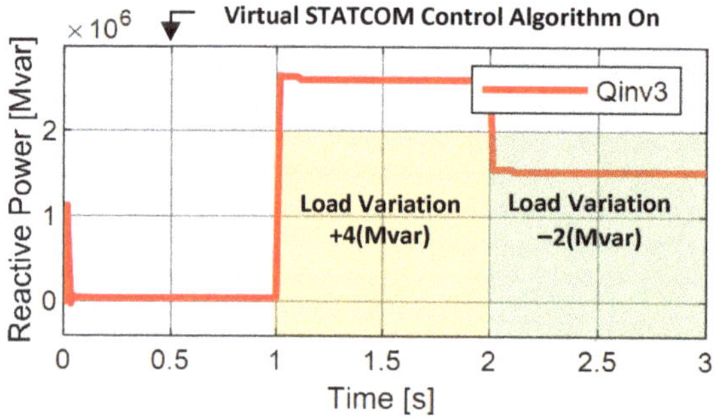

(**c**)

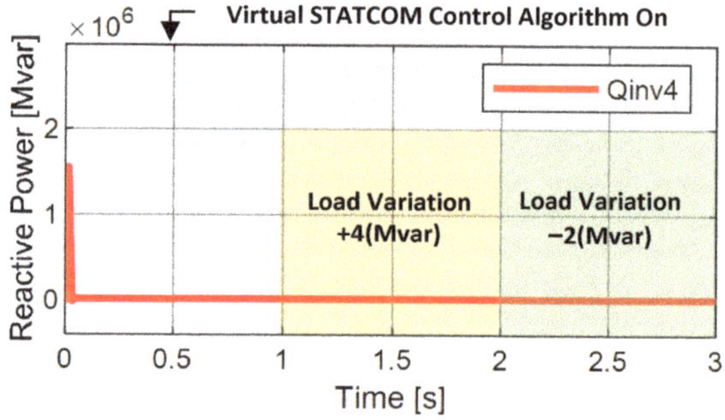

(**d**)

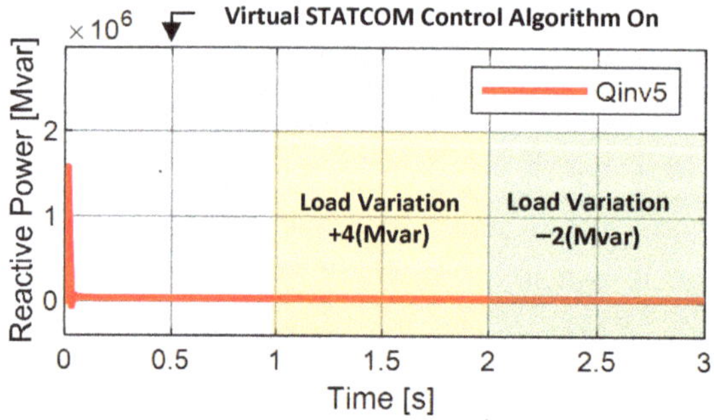

(**e**)

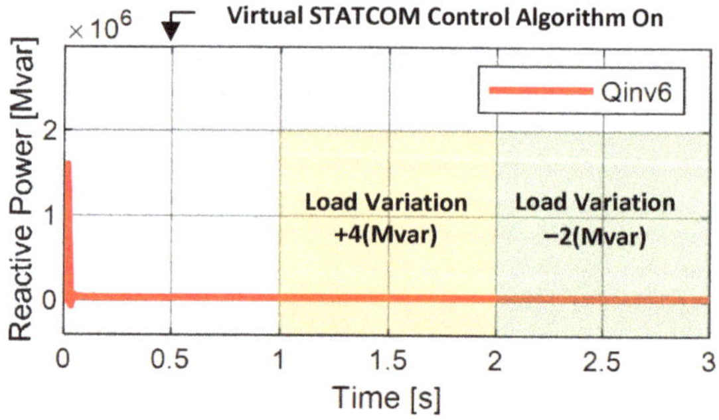

(**f**)

Figure 11. *Cont.*

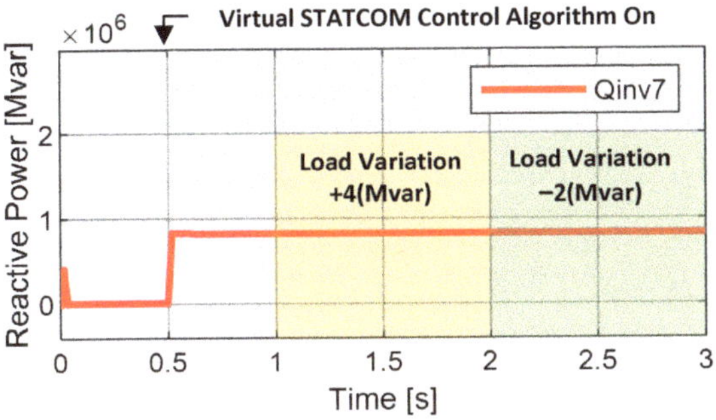

(g)

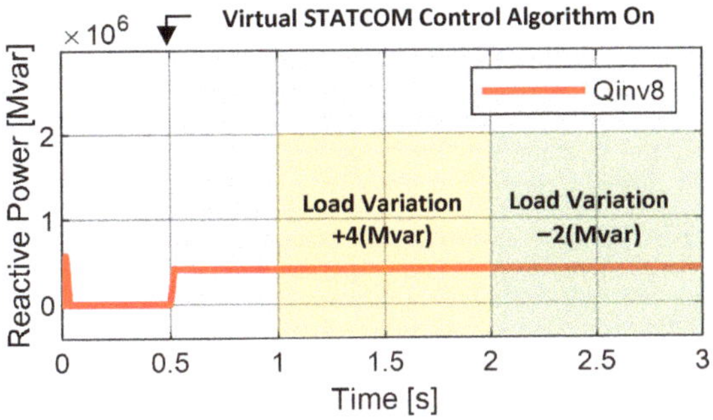

(h)

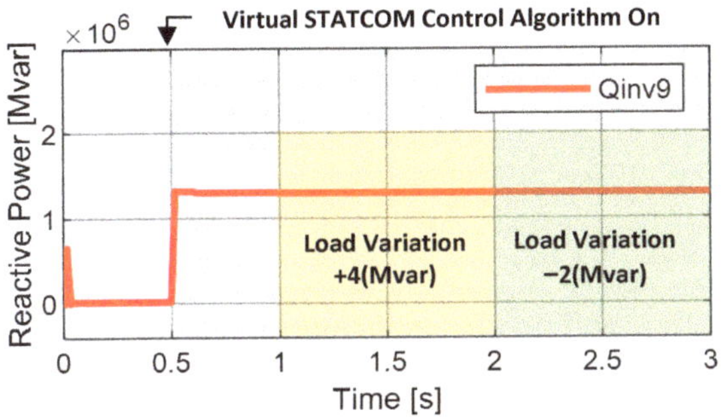

(i)

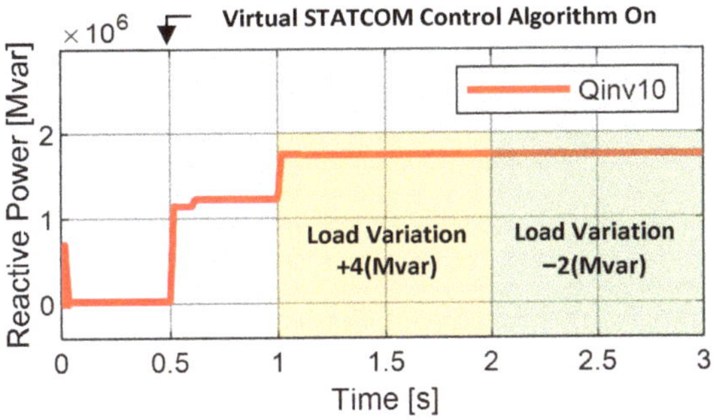

(j)

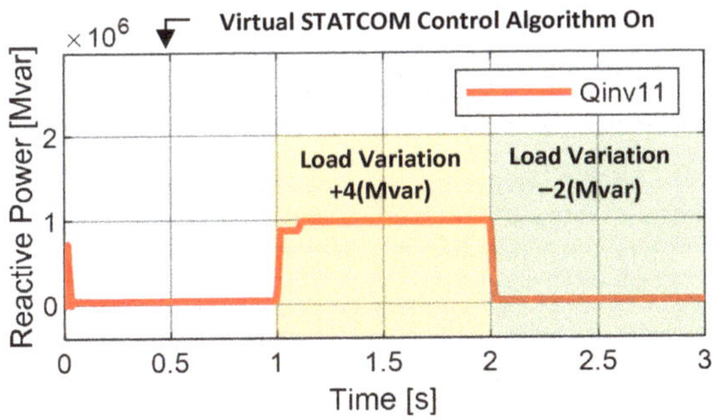

(k)

Figure 11. 11 Reactive Power Compensation Amount of Virtual STATCOM Inverter 1: (**a**) Inverter 1, (**b**) Inverter 2, (**c**) Inverter 3, (**d**) Inverter 4, (**e**) Inverter 5, (**f**) Inverter 6, (**g**) Inverter 7, (**h**) Inverter 8, (**i**) Inverter 9, (**j**) Inverter 10, (**k**) Inverter 11.

In the time of 0~0.5 (s), the Virtual STATCOM control algorithm does not operate, so the Virtual STATCOM inverters are not outputting the reactive power. The control algorithm operates at 0.5 (s), and the reactive power compensation is carried out in consideration of the distance of the inverter to the maximum reactive power output. The reactive power compensation is carried out by dividing reactive power 5.8 (Mvar) generated in PCC by inverters 1, 2, 7, 8, 9, and 10. Inverter 1 compensated 0.4 (Mvar) reactive power, Inverter 2 compensated 1.35 (Mvar), Inverter 7 compensated 0.9 (Mvar), Inverter 8 compensated 0.45 (Mvar), Inverter 9 compensated 1.4 (Mvar), and Inverter 10 compensated 1.2 (Mvar).

In Load Variation 1 (s), inverters 1, 2, 3, 7, 8, 9, 10 and 11 divided PCC reactive power of 9.8 (Mvar), so Inverter 1 compensated 0.4 (Mvar), Inverter 2 compensated 1.35 (Mvar), Inverter 3 compensated 2.5 (Mvar), Inverter 7 compensated 0.9 (Mvar), Inverter 8 compensated 0.45 (Mvar), Inverter 9 compensated 1.4 (Mvar), Inverter 10 compensated 1.8 (Mvar), and Inverter 11 compensated 1 (Mvar). In the case of a load variation of 2 (s), inverters 1, 2, 3, 7, 8, 9 and 10 divided PCC reactive power of 7.8 (Mvar), so Inverter 1 compensated 0.4 (Mvar), Inverter 2 compensated 1.35 (Mvar), Inverter 3 compensated 1.5 (Mvar), Inverter 7 compensated 0.9 (Mvar), Inverter 8 compensated 0.45 (Mvar), Inverter 9 compensated 1.4 (Mvar), and Inverter 10 compensated 1.8 (Mvar).

5. Conclusions

This paper proposed a Virtual STATCOM Configuration and Control method that operates like a single STATCOM based on multiple DGs connected to the system. The conventional STATCOM has the following demerits: as it is installed in the power transmission/transformation system, it is hard to operate while considering the power quality of the distribution system; when installing STATCOM in the distribution system, it is hard to select the installation location because of the complexity of the distribution system; and when proper positioning is not considered, the economic feasibility of STATCOM will be reduced. The Virtual STATCOM solves the problem of installation location selection with the configuration that includes a DG inverter connecting to the grid system, and enhances the economic feasibility by using the equipment connected to the existing system.

Virtual STATCOM is controlled using the operating principle of the existing STATCOM. It improves voltage stability through reactive power compensation. Multiple DGs connected to the grid were integrated and controlled through an LP-based algorithm. As a result of the simulation, we confirmed that the voltage drop occurred because of the line impedance and load impedance components before compensation through the Virtual STATCOM integrated control algorithm. As a result of the compensation using the Virtual STATCOM algorithm, we confirmed that the voltage stability was improved through the reactive power compensation of the PCC. To verify the performance, a simulation was conducted based on a real-time simulator. The off-line simulation results and the real-time simulation results are the same, so we verified that there is no problem in an environment similar to the real one.

Author Contributions: Conceptualization, T.-S.P. and S.-H.P.; methodology, B.-H.A. and J.-D.P.; software, S.-H.P. and T.-H.K.; validation, J.-H.P. and M.-J.L.; formal analysis, S.-H.P.; writing—original draft preparation, S.-H.P.; review and editing, T.-S.P. and S.-H.P. All authors have read and agreed to the published version of the manuscript.

Funding: This work was supported by the Gwangju Jeonnam Local Energy Cluster Manpower Training of the Korea Institute of Energy Technology Evaluation and Planning (KETEP) grant funded by the Korean government's Ministry of Knowledge Economy (20214000000560).

Conflicts of Interest: The authors declare no conflict of interest.

References

1. Khalilpour, K.R.; Vassallo, A. *Community Energy Networks with Storage: Modeling Frameworks for Distributed Generation*; Springer: Berlin/Heidelberg, Germany, 2016.
2. Amini, M.H.; Nabi, B.; Haghifam, M.-R. Load management using multi-agent systems in smart distribution network. In Proceedings of the 2013 IEEE Power & Energy Society General Meeting, Vancouver, BC, Canada, 21–25 July 2013; pp. 1–5.
3. Hung, D.Q.; Mithulananthan, N.; Bansal, R. Analytical strategies for renewable distributed generation integration considering energy loss minimization. *Appl. Energy* **2013**, *105*, 75–85. [CrossRef]
4. Rueda-Medina, A.C.; Padilha-Feltrin, A. Distributed generators as providers of reactive power support—A market approach. *IEEE Trans. Power Syst.* **2012**, *28*, 490–502. [CrossRef]
5. Adefarati, T.; Bansal, R. Reliability assessment of distribution system with the integration of renewable distributed generation. *Appl. Energy* **2017**, *185*, 158–171. [CrossRef]
6. Mithulananthan, N.; Hung, D.Q.; Lee, K.Y. *Intelligent Network Integration of Distributed Renewable Generation*; Springer: Berlin/Heidelberg, Germany, 2017.

7. Hingorani, N.G.; Gyugyi, L.; El-Hawary, M. *Understanding FACTS: Concepts and Technology of Flexible AC Transmission Systems*; IEEE Press: New York, NY, USA, 2000; Volume 1.
8. Hingorani, N.G. FACTS-flexible AC transmission system. In Proceedings of the International Conference on AC and DC Power Transmission, London, UK, 17–20 September 1991; IET: London, UK, 1991; pp. 1–7.
9. Hingorani, N.G. Flexible AC transmission. *IEEE Spectr.* **1993**, *30*, 40–45. [CrossRef]
10. Hingorani, N.G. High power electronics and flexible AC transmission system. In Proceedings of the American Power Conference, Chicago, IL, USA, 18–20 April 1988.
11. Rao, P.; Crow, M.; Yang, Z. STATCOM control for power system voltage control applications. *IEEE Trans. Power Deliv.* **2000**, *15*, 1311–1317. [CrossRef]
12. Tuzikova, V.; Tlusty, J.; Muller, Z. A novel power losses reduction method based on a particle swarm optimization algorithm using STATCOM. *Energies* **2018**, *11*, 2851. [CrossRef]
13. Bhat, M.V.; Manjappa, N. Flower pollination algorithm based sizing and placement of DG and D-STATCOM simultaneously in radial distribution systems. In Proceedings of the 2018 20th National Power Systems Conference (NPSC), Tiruchirappalli, India, 14–16 December 2018; pp. 1–5.
14. Casey, L.F.; Schauder, C.; Cleary, J.; Ropp, M. Advanced inverters facilitate high penetration of renewable generation on medium voltage feeders-impact and benefits for the utility. In Proceedings of the 2010 IEEE Conference on Innovative Technologies for an Efficient and Reliable Electricity Supply, Waltham, MA, USA, 27–29 September 2010; pp. 86–93.
15. Turitsyn, K.; Sulc, P.; Backhaus, S.; Chertkov, M. Options for control of reactive power by distributed photovoltaic generators. *Proc. IEEE* **2011**, *99*, 1063–1073. [CrossRef]
16. Smith, J.; Sunderman, W.; Dugan, R.; Seal, B. Smart inverter volt/var control functions for high penetration of PV on distribution systems. In Proceedings of the 2011 IEEE/PES Power Systems Conference and Exposition, Phoenix, AZ, USA, 20–23 March 2011; pp. 1–6.
17. Schauder, C. *Advanced Inverter Technology for High Penetration Levels of PV Generation in Distribution Systems*; National Renewable Energy Lab (NREL): Golden, CO, USA, 2014.
18. Siavashi, E.M. Smart PV Inverter Control for Distribution Systems. Ph.D. Thesis, The University of Western Ontario, London, ON, Canada, 2015.
19. Varma, R.K.; Rahman, S.A.; Mahendra, A.; Seethapathy, R.; Vanderheide, T. In Novel nighttime application of PV solar farms as STATCOM (PV-STATCOM). In Proceedings of the 2012 IEEE Power and Energy Society General Meeting, San Diego, CA, USA, 22–26 July 2012; pp. 1–8.
20. Varma, R.K.; Rahman, S.A.; Vanderheide, T. New control of PV solar farm as STATCOM (PV-STATCOM) for increasing grid power transmission limits during night and day. *IEEE Trans. Power Deliv.* **2014**, *30*, 755–763. [CrossRef]
21. Varma, R.K.; Maleki, H. PV Solar System Control as STATCOM (PV-STATCOM) for Power Oscillation Damping. *IEEE Trans. Sustain. Energy* **2019**, *10*, 1793–1803. [CrossRef]
22. Varma, R.K.; Khadkikar, V.; Seethapathy, R. Nighttime application of PV solar farm as STATCOM to regulate grid voltage. *IEEE Trans. Energy Convers.* **2009**, *24*, 983–985. [CrossRef]
23. Wandhare, R.G.; Agarwal, V. Novel stability enhancing control strategy for centralized PV-grid systems for smart grid applications. *IEEE Trans. Smart Grid* **2014**, *5*, 1389–1396. [CrossRef]
24. Shah, R.; Mithulananthan, N.; Lee, K.Y. Large-scale PV plant with a robust controller considering power oscillation damping. *IEEE Trans. Energy Convers.* **2012**, *28*, 106–116. [CrossRef]
25. Luo, L.; Gu, W.; Zhang, X.-P.; Cao, G.; Wang, W.; Zhu, G.; You, D.; Wu, Z. Optimal siting and sizing of distributed generation in distribution systems with PV solar farm utilized as STATCOM (PV-STATCOM). *Appl. Energy* **2018**, *210*, 1092–1100. [CrossRef]
26. Kundur, P. Power system stability. In *Power System Stability and Control*; CRC press: Boca Raton, FL, USA, 2007; p. 10.
27. Hossain, J.; Mahmud, A. *Large Scale Renewable Power Generation*; Springer: Singapore, 2014.
28. Tabatabaei, N.M.; Aghbolaghi, A.J.; Bizon, N.; Blaabjerg, F. *Reactive Power Control in AC Power Systems*; Springer: Cham, Switzerland, 2017.
29. Arulampalam, A.; Ekanayake, J.B.; Jenkins, N. Application study of a STATCOM with energy storage. *IEE Proc.-Gener. Transm. Distrib.* **2003**, *150*, 373–384. [CrossRef]
30. Luenberger, D.G.; Ye, Y. *Linear and Nonlinear Programming*; Springer: New York, NY, USA, 2008; Volume 116.

energies

MDPI

Article

Co-Optimizing Battery Storage for Energy Arbitrage and Frequency Regulation in Real-Time Markets Using Deep Reinforcement Learning

Yushen Miao [1], Tianyi Chen [1], Shengrong Bu [2,*], Hao Liang [3] and Zhu Han [4]

1 James Watt School of Engineering, University of Glasgow, Glasgow G12 8QQ, UK; y.miao.1@research.gla.ac.uk (Y.M.); t.chen.1@research.gla.ac.uk (T.C.)
2 Department of Engineering, Brock University, St. Catharines, ON L2S 3A1, Canada
3 Department of Electrical and Computer Engineering, University of Alberta, Edmonton, AB T6G 2R3, Canada; hao2@ualberta.ca
4 Department of Electrical and Computer Engineering, University of Houston, Houston, TX 77204, USA; zhan2@uh.edu
* Correspondence: sbu@brocku.ca

Abstract: Battery energy storage systems (BESSs) play a critical role in eliminating uncertainties associated with renewable energy generation, to maintain stability and improve flexibility of power networks. In this paper, a BESS is used to provide energy arbitrage (EA) and frequency regulation (FR) services simultaneously to maximize its total revenue within the physical constraints. The EA and FR actions are taken at different timescales. The multitimescale problem is formulated as two nested Markov decision process (MDP) submodels. The problem is a complex decision-making problem with enormous high-dimensional data and uncertainty (e.g., the price of the electricity). Therefore, a novel co-optimization scheme is proposed to handle the multitimescale problem, and also coordinate EA and FR services. A triplet deep deterministic policy gradient with exploration noise decay (TDD–ND) approach is used to obtain the optimal policy at each timescale. Simulations are conducted with real-time electricity prices and regulation signals data from the American PJM regulation market. The simulation results show that the proposed approach performs better than other studied policies in literature.

Keywords: battery energy storage; energy arbitrage; frequency regulation; real-time market; deep reinforcement learning

Citation: Miao, Y.; Chen, T.; Bu, S.; Liang, H.; Han, Z. Co-Optimizing Battery Storage for Energy Arbitrage and Frequency Regulation in Real-Time Markets Using Deep Reinforcement Learning. *Energies* **2021**, *14*, 8365. https://doi.org/10.3390/en14248365

Academic Editors: Victor Becerra and Ahmed Rachid

Received: 29 October 2021
Accepted: 7 December 2021
Published: 11 December 2021

1. Introduction

With wider integration of renewable resources, energy storage has become a significant technology to help eliminate uncertainties associated with renewable energy generation, in order to maintain stability and improve flexibility of power networks. Among different kinds of energy storage technologies, battery energy storage systems (BESSs) have played an irreplaceable role in energy storage, grid synchronization, and other operation-assistance services [1,2] due to the following advantages: (1) BESSs can be flexibly configured depending on the power and energy requirements of system applications [3]; (2) BESSs have an instantaneous response nature [4,5]; (3) BESSs are not limited by external conditions such as geographical resources. Various research has been carried out related to battery energy storage systems planning and design for different applications. Optimized planning was proposed for a battery energy storage system considering battery degradation to reduce the operational costs of the nanogrid and microgrid [6]. In [7], the authors identified the optimal conditions for wireless charging of electric vehicles when they were in motion to reduce energy consumption. A model was presented for a residential energy management system to dispatch battery energy storage in a market-based setting [8]. A privacy-aware framework was presented for utility-driven demand-side management with a realistic energy storage system model [9]. However, the economic viability of using BESSs to provide various services with a large scale is questionable due to their high investment costs [10].

One of the most discussed revenue sources for BESSs is to provide energy arbitrage (EA) services in a real-time electricity market by deliberately charging at low prices and discharging at higher prices to gain profit [11,12]. EA using BESSs was studied in [13], where the electricity price was assumed to be known before making storage decisions. More recent research took electricity price uncertainty into consideration, and thus many forecast methods were proposed to improve the quality of electricity price prediction, and a reinforcement learning method was proposed to maximize the profit of EA based on historical prices [14]. A stochastic dynamic programming method was used to optimize the BESS based on the forecast electricity price [15]. Neural networks were used to address the price prediction uncertainty by introducing a scenario-based optimal control framework [16]. Different models were presented in [17] to process the various price signals to optimize the price forecast.

In order to further increase the revenue of BESSs, some research work has considered a battery to provide EA and frequency regulation (FR) services simultaneously [18], since FR is a significant income source for energy storage [19–23]. For FR, BESSs are used to regulate the frequency of the power grid by charging or discharging based on the regulation signals sent by the power grid operator [5,19,24,25]. A comprehensive evaluation for stacked revenue by using the grid-connected BESS was introduced to provide EA and FR services [26]. A linear programming method was used to maximize the potential revenue of electrical energy storage from participation in EA and FR in the day-ahead market [27]. Co-optimizing EA and FR services simultaneously is considered a multitimescale problem, and a dynamic programming approach was proposed to solve the co-optimization problem [19,20]. These two existing works on co-optimizing EA and FR services assumed that the electricity prices, regulation signals, or their distributions were known in advance. However, the distributions or the values are hard to attain in the real-time market. Furthermore, these two works did not consider the degradation cost of the BESS, a key factor in energy operational planning, without which there might be aggressive charging or discharging of the BESS [4].

Deep reinforcement learning (DRL), combined with deep neural networks (DNNs) and reinforcement learning (RL) techniques, can be powerful tools for addressing BESS-related decision-making problems using the trial-and-error mechanism [28,29]. Compared to model-based methods, such as MILP methods, DRL approaches have the following advantages: the ability to learn from historical data, to be self-adaptable, and to learn a good control policy even under a very complex environment [12]. A novel continuous DRL algorithm was used for energy management of the hybrid electric vehicles [30]. An expert-assistance deep deterministic policy gradient (DDPG) strategy was introduced to minimize the energy consumption and optimize the power allocation of the hybrid electric buses [31]. A multiphysics-constrained fast-charging strategy was proposed for lithium-ion batteries in [32] based on an environmental perceptive DDPG. However, DDPG is not effective in avoiding overestimation in the actor–critic setting [33,34].

To address the above issues, a novel co-optimization scheme considering the degradation of the battery cell in the BESS is proposed for the multitimescale problem of co-optimizing EA and FR services. A novel deep reinforcement learning (DRL) approach, a triplet deep deterministic policy gradient with exploration noise decay (TDD–ND), is proposed to handle the uncertainty of the real-time electricity prices and frequency regulation signals in the multitimescale co-optimization problem due to the following reasons: (1) TDD–ND does not rely on the knowledge of probability distributions; (2) TDD–ND can be used to solve the problem with continuous action space directly by using deterministic policy in an actor–critic algorithm [34–36]; (3) The TDD–ND algorithm takes the weighted action value of triplet critics, which overcomes estimation bias in the deep deterministic policy gradient (DDPG) algorithm and the twin delayed deep deterministic policy gradient

(TD3) algorithm [34]; (4) The TDD–ND algorithm adopts the exploration ND policy, which improves the exploration at the beginning of the training compared to DDPG and TD3.

The main contributions of this paper are as follows:

1. A novel co-optimization scheme is proposed to handle the multitimescale problem. The BESS decides an optimal EA action every five minutes to maximize its revenue due to the total amount of energy change, and every two seconds the BESS decides an optimal FR action to maximize the total reward including the revenue due to energy change and FR settlement reward. Based on the FR action, the EA action has to be adjusted based on the power constraints of the BESS to maximize the total revenue of the day on the two-second level.
2. The TDD–ND algorithm is proposed to solve the co-optimization problem. To the best of our knowledge, the TDD algorithm [34] is for the first time used for energy storage. Our proposed method combines the TDD algorithm with ND policy to improve the exploration during the training, and thus to achieve the higher total revenue.
3. Real-time data are used to evaluate the performance of the proposed TDD–ND co-optimization approach. Simulation results show that our proposed DRL approach with the co-optimization scheme performs better than studied policies.

The rest of this paper is organized as follows. Section 2 explains the Pennsylvania New Jersey Maryland (PJM)'s frequency regulation market. Section 3 presents the nested system model used to formulate the co-optimizing problem. Our proposed TDD–ND approach is described in Section 4. The simulation results are discussed in Section 5. The conclusion is made in Section 6.

2. PJM Frequency Regulation Market

In the PJM frequency regulation market, generators and other devices (e.g., energy storage) can provide grid ancillary services in exchange for regulation credits [37]. PJM sends the regulation (RegD) signal to the resources wishing to provide regulation services every two seconds. Afterwards, PJM tracks the response from each resource and computes a performance score for each resource every two seconds based on the RegD signal and regulation response. For every five minutes, the market also calculates the average performance score within the five-minute period. The performance score is a weighted sum of correlation, delay, and precision [38,39]. A BESS typically has the nature of the instantaneous response and hence the scores of correlation and delay are close to 1. Therefore, the average performance score SC of a BESS within a five-minute period can be calculated based on the precision score as follows [19]:

$$SC = \frac{\sum_{t=0}^{150\Delta t} SC_t}{150}, \quad (1)$$

$$SC_t = |1 - \left|\lambda_t^F + rd_t\right| / ar|, \quad (2)$$

where λ_t^F, rd_t, ar are denoted as the regulation response power taken by BESS response to the RegD signal at time t, the RegD signal at time t, and the maximum power capacity assigned for FR, respectively. Δt is set to two seconds because the RegD signal rd_t is sent every two seconds. SC_t denotes the two-second performance score. When the BESS is 100% following the RegD signal, $|\lambda_t^F + rd_t| = 0$ and $SC_t = 1$.

Every five minutes, the PJM market determines the eligibility of the resource for regulation based on its average performance score SC, and calculates the amount of the regulation credit settlement received by the eligible resource. If the average performance score is less than 40%, the resource will lose its regulation qualification and regulation credits during that time period [37]. The five-minute regulation credit settlement R^C can be calculated as follows [37]:

$$R^C = \begin{cases} SC \cdot ar \cdot P^C, & SC \geq 0.4, \\ 0, & SC < 0.4, \end{cases} \tag{3}$$

where P^C, in \$/MW·5min, is the five-minute regulation clearance price for 1 MW regulation capacity.

3. Nested System Model and Problem Formulation

The system model, illustrated in Figure 1, consists of two main parts, i.e., the power grid and the BESS including a battery and an energy management system (EMS). The BESS participates in the energy and regulation market. The power grid sends the real-time electricity locational marginal price (LMP), FR signal, and FR market clearance price to the EMS in the BESS. The EMS then generates the operation signal to the battery to take action. At the same time, the battery sends feedback with its real-time status to the EMS. Based on the real-time status of the battery and the information from the power grid, the EMS generates a new operation signal to the battery.

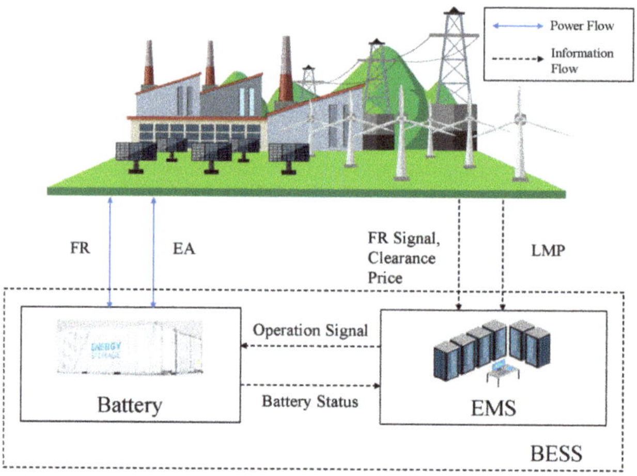

Figure 1. The configuration of the system model.

The BESS co-optimizes EA and FR services to maximize its total reward within a one-day time horizon in a real-time PJM market: EA acts every five minutes, and FR responds every two seconds [19]. Due to the nature of the problem, the timescale is divided into two dimensions: a large timescale with five-minute intervals and a small timescale with two-second intervals, where two-second intervals are nested in the five-minute timescales. The two optimization problems are formulated as two nested MDP submodels in the two following subsections, respectively.

3.1. The Five-Minute MDP Submodel Formulation

The one-day horizon of five-minute submodel T^A, decomposed into 288 five-minute increments (i.e., $\Delta T = 5\,\text{min}$) illustrated in Figure 2, is denoted as $\boldsymbol{T}^A = \{0, \Delta T, 2\Delta T, 3\Delta T, \ldots 287\Delta T\}$. The BESS takes a charging or discharging action every five minutes based on its current state to maximize the cumulative reward within the one-day horizon. The state, action, and reward are defined as follows.

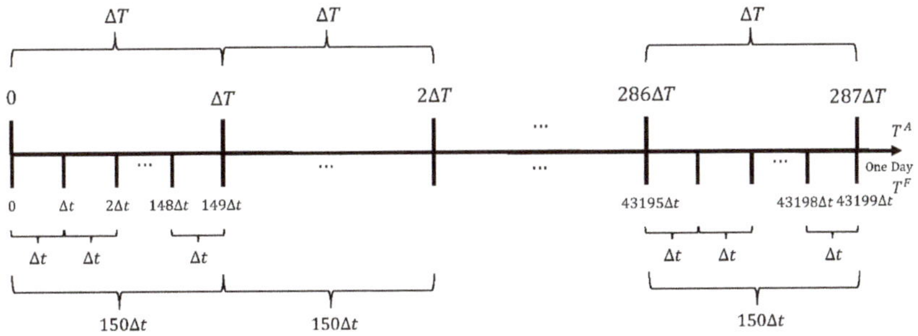

Figure 2. The nested timescales in a day.

3.1.1. State

The state of the BESS at time T can be defined as $S_T^A = (E_T, P_T^A)$, where E_T is the BESS energy level, and P_T^A is the real-time electricity locational marginal price (LMP) at time T.

3.1.2. Action

The action in the five-minute submodel, denoted as λ_T, is the total amount of power change due to EA and FR at time T within the five-minute interval. $\lambda_T > 0$ represents that the BESS is charging, while $\lambda_T < 0$ implies that the BESS is discharging. The optimal action at time T is denoted as λ_T^*. The action space should not exceed the maximum power capacity of BESS B:

$$|\lambda_T| \leq B. \tag{4}$$

The total amount of energy stored in the BESS at time T should be within its maximum energy capacity $E^{\max}$:

$$0 \leq E_T + \lambda_T \cdot \Delta T \leq E^{\max}. \tag{5}$$

After taking action λ_T, state S_T^A is converted to state $S_{T+\Delta T}^A$ at time $T + \Delta T$. The real-time price P_T^A is updated to $P_{T+\Delta T}^A$, and the energy level E_T evolves to $E_{T+\Delta T}$, which can be calculated as follows:

$$E_{T+\Delta T} = \begin{cases} E_T + \Delta T \cdot \lambda_T \cdot \eta^c, & \lambda_T \geq 0, \\ E_T + \Delta T \cdot \lambda_T / \eta^d, & \lambda_T < 0, \end{cases} \tag{6}$$

where η^c and η^d denote the charging and discharging efficiency, respectively.

3.1.3. Degradation Cost and Reward

The degradation cost of the BESS is a key factor in energy operational planning [4] as the battery cells degrade for repeated charge/discharge cycles. The degradation cost of the BESS can be calculated as follows [4]:

$$f_T(b) = c_b |\lambda_T| \cdot \Delta T, \tag{7}$$

where c_b is the degradation cost coefficient, and can be calculated as follows [4]:

$$c_b = \frac{P_{cell}}{2N \cdot (SOC_{\max} - SOC_{\min})}, \tag{8}$$

where P_{cell} is the price of the battery cell in the BESS and N is the number of cycles that the BESS could be operated within the state of charge (SoC) constraint $[SOC_{\min}, SOC_{\max}]$.

After taking the action, the BESS will receive a reward. In order to avoid conservative actions caused by the negative reward in the learning process, an average electricity price

$\bar{P}^A$ is introduced in the reward $R_T^A(S_T^A, \lambda_T)$ for performing λ_T action in state S_T^A based on the basic principle of EA [14] as follows:

$$R_T^A\left(S_T^A, \lambda_T\right) = \left(\bar{P}^A - P_T^A\right) \cdot \lambda_T \cdot \Delta T - f_T(b). \quad (9)$$

3.2. The Two-Second MDP Submodel Formulation

The BESS needs to respond to the updated RegD signal every two seconds. Because two-second intervals are nested in the five-minute horizon, the time horizon of one day within every two-second increment is denoted as $\boldsymbol{T}^F = \{0, \Delta t, 2\Delta t, 3\Delta t, \ldots (150 \cdot 288 - 1)\Delta t\}$, where $\Delta t = 2\,\text{s}$, shown in Figure 2. The BESS takes a charging or discharging action every two seconds based on its current state to maximize the cumulative reward within the one-day horizon. The state, action, and reward are defined as follows.

3.2.1. State

The state at time t is denoted as $S_t^F = (E_t, rd_t)$, where E_t is the energy level of the BESS and rd_t is the received RegD signal at that time.

3.2.2. Action

The action is the regulation response power, denoted as λ_t^F at time t, which is constrained by Equation (4). The action space also should not go beyond the maximum power capacity of BESS B. After performing an action λ_t^F at time t, state S_t^F will transfer to state $S_{t+\Delta t}^F$ at time $t + \Delta t$, and the energy level E_t will be updated to $E_{t+\Delta t}$ based on λ_t denoted as the total amount of power change at time t due to EA and FR:

$$E_{t+\Delta t} = \begin{cases} E_t + \Delta t \cdot \lambda_t \cdot \eta^c, & \lambda_t \geq 0, \\ E_t + \Delta t \cdot \lambda_t / \eta^d, & \lambda_t < 0. \end{cases} \quad (10)$$

3.2.3. Reward

Based on the PJM market regulation policy, the reward $R_t(S_t^F, \lambda_t^F)$ by performing action λ_t^F at state S_t^F can be calculated as

$$R_t\left(S_t^F, \lambda_t^F\right) = R_t^A + R_t^F - f_t(b), \quad (11)$$

where $f_t(b) = c_b|\lambda_t| \cdot \Delta t$ according to Equation (7) and R_t^A is the reward due to the total amount of energy change caused by both EA and FR within the two-second interval:

$$R_t^A = \left(\bar{P}^A - P_T^A\right) \cdot \lambda_t \cdot \Delta t. \quad (12)$$

Instead of calculating the FR settlement at the end of every five minutes, in the two-second submodel, we need to evaluate the FR reward every two seconds once choosing an action λ_t. Based on Equation (3), R_t^F is the equivalent real-time regulation settlement reward within the two-second interval:

$$R_t^F = \begin{cases} SC_t \cdot B \cdot (P^C/150), & SC_t \geq 0.4, \\ 0, & SC_t < 0.4, \end{cases} \quad (13)$$

assuming that the maximum power capacity ar assigned for FR is the power capacity of the BESS B.

3.3. Proposed Co-Optimization Scheme

Solving the co-optimizing problem for EA and FR is to find the optimal action selection policy for the BESS to obtain the maximum expected reward within a day. A co-optimization scheme is proposed to handle the multitimescale problem and coordinate the EA and FR services, which is illustrated in Figure 3. Once λ_t^F is derived, λ_t can be calculated as follows:

$$\lambda_t = \begin{cases} B + \lambda_t^F, & \lambda_T^* - \lambda_t^F > B, \\ \lambda_T^*, & |\lambda_T^* - \lambda_t^F| \le B, \\ -B + \lambda_t^F, & \lambda_T^* - \lambda_t^F < -B, \end{cases} \tag{14}$$

where λ_t is not always equal to λ_T^*, due to the power constraint (Equation (4)) of the BESS. The first case shows that when the optimal action for FR λ_t^F is discharging (i.e., $\lambda_t^F < 0$) and the best action for EA is charging, the action for EA will be charging with the highest power capacity B. In this case, the charging value of λ_T^* was set too high, and λ_t is less than λ_T^*. For the second case, the λ_t is set to λ_T^*. For the third case, when the optimal action for FR λ_t^F is charging and the best action for EA is discharging, the action for EA is discharging with the highest power capacity $-B$. In this case, the discharging value of λ_T^* was set too low, and λ_t is greater than λ_T^*.

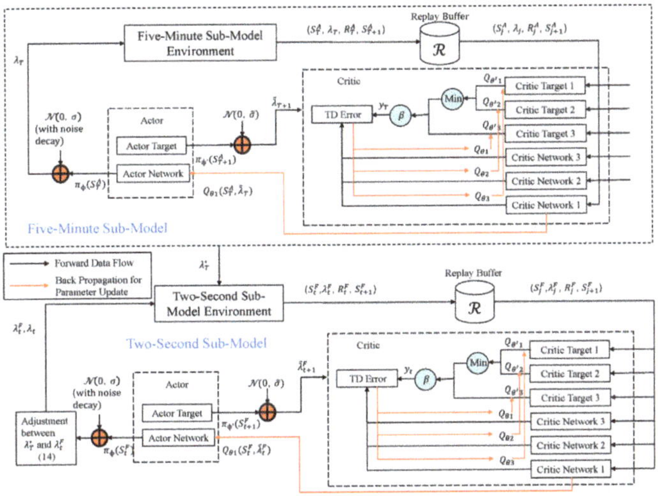

Figure 3. The TDD–ND approach for the proposed co-optimization scheme.

4. Proposed Triplet Deep Deterministic Policy Gradient with Exploration Noise Decay Approach

A novel DRL approach, combining TDD [34] and ND, is proposed to address the co-optimization problem. TDD–ND is a model-free, off-policy actor–critic algorithm, in which the triplet critics are used to limit estimation bias, and the exploration ND policy is used to improve the exploration in the algorithm.

4.1. Triplet Deep Deterministic Policy Gradient Algorithm

The TDD algorithm [34] is an off-line RL algorithm which can be applied to solve the optimization problem with continuous state space as well as continuous actions [35,36]. TDD includes a single actor network (i.e., a deterministic policy network) π_ϕ and its actor target network $\pi_{\phi'}$. In addition, TDD adopts three critic networks Q_{θ_1}, Q_{θ_2}, and Q_{θ_3} for Q-value estimation. $Q_{\theta'_1}$, $Q_{\theta'_2}$ and $Q_{\theta'_3}$ represent three target networks, corresponding to critic networks Q_{θ_1}, Q_{θ_2}, and Q_{θ_3}, respectively. The target value y_t can be updated using the weighted minimum Q-value of target Q-networks $Q_{\theta'_1}$ and $Q_{\theta'_2}$, combined with the weighted value of $Q_{\theta'_3}$ as follows [34]:

$$y_t = r_t + \gamma\left[\beta \min_{j\in\{1,2\}} Q_{\theta'_j}(s_{t+1}, \tilde{a}_{t+1}) + (1-\beta)Q_{\theta'_3}(s_{t+1}, \tilde{a}_{t+1})\right], \tag{15}$$

where $\beta \in (0,1)$ is the weight of the pair of critics, $\gamma \in [0,1]$ is a discount factor, and $\tilde{a}_{t+1}$ is the clipped target action, calculated as follows:

$$\tilde{a}_{t+1} \leftarrow \pi_{\phi'}(s_{t+1}) + \epsilon, \quad \epsilon \sim \text{clip}(\mathcal{N}(0,\tilde{\sigma}), -c, c), \tag{16}$$

where ϵ is the clipped Gaussian noise with standard deviation of $\tilde{\sigma}$, and c is the edge value. The parameters of the critic networks will be updated by minimizing the following loss:

$$\mathcal{L}(\theta) = \mathbb{E}_{(s_t,a_t,r_t,s_{t+1})\sim\mathcal{R}}\left[(Q_\theta(s_t,a_t) - y_t)^2\right], \tag{17}$$

where $\mathcal{R}$ is a replay buffer to store and relay experience transactions (s_t, a_t, r_t, s_{t+1}) including states, actions, rewards, and next states. The deterministic policy network in actor is updated using sampled policy gradient which is shown as follows:

$$\nabla_\phi J(\phi) = N^{-1} \sum \nabla_a Q_{\theta_1}(s_t, a_t)\Big|_{a=\pi_\phi(s_t)} \nabla_\phi \pi_\phi(s_t), \tag{18}$$

4.2. Proposed TDD–ND Co-Optimization Approach

The ND policy is combined with the TDD algorithm to address the co-optimization problem. For the ND policy, the standard deviation of the exploration noise ϵ is set to the maximum value σ_{max} at the beginning of the training, gradually reduced to the minimum value σ_{min} with a decay of σ_{decay} with the increase of the number of the training episodes, and kept at the minimum value σ_{min} for the rest of the training. A TDD–ND algorithm for five-minute submodel optimization is presented in Algorithm 1.

Algorithm 1: The TDD–ND training process for five-minute submodel optimization

Initialize the actor network π_ϕ, the actor target network $\pi_{\phi'} \leftarrow \pi_\phi$, the size R of replay buffer $\mathcal{R}$, and the mini-batch size m.
Initialize the three critic networks Q_{θ_1}, Q_{θ_2} and Q_{θ_3}, and three critic target networks $Q_{\theta'_1} \leftarrow Q_{\theta_1}$, $Q_{\theta'_2} \leftarrow Q_{\theta_2}$ and $Q_{\theta'_3} \leftarrow Q_{\theta_3}$.

for episode $i \leftarrow 0$ to I **do**
  for $t \in T^A$ **do**
    Based on the state of the BESS S_T^A including E_T and P_T^A, choose action λ_T, observe reward R_T^A and next state of the BESS S_{T+1}^A.
    Store transition $(S_T^A, \lambda_T, R_T^A, S_{T+1}^A)$ in $\mathcal{R}$.
    Sample a batch of transitions $\left(S_j^A, \lambda_j, R_j^A, S_{j+1}^A\right)$ from $\mathcal{R}$.
    From the next state of the BESS S_{T+1}^A, the actor target plays the next charging or discharging action of the BESS λ_{T+1} via Equation (16).
    Select Gaussian noise $\epsilon \sim \mathcal{N}(0,\sigma)$ to this next action λ_{T+1}. Decrease σ from σ_{max} to σ_{min} with the decay of σ_{decay} as the increasing of the episode.
    Calculate the estimated target value via Equation (15).
    Update parameters of the three critic networks by minimizing the loss defined by Equation (17).
    Update the weights of the critic target networks by:
    $\boldsymbol{\theta}'_i \leftarrow \tau\boldsymbol{\theta}_i + (1-\tau)\boldsymbol{\theta}'_i$, $i = 1,2,3$ every 2 iterations, where $\tau \ll 1$ is the target update parameter.
    Update the actor network by performing gradient every 2 iterations based on Equation (18).
    Update the weights of the actor target networks by:
    $\boldsymbol{\phi}' \leftarrow \tau\boldsymbol{\phi} + (1-\tau)\boldsymbol{\phi}'$ every 2 iterations.
  end for
end for

The flow chart of the proposed TDD–ND co-optimization approach is illustrated in Figure 3. The TDD–ND algorithm is used to train the neural networks for five-minute submodel optimization. The best actions of the five-minute submodel λ_T^* are then input into the two-second submodel environment. The TDD–ND algorithm is then used to train neural networks for two-second submodel optimization. For each training iteration, after action λ_t^F is chosen, λ_t is calculated based on Equation (14), and the reward $R_t(S_t^F, \lambda_t^F)$ will be calculated using Equation (11) to maximize the accumulated reward within the one-day horizon. After each time step, a mini-batch of m transitions is sampled uniformly from a replay buffer $\mathcal{R}$.

5. Experimental Results

The performance of the proposed co-optimization approach is evaluated in a real-world scenario. The values of the parameters used in the simulations are listed in Table 1. Some of the parameters are varied in the simulation and will be noted accordingly. The parameter settings for the TDD–ND algorithm are listed in Table 2.

Table 1. The value of the parameters used in the simulation.

Parameters	Value
P_T^A	PJM historical real-time LMP from 00:00:00 AM to 11:55:00 PM, 30 July 2021 [40]
P^C	PJM historical real-time clearance price for FR from 00:00:00 AM to 11:55:00 PM, 30 July 2021 [40]
rd_t	Historical real-time RegD signal from 00:00:00 AM to 11:59:58 PM, 30 July 2021 [41]
$E^{\max}$	5 MWh
B, ar	1 MW
P_{cell}	8×10^4 \$/MWh
C_b	4/MWh

Table 2. TDD–ND parameter settings.

Parameters	Value
α_{actor}	8×10^{-4}
α_{critic}	8×10^{-5}
γ	0.99
σ_{max}	1
σ_{min}	0.01
σ_{decay}	3×10^{-3}
R	1×10^6
m	1×100

5.1. Performance Evaluation of the Proposed TDD–ND Algorithm

Based on the principle of EA, the BESS charges at low electricity prices and discharges at high electricity prices. The average price works as a simple indicator to determine whether the price P_T^A is low or high compared to the historical values. The operations of the BESS in a day are illustrated in Figure 4. The figure shows that when the P_T^A is lower than the average price, the BESS actions are mainly larger than 0, which means the BESS is charging. However, when the P_T^A is higher than the average price, the BESS operations are discharging to gain profits. The figure matches well with the principle of EA.

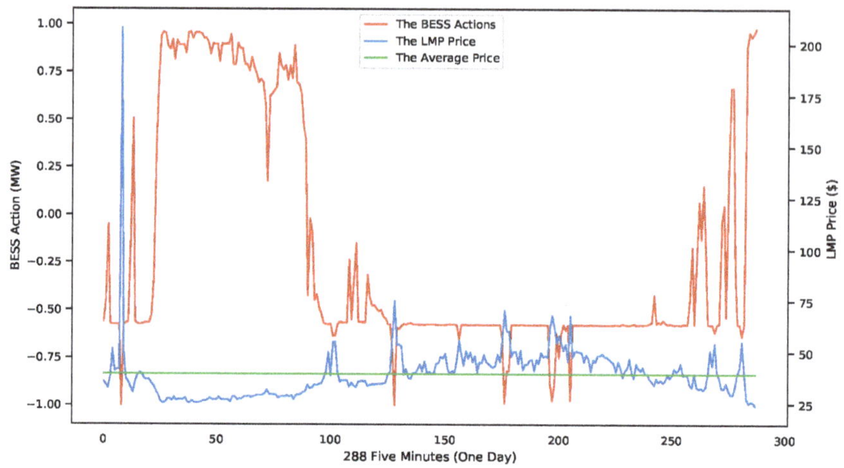

Figure 4. The BESS operation in a one-day period after five-minute submodel optimization.

The performance of the TDD–ND algorithm for co-optimizing EA and FR is studied by comparing it with another widely-used DRL algorithm, the deep Q-learning (DQL) algorithm. TDD–ND and DQL algorithms were used to train the five-minute and two-second submodels for 500 times (500 episodes). During the training using TDD–ND, the total revenue of a day was validated after every 10 episodes without adding exploration noise ϵ to see whether the results were close to the training results. The learning curves of the TDD–ND algorithm and the DQL algorithm are illustrated in Figures 5 and 6, respectively. These two figures show that the TDD–ND algorithm has a much better performance than the DQL algorithm in terms of the average performance score and the total reward. The reason is that the TDD–ND algorithm can choose more accurate continuous actions rather than using discretized actions in DQL, and can thus obtain a higher average performance score and total reward.

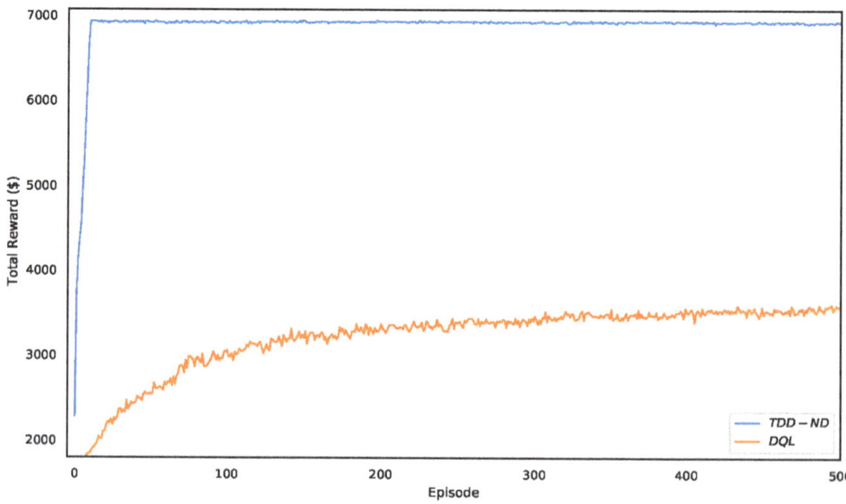

Figure 5. The learning curve of the average performance score of the day trained by TDD–ND and DQL.

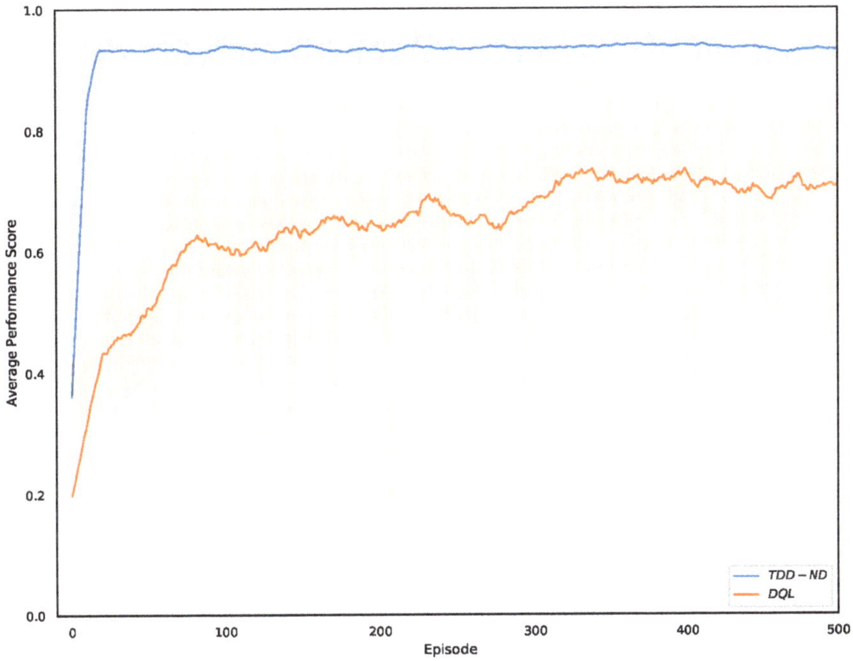

Figure 6. The learning curves of TDD–ND and DQL of the total revenue within a day.

The impact of different levels of power capacity B and energy capacity $E^{\max}$ on the performance of the TDD–ND algorithm and the DQL algorithm are studied. After training, the TDD–ND test results are slightly higher than their training values without the exploration noise. Figure 7 shows that the proposed TDD–ND algorithm always performs better than the DQL algorithm. The reason is that the DQL algorithm chooses discretized actions rather than continuous actions to take, and thus negatively impacts the total revenue. The figure also shows that the total revenue using both algorithms increases with power capacity B in the similar trend. For both algorithms, the total revenue increases sharply with B when B is between 0.5 and 1.0. The reason is that when B is 0.5, the SC_t is smaller than 0.4 in most time slots, and thus the regulation settlement reward R_t^F becomes 0. When B increases to 1, the SC_t is greater than 0.4 in many more time slots. Therefore, the total revenue is significantly increased. Between $B = 1$ and $B = 2.5$, the improvement in total revenue approximately follows the increase of B, since R_t^F is the dominant factor in the total revenue, and is a linear function of B.

How energy capacity $E^{\max}$ impacts the total revenue using the proposed TDD–ND algorithm and the DQL algorithm is shown in Figure 8. The figure shows that the TDD–ND algorithm generates more total revenue than the DQL algorithm under each of the $E^{\max}$ settings. Compared to the impact of power capacity B, the increase of energy capacity $E^{\max}$ only makes a slight change to the total revenue. For both algorithms, the total revenue rises slowly with the increase of $E^{\max}$ between $E^{\max} = 2.5$ and 12.5, as the energy capacity increasing only improves R_t^A but R_t^F dominates the total revenue in Equation (11) when $B = 1$ MW. Compared to the TDD–ND algorithm, the DQL algorithm has a slightly higher improvement rate of the total reward with the increase of $E^{\max}$, since higher $E^{\max}$ allows the DQL algorithm to choose better discretized actions for EA, and thus a higher improvement rate of R_t^A compared to the TDD–ND algorithm with continuous actions.

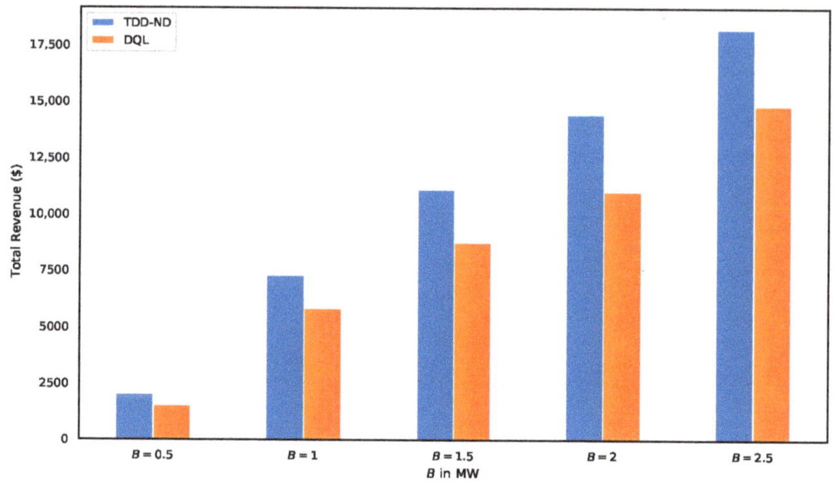

Figure 7. The total revenue of the day with different levels of power capacity B.

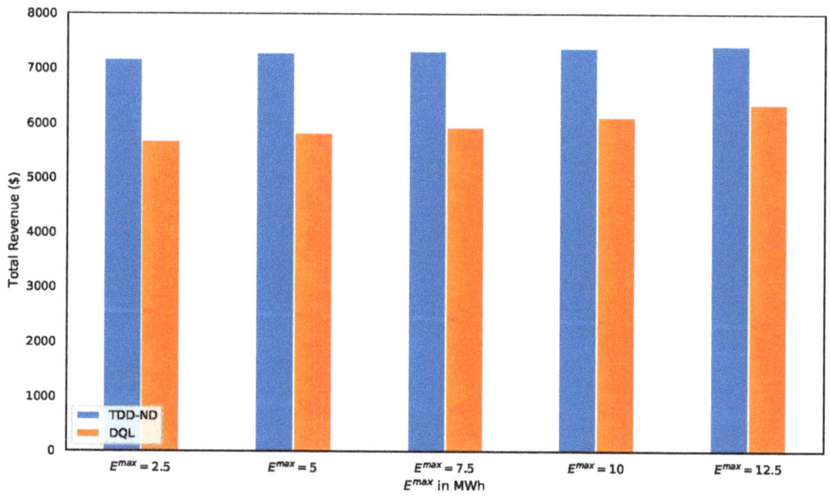

Figure 8. The total revenue of the day with different levels of energy capacity $E^{\max}$.

5.2. Performance Comparison of Various Schemes

To demonstrate the effectiveness of our proposed TDD–ND co-optimization scheme, the following methods are compared: (1) Pure-EA scheme, in which the BESS only provides the EA service; (2) Pure-FR scheme, in which the BESS only provides the FR service; (3) Rule-based co-optimization scheme, in which the BESS provides the EA and FR services. The rule is as follows: The action λ_t is set to λ_T^* to maximize the reward due to the total amount of energy change caused by both EA and FR; (4) TDD–ND co-optimization scheme is our proposed TDD–ND algorithm and co-optimization scheme.

The total revenue using each scheme with different settings of B and $E^{\max}$ is illustrated in Figures 9 and 10, respectively. Figure 9 shows that the TDD–ND co-optimization scheme generates much more total revenue than the other three schemes at every setting of B, as the TDD–ND co-optimization scheme tries to maximize the total accumulated reward. The total revenue using the pure-EA scheme is very small, since FR is much more profitable than EA. For the TDD–ND scheme, the rule-based scheme and the pure-FR scheme, the regulation settlement reward R_t^F increases with B, and thus the total revenue increases with B. For the pure-EA scheme, the higher B allows the BESS to charge more when P_T^A is low and discharge more when P_T^A is high. The increasing rates of the total revenue using the TDD–ND co-optimization scheme and the rule-based co-optimization scheme from $B = 0.5$ to $B = 1$ are higher than those from $B = 1$ to $B = 2.5$. The reason is that when $B = 0.5$, the performance score is smaller than 0.4 in most time slots, and thus the regulation settlement reward becomes 0. Therefore, the total revenue of the rule-based co-optimization scheme is close to that of the pure-FR scheme. When B increases to 1, the performance score is greater than 0.4 in many more time slots, and with the coordination of EA, the total revenue is significantly increased. When B is between 1 and 1.5, the total revenue of the pure-FR scheme is much lower than that of the rule-based co-optimization scheme and that of the TDD–ND cop-optimization scheme. The reason is that the pure-FR scheme cannot follow rd_t signals closely due to the limitations of the energy capacity, while the rule-based scheme can coordinate the energy capacity for EA and FR. When B reaches 2 or higher, the rule-based scheme has similar total revenue to the pure-FR scheme, as the setting of B allows the rule-based scheme to follow rd_t signal closely.

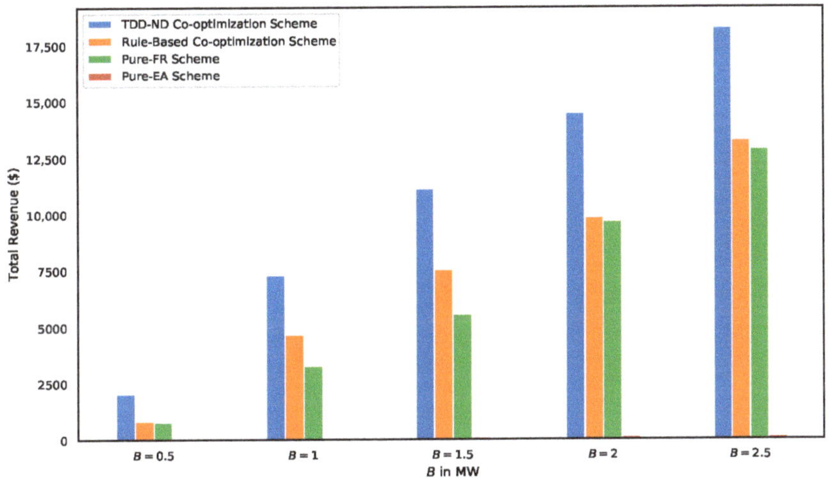

Figure 9. The comparison of the total revenues between using our proposed TDD–ND co-optimization scheme, rule-based co-optimization scheme, pure-FR scheme, and pure-EA scheme under different levels of power capacity B.

The total revenue of each of the four schemes under different settings of energy capacities $E^{\max}$ is presented in Figure 10. The total revenue of the proposed TDD–ND co-optimization scheme is much higher than those of the other three schemes. FR is much more profitable than EA under all of the $E^{\max}$ settings. The total revenue of the TDD–ND scheme and the rule-based scheme increases slightly with $E^{\max}$, because the increase of $E^{\max}$ only improves the value of R_t^A, which is a small portion of the total revenue R_t. For the pure-EA scheme, the total revenue increases with energy capacity $E^{\max}$, as higher energy capacity allows the BESS to charge more when P_T^A is low and discharge more when P_T^A is high.

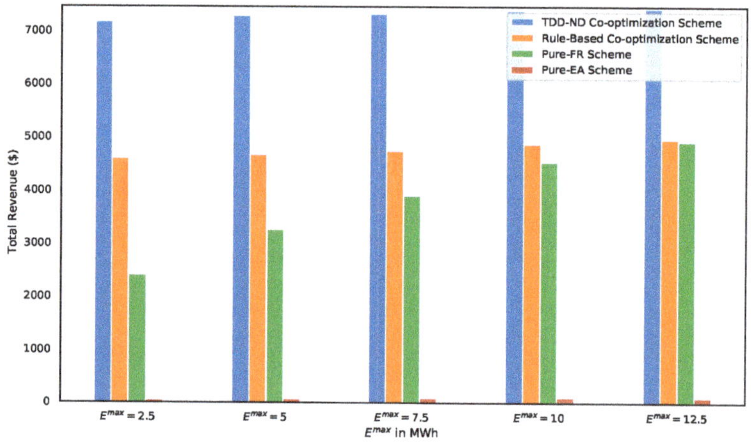

Figure 10. The comparison of the revenues between following the proposed TDD–ND co-optimization, rule-based co-optimization, and pure-FR and pure-EA schemes under different settings of energy capacity E^{max}.

6. Conclusions

A battery energy storage system (BESS) providing both energy arbitrage (EA) and frequency regulation (FR) services simultaneously to maximize its total revenue within a day was considered. The BESS takes an EA action every five minutes and an FR action every two seconds. The multitimescale co-optimization problem was formulated as two nested Markov decision process (MDP) submodels. A novel co-optimization scheme was proposed to handle the multitimescale problem and to coordinate the EA and FR services to maximize the total revenue. The novel deep reinforcement learning (DRL) algorithm, triplet deep deterministic policy gradient with exploration noise decay (TDD–ND), was proposed to determine the best actions to take to maximize the accumulated reward within the one-day horizon. The proposed TDD–ND algorithm achieved 22.8% to 32.9% higher total revenue than the deep Q-learning (DQL) algorithm under various power capacity settings of the BESS when its energy capacity was 5 MWh, and achieved 16.7% to 26.6% higher total revenue under various energy capacity settings when the power capacity was 1 MW. Additionally, our proposed TDD–ND co-optimization scheme achieved 37.7% to 148.8%, 41.8% to 156.3%, and 3507.8% to 15,583.2% higher total revenues compared to the rule-based co-optimization scheme, the pure-FR scheme, and the pure-EA scheme, respectively, under various power capacity settings when the energy capacity of the BESS was 5 MWh. When the power capacity was set to 1 MW, the proposed TDD–ND co-optimization scheme achieved total revenues 49.6% to 56.2%, 51.0% to 198.4%, and 7156.2% to 12,777.0% higher than the rule-based co-optimization scheme, the pure-FR scheme, and the pure-EA scheme, respectively, under the various energy capacity settings. In the future, investigation can be carried out on the use of the co-optimization methods in multivector energy systems considering different timescales.

Author Contributions: Conceptualization, Y.M.; methodology, Y.M., T.C. and S.B.; software, Y.M., and T.C.; simulation, Y.M.; validation, Y.M.; formal analysis, Y.M.; investigation, Y.M., T.C. and S.B.; resources, Y.M. and S. B.; data curation, Y.M.; writing—original draft preparation, Y.M.; writing—review and editing, Y.M., S.B., H.L. and Z.H.; visualization, Y.M.; supervision, S.B.; project administration, Y.M., T.C. and S.B. All authors have read and agreed to the published version of the manuscript.

Funding: This work was supported by start-up funds provided by Brock University, NSF CNS-2128368, CNS-2107216, Toyota and Amazon.

Conflicts of Interest: The authors declare no conflicts of interest.

Nomenclature

ar	The maximum regulation capacity in MW assigned by PJM
B	The maximum power capacity of the BESS in MW
c_b	The linearized battery degradation cost co-efficient
E_T	Energy level of the BESS in MWh at time T in five-minute submodel
E_t	Energy level of the BESS in MWh at time t in two-second submodel
E^{max}	The maximum energy capacity of the BESS in MWh
$f(b)$	The degradation cost
m	The mini-batch size
N	The number of cycles that the BESS
P_T^A	The real-time electricity price at time T
$\bar{P}^A$	The average value of electricity prices in the past day
P_{cell}	The price of the battery cell in the BESS
RegD	Dynamic signal for fast regulation, which is a measure of the imbalance between sources and uses of power in MW in the grid
rd_t	The regulation signal (RegD) sent by PJM at time t to the BESS to provide regulation service
R^C	The five-minute regulation settlement
R_T^A	The reward for performing an action λ_T state S_T^A in five-minute submodel
R_t	The reward for performing action λ_t^F at state S_t^F in two-second submodel
R_t^F	The real-time regulation settlement reward within the two-second interval
SC	Average performance score within a five-minute period indicating the performance of FR
SC_t	The two-second performance score at time t
S_T^A	The state of five-minute submodel at time T
S_t^F	The state of two-second submodel at time t
T	The time indicator in five-minute submodel
t	The time indicator in two-second submodel
$\boldsymbol{T}^A$	The one-day horizon of five-minute submodel
$\boldsymbol{T}^F$	The one-day horizon of two-second submodel
ΔT	The five-minute time interval
Δt	The two-second time interval
λ_T	The action of the total amount of power change in MW due to EA and FR at time T in five-minute submodel
λ_t^F	The action in MW of BESS response to the RegD signal at time t in two-second submodel
λ_T^*	The optimal action of five-minute submodel at time T
λ_t	The total amount of power change at time t due to EA and FR in two-second submodel
η^c	The charging efficiency of the BESS
η^d	The discharging efficiency of the BESS,
α_{actor}	learning rate for actor
α_{critic}	learning rate for critic
σ_{max}	The maximum standard deviation value in the exploration noise
σ_{min}	The minimum standard deviation value in the exploration noise
σ_{decay}	The decay of standard deviation value in the exploration noise decay policy
γ	The discount factor for future rewards
$\mathcal{R}$	Replay buffer
R	The size of replay buffer
ϵ	Clipped Gaussian noise
π_ϕ	The actor network in TDD–ND
$\pi_{\phi'}$	The actor target network in TDD–ND
Q_θ	Critic networks in TDD–ND
$Q_{\theta'}$	Target networks in TDD–ND

References

1. Chatzinikolaou, E.; Rogers, D.J. A comparison of grid-connected battery energy storage system designs. *IEEE Trans. Power Electron.* **2017**, *32*, 6913–6923. [CrossRef]
2. He, G.; Chen, Q.; Kang, C.; Pinson, P.; Xia, Q. Optimal bidding strategy of battery storage in power markets considering performance-based regulation and battery cycle life. *IEEE Trans. Smart Grid* **2016**, *7*, 2359–2367. [CrossRef]
3. Rosewater, D.; Baldick, R.; Santoso, S. Risk-Averse Model Predictive Control Design for Battery Energy Storage Systems. *IEEE Trans. Smart Grid* **2020**, *11*, 2014–2022. [CrossRef]
4. Shi, Y.; Xu, B.; Wang, D.; Zhang, B. Using battery storage for peak shaving and frequency regulation: Joint optimization for superlinear gains. *IEEE Trans. Power Syst.* **2018**, *33*, 2882–2894. [CrossRef]
5. Meng, L.; Zafar, J.; Khadem, S.K.; Collinson, A.; Murchie, K.C.; Coffele, F.; Burt, G.M. Fast Frequency Response from Energy Storage Systems—A Review of Grid Standards, Projects and Technical Issues. *IEEE Trans. Smart Grid* **2020**, *11*, 1566–1581. [CrossRef]
6. Arévalo, P.; Tostado-Véliz, M.; Jurado, F. A novel methodology for comprehensive planning of battery storage systems. *J. Energy Storage* **2021**, *37*, 102456. [CrossRef]
7. Mohamed, N.; Aymen, F.; Ben Hamed, M.; Lassaad, S. Analysis of battery-EV state of charge for a dynamic wireless charging system. *Energy Storage* **2020**, *2*, 1–10. [CrossRef]
8. Antoniadou-Plytaria, K.; Steen, D.; Tuan, L.A.; Carlson, O.; Fotouhi Ghazvini, M.A. Market-Based Energy Management Model of a Building Microgrid Considering Battery Degradation. *IEEE Trans. Smart Grid* **2021**, *12*, 1794–1804. [CrossRef]
9. Avula, R.R.; Chin, J.X.; Oechtering, T.J.; Hug, G.; Mansson, D. Design Framework for Privacy-Aware Demand-Side Management with Realistic Energy Storage Model. *IEEE Trans. Smart Grid* **2021**, *12*, 3503–3513. [CrossRef]
10. Arias, N.B.; Lopez, J.C.; Hashemi, S.; Franco, J.F.; Rider, M.J. Multi-Objective Sizing of Battery Energy Storage Systems for Stackable Grid Applications. *IEEE Trans. Smart Grid* **2020**, *3053*, 1–14. [CrossRef]
11. Sioshansi, R.; Denholm, P.; Jenkin, T.; Weiss, J. Estimating the value of electricity storage in PJM: Arbitrage and some welfare effects. *Energy Econ.* **2009**, *31*, 269–277. [CrossRef]
12. Cao, J.; Harrold, D.; Fan, Z.; Morstyn, T.; Healey, D.; Li, K. Deep Reinforcement Learning-Based Energy Storage Arbitrage with Accurate Lithium-Ion Battery Degradation Model. *IEEE Trans. Smart Grid* **2020**, *11*, 4513–4521. [CrossRef]
13. Eyer, J.M.; Iannucci, J.J.; Corey, G.P.; SANDIA. Energy storage benefits and markets analysis handbook. In *A Study for DOE Energy Storage Systems Program*; Sandia National Laboratories: Albuquerque, NM, USA; Livermore, CA, USA, 2004; p. 105.
14. Wang, H.; Zhang, B. Energy storage arbitrage in real-time markets via reinforcement learning. In Proceedings of the IEEE Power and Energy Society General Meeting, Portland, OR, USA, 5–9 August 2018; pp. 1–5.
15. Abdulla, K.; De Hoog, J.; Muenzel, V.; Suits, F.; Steer, K.; Wirth, A.; Halgamuge, S. Optimal operation of energy storage systems considering forecasts and battery degradation. *IEEE Trans. Smart Grid* **2018**, *9*, 2086–2096. [CrossRef]
16. Chen, Y.; Hashmi, M.U.; Deka, D.; Chertkov, M. Stochastic Battery Operations using Deep Neural Networks. In Proceedings of the 2019 IEEE Power and Energy Society Innovative Smart Grid Technologies Conference, ISGT 2019, Washington, DC, USA, 18–21 February 2019; pp. 7–11.
17. Krishnamurthy, D.; Uckun, C.; Zhou, Z.; Thimmapuram, P.R.; Botterud, A. Energy storage arbitrage under day-ahead and real-time price uncertainty. *IEEE Trans. Power Syst.* **2017**, *33*, 84–93. [CrossRef]
18. *Grid Energy Storage*; Tech. Rep.; Dept. Energy: Washington, DC, USA, 2013.
19. Cheng, B.; Powell, W.B. Co-optimizing battery storage for the frequency regulation and energy arbitrage using multi-scale dynamic programming. *IEEE Trans. Smart Grid* **2018**, *9*, 1997–2005. [CrossRef]
20. Cheng, B.; Asamov, T.; Powell, W.B. Low-rank value function approximation for co-optimization of battery storage. *IEEE Trans. Smart Grid* **2018**, *9*, 6590–6598. [CrossRef]
21. Walawalkar, R.; Apt, J.; Mancini, R. Economics of electric energy storage for energy arbitrage and regulation in New York. *Energy Policy* **2007**, *35*, 2558–2568. [CrossRef]
22. Perekhodtsev, D. Two Essays on Problems of Deregulated Electricity Markets. Ph.D. Thesis, Carnegie Mellon University, Pittsburgh, PA, USA, 2004; p. 94.
23. Engels, J.; Claessens, B.; Deconinck, G. Optimal Combination of Frequency Control and Peak Shaving with Battery Storage Systems. *IEEE Trans. Smart Grid* **2019**, *11*, 3270–3279. [CrossRef]
24. Sun, Y.; Bahrami, S.; Wong, V.W.; Lampe, L. Chance-constrained frequency regulation with energy storage systems in distribution networks. *IEEE Trans. Smart Grid* **2020**, *11*, 215–228. [CrossRef]
25. Jiang, T.; Ju, P.; Wang, C.; Li, H.; Liu, J. Coordinated Control of Air-Conditioning Loads for System Frequency Regulation. *IEEE Trans. Smart Grid* **2021**, *12*, 548–560. [CrossRef]
26. Tian, Y.; Bera, A.; Benidris, M.; Mitra, J. Stacked Revenue and Technical Benefits of a Grid-Connected Energy Storage System. *IEEE Trans. Ind. Appl.* **2018**, *54*, 3034–3043. [CrossRef]
27. Nguyen, T.T.; Nguyen, N.D.; Nahavandi, S. Deep reinforcement learning for multi-agent systems: A review of challenges, solutions and applications. *IEEE Trans. Cybern.* **2018**, *50*, 3826–3839. [CrossRef] [PubMed]
28. Chen, T.; Bu, S.; Liu, X.; Kang, J.; Yu, F.R.; Han, Z. Peer-to-Peer Energy Trading and Energy Conversion in Interconnected Multi-Energy Microgrids Using Multi-Agent Deep Reinforcement Learning. *IEEE Trans. Smart Grid* **2021**. [CrossRef]

29. Wang, B.; Li, Y.; Ming, W.; Wang, S. Deep Reinforcement Learning Method for Demand Response Management of Interruptible Load. *IEEE Trans. Smart Grid* **2020**, *11*, 3146–3155. [CrossRef]
30. Wu, J.; Wei, Z.; Li, W.; Wang, Y.; Li, Y.; Sauer, D.U. Battery Thermal-and Health-Constrained Energy Management for Hybrid Electric Bus Based on Soft Actor-Critic DRL Algorithm. *IEEE Trans. Ind. Inform.* **2021**, *17*, 3751–3761. [CrossRef]
31. Wu, J.; Wei, Z.; Liu, K.; Quan, Z.; Li, Y. Battery-Involved Energy Management for Hybrid Electric Bus Based on Expert-Assistance Deep Deterministic Policy Gradient Algorithm. *IEEE Trans. Veh. Technol.* **2020**, *69*, 12786–12796. [CrossRef]
32. Wei, Z.; Quan, Z.; Wu, J.; Li, Y.; Pou, J.; Zhong, H. Deep Deterministic Policy Gradient-DRL Enabled Multiphysics-Constrained Fast Charging of Lithium-Ion Battery. *IEEE Trans. Ind. Electron.* **2021**. [CrossRef]
33. Fujimoto, S.; Van Hoof, H.; Meger, D. Addressing Function Approximation Error in Actor-Critic Methods. In Proceedings of the 35th International Conference on Machine Learning, ICML 2018, Stockholm, Sweden, 10–15 July 2018; Volume 4, pp. 2587–2601.
34. Wu, D.; Dong, X.; Shen, J.; Hoi, S.C. Reducing Estimation Bias via Triplet-Average Deep Deterministic Policy Gradient. *IEEE Trans. Neural Netw. Learn. Syst.* **2020**, *31*, 4933–4945. [CrossRef]
35. Lillicrap, T.P.; Hunt, J.J.; Pritzel, A.; Heess, N.; Erez, T.; Tassa, Y.; Silver, D.; Wierstra, D. Continuous control with deep reinforcement learning. In Proceedings of the 4th International Conference on Learning Representations, ICLR 2016—Conference Track Proceedings, San Juan, PR, USA, 2–4 May 2016.
36. Shi, Q.; Lam, H.K.; Xuan, C.; Chen, M. Adaptive neuro-fuzzy PID controller based on twin delayed deep deterministic policy gradient algorithm. *Neurocomputing* **2020**, *402*, 183–194. [CrossRef]
37. PJM. *PJM Manual 11: Services, Ancillary Operations, Market Operations, Real-Time Market*; PJM: Valley Forge, PA, USA, 2019.
38. Croop, D. *PJM—Performance Scoring, Regulation Market Issues Senior Task Force*; PJM: Valley Forge, PA, USA, 2016.
39. Operations, B.; Pilong, C. *PJM Manual 12: Regulation*; PJM: Valley Forge, PA, USA, 2016.
40. PJM. *Real-Time Five Minute LMPs*; PJM: Valley Forge, PA, USA, 2019.
41. PJM. *Ancillary Services*; PJM: Valley Forge, PA, USA, 2019.

energies

Article

Multi-Objective Optimal Integration of Solar Heating and Heat Storage into Existing Fossil Fuel-Based Heat and Power Production Systems

Guangxuan Wang [1,2,*] and Julien Blondeau [1,2]

1 Thermo and Fluid Dynamics (FLOW), Faculty of Engineering, Vrije Universiteit Brussel (VUB), Pleinlaan 2, 1050 Brussels, Belgium; julien.blondeau@vub.be

2 Brussels Institute for Thermal-Fluid Systems and Clean Energy (BRITE), Vrije Universiteit Brussel (VUB) and Université Libre de Bruxelles (ULB), 1050 Brussels, Belgium

* Correspondence: guangxuan.wang@vub.be

Abstract: Increasing the share of Renewable energy sources in District Heating (DH) systems is of great importance to mitigate their CO_2 emissions. The combined integration of Solar Thermal Collectors (STC) and Thermal Energy Storage (TES) into existing Combined Heat and Power (CHP) systems can be a very cost-effective way to do so. This paper aims at finding the optimal design of STC and TES systems integrated in existing CHP's considering two distinct objectives: economic profitability and environmental impact. To do so, we developed a three-stage framework based on Pareto-optimal solutions generated by multi-objective optimization, a Technique for Order Preference by Similarity to Ideal Solution (TOPSIS)-entropy method to select the optimal solution, followed by the definition of final Operation strategy. We proposed relevant improvement of the state-of-the-art models used in similar analysis. We also applied the proposed methodology to the case of a representative, 12 MW_{th} CHP plant. Our results show that, while the addition of TES or STC alone results in limited performances and/or higher costs, both the cost and the CO_2 emissions can be reduced by integrating the optimal combination of STC and TES. For the selected, optimal solution, carbon emissions are reduced by 10%, while the Annual Total Cost (*ATC*) is reduced by 3%. It also improves the operational flexibility and the efficiency by peak load shaving, load valley filling and thus by decreasing the peak load boiler operation. Compared to the addition of STC alone, the use of TES results in an increased efficiency, from 88% to 92%. The optimal share of STC is then increased from 7% to 10%.

Keywords: combined heat and power; thermal storage; solar heating; multi-objective optimization; decision making

Citation: Wang, G.; Blondeau, J. Multi-Objective Optimal Integration of Solar Heating and Heat Storage into Existing Fossil Fuel-Based Heat and Power Production Systems. *Energies* **2022**, *15*, 1942. https://doi.org/10.3390/en15051942

Academic Editors: Francesco Calise and Hee-Je Kim

Received: 13 December 2021
Accepted: 1 March 2022
Published: 7 March 2022

1. Introduction

The share of Renewable Energy Sources (RES) in energy systems is growing rapidly to accelerate the energy transition and tackle climate change. However, their penetration in the heating and cooling sector, which accounts for more than 50% of the final energy demand in EU [1], is only about 22% in Europe [2]. One of the key solutions to increase this share on the short term and in a cost-effective way is to integrate renewable energy in existing District Heating (DH) systems. Many initiatives are currently taken to do so. According to IRENA [3], Denmark has the ambition to increase the share of RES in their DH systems up to 73% in 2030 (vs. 42% in 2014). China targets a 24% share by 2030.

In terms of renewable resources, biomass, solar heating and geothermal energy are the options with largest potential to reach higher shares of RES in DH's [3], also at large scale [4–6]. Among these options, solar collectors present the additional challenge of being an intermittent source of energy, therefore potentially requiring additional heat storage, which represents an interesting optimization problem.

Hot water Thermal Energy Storage (TES) can provide load shifting and is 100 times cheaper than electricity storage for the same energy capacity [7]. Thanks to TES, up to 25% RES would be integrated in energy systems without significantly affecting its efficiency [8]. Hybrid systems composed of Combined Heat and Power (CHP) units integrated with RES and TES is therefore considered as a first step towards the 4th generation district heating systems [8].

TES alone can also be added to existing CHP's. It contributes to peak load shaving, it can store energy when the demand is low and deliver it when the demand is high. With an optimal operation, it can also increase the energy efficiency, which in turn results in lower CO_2 emissions [9]. In many studies, the optimization of a combined CHP-TES system often focuses on economic aspects, taking into account the cost of CO_2 emissions. Benalcazar [10] proposed an optimization method based on the economic performance for the optimal sizing of how water TES, integrated in an existing CHP plant considering specific investment cost and different carbon prices. His analysis showed that the integration of the TES units can save operational cost and decrease the use of the heat-only boiler, which reduces fuel consumption and decreases CO_2 emissions. Mugnini et al. [11] assessed possible energy flexibility strategies to improve the performance of such system. Their results revealed that a hot water tank can increase the CHP working hours and primary energy savings. Lai et al. [12] developed an operation optimization model based on Particle swarm optimization method, to investigate the flexibility and thermodynamic performance of a CHP unit integrated with an integrated heat storage tank. Their results show that such an integration led to an increased range of operational conditions of CHP units.

Although the addition of Solar Thermal Collectors (STC) can lead to larger CO_2 emission savings than TES alone, they generally increase the production costs of existing systems [3]. The question of the optimal combination of STC and TES in terms of both economic and environmental impacts may therefore be raised. This corresponds to a multi-objective optimization problem with two design variables, i.e., the sizes of the STC and the TES systems.

Single economic objective optimization does not provide alternative solutions to deal with conflicting objectives [13]. Therefore, recent research efforts focused on multi-objective optimization of energy systems. Multi-objective optimization is used to find a trade-off between two or more conflicting objectives to support decision making. Ren et al. [14] proposed a multi-objective linear programming method for operational strategy of a Distributed Energy System (DES). Their model was based on trade-off analysis of economic and environmental optimization. Fazlollahi et al. [15] developed a multi-objective, multi-period optimization for sizing and operating a DH system with the objectives of maximizing the system efficiency and minimizing the CO_2 emissions and the Annual Total Cost (*ATC*), the annualized value of the total cost over the lifetime of the project. Luo et al. [16] developed a framework for the optimization of DES integrated with Genetic Algorithm for multi-objective optimization, and multi-criteria evaluated by Technique for Order Performance by Similarity to an Ideal Solution (TOPSIS) method. Karmellos et al. [17] presented a multi-objective Mixed-integer linear programming (MILP) model for the optimal design and operation of DES by ε-constraint method, with minimizing the *ATC*, and the total carbon emission as objective function. Franco and Versace [18] carried out a multi-objective strategy considering energetic and economic objectives to investigate design and operation strategy of a CHP-TES to DH network.

As discussed above, various approaches have therefore been applied to determine the economically and/or environmentally optimal design and operation of RES and TES integrated into existing CHP systems. However, the proposed methodologies can be further improved in the following respects. First, the modelling of the CHP systems could be more accurate. An increased accuracy of the techno-economic models for the following aspects could lead to more accurate results [19]: Piece-wise Linear Investment functions allowing for a non-linear evolution of the investment costs, and account for part-load efficiencies, start-up costs, CHP acceptable operation ranges and maximum ramp rates, which significantly affect technical and economic performances. Secondly, few recent works [15,16] integrated multi-objective optimization models with decision making methods to optimize the capacity and the operation strategy of TES and RES integrated to existing CHP systems. Moreover, limited research was carried out on the effect of fluctuating investment cost on the sizing and the operation of the system.

In this work, we therefore aim at integrating multi-objective optimization and decision-making methods featuring advanced techno-economic models, and to apply them to 1the STC and TES systems.

Our main objectives are the development a comprehensive methodology to allow decision-makers to determine the optimal design of hybrid heat and power production systems and to assess the economic and environmental impact of the optimal integration of STC and TES systems into existing, conventional CHP systems.

The main innovative features of this work can be summarized as follows:

- The techno-economic models of the sub-systems features Piece-Wise Linear Investment function, part-load efficiencies, start-up costs, maximum ramp rates and CHP acceptable operation ranges.
- The variation of the economic and technical parameters, such as ambient temperature, electricity and fuel prices, is considered.
- Pareto-optimal solutions are generated using multi-objective optimization, from which the optimal solution is picked using the TOPSIS-entropy method, an effective method to make decisions processes more reliable and accurate.

The paper is organized as follows. Section 2 describes the proposed methodologies. The case of a hybrid energy system is defined in Section 3, including the input data and the investigated scenarios. Section 4 presents and discusses the results of the case study. Lastly, conclusions are drawn in Section 5.

2. Materials and Methods

This Section describes the proposed methodology for the optimization of the studied hybrid systems. Their general structure is defined in Section 2.1. The multi-objective optimization model is then described in Section 2.2. Finally, the Decision-making method is presented in Section 2.3.

Figure 1 illustrates the flow chart of the optimization framework used in this study. The framework consists of three stages:

1. Pareto-optimal solutions, generated by multi objective optimization model. It involves a trade-off analysis between economic and environmental aspects.
2. Optimal solution selected among the Pareto solutions using a decision-making tool. The optimal solution with the maximum relative quality ranking is picked up as the final optimal solution.
3. Final design and operation strategy. The hourly operation strategy of the optimal solution is further described to assess performance of each unit with optimal capacity. More detailed description will be introduced in the Section 2.2.

2.1. Hybrid Heat and Power Production System

The components of the existing CHP system considered in this study are a 12 MW_{th} coal-fired CHP with Extraction Condensing (EC) steam turbine, and two gas-fired Heat Only Boilers (HOB) with a thermal power of 5 MW_{th}, respectively, also connected to the system. The sizes of the STC and of the TES that will be added to the existing system are the design variables. A schematic of the system is given in Figure 2. Electricity is produced by the CHP and sold to the local market. The generated heat can be sent to both the heat clients and the TES system. HOBs are only in operation when the heat production is insufficient. The constraints related to each unit described in Section 2.2.2, and the input data described in Section 3.1.

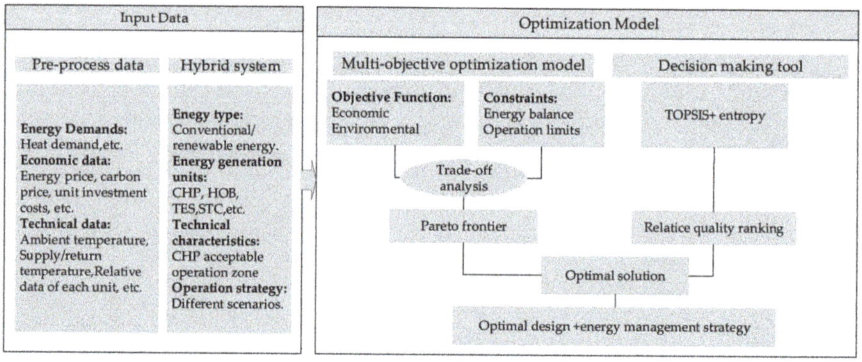

Figure 1. Optimization Framework of hybrid energy system.

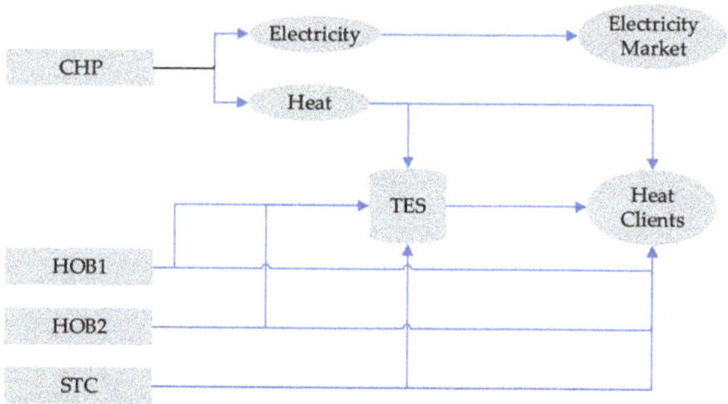

Figure 2. Schematic of Hybrid energy system.

2.2. Optimization Model

2.2.1. Objective Function

In this study, the weighting method [20] will be used to solve the multi-objective optimization problem through MILP method. The two objective functions to be minimized are the *ATC* and the total carbon emissions. A general objective function is then defined as shown in Equation (1):

$$min\left\{F_{obj} = \alpha\cdot\frac{ATC}{ATC_{min}} + (1-\alpha)\cdot\frac{Carbon}{Carbon_{min}}\right\} \tag{1}$$

where α is a weighting factor and where the objectives functions are normalized using the corresponding single optimization value ATC_{min} and $Carbon_{min}$.

The *ATC* of the studied hybrid systems is defined in Equation (2) as the sum of the annualized investment cost C_{inv}, the maintenance cost C_M and the operating cost C_O, minus the revenues R_S from the electricity production [14]. The total CO_2 emissions are calculated for all heat and power generation systems by multiplying the fuel consumption with the corresponding CO_2 emission factor, as shown in Equation (3).

$$ATC = C_{inv} + C_M + C_O - R_S \tag{2}$$

$$Carbon = \sum_u \sum_t CF_u * FC_{u,t} \tag{3}$$

The total investment cost C_{inv} is calculated following Equation (4). The annualized investment cost $c_{u,t}^{inv}$, c_t^{invTES} and c_t^{invSTC} (Equations (5)–(7)) are calculated using the annuity factor defined in Equation (8), which considers the discount rate i and the lifetime n_u for each unit, respectively. The maintenance costs are defined as a fixed proportion of the investment costs and are calculated following the same structures, see Equation (9). Operational costs consist of fuel consumption and unit startup costs, see Equation (10).

$$C_{inv} = \sum_u \sum_t (c_{u,t}^{inv} * Cap_u + c_t^{invTES} * V + c_t^{invSTC} * A) \tag{4}$$

$$c_{u,t}^{inv} = \frac{a_u * I_u^{inv}}{8760} \tag{5}$$

$$c_t^{invTES} = \frac{a_u * I^{invTES}}{8760} \tag{6}$$

$$c_t^{invSTC} = \frac{a_u * I^{invSTC}}{8760} \tag{7}$$

$$a_u = \frac{i * (1+i)^{n_u}}{(1+i)^{n_u} - 1}, \ \forall \text{ u} \in \text{units, t} \in \text{periods} \tag{8}$$

$$C_M = \sum_u \sum_t \left(c_{u,t}^{M} * Cap_u + c_t^{MTES} * V + c_t^{MSTC} * A\right) \tag{9}$$

$$C_O = \sum_u \sum_t (c_{u,t}^{fuel} * FC_{u,t} + \delta_{u,t} * SC_u) \tag{10}$$

The revenues R_S consist in selling electricity to the grid, see Equation (11).

$$R_S = \sum_t P * El_t \tag{11}$$

2.2.2. Constraints for System Design and Operation

All units are subject to some constraints. The following constraints are considered in this study for the design and the operation of each sub-system.

Common Types of Constraints

Common types of constraints are applied to all units:

- Minimum and maximum loads

For all units, the production must be within the minimum and maximum loads, see Equation (12):

$$PLR_{u,t}^{lb} * Q_u^{norm} \leq Q_{u,t} \leq PLR_{u,t}^{ub} * Q_u^{norm} \ \forall \text{ u} \in \text{units, t} \in \text{periods} \tag{12}$$

- Ramping rate limits

Maximum ramping rates (up and down) of the thermal units should be defined for stability, integrity and safety reasons, see Equation (13) and (14):

$$Q_{u,t+1} - Q_{u,t} \leq ramp_{up_{u,t}} * Q_u^{norm} \ \forall \text{ u} \in \text{units, t} \in \text{periods} \tag{13}$$

$$Q_{u,t+1} - Q_{u,t} \leq ramp_{down_{u,t}} * Q_u^{norm} \ \forall \text{ u} \in \text{units, t} \in \text{periods} \tag{14}$$

- State of units

In the framework of MILP optimization, the binary variable $I_{u,t}^{use}$ is used to illustrate the state of the units. It is constrained by the variables $T_{u,t}^{ON}$ and $T_{u,t}^{OFF}$ defining whether the unit is on or off at time t, see Equations (15) and (16).

$$I_{u,t+1}^{use} - I_{u,t}^{use} \leq T_{u,t+1}^{ON} \ \forall \text{ u} \in \text{units, t} \in \text{periods} \tag{15}$$

$$I_{u,t}^{use} - I_{u,t+1}^{use} + T_{u,t+1}^{ON} \leq T_{u,t+1}^{OFF} \ \forall \text{ u} \in \text{units, t} \in \text{periods} \tag{16}$$

- Minimum uptime and downtime

Minimum uptime and downtime constraints are considered for safety and reliability reasons. If a unit is in service, the duration should be at least equal to the defined minimum uptime, as indicated in Equation (17). Similarly, if a unit is not in running, the period should be at least equal to the defined minimum downtime as expressed in Equation (18).

$$\sum_{t-min\_uptime}^{t} T_{u,t}^{ON} \leq I_{u,t}^{use}\ \forall\, \mathrm{u} \in \text{units},\ \mathrm{t}\ \in (min\_uptime,\ \text{periods}), \quad (17)$$

$$\sum_{t-min\_downtime}^{t} T_{u,t}^{OFF} \leq 1 - I_{u,t}^{use}\ \forall\, \mathrm{u} \in \text{units},\ \mathrm{t}\ \in (min\_downtime,\ \text{periods}), \quad (18)$$

Extraction Condensation CHP Unit

The CHP plant considered in this study is of the Extraction Condensation type (EC): heat is provided to the heat clients through the condensation of steam extracted from the turbine. EC-based CHP can be operated in a flexible way, since steam can usually be extracted from more than one pressure stage, and power may vary with same heat production. In this study, the total energy efficiency of the CHP plant is considered as constant, as suggested in [17,19].

For EC-based CHP's, the power loss coefficient β, defined as the power generation reduction caused by heat extraction, is a key characteristic [21]. It can be calculated based on the ambient and DH temperatures [22], see Equations (19) and (20):

$$\beta = 1 - \frac{T_0}{T_M} \quad (19)$$

$$T_M = \frac{T^{supply} - T^{return}}{\ln\left(\frac{T^{supply}}{T^{return}}\right)} \quad (20)$$

The acceptable operation zone of the CHP plant in terms of heat extraction and electrical power for a given supply and return temperature is illustrated in Figure 3. It is elaborated as suggested by [22]. The blue line BC is the backpressure line, exhibiting a slope σ. It represents the operation points with the designed power to heat ratio [21]. The maximum load line CD is parallel to the minimum load line AB (both with a slope β). Any line with the same slope corresponds to a constant fuel consumption.

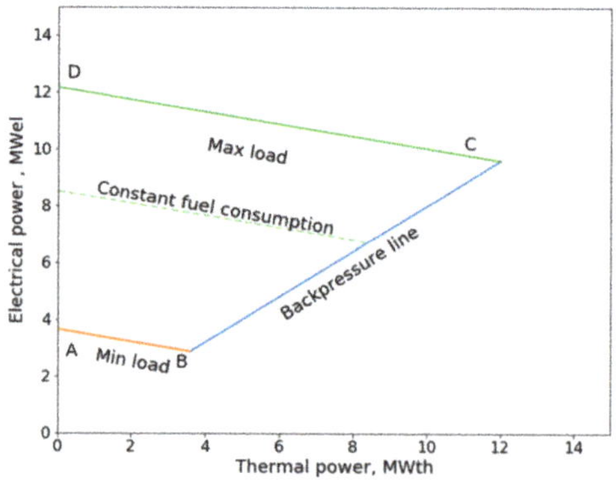

Figure 3. Cogeneration of Power and heat extraction for EC based CHP plant.

Equation (21) and (22) express the corresponding constraints on the operational loads. The fuel consumption can be calculated using Equation (23).

$$P_t^{EC} = PLR_t^{EC} * I_{EC,t}^{use} * \left(\sigma^{EC} + \beta_t\right) * Q_{EC}^{norm} - Q_t^{EC} * \beta_t \ \forall \text{ t } \in \text{periods} \quad (21)$$

$$P_t^{EC} \geq \sigma^{EC} * Q_t^{EC} \ \forall \text{ t } \in \text{periods} \quad (22)$$

$$FC_t^{EC} = (Q_t^{EC} + P_t^{EC}) / \eta_{EC} \ \forall \text{ t } \in \text{periods} \quad (23)$$

Heat Only Boiler

The part load efficiency of the gas-fired Heat Only Boiler (HOB) is considered according to the model proposed by [19]. Equation (24) provides an expression for the consumption of the HOB boiler, which leads to the non-linear evolution of its efficiency, as shown in Figure 4.

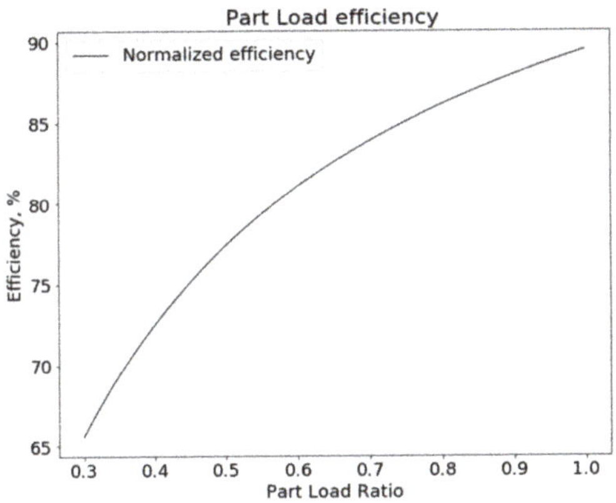

Figure 4. Normalized efficiency for HOB, according to [19].

$$FC_t^{HOB} = 0.4576 * Q_{HOB}^{norm} * I_{HOB,t}^{use} + 0.6599 * Q_t^{HOB} \ \forall \text{ t } \in \text{periods} \quad (24)$$

Equations (25) and (26) state that the two HOB (HOB1 and HOB2) are only in operation when the heat demand cannot be satisfied by the main steam boiler (peak load), during maintenance or when the demand exceeds the CHP output.

$$Q_t^{EC} \geq I_{HOB1,t}^{use} * Q_{CHP}^{norm} \quad (25)$$

$$Q_t^{EC} + Q_t^{HOB} \geq I_{HOB2,t}^{use} * (Q_{CHP}^{norm} + Q_{HOB}^{norm}) \quad (26)$$

Thermal Energy Storage Unit

The Thermal Energy Storage (TES) system is modeled based on the formulation proposed by Wang et al. [23]. Upper and lower boundary conditions are defined for the charging and discharging rates, see Equations (27) and (28). To ensure the availability of the TES, the stored thermal energy at the end of the schedule period is assumed to be equal to its initial value, see Equation (29). The energy balance of the system, taking into account the storage efficiency and the charging and discharging efficiencies, is expressed in Equation (30). The binary variable $z(t)$ is used to illustrate the TES charging and discharging operation status, see Equation (31). Equations (32) and (33) express the constraints that the stored thermal energy must be lower than the maximum storage capacity of the system, which depends on TES volume and DH supply and return temperatures.

$$0 \leq HS_t^{chr} \leq \alpha_t^{chr} \cdot Cap_t^{TES} \ \forall \text{ t } \in \text{periods} \quad (27)$$

$$0 \leq HS_t^{dis} \leq \alpha_t^{dis} \cdot Cap_t^{TES} \; \forall \text{ t } \in \text{periods} \tag{28}$$

$$Q_1^{TES} = Q_{end}^{TES} \; \forall \text{ t } \in \text{periods} \tag{29}$$

$$Q_t^{TES} = \eta_s^{TES} Q_{t-1}^{TES} + \eta_{chr}^{TES} HS_t^{chr} - HS_t^{dis} / \eta_{dis}^{TES} \; \forall \text{t } \in \text{periods} \tag{30}$$

$$z(t) = \begin{cases} 1, & HS_t^{dis} = 0 \\ 0, & HS_t^{chr} = 0 \end{cases} \; \forall \text{ t } \in \text{periods} \tag{31}$$

$$Cap_t^{TES} = \frac{\rho V c}{3600} \{\min((T_t^{supply} - 5), 98) - (T_t^{return} + 5)\} \; \forall \text{ t } \in \text{periods} \tag{32}$$

$$Q_t^{TES} \leq Cap_t^{TES} \; \forall \text{ t } \in \text{periods} \tag{33}$$

The investment costs for the TES system, including the auxiliary equipment, can be estimated following [10,24]. Due to limitation of linear programming method, the exponential decay function of specific investment cost should be replaced with piecewise linear approximation for further use [10]. The specific investment costs curve according to [24], a piecewise linear function is used to make it compatible with MILP, see in Figure 5.

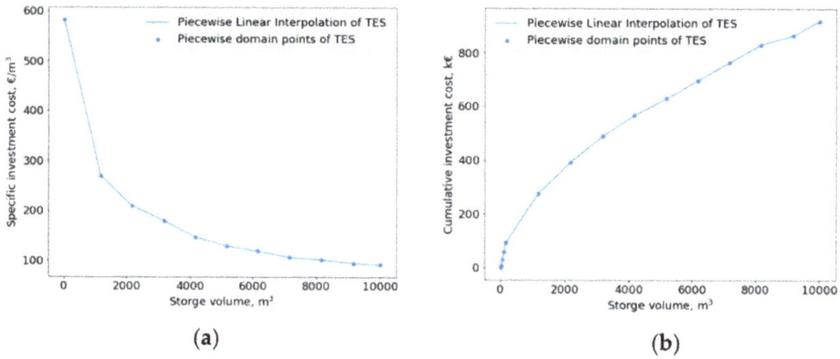

Figure 5. Investment costs for TES systems: (**a**) Specific investment cost according to [24]; (**b**) Cumulative investment cost.

Solar Thermal Collectors Unit

Equations (34) shows how the heating performance of STC can be calculated based on the collector area, solar irradiance and DHN temperature according to the formulation from [25].

$$\begin{aligned} Q_t^{STC} = A * ((0.839 * \, Q_t^{SR} - 2.46 * \left(T_t^{aver} - T_t^{ambient}\right) \\ -0.0197 * \left(T_t^{aver} - T_t^{ambient}\right)^2) \; \forall \text{ t } \in \text{periods} \end{aligned} \tag{34}$$

The specific investment costs of ground mounted solar collector field (shown in Figure 6a) are estimated based on [25]. PLI is also applied, cumulative investment cost see Figure 6b.

Energy Balance

Equation (35) expresses that the heat demand from the local distribution network must always be fulfilled for each time interval.

$$Q_t^{CHP} + Q_t^{HOB} + Q_t^{STC} + \eta_{dis}^{TES} HS_t^{dis} - HS_t^{chr} \geq Q_t^{demand} \; \forall \text{ t } \in \text{periods} \tag{35}$$

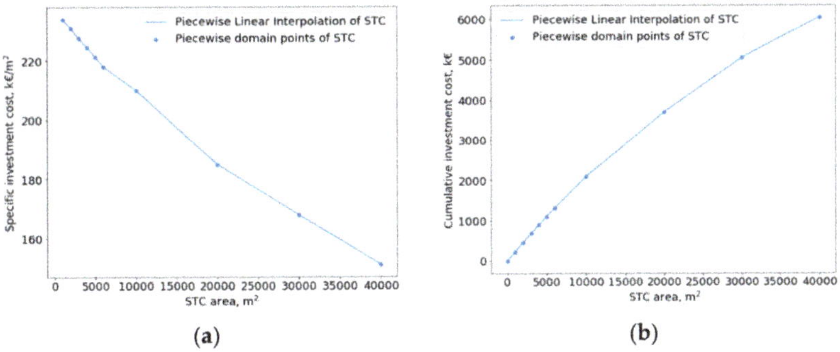

Figure 6. Investment costs for STC: (**a**) Specific investment cost according to [25]; (**b**) Cumulative investment cost.

2.3. Decision-Making Method

Figure 7 illustrates the flow chart of the decision-making process used in this study. The decision-making process is to provide a quantitative evaluation on the system with several conflicting objectives, to support the Decision makers the most potential solutions by considering the important criteria [13]. To evaluate performance of the multi-objective optimization model, a decision-making method combining Technique for Order Preference by Similarity to Ideal Solution (TOPSIS) with Entropy method (EM) is investigated. TOPSIS is a classic method to find optimal solution within finite objectives [26], while EM aims at determining the weights of each objective in an automatic way, in order to minimize man-made error [27]. As illustrated in Figure 7, The TOPSIS-entropy method selects the solution with the maximal relative quality as the optimal solution.

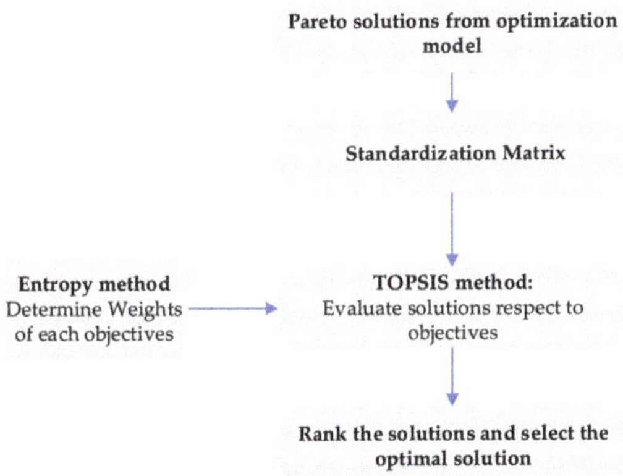

Figure 7. Flowchart of Decision-making process.

The different steps followed to the TOPSIS-entropy method are as follows [28]:

- Weights calculation:

The first step is matrix normalization, see Equation (36). This ensures that all indicators are positive and present comparable ranges. Then, standardized value p_{ij} and entropy value of objective e_j are calculated using Equations (37) and (38). Finally, weighting values of the objectives ω_j are defined, see Equation (39).

$$y_{ij} = \frac{\max(x_{ij}) - x_{ij}}{\max(x_{ij}) - \min(x_{ij})}\ \forall\, \mathrm{i} \in [1 \ldots m],\ \mathrm{i} \in (1 \ldots \mathrm{n}) \quad (36)$$

$$p_{ij} = \frac{y_{ij}}{\sqrt{\sum_{i=1}^{m} y_{ij}}} \forall \mathrm{i} \in [1 \ldots m], \mathrm{i} \in (1 \ldots \mathrm{n}) \tag{37}$$

$$e_j = -\frac{1}{\ln(m)} \sum_{i=1}^{m} p_{ij} \ln(p_{ij}) \forall \mathrm{j} \in (1 \ldots \mathrm{n}) \tag{38}$$

$$\omega_j = \frac{1 - e_j}{\sum_{j=1}^{n} (1 - e_j)} \tag{39}$$

- TOPSIS method

The weighted matrix used as a basis for the TOPSIS method is described in. Equation (40) The positive ideal solution V^+ and negative ideal solution V^- are then determined using Equations (41) and (42). The distances D^+ and D^- between any evaluated result and the two positive and negative solution V^+ and V^- are calculated using Equations (43) and (44). Finally, the relative quality C_i is defined to compare the distance between any result and the two ideal solutions, see Equation (45). The optimal solution corresponds to the highest value of C_i.

$$v_{ij} = \omega_{ij} * y_{ij} \forall \mathrm{i} \in [1 \ldots m], \mathrm{i} \in (1 \ldots \mathrm{n}) \tag{40}$$

$$V^+ = \{(max\ V_{ij} | \mathrm{j} \in J_1), (min\ V_{ij} | \mathrm{j} \in J_2)\} \forall \mathrm{i} \in (1 \ldots \mathrm{n}) \tag{41}$$

$$V^- = \{(min\ V_{ij} | \mathrm{j} \in J_1), (max\ V_{ij} | \mathrm{j} \in J_2)\} \forall \mathrm{i} \in (1 \ldots \mathrm{n}) \tag{42}$$

$$D^+ = \sqrt{\sum_{j=1}^{n} (V_{ij} - V_j^+)^2} \forall \mathrm{i} \in (1 \ldots \mathrm{n}) \tag{43}$$

$$D^- = \sqrt{\sum_{j=1}^{n} (V_{ij} - V_j^-)^2} \forall \mathrm{i} \in (1 \ldots m) \tag{44}$$

$$C_i = \frac{D^-}{D^+ - D^-} \forall \mathrm{i} \in (1 \ldots m) \tag{45}$$

3. Case Study

In this Section, the proposed optimization framework is applied to a specific case of hybrid system described in Section 2.1. The considered input data is described in Section 3.1. Four different scenarios are defined in Section 3.2.

3.1. Input Data

The considered heat demand, ambient temperature, supply and return temperatures are field data retrieved in 2017 from an existing DH plant located in France. The hourly market electricity price was imported from EPEX SPOT for the year 2017 [29] and local climate data of solar radiation were retrieved from PVGIS for the same year [30]. The impact of the uncertainty on the main parameters will be investigated in Section 4.4.

However, the annual profile with 8760 time steps makes is difficult to solve due to complexity of optimization model [31–33]. To limit the computational time without significantly impacting the simulation results [32], the approach of typical periods is used in this work [33], a full year period was discretized into 12 typical days, each of them being representative for a month.

Figure 8 shows that the heat demand varies from 1.44 MW_{th} to 16.9 MW_{th}. The average heat load and heat load variations are larger in the winter. The corresponding supply and return temperatures range from 75 to 90 °C and from 55 to 75 °C, respectively. The average electricity price is also slightly higher in the winter, as illustrated in Figure 9. For each typical day, the heating load peak is observed between 8 and 10 AM and between 8 and 10 PM, while the peak of electricity price is observed between 8 and 9 AM and between 6 and 7 PM. The hourly variations of ambient temperature and solar irradiance are also presented in Figure 10. The heat collected by the STC was calculated following the methodology of Schmidt (Ref. [10]). The considered tilt angle of the collector was 40°.

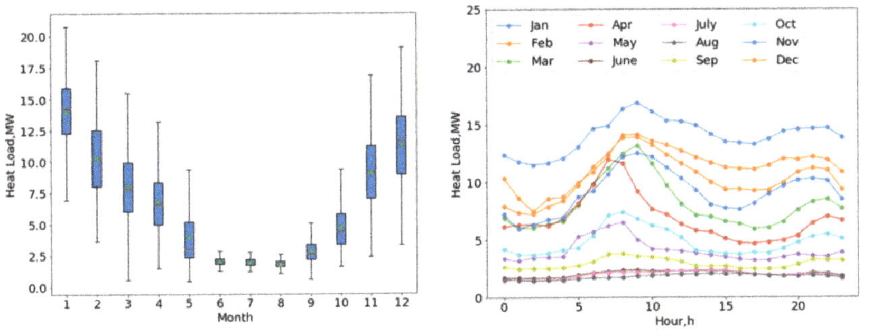

(a) (b)

Figure 8. Heat Load in: (**a**) one year; (**b**) monthly typical days.

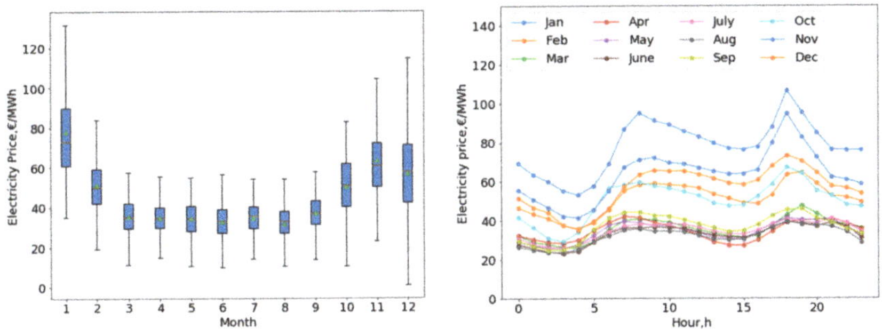

(a) (b)

Figure 9. Electricity price in: (**a**) one year; (**b**) monthly typical days [29].

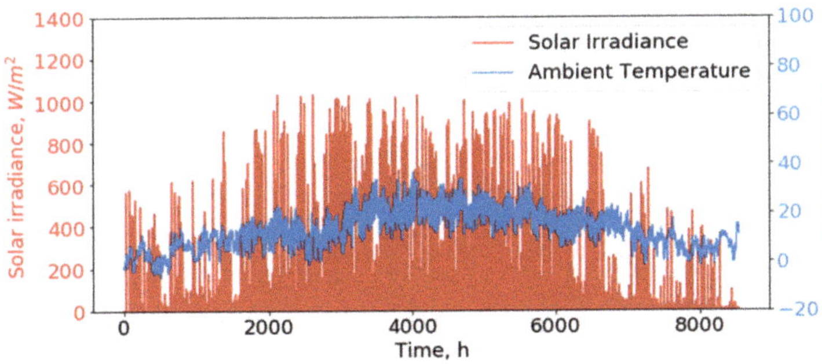

Figure 10. Hourly variation of ambient temperature and solar irradiance over the year [30].

The considered technical parameters of the heating and the thermal storage systems are provided in Tables 1 and 2. The related economic parameters, including investment and maintenance costs, lifetime and startup costs are presented in Table 3. The startup costs actually vary with the size of the units, but the constant values from [34] were considered here as they match the considered range of unit size.

CO_2 emissions are calculated based on the specific emission factors to the primary consumption [1]: 350 kg CO_2/MWh for coal and 200 kg CO_2/MWh for natural gas [35]. The CO_2 price in France in 2017 is collected from World Bank Carbon Pricing dashboard 30.5 € per tonne [36]. The fuel costs are assumed constant: 28 €/MWh for Coal and 32 €/MWh for natural gas [34].

The following assumption were also made on the efficiency of the systems: the effect of incidence angle was neglected in the design of the STC system and the CHP operation efficiency was considered constant.

Table 1. Technical parameters of each unit.

Units	Capacity	Minimum Part Load Ratio	Min Uptime	Min Downtime	Ramp-Up Rate %/h	Ramp-Down Rate %/h	Norm Efficiency
CHP	12 MW	0.3	10	7	30	30	0.883
HOB1	5 MW	0.3	2	2	100	100	0.9
HOB2	5 MW	0.3	2	2	100	100	0.9
TES	0–6000 m^3	0	4	4	100	100	-
STC	0–40,000 m^2	0	1	1	100	100	-

Table 2. Technical parameters of TES [23].

Unit	Charging Ratio	Discharging Ratio	Storage Efficiency per Hour	Charging Efficiency	Discharging Efficiency
TES	0.4	0.4	0.998	0.95	0.95

Table 3. Economic data of each unit [34].

Units	Investment Cost	Maintenance Cost	Startup Cost per Time	Lifetime
CHP	1154 €/kW	43.2 €/(kW·year)	5000 €	25
HOB	62.9 €/kW	1.26 €/(kW·year)	1290 €	17
TES	See Figure 5	-	-	25
STC	See Figure 6	-	-	30

3.2. Scenarios

Different scenarios are investigated in this study to analyze the impact of the integration of TES and STC systems into the existing CHP system. Scenario 1 covers the operation of the existing system without the addition of TES nor STC. Both systems are then considered and optimized in Scenario 2. In Scenarios 3 and 4, the addition of TES or STC alone is studied. The scenarios are summarized in Table 4.

Table 4. Overview of each Scenario.

Scenario Name	CHP	HOB1	HOB2	TES	STC
1	•	•	•		
2	•	•	•	•	•
3	•	•	•	•	
4	•	•	•		•

4. Results and Discussion

In this Section, the results of the optimization performed for the 4 Scenarios of the case study are presented and discussed. In this study, the MILP model implemented in Python and solved with IBM DOcplex toolbox. Section 4.1 gives the results of the Pareto Frontiers for all Scenarios. In Section 4.2, the optimal solutions chosen using the TOPSIS-entropy method are presented. Section 4.3 gives more detailed information on the hourly operation strategy, and Section 4.4 presents a sensitivity analysis on the main parameters.

4.1. Pareto Frontiers

Figure 11 shows the Pareto frontiers computed for Scenarios 2 to 4 in terms of *ATC* and CO_2 emissions, expressed relatively to the reference Scenario 1. In Scenario 1, the *ATC* is 4380 k€/year and the CO_2 emissions are 43,435 ton/year.

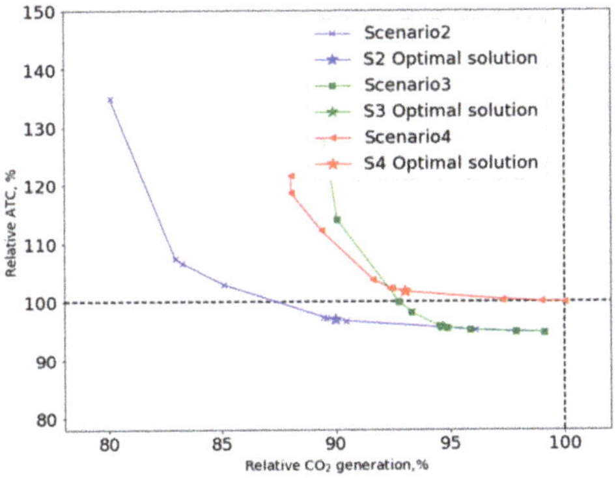

Figure 11. Relative Pareto frontiers for Scenarios relative to Scenario 1.

Considering the addition of TES only (Scenario 3) quickly allows for a reduction of the system cost (5%) and a moderate reduction of the CO_2 emissions (up to 1% less). This is due to the larger consumption of the produced heat hen heat storage is available (see Section 4.3). When even lower CO_2 emissions must be reached with TES only, the price of the system however increases very sharp: it becomes higher than the reference cost for CO_2 emissions reductions around 7% and increases exponentially afterwards: +30% compared to the reference scenario for a 10% CO_2 emissions reduction.

On the other hand, considering STC only (Scenario 4) immediately leads to a higher cost, because of the higher price of STC. This system however becomes cheaper than TES only to reach CO_2 emissions reductions larger than 8%, although the cost also increases very rapidly for larger emission savings.

Figure 11 also shows that, combining TES and STC (Scenario 2) allows decreasing the *ATC* while reaching much larger CO_2 emissions reduction. The combination of these systems indeed leads to the valorization of a higher amount of heat collected by the STC system by storing it and making it available when the demand is high. Parity with the reference scenario in terms of costs is reached for 12% less CO_2 emissions. For even lower emissions, the cost increase is also significantly reduced compared to TES or STC only: 17% less CO_2 is emitted for cost increase of 7%. The installation of the additional equipment is made more cost-effective by their complementarity.

4.2. TOPSIS-Entropy Method Analysis

The TOPSIS-entropy method was used to choose the optimal solution on the Pareto frontiers obtained from the MILP optimizations. The weights calculated by the entropy method for each Scenario are very similar: around 0.54 and 0.46 for *ATC* and CO_2 emissions respectively. The solutions presenting the maximum relative quality for each Scenario are given in Table 5 and illustrated on the Pareto frontiers of Figure 11. The results generated for 12 representative days, see Section 3.1, have been extrapolated to a full year. Compared to the reference Scenario 1, the identified, optimal solution for Scenario 2 corresponds to a decrease of both the CO_2 emissions (10%) and the *ATC* (3%), which is an excellent trade-off. The total efficiency of the system increases from 87% to 92%, while the share of renewable energy in the system reaches 10%. The optimal solution for TES only (Scenario 3) is selected before the strong increase of the system cost on the Pareto frontier. The CO_2 emissions are reduced by 5% while the cost is reduced by 4%. When only the STC system is installed (Scenario 4), the optimal solution is again selected before the strong increase of the cost. In that case, the cost increases with 2%, for a CO_2 emissions reduction of 7%. The *ATC* is higher than for Scenario 2 because of the lower proportion of renewable heat that can be consumed. The addition of TES indeed allows for the displacement of the consumption of the heat collected by the STC to the periods when the demand is higher.

Table 5. Optimal solutions by TOPSIS-entropy method for each scenarios.

Scenario Name	TES/m^3	STC/m^2	CO_2 (ton/year)	*ATC* (k€/year)	HOB Operation Hour (h/year)	Total Efficiency, %	Share of RES, %
1	-	-	43,435	4380	1520	87	0
2	1382	22,399	39,268	4258	730	92	10
3	742	-	41,184	4198	730	88	0
4	-	17,649	40,363	4471	1460	90	7

4.3. Hourly Operation Strategy

In this Section, the hourly operation of the optimal systems selected for the 4 scenarios during the typical days are presented and discussed. Figure 12 shows the hourly heat production per system and the hourly operation modes of the CHP on its operational map, over the 12 typical days in a row. It is obvious that integration of TES and STC significantly improve system operation flexibility and efficiency.

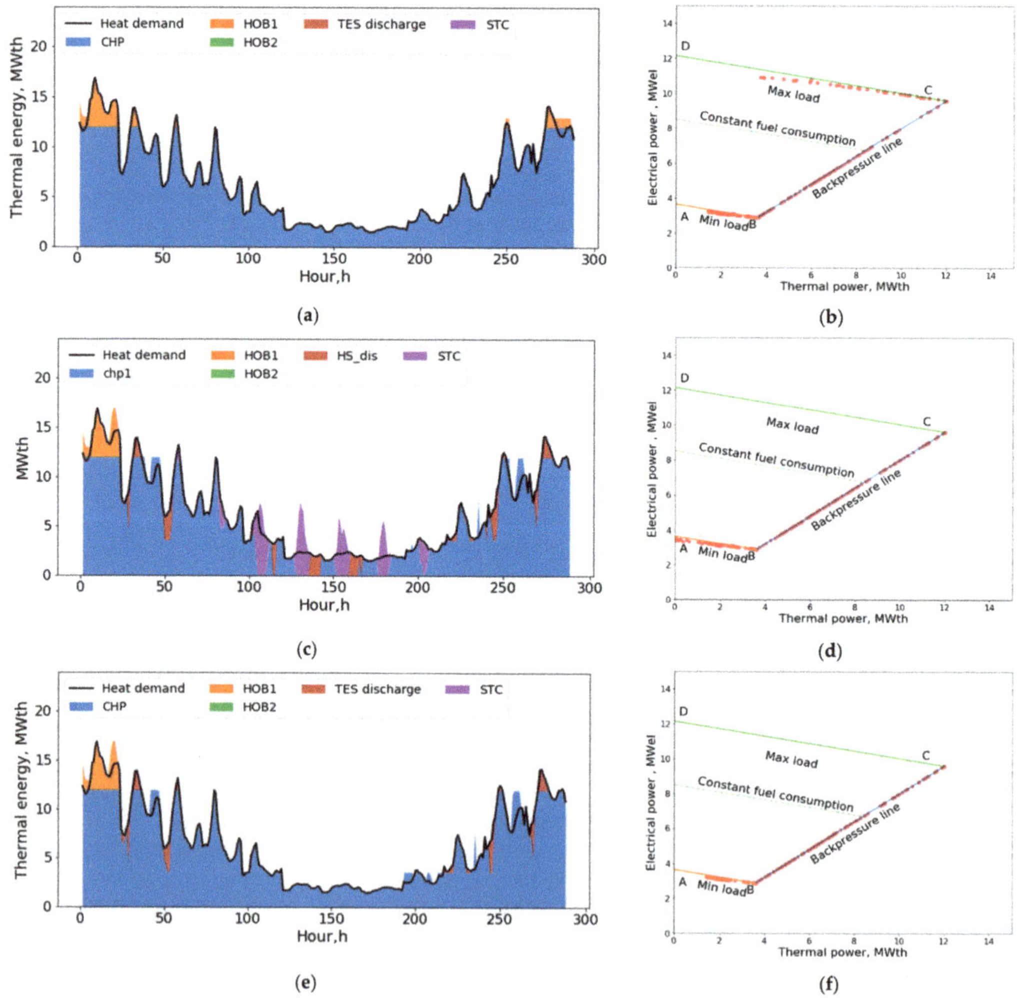

Figure 12. *Cont.*

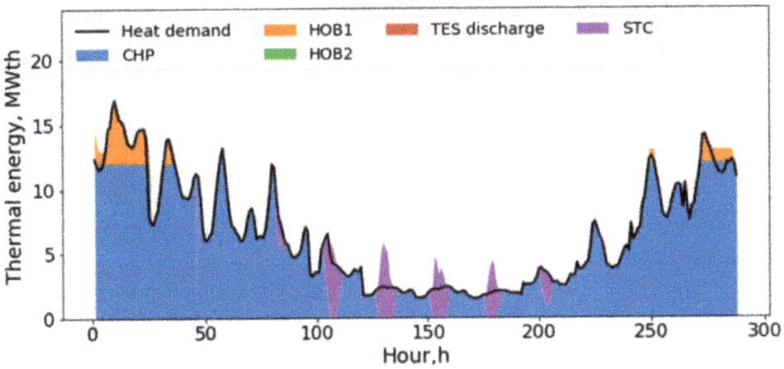

(g)

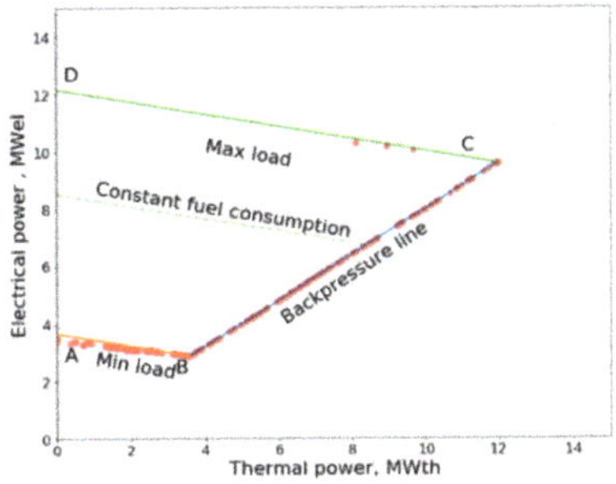

(h)

Figure 12. Hourly heat production for all Scenarios over the 12 typical days in a row: (**a**) Hourly Heat production of Scenario 1; (**b**) CHP operation field of Scenario 1; (**c**) Hourly Heat production of Scenario 2; (**d**) CHP operation field of Scenario 2; (**e**) Hourly Heat production of Scenario 3; (**f**) CHP operation field of Scenario 3; (**g**) Hourly Heat production of Scenario 4; (**h**) CHP operation field of Scenario 4.

First, it can be observed that the TES and the STC systems play an important role in increasing the system flexibility. The system with TES is more flexible as it can achieve peak load shaving and load valley filling, which results in a decrease of the HOB operation [10]. The HOB's are operated up to 52% less when TES is implemented (Scenarios 2 and 3), as shown in Table 5. In the winter period, the thermal energy supplied by HOB during peak load time in Scenarios 1 is replaced by TES in Scenarios 2 and 3. During the summer period, the daytime heat demand is covered by STC in Scenarios 2 and 4. As expected, the excess heat produced by STC in Scenario 2 is stored in TES and discharged later on when STC cannot meet the demand.

Secondly, the operational flexibility of the CHP unit is also increased by the integration of TES and STC. Figure 12 shows that the CHP operates more on the left-hand part of the A-B line (minimum fuel consumption) in Scenario 2 than in Scenario 1, which helps increasing the electrical power generation for the same fuel consumption and, hence, the same CO_2 emissions.

The detailed performances of the TES system are also shown in Figure 13. TES enables the operation of the system associated with electricity prices, to maximize the profits from selling electricity without affect heat supply safety. Furthermore, TES is more active in the summer period when used in combination with STC. The optimal TES size in Scenario 2 is therefore twice larger than for Scenario 3, see Table 5.

4.4. Sensitive Analysis

The input parameters used in the case study above are prone to uncertainty. To assess the impact of these uncertainties, a sensitivity analysis is carried out for the optimal solution of Scenario 2, for the following key input parameters [37]: heat demand, carbon price, electricity price, fuel price and interest rate.

Figure 14 shows the performance of the optimal solutions in terms of *ATC*, CO_2 emissions and size of TES and STC, with the variation of the inputs (±20%). It must be noted that each of these results corresponds to an optimal solution computed for the new set of input parameters, and not to the results obtained for the reference optimal solution for other input parameters.

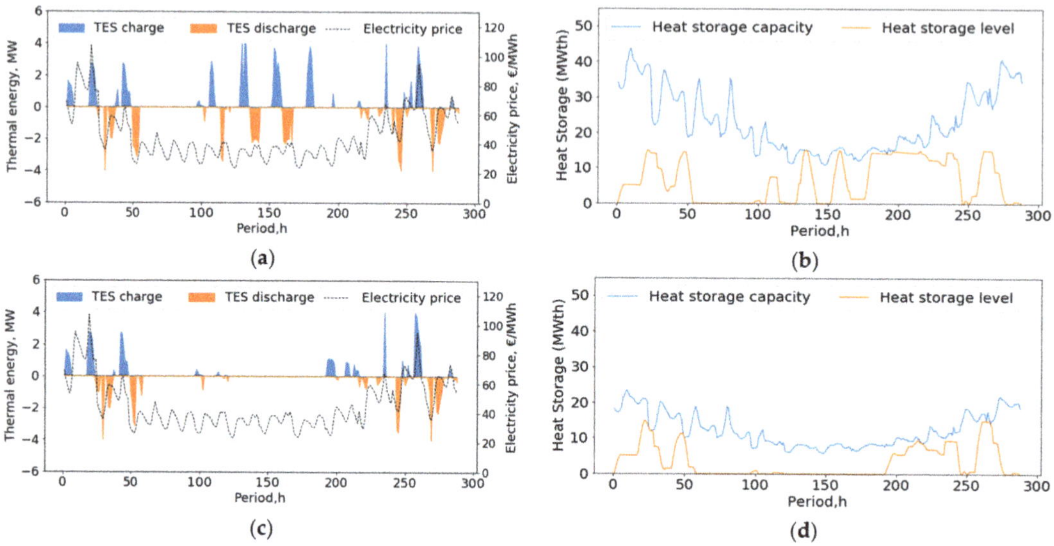

Figure 13. Charging or discharging load and heat storage levels of TES over the 12 typical days in a row: (**a**) TES charging and discharging load of Scenario 2; (**b**) TES heat storage levels of Scenario 2; (**c**) TES charging and discharging load of Scenario 3; (**d**) TES heat storage levels of Scenario 3.

Although they affect the optimal sizes of the TES and STC systems, the influence of economic parameters on the amount of CO_2 emitted in the optimal cases is negligible compared to the influence of the heat demand, see Figure 14a. As far as *ATC* is concerned (Figure 14b), heat demand and fuel price have a similar impact: *ATC* increased by 15% when the input parameter increases by 20%. The price of electricity exhibits the opposite trend: the *ATC* is 12.5% higher when it decreases with 20%, due to the lower revenues from the CHP unit. The price of CO_2 and the interest rate should not be neglected, although they have a lower impact on *ATC*.

In terms of impact on the systems design in Figure 14c,d, the TES volume is less sensitive to the uncertainties on the inputs than the STC area, which significantly varies with heat demand, fuel price and interest rate. Although this study showed that their combination with TES can result in a profitable reduction of CO_2 emissions, this shows that economic aspects can be a limiting factor in the implementation of STC.

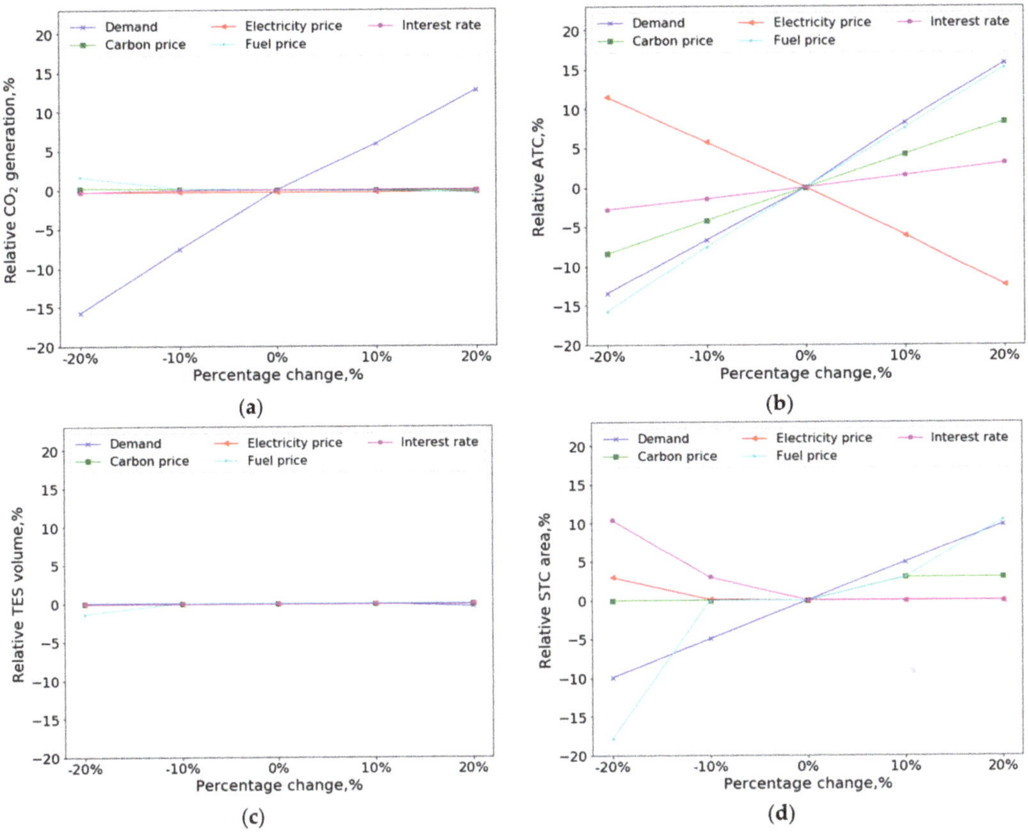

Figure 14. Sensitivity of the main results to the variation of key input parameters: (**a**) CO_2 generation; (**b**) *ATC*; (**c**) TES volume; (**d**) STC area.

5. Conclusions and Future Work

A framework for the multi-objective optimization of the integration of Solar Thermal Collectors (STC) and Thermal Energy Storage (TES) systems in existing fossil-fuel based heat and power production systems was presented. The proposed method was applied to the representative case of a medium-scale CHP system coupled to a District Heating network. As a comparison, the integration of TES or STC alone was also considered.

The proposed TOPSIS-entropy method has been proved to be efficient to select the optimal design in terms of trade-off between cost and CO_2 reduction. Our results show that, while the addition of TES or STC alone results in limited economic and environmental performances and exhibits a rapid increase of the cost with the targeted CO_2 emission reduction, the optimal combination of TES and STC can lead to a reduction of both the cost and the CO_2 emissions: respectively 3% and 10% in the studied case. For larger CO_2 emissions savings, beyond the optimal trade-off, the additional cost remains limited compared to the other solutions. The integration of TES and STC also significantly improves the system flexibility and efficiency. TES allows for peak load shaving and load valley filling, resulting in up to 52% less operation of peak units. For the optimal design, the total efficiency of the system increases from 87% to 92%. The share of renewable energy reaches 10% when both TES and STC are integrated, compared to 7% with STC alone.

The operational flexibility of the CHP unit itself is also increased by the integration of TES and STC, which helps increasing the electrical power generation.

A sensitivity analysis shows that only the heat demand has a significant impact on the environmental performance, while both the heat demand and the fuel price have a significant influence on the economic performances. Furthermore, the optimal TES volume is less sensitive to the uncertainties on the inputs than the STC surface, that is more impacted by the economic parameters.

In future works, the same methodology could be applied to the integration of more renewable energy sources such as heat pumps and power-to-X based on renewable electricity production. The case of a new CHP unit will also be investigated instead of a retrofit, taking into account the optimal sizing of this unit and the related investment costs. Moreover, the impact of the uncertainties on the input data could be studied using advanced Uncertainty Quantification and Robust Design techniques.

Author Contributions: G.W.; conceptualization, data curation, methodology, software, formal analysis, investigation, writing—original draft, writing—review and editing, visualization. J.B.; conceptualization, methodology, investigation, writing—review and editing, supervision, project administration, funding acquisition. All authors have read and agreed to the published version of the manuscript.

Funding: The research received no external funding.

Informed Consent Statement: Informed consent was obtained from all subjects involved in the study.

Conflicts of Interest: The authors declare that they have no known competing financial interests or personal relationships that could have appeared to influence the work reported in this paper.

Nomenclature

Abbreviations	
ATC	Annual total cost
CHP	Combined heat and power
DES	Distributed energy system
DH	District heating
EC	Extraction condensation steam turbine
HOB	Heat-only boiler
MILP	Mixed-integer linear programming
RES	Renewable energy source
STC	Solar thermal collector
TES	Thermal energy storage
TOPSIS	Technique for Order Preference by Similarity to Ideal Solution
Indices and sets	
u	Unit index, u ∈ units
t	Time index, t ∈ periods
Parameters	
Cap_u	Maximum capacity of each unit, MW.
c_u^{inv}	Specific investment cost per hour, €/MW.
c_t^{invTES}	Specific investment cost per hour for TES, €/m^3.
c_t^{invSTC}	Specific investment cost per hour for STC, €/m^2.
a_u	Annuity factor.
I_u^{inv}	Specific Investment cost per unit, €/MW.
I^{invTES}	Specific Investment cost per volume for TES €/m^3.
I^{invSTC}	Specific Investment cost per area for STC €/m^2.
$c_{u,t}^M$	Specific maintenance cost, €/(MW·h).
c_t^{MTES}	Specific maintenance cost for TES, €/(m^3·h).
c_t^{MSTC}	Specific maintenance cost for STC, €/(m^2·h).
$c_{u,t}^{fuel}$	Fuel cost, €/MWh
SC_u	Start-up cost per time, €.
CF_u	Carbon emission factor for each fuel, kg CO_2/MWh
El_t	Electricity price, €/MWh.
$ramp\_up_{u,t}$	Maximum ramp up rate.
$ramp\_down_{u,t}$	Maximum ramp down rate.

$PLR_{u,t}^{lb}$	Minimum Part load ratio.
$PLR_{u,t}^{ub}$	Maximum Part load ratio.
Q_{u}^{norm}	Norm Heat capacity, MW.
$Q_{u}^{initial}$	Initial generation for each units, MW.
η_{EC}	Total efficiency of EC CHP
η_{s}^{TES}	Storage efficiency of TES.
η_{chr}^{TES}	Charging efficiency of TES.
η_{dis}^{TES}	Discharging efficiency of TES.
T_{t}^{supply}	Supply temperature of DH network, K.
T_{t}^{return}	Return temperature of DH network, K.
T_{t}^{aver}	Mean panel temperature, K.
$T_{t}^{ambient}$	Ambient temperature, K.
T_0	Reference temperature, K.
T_M	Average temperature, K.
Q_{t}^{SR}	Solar irradiance, MW.
α_{t}^{chr}	Maximum charging ratio of TES.
α_{t}^{dis}	Maximum discharging ratio of TES.
σ^{EC}	Power to heat ratio of EC CHP.
β	Power loss coefficient
y_{ij}	Value of positive matrix Y.
e_j	Entropy value of objective
p_{ij}	Standardized value in normalized matrix P.
J_1, J_2	Benefit and cost indicators.
ω_j	Weighting value of the objective
V^+	Positive ideal solution
V^-	Negative ideal solution
C_i	Relative quality
Positive Variables	
A	STC area, m^2.
V	Thermal storage tank volume, m^3.
C_{inv}	Annualization of investment, k€.
C_M	Maintenance cost, k€.
C_O	Operation cost, k€.
Cap_{t}^{TES}	Maximum storage capacity of TES, MW.
$FC_{u,t}$	Fuel consumption, MW
HS_{t}^{chr}	Thermal energy charging amount at time t, MW.
HS_{t}^{dis}	Thermal energy discharging amount at time t, MW.
P_{t}^{EC}	Power production of EC CHP, MW.
PLR_{t}^{EC}	Part load ratio of EC CHP.
$Q_{u,t}$	Heat production, MW.
R_S	Revenue from selling electricity back into grid, k€.
Binary Variables	
$\delta_{u,t}$	Binary variables, $\delta_{u,t} = 1$ when units turn on.
$T_{u,t}^{ON}$	Binary variables of ON status
$T_{u,t}^{OFF}$	Binary variables of OFF status
$I_{u,t}^{use}$	Binary variables of In_Use status.
z_t	TES charging/discharging operation status

References

1. European Commission. Mapping and Analyses of the Current and Future (2020–2030). Heating/Cooling Fuel Deployment—Executive Summary. Available online: https://energy.ec.europa.eu/mapping-and-analyses-current-and-future-2020-2030-heatingcooling-fuel-deployment-fossilrenewables-1_en (accessed on 6 December 2021).
2. Eurostat. Renewable Energy Statistics. Available online: https://ec.europa.eu/eurostat/statistics-explained/index.php?title=Renewable_energy_statistics (accessed on 6 December 2021).
3. International Renewable Energy Agency. Renewable Energy in District Heating and Cooling. Available online: https://www.irena.org/publications/2017/Mar/Renewable-energy-in-district-heating-and-cooling (accessed on 6 December 2021).
4. Danish Energy Agency. Technology Data: Generation of Electricity and District Heating. Available online: https://ens.dk/sites/ens.dk/files/Analyser/technology_data_catalogue_for_el_and_dh.pdf (accessed on 6 December 2021).

5. Verkis Svartsengi. Power Plant. Available online: https://www.verkis.com/projects/energy-production/geothermal-energy/nr/936 (accessed on 6 December 2021).
6. International Energy Agency (IEA). Large Biomass CHP Plant in Stockholm, Sweden. Available online: https://www.ieabioenergy.com/wp-content/uploads/2018/02/8-LargeCHP-Va%CC%88rtaverket_SE_Final.pdf (accessed on 6 December 2021).
7. Lund, H.; Østergaard, P.A.; Connolly, D.; Ridjan, I.; Mathiesen, B.V.; Hvelplund, F.; Thellufsen, J.Z.; Sorknsæs, P. Energy Storage and Smart Energy Systems. *Int. J. Sustain. Energy Plan. Manag.* **2016**, *11*, 3–14. [CrossRef]
8. Lund, H.; Werner, S.; Wiltshire, R.; Svendsen, S.; Thorsen, J.E.; Hvelplund, F.; Mathiesen, B.V. 4th Generation District Heating (4GDH). Integrating Smart Thermal Grids into Future Sustainable Energy Systems. *Energy* **2014**, *68*, 1–11. [CrossRef]
9. Cabeza, L.F.; Miró, L.; Oró, E.; De Gracia, A.; Martin, V.; Krönauer, A.; Rathgeber, C.; Farid, M.M.; Paksoy, H.O.; Martínez, M.; et al. CO_2 Mitigation Accounting for Thermal Energy Storage (TES) Case Studies. *Appl. Energy* **2015**, *155*, 365–377. [CrossRef]
10. Benalcazar, P. Optimal Sizing of Thermal Energy Storage Systems for CHP Plants Considering Specific Investment Costs: A Case Study. *Energy* **2021**, *234*, 121323. [CrossRef]
11. Mugnini, A.; Comodi, G.; Salvi, D.; Arteconi, A. Energy Flexible CHP-DHN Systems: Unlocking the Flexibility in a Real Plant. *Energy Convers. Manag. X* **2021**, *12*, 100110. [CrossRef]
12. Lai, F.; Wang, S.; Liu, M.; Yan, J. Operation Optimization on the Large-Scale CHP Station Composed of Multiple CHP Units and a Thermocline Heat Storage Tank. *Energy Convers. Manag.* **2020**, *211*, 112767. [CrossRef]
13. Savic, D. Single-Objective vs. Multiobjective Optimisation for Integrated Decision Support. In Proceedings of the First Biennial Meeting of the International Environmental Modelling and Software Society, Lugano, Switzerland, 24–27 June 2002; pp. 7–12.
14. Ren, H.; Zhou, W.; Nakagami, K.; Gao, W.; Wu, Q. Multi-Objective Optimization for the Operation of Distributed Energy Systems Considering Economic and Environmental Aspects. *Appl. Energy* **2010**, *87*, 3642–3651. [CrossRef]
15. Fazlollahi, S.; Becker, G.; Ashouri, A.; Maréchal, F. Multi-Objective, Multi-Period Optimization of District Energy Systems: IV—A Case Study. *Energy* **2015**, *84*, 365–381. [CrossRef]
16. Luo, Z.; Yang, S.; Xie, N.; Xie, W.; Liu, J.; Souley Agbodjan, Y.; Liu, Z. Multi-Objective Capacity Optimization of a Distributed Energy System Considering Economy, Environment and Energy. *Energy Convers. Manag.* **2019**, *200*, 112081. [CrossRef]
17. Karmellos, M.; Mavrotas, G. Multi-Objective Optimization and Comparison Framework for the Design of Distributed Energy Systems. *Energy Convers. Manag.* **2019**, *180*, 473–495. [CrossRef]
18. Franco, A.; Versace, M. Multi-Objective Optimization for the Maximization of the Operating Share of Cogeneration System in District Heating Network. *Energy Convers. Manag.* **2017**, *139*, 33–44. [CrossRef]
19. Wirtz, M.; Hahn, M.; Schreiber, T.; Müller, D. Design Optimization of Multi-Energy Systems Using Mixed-Integer Linear Programming: Which Model Complexity and Level of Detail Is Sufficient? *Energy Convers. Manag.* **2021**, *240*, 114249. [CrossRef]
20. Wu, Q.; Ren, H.; Gao, W.; Ren, J. Multi-Objective Optimization of a Distributed Energy Network Integrated with Heating Interchange. *Energy* **2016**, *109*, 353–364. [CrossRef]
21. Verbruggen, A.; Dewallef, P.; Quoilin, S.; Wiggin, M. Unveiling the Mystery of Combined Heat & Power (Cogeneration). *Energy* **2013**, *61*, 575–582. [CrossRef]
22. Mollenhauer, E.; Christidis, A.; Tsatsaronis, G. Evaluation of an Energy- and Exergy-Based Generic Modeling Approach of Combined Heat and Power Plants. *Int. J. Energy Environ. Eng.* **2016**, *7*, 167–176. [CrossRef]
23. Wang, H.; Yin, W.; Abdollahi, E.; Lahdelma, R.; Jiao, W. Modelling and Optimization of CHP Based District Heating System with Renewable Energy Production and Energy Storage. *Appl. Energy* **2015**, *159*, 401–421. [CrossRef]
24. Sveinbjörnsson, D.; Jensen, L.L.; Trier, D.; Bava, F.; Hassine, I.B.; Jobard, X. Fifth Generation, Low Temperature, High Exergy District Heating and Cooling Networks: D2.3 Large Storage Systems for DHC Networks. Available online: https://ec.europa.eu/research/participants/documents/downloadPublic?documentIds=080166e5c2089739&appId=PPGMS (accessed on 6 December 2021).
25. Schmidt, T.; Miedaner, O. Solar District Heating Guidelines. Available online: https://www.solarthermalworld.org/sites/default/files/story/2015-04-03/sdh-wp3-d31-d32_august2012_0.pdf (accessed on 6 December 2021).
26. Chen, P. Effects of the Entropy Weight on TOPSIS. *Expert Syst. Appl.* **2021**, *168*, 114186. [CrossRef]
27. Li, X.; Wang, K.; Liuz, L.; Xin, J.; Yang, H.; Gao, C. Application of the Entropy Weight and TOPSIS Method in Safety Evaluation of Coal Mines. *Procedia Eng.* **2011**, *26*, 2085–2091. [CrossRef]
28. Ding, L.; Shao, Z.; Zhang, H.; Xu, C.; Wu, D. A Comprehensive Evaluation of Urban Sustainable Development in China Based on the TOPSIS-Entropy Method. *Sustainability* **2016**, *8*, 746. [CrossRef]
29. Epexspot. EPEX SPOT 2021. Available online: https://www.epexspot.com/en (accessed on 6 December 2021).
30. European Commission. JRC Photovoltaic Geographical Information System (PVGIS)—Commission. Available online: https://re.jrc.ec.europa.eu/pvg_tools/en/#MR (accessed on 6 December 2021).
31. Benalcazar, P. Sizing and Optimizing the Operation of Thermal Energy Storage Units in Combined Heat and Power Plants: An Integrated Modeling Approach. *Energy Convers. Manag.* **2021**, *242*, 114255. [CrossRef]
32. Morvaj, B.; Evins, R.; Carmeliet, J. Optimising Urban Energy Systems: Simultaneous System Sizing, Operation and District Heating Network Layout. *Energy* **2016**, *116*, 619–636. [CrossRef]
33. Fazlollahi, S.; Becker, G.; Maréchal, F. Multi-Objectives, Multi-Period Optimization of District Energy Systems: II-Daily Thermal Storage. *Comput. Chem. Eng.* **2014**, *71*, 648–662. [CrossRef]

34. Limpens, G.; Moret, S.; Jeanmart, H.; Maréchal, F. EnergyScope TD: A Novel Open-Source Model for Regional Energy Systems. *Appl. Energy* **2019**, *255*, 113729. [CrossRef]
35. Quaschning, V. *Understanding Renewable Energy Systems*; Routledge: Abingdon, UK, 2005; Volume 67, ISBN 9781844071289.
36. Worldbank. Carbon Pricing Dashboard, Up-to-Date Overview of Carbon Pricing Initiatives. Available online: https://carbonpricingdashboard.worldbank.org/map_data (accessed on 6 December 2021).
37. Akbari, K.; Nasiri, M.M.; Jolai, F.; Ghaderi, S.F. Optimal Investment and Unit Sizing of Distributed Energy Systems under Uncertainty: A Robust Optimization Approach. *Energy Build.* **2014**, *85*, 275–286. [CrossRef]

energies

Article

Design of a Neural Super-Twisting Controller to Emulate a Flywheel Energy Storage System

Daniel A. Magallón [1], Carlos E. Castañeda [1,*], Francisco Jurado [2] and Onofre A. Morfin [3]

1 Centro Universitario de los Lagos, Universidad de Guadalajara, Lagos de Moreno 47460, Mexico; daniel.magallon6532@academicos.udg.mx
2 Tecnológico Nacional de México/I.T. La Laguna, Torreón 27000, Mexico; fjuradoz@lalaguna.tecnm.mx
3 Universidad Autónoma de Ciudad Juárez, Ciudad Juárez 32310, Mexico; omorfin@uacj.mx
* Correspondence: ccastaneda@lagos.udg.mx

Abstract: In this work, a neural super-twisting algorithm is applied to the design of a controller for a flywheel energy storage system (FESS) emulator. Emulation of the FESS is achieved through driving a Permanent Magnet Synchronous Machine (PMSM) coupled to a shaft to shaft DC-motor. The emulation of the FESS is carried out by controlling the velocity of the PMSM in the energy storage stag and then by controlling the DC-motor velocity in the energy feedback stage, where the plant's states of both electrical machines are identified via a neural network. For the neural identification, a Recurrent Wavelet First-Order Neural Network (RWFONN) is proposed. For the design of the velocity controller, a super-twisting algorithm is applied by using a sliding surface as the argument; the latter is designed based on the states of the RWFONN, in combination with the block control linearization technique to the control of the angular velocity from both machines in their respective operation stage. The RWFONN is trained online using the filtered error algorithm. Closed-loop stability analysis is included when assuming boundedness of the synaptic weights. The results obtained from Matlab/Simulink validate the performance of the proposal in the control of an FESS.

Keywords: wavelet neural network; block control form; filtered error algorithm; neural super-twisting control; flywheel energy storage system

Citation: Magallón, D.A.; Castañeda, C.E.; Jurado, F.; Morfin, O.A. Design of a Neural Super-Twisting Controller to Emulate a Flywheel Energy Storage System. *Energies* **2021**, *14*, 6416. https://doi.org/10.3390/en14196416

Academic Editors: Victor Becerra and Ahmed Rachid

Received: 3 September 2021
Accepted: 29 September 2021
Published: 7 October 2021

1. Introduction

For many years, flywheels made of stone or other primitive materials have been used for the operation of different mechanisms. Currently, there are mechanisms built from metal flywheels, powered by electric motors, which can work in both motor and generator mode and whose main function is to store energy in the flywheel; because of this property, these mechanisms are known as Flywheel Energy Storage Systems (FESSs) [1]. Induction motors and synchronous motors are some types of electrical machines used in the application of FESSs. For the proper functioning of the electrical grid, it is essential to have a balance between consumption and supply of electrical energy. Due to its rapid response of operation, an FESS has the ability to reduce some of the problems caused by power variations. The control and comparison between different electrical machines used in FESSs are shown in [2]. The FESSs have been studied by different authors showing the performance from different control strategies. Furthermore, in several published works, a complete FESS has been emulated through the use of electric motors where the main operation of the flywheel is the storage of mechanical energy in a rotating flywheel, which can then be converted to electrical energy using an electrical machine with power inverters [3]. An FESS has several advantages in contrast with other energy storage systems, including long useful life, and an FESS has multiple applications such as in wind generators, electric vehicles, etc. [4]. In [5], a vector control strategy was presented to emulate a FESS using a DC permanent magnet machine, where the machine was coupled to a PMSM to control the angular velocity of the flywheel in the charging and discharging stages. In [6], an integral sliding mode controller with a super-twisting algorithm was presented. In that work, the rotor side converter in the doubly-fed induction generator is used for controlling a wind generator, where the sliding surfaces are generated such that these are compatible with the errors in the stator active and reactive powers. In [7], emulation of an FESS using reconfigurable hardware test-bed of power converters was presented, where the emulation was carried out to the acceleration, deceleration and standby modes. The flywheel's model is described in the dq domain with relations among voltage, current and rotating speed. In [8], an adaptive control for high-speed FESSs was proposed where the controller has the task of controlling the DC-link voltage for the FESS and controlling the active power exchange between the FESS and the grid, whose advantage of inertia emulation is that the inertia and damping can be altered in real-time. In [9], under the assumption that all the states are measurable, a sliding mode control system for a 4-quadrant DC–DC converter was designed. Furthermore, a low voltage prototype for a flywheel application was constructed.

Recently, a control scheme applied to an FESS for the improvement of the dynamic performance of the utility grid, based on a PMSM incorporated into a multi-machine system, was presented in [10]. In that work, the speed and voltages generated were monitored by the FESS in order to generate the real and reactive power. Such control strategy was designed against three-phase faults.

On the one hand, considering that the fundamental parts of an energy storage system are the electric motors, the PMSM is a kind of synchronous motor that features low inertia, low noise, high power density and high efficiency, simple structure, and easy control. PMSMs have been used for the control of robotic movements, electric and aerospace propulsions, etc. [11]. The nonlinear mathematical model of a PMSM is strongly coupled. In addition, its parameters are not fixed but are kept within an operating range depending on the work environment. Angular velocity and torque can fluctuate in some parameter regions [12]. Recently, a feedback technique of dynamic surface control for a chaotic PMSM, based on nonlinear mappings using neural networks (NNs) to approximate the system dynamics as well as to estimate the unknown parameters, has been proposed in [13]. In [14], NNs were implemented in a PMSM for the detection of damage or failure in motor bearings by analyzing mechanical vibrations, comparing different types of NNs and the effectiveness of failure detection.

On the other hand, DC machines are generators that convert mechanical energy into DC electrical energy and motors that convert DC electrical energy into mechanical energy. Most DC machines are similar to alternating current machines in that they have AC voltages and currents inside them. DC machines have a DC output only because there is a mechanism that converts internal AC voltages into DC voltages at their terminals; since this mechanism is called a commutator, DC machinery is also known as a commutator or commutator machinery [15].

Implementation of the sliding-mode control to DC motors has been studied by different authors. In [16], an adaptive neural controller in discrete-time was proposed, where a NN was used to identify the plant model, using the Kalman extended filter training algorithm, when applying the block control in sliding mode for angular velocity control. A speed control scheme for a DC motor, based on state observers of sliding modes, was presented in [17]. In that work, the state observers estimate the angular velocity and the load torque; thus, in this way, it was possible to apply the controller, designed from the application of the exact linearization technique in combination with the super-twisting control.

The application of super-twisting control has been recently studied in [18], introducing the control to roll suppression of marine vessels subjected to harmonic wave excitations, where some of the main control objectives were to reduce the phenomenon of vibration over dynamic behaviors such as chaos, due to parametric variation.

The main contribution of this work is the proposal of a methodology to emulate a flywheel energy storage system by driving a motor-generator group formed by two machines, namely, a PMSM coupled with a DC-motor. The state variables of the system are identified via recurrent wavelet NNs of first-order trained online by the filtered error algorithm. By using these NNs, it is possible to involve unmodeled dynamics and parameter variations, avoiding uncertainty in the machine parameters. The PMSM velocity controller is designed to emulate the stored kinetic energy of the flywheel, while the DC-motor velocity controller is used to emulate the energy that is discharged towards the utility grid. Both controllers are synthesized based on the proposed NN model transformed into the block controllable form to define a sliding surface that is steered to the origin in finite time by the action of the super-twisting control algorithm.

This paper is organized as follows. In Section 2, the mathematical background is presented, which is related to the kinetic energy storage, identification of the nonlinear system via recurrent wavelet first order neural network (RWFONN) using the filtered error training algorithm, and the super-twisting control algorithm; Section 3 presents the proposed methodology, where the procedures of charge/discharge are explained in detail; in Section 4, the simulation results are shown, from which the proposed methodology is validated; some important discussions are presented in Section 5; conclusion and some remarks about the application of the Neural Super-Twisting Control (NSTC) are drawn in Section 6. Furthermore, in Appendices A and B, the boundedness of the identification error and the complete closed-loop stability analysis are shown, respectively.

2. Mathematical Background

This section describes the energy storage in an FESS, the nonlinear mathematical representation of a permanent magnet synchronous motor and a DC-motor, the mathematical model of the dynamic behavior for an RWFONN and summarizes the filtered error algorithm, the block control linearization technique, and the super-twisting control algorithm.

2.1. FESS

The main function of FESSs is to store rotational kinetic energy, which can then be converted into electrical energy by means of an electric machine. In order to generate rotational kinetic energy, the flywheel must be driven by an electric machine that meets the requirement of operating in both motor and generator mode, allowing to convert rotational kinetic energy into electric energy and vice versa [4]. The main components from an FESS are the steering wheel, a vacuum containment to avoid loss by air friction, magnetic or metallic bearings, an electric machine capable of operating in both operation modes, i.e., motor and generator mode, and a three-phase converter to connect the FESS with the utility-grid [1]. The kinetic energy stored in the flywheel can be approximated by the following equation [19]

$$E_k = \frac{1}{2} J_m \omega_m^2, \tag{1}$$

where E_k is the stored kinetic energy in J, J_m is the inertial moment in N·m·s^2, and ω_m is the rated angular velocity in rad/s. In order to emulate the FESS deceleration, it is considered the next equation

$$\frac{d}{dt}\omega_m = \frac{-B_m \omega_m}{J_m}, \tag{2}$$

which defines the mechanical behavior of the flywheel when the energy is delivered to the utility grid [20], with B_m the frictional coefficient in N·m·s.

2.2. Nonlinear Systems

A class of nonlinear multi-input/multi-output system given in the general form

$$\dot{\chi} = \mathbf{f}(\chi, t) + \mathbf{B}(\chi, t)\mathbf{u} + \mathbf{g}(\chi, t) \tag{3}$$

where $\chi \in \mathbf{X} \subset \mathbb{R}^n$ is the state vector of the system, $\mathbf{f}(\chi, t)$ is the nonlinear system function vector; $\mathbf{u} \in \mathbf{U} \subset \mathbb{R}^m$ is the control input vector, $\mathbf{B}(\chi, t)$ is a non-singular matrix that characterizes the system, and $\mathbf{g}(\chi, t)$ is a smooth function vector that represents external disturbances and parametric variations. It must be noticed that in the present work, both PMSM and DC-motor mathematical models match with that given in the general form (3). It is important to mention that due to the purpose of this work, these mathematical models are not shown here.

2.3. RWFONN

First-order NNs are used for system identification as well as to design neural controllers for some electrical, mechanical or electromechanical systems. A novel neural identification scheme, namely a Recurrent Morlet Wavelet Neural Network (RMWNN) also known as RWFONN, based on a recurrent high-order neural network (RHONN) structure was proposed in [21]. In the NN propose in this work, the classical sigmoid activation function has been replaced by a Morlet wavelet function. The general structure from an RWFONN is given by [22]

$$\dot{x}_j^i = -a_j^i x_j^i + (w_{jk}^i)^\top \psi_{jk}^i \tag{4}$$

where x_j^i is the state of the i-th neuron, $a_j^i > 0$ for $i = 1, 2, \ldots, n$ is part of the underlying network architecture and it is fixed during the training process, w_{jk}^i is the k-th adjustable synaptic weight connecting the j-th state to the i-th neuron, and ψ_{jk}^i is a Morlet wavelet activation function. The dynamic behavior of the whole neural network can be described by expressing Equation (4) in the vector notation

$$\dot{\mathbf{x}} = \mathbf{A}\mathbf{x} + \mathbf{w}^\top \boldsymbol{\psi}, \tag{5}$$

where $\mathbf{x} = \left[x_j^1 \; x_j^2 \ldots x_j^n\right]^\top \in \mathbb{R}^n$ is the state vector of the NN, $\mathbf{w} = \left[w_j^1 \; w_j^2 \ldots w_j^n\right]^\top \in \mathbb{R}^{L \times n}$ is the synaptic weights vector, with L representing the total number of weights used to identify the plant behavior, $\mathbf{A} = \mathrm{diag}\left[-a_j^1 \; - a_j^2 \ldots \; - a_j^n\right] \in \mathbb{R}^{n \times n}$ is the parameter (Hurwitz) matrix, and $\boldsymbol{\psi}$ is the regressor vector. In this work, $\psi(\chi) = e^{(-\chi^2/\beta)}\cos(\lambda\chi)$ is the Morlet wavelet function used here as the activation function where χ is the state from the plant (Equation (3)) and the parameters β, λ are the expansion and dilation terms [22], respectively. Thus, the nonlinear system (3) is identified online by using an RWFONN (Equation (5)), where the synaptic weights vector $\mathbf{w}$ is adjusted via a filtered error algorithm.

2.4. Filtered Error Training Algorithm

For the application of the filtered error algorithm, this study begins under the assumption of an unknown system (3), which can be modeled (identified) by using an RWFONN structure (5). The synaptic weights are adjusted according to the following update law [23,24]

$$\dot{w}_{jk}^i = -\Gamma_{jk}^i \psi_{jk}^i \xi_{jk}^i, \tag{6}$$

where $\Gamma \in \mathbb{R}^{L \times L}$ is a definite positive gain matrix, ψ is the Morlet wavelet activation function, and ξ is the identification error, i.e., the difference between the states from the RWFONN (Equation (5)) and those from the plant (3), namely, $\xi_j^i = x_j^i - \chi_j^i$. The update law (Equation (6)) corresponds to the filtered error training algorithm [23]. Since the RWFONN is given in a generalized way from the structure for a RHONN, the following theorem establishes that identification and parametric errors for the RWFONN are bounded.

Theorem 1. *Consider the RWFONN model whose weights are adjustable according to Equation (6) for each $i = 1, 2, \ldots, n$;*

1. ξ_j^i, $w_j^i \in \mathcal{L}_\infty$ (*i.e.*, ξ_j^i *and* w_j^i *are uniformly bounded*);
2. $\lim_{t\to\infty} \xi_i(t) = 0$.

Proof. See [24,25]. □

Appendix A presents the boundedness of the identification error ξ_j^i given by the synaptic weights w_j^i.

2.5. Nonlinear Block Controllable Form

One of the main characteristics of the block control linearization technique is to transform a nonlinear system into an equivalent form expressed by first-order subsystems consisting of r blocks, which can be solved independently one from each other. Such equivalent form is commonly known as the block controlled form whose representation is described by [26]

$$\begin{aligned} \dot{\mathbf{x}}_1 &= \mathbf{f}_1(\mathbf{x}_1, t) + \mathbf{B}_1(\mathbf{x}_1, t)\mathbf{x}_2 + \mathbf{g}_1(\mathbf{x}_1, t), \\ \dot{\mathbf{x}}_{\hat{i}} &= \mathbf{f}_{\hat{i}}(\bar{\mathbf{x}}_{\hat{i}}, t) + \mathbf{B}_{\hat{i}}(\bar{\mathbf{x}}_{\hat{i}}, t)\mathbf{x}_{\hat{i}+1} + \mathbf{g}_{\hat{i}}(\bar{\mathbf{x}}_{\hat{i}}, t),\ \hat{i} = 2, \ldots, r-1, \\ \dot{\mathbf{x}}_r &= \mathbf{f}_r(\bar{\mathbf{x}}_r, t) + \mathbf{B}_r(\bar{\mathbf{x}}_r, t)\mathbf{u} + \mathbf{g}_r(\bar{\mathbf{x}}_r, t), \end{aligned} \tag{7}$$

where $\mathbf{f}(\mathbf{x}, t)$ is a smooth and bounded mapping, $\mathbf{x} = [x_1\ x_2 \ldots x_r]^\top$ is the state vector decomposed, $\bar{\mathbf{x}}_{\hat{i}} = [x_1 \ldots x_{\hat{i}}]^\top$, for $\hat{i} = 2, \ldots, r$, $\mathbf{x}_{\hat{i}}$ is a $n_{\hat{i}} \times 1$ vector, and the subscripts $(n_1, n_2, \ldots, n_r)$ define the structure of the system. The matrix $\mathbf{B}_{\hat{i}}$, since the fictitious $\mathbf{x}_{\hat{i}+1}$ for each $\hat{i}$-th block has full rank, $\text{rank}(\mathbf{B}_{\hat{i}}) = n_{\hat{i}}$, $\forall \mathbf{x} \in \mathbf{X} \subset \mathbb{R}^n$ and $t \in [0, \infty]$, $\hat{i} = 1, \ldots, r$. Therefore, in this work, the block control technique is applied to the NN structure (5) to define a sliding surface, which is the argument of the super-twisting controller.

2.6. Super-Twisting Control Algorithm

The first-order sliding mode has long been used as a robust control technique to cancel nonlinear terms and uncertainties due to external disturbances in a system, but its main disadvantage is the presence of the chattering effect, which induces vibrations in the controlled system. In order to reduce the chattering effect, high-order sliding modes have been used. A particular case is the super-twisting algorithm (STA), which has been playing a very important role in sliding mode controllers. The super-twisting control (STC) algorithm of second-order can be applied to systems where the control appears in the first derivative of the sliding surface. The application of the STC allows to cancel perturbations and nonlinearities, forcing the sliding surface toward zero in finite time. Furthermore, the states that define the sliding surface are directed to zero with asymptotic motion. The dynamics of the sliding surface are known as sliding mode movement, and it is characterized by being a reduced-order system [27,28].

The sliding surface for the NSTC is defined from the nonlinear system once the latter resembles the block controllable form (7) for the variable on which the control input of the system appears. Therefore, the sliding surface is [29]

$$\mathbf{s} = \mathbf{x}_r, \tag{8}$$

where the dynamics of the sliding surface (Equation (8)), involving the STC algorithm and the system (7), take the form

$$\begin{aligned} \dot{\mathbf{s}} &= \mathbf{f}_r(\bar{\mathbf{x}}_r, t) - \mathbf{B}_r(\bar{\mathbf{x}}_r, t)(\boldsymbol{\lambda}_s|\mathbf{s}|^{\frac{1}{2}}\text{sign}(\mathbf{s}) + \mathbf{u}_s) + \mathbf{g}_r(\bar{\mathbf{x}}_r, t), \\ \dot{\mathbf{u}}_s &= \boldsymbol{\alpha}_s\, \text{sign}(\mathbf{s}), \end{aligned} \tag{9}$$

with $\boldsymbol{\lambda}_s = \begin{bmatrix} \lambda_{s11} & 0 \\ 0 & \lambda_{s22} \end{bmatrix}$ and $\boldsymbol{\alpha}_s = \begin{bmatrix} \alpha_{s11} & 0 \\ 0 & \alpha_{s22} \end{bmatrix}$ diagonal matrices. Then, the super-twisting control law provides finite-time convergence to zero, simultaneously $\mathbf{s}$ and $\dot{\mathbf{s}}$, where the term $\boldsymbol{\lambda}_s|\mathbf{s}|^{\frac{1}{2}}\text{sign}(\mathbf{s})$ commutes its gain at high frequency for giving robustness to annul the nonlinearities $\mathbf{f}_r(\bar{\mathbf{x}}_r, t)$, and to cancel the external disturbances $\mathbf{g}_r(\bar{\mathbf{x}}_r, t)$, while the term $\mathbf{u}_s$ reduces the chattering effect and the steady state error [28].

In order to ensure the stability of the system with the super-twisting control law (Equation (9)), the control matrix gains $\boldsymbol{\lambda}$ and $\boldsymbol{\alpha}$ are holding to the following restrictions:

$$k_i > 2\delta_i, \;\; k_i > \frac{1}{2}\frac{k_i^2(\delta_i - k_i)}{k_i - 2\delta_i} \tag{10}$$

where the disturbance norm is defined as $|\mathbf{f}_r(\bar{\mathbf{x}}_r, t) + \mathbf{g}_r(\bar{\mathbf{x}}_r, t)| = \delta|s|^{1/2}$, for $\delta > 0$. The inequalities in Equation (10) are defined in the stability analysis presented in Appendix B.

3. Proposed Methodology

This section describes the procedure for emulating the FESS through a group motor-generator formed by two electrical machines mechanically coupled. The angular velocity of the PMSM, when it is working in motor mode, is controlled in order to emulate the storing of energy, whereas the velocity for the DC-machine is controlled emulating the discharge of the flywheel where the PMSM is now working in generator mode. The FESS's emulator scheme is shown in Figure 1. The procedure followed to the emulation of the FESS is explained next:

- The charging procedure (storing kinetic energy in the flywheel). This procedure consists of the PMSM in the *dq* coordinate frame [30] working in motor mode to the control the angular velocity through the RWFONN using the block control linearization technique. In this scenario, the kinetic energy is stored in the flywheel (dotted red block in Figure 1);
- Discharging procedure (releasing energy). In this stage, the stored energy is now transferred to the PMSM working in generator mode. This scenario is considered in the case when there exists an electrical failure in the utility grid or when the flywheel is required to compensate with active power to the utility grid to solve a problem of the power management office, and the velocity controller for the DC-motor emulates the deceleration of the flywheel through the use of Equation (2). The flywheel energy discharge is emulated by the DC-machine (dotted blue block in Figure 1).

In this work, the sliding surface defined in Equation (8) is designed through the transformation of the RWFONN to the block controllable form involving the dynamics of the tracking error, such that this transformation resembles the structure described in Equation (7). By applying the super-twisting algorithm to the system (7) and using the RWFONN (Equation (5)), trained with the filtered error halgorithm (6), the result is the NSTC system, which is uniformly ultimately bounded (UUB). The stability proof is carried out in Appendix B.

Note that from Figure 1, the dotted blue block contains the motor-side converter gray block, where such block represents a future real-time implementation, while the dotted red block contains the gray blocks of the SVPWM and 3-phase inverter, which also represent a future real-time implementation.

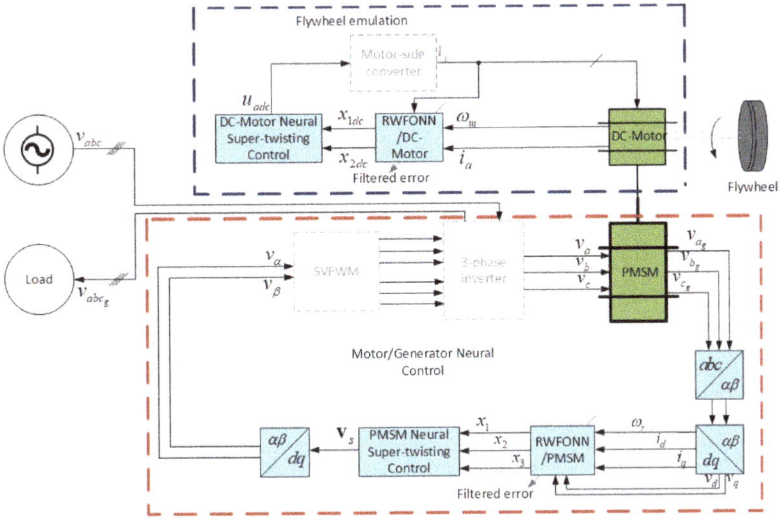

Figure 1. FESS's emulator scheme.

Additionally, Figure 2 shows a general flowchart of the proposed algorithm applied to the control of both electrical machines, namely, the PMSM and the DC-motor. Both identification and control processes are indicated separately. Note that in this figure, there are particular notations for the PMSM (ξ_i, x_1, $\mathbf{u}_{1s}$, and ε_1) and for the DC-motor (ξ_{idc}, x_{1dc}, u_{sdc}, and ζ_1). The particular notations and the processes of identification and control will be explained in the following subsections.

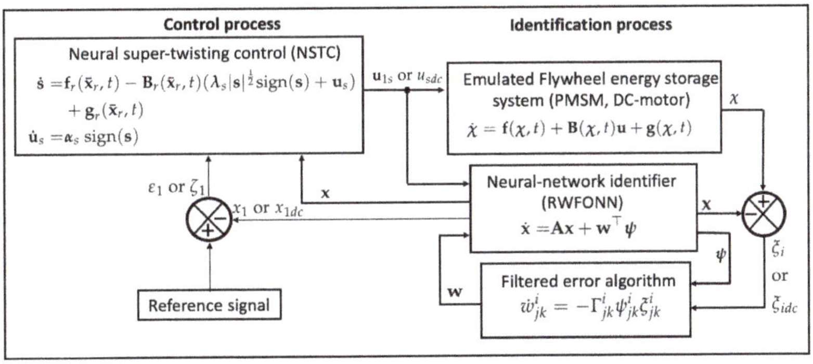

Figure 2. General flowchart of the proposed control algorithm.

3.1. Flywheel Storage System

In this section, the mathematical analysis of the stored kinetic energy from the flywheel, emulated by the DC-motor, is developed. Under the assumption that the mass is uniformly distributed and the axis of rotation passes through the mass center, the mass moment of inertia for a solid disk is given by [31]

$$J_m = \frac{1}{2}mr^2, \tag{11}$$

where J_m is the mass moment of inertia and m, r are the mass and radius of the disk, respectively. Then, for the numerical calculation of the mass moment of inertia of the disk, when considering $m = 2$ kg and $r = 0.3$ m, Equation (11) yields

$$J_m = 0.09\ \mathrm{kg} \cdot \mathrm{m}^2. \tag{12}$$

Now, the kinetic energy storage equation is calculated when substituting Equation (12) into (1). Therefore, considering the angular velocity as $\omega_m = 130$ rad/s, the energy stored in the flywheel results

$$E_k = 760.5\ J \tag{13}$$

when considering that the deceleration caused by the kinetic energy stored in the flywheel is emulated by the DC-motor and its drive. To calculate the deceleration of the emulated flywheel, from the knowledge of the mass moment of inertia in Equation (12), replacing J_m in Equation (2), it is then possible to calculate the reference angular velocity ω_{mref} for the DC-motor and, consequently, to be able to apply the NSTC.

3.2. Nonlinear Identification and NSTC Applied to PMSM

To convert the electrical energy applied to the PMSM into kinetic energy, the PMSM is activated in motor mode with the inputs v_d and v_q. In the following subsections, the neural identification and the NSTC applied to the PMSM in motor mode are presented.

3.2.1. PMSM Neural Identification

In this work, the mathematical model of the PMSM is taken from [30], which consists of a third order nonlinear system given in the general form (3), which is modeled in a dq coordinated frame rotating at the rotor velocity. Then, the RWFONN with structure (5), proposed to identify the dynamic system of the PMSM mathematical model, is defined as

$$\begin{aligned}
\dot{x}_1 &= -a_1 x_1 + b_1 w_1 \psi_1(\chi_1) + x_2 + x_3, \\
\dot{x}_2 &= -a_2 x_2 + b_2 w_2 \psi_2(\chi_2) + v_d, \\
\dot{x}_3 &= -a_3 x_3 + b_3 w_3 \psi_3(\chi_3) + v_q,
\end{aligned} \tag{14}$$

where the RWFONN's states x_1, x_2, and x_3 identify the states $\omega_r = \chi_1$ (angular velocity), $i_d = \chi_2$ (d-current), and $i_q = \chi_3$ (q-current) from the PMSM, respectively. a_1, a_2, a_3, b_1, b_2, and b_3 are positive fixed parameters; w_1, w_2, w_3 are the synaptic weights; $\psi_1(\chi_1) = e^{(-\chi_1^2/\beta_1)}\cos(\lambda_1\chi_1)$, $\psi_2(\chi_2) = e^{(-\chi_2^2/\beta_2)}\cos(\lambda_2\chi_2)$, and $\psi_3(\chi_3) = e^{(-\chi_3^2/\beta_3)}\cos(\lambda_3\chi_3)$ are Morlet wavelet activation functions. The RWFONN's inputs are the same as PMSM, v_d and v_q. The filtered error algorithm used for training the RWFONN (Equation (14)) is performed as $w_1 = \gamma_1\psi_1(\chi_1)\xi_1$, $w_2 = \gamma_2\psi_2(\chi_2)\xi_2$ and $w_3 = \gamma_3\psi_3(\chi_3)\xi_3$, with γ_1, γ_2, γ_3 adjustable parameters, where $\xi_1 = x_1 - \chi_1$, $\xi_2 = x_2 - \chi_2$ and $\xi_3 = x_3 - \chi_3$ are the respective identification errors. According to Equation (5), $i = j = k = 1, 2, 3$.

It should be noticed that the proposed structure (14) has only one neuron for each state, which represents a low computational burden in contrast with that from structures for high-order neural networks (HONNs). The parameters from the PMSM mathematical model are mentioned in [30], and the values are shown in Table 1 [32].

3.2.2. NSTC Application to PMSM

In order to control the angular velocity ω_r of the PMSM, a first step toward that goal consists of transforming the RWFONN (Equation (14)) into the form (7) by using a recursive transformation of the tracking error ε_i [29]. For this, the structure of the system (14) is rewritten in the vector form

$$\begin{aligned}
\dot{x}_1 &= -a_1 x_1 + b_1 w_1 \psi_1(\chi_1) + \mathbf{a}_1 \mathbf{x}_2, \\
\dot{\mathbf{x}}_2 &= \mathbf{A}_2 \mathbf{x}_2 + \mathbf{A}_3 \mathbf{b}_1 + \mathbf{B}_2 \mathbf{v},
\end{aligned} \tag{15}$$

with $\mathbf{a}_1 = \begin{bmatrix} 1 & 1 \end{bmatrix}$, $\mathbf{A}_2 = \begin{bmatrix} -a_2 & 0 \\ 0 & -a_3 \end{bmatrix}$, $\mathbf{A}_3 = \begin{bmatrix} b_2 w_2 \psi_2(\chi_2) & 0 \\ 0 & b_3 w_3 \psi_3(\chi_3) \end{bmatrix}$, $\mathbf{b}_1 = \begin{bmatrix} 1 & 1 \end{bmatrix}^\top$, $\mathbf{B}_2 = \begin{bmatrix} 1 & 0 \\ 0 & 1 \end{bmatrix}$, $\mathbf{x}_2 = \begin{bmatrix} x_2 & x_3 \end{bmatrix}^\top$, and $\mathbf{v} = \begin{bmatrix} v_d & v_q \end{bmatrix}^\top$. Now, by applying the recursive transformation to the structure (15) to resemble it into the block control form, the tracking error ε_1 is defined by

$$\varepsilon_1 = \omega_{rref} - x_1, \tag{16}$$

where ω_{rref} is the reference angular velocity to the PMSM, and x_1 represents the state of the neural network (15). Therefore, the tracking error dynamics are given as

$$\dot{\varepsilon}_1 = \dot{\omega}_{rref} - \dot{x}_1. \tag{17}$$

Table 1. Values and parameters of PMSM [32] and DC-motor [33].

PMSM Parameters	Value
Stator Resistance R	1.4 Ω
Inductance L_d	6.6 mH
Inductance L_q	5.8 mH
Inertial Moment J	0.00176 kg·m^2
Damping Coefficient B	0.00038818 N·m ·s/rad
Flux Linkage λ_{af}	0.1546 V·s/rad
Pair Poles P	3
DC-motor Nameplate Data and Parameters	**Value**
Field Voltage	120 V
Field Current	0.5 A
Armature Voltage	120 V
Armature Current	3.0 A
Rotor Velocity	1750 rpm
Armature Resistance R_a	12.5 Ω
Armature Inductance L_a	0.075 H
Motor Constant K_m	2.602
Inertial Moment J_m	0.0036 N·m·s^2
Frictional Coefficient B_m	0.002 N·m·s

Substituting $\dot{x}_1$ Equation (15) into (17), and assigning the desired dynamics $k_1\varepsilon_1$, it yields

$$\dot{\varepsilon}_1 = \dot{\omega}_{rref} - (-a_1 x_1 + b_1 w_1 \psi_1(\chi_1) + \mathbf{a}_1 \mathbf{x}_2) = -k_1 \varepsilon_1, \tag{18}$$

where k_1 is adjusted to vanish the tracking error. In this way, it is possible to synthesize a control law through the vector state $\mathbf{x}_2$, where the reference vector is defined as

$$\mathbf{x}_{2ref} = \mathbf{a}_1^{\text{pinv}} [\dot{\omega}_{rref} + a_1 \omega_{rref} - a_1 \varepsilon_1 - b_1 w_1 \psi_1(\chi_1) + k_1 \varepsilon_1], \tag{19}$$

where the superscript "pinv" denotes the pseudo-inverse. Furthermore, the second error in vector form is defined as

$$\boldsymbol{\varepsilon}_2 = \mathbf{x}_{2ref} - \mathbf{x}_2 \tag{20}$$

with dynamics

$$\dot{\boldsymbol{\varepsilon}}_2 = \dot{\mathbf{x}}_{2ref} - \dot{\mathbf{x}}_2. \tag{21}$$

Substituting Equation (15) into (21) yields

$$\dot{\varepsilon}_2 = \dot{\mathbf{x}}_{2ref} - \mathbf{A}_2\mathbf{x}_{2ref} + \mathbf{A}_2\varepsilon_2 - \mathbf{A}_3\mathbf{b}_1 - \mathbf{B}_2\mathbf{v}. \tag{22}$$

Moreover, using Equations (18) and (22), the representation in the block control form from Equation (15), in terms of the tracking error ε_i, is obtained as

$$\begin{aligned}
\dot{\varepsilon}_1 =& -k_1\varepsilon_1 + \mathbf{a}_1\varepsilon_2,\\
\dot{\varepsilon}_2 =& \mathbf{a}_1^{\text{pinv}}[\ddot{\omega}_{rref} + a_1\dot{\omega}_{rref} - b_1 w_1 \psi_1(\chi_1)]\\
& - \mathbf{A}_2\left(\mathbf{a}_1^{\text{pinv}}[\ddot{\omega}_{rref} + a_1\dot{\omega}_{rref} - b_1 w_1 \psi_1(\chi_1)]\right)\\
& + 0.5\begin{bmatrix} a_1k_1 - k_1^2 + a_1a_2k_1 - a_2k_1^2 \\ a_1k_1 - k_1^2 + a_1a_3k_1 - a_3k_1^2 \end{bmatrix}\varepsilon_1 + \mathbf{A}_2\varepsilon_2 - \mathbf{A}_3\mathbf{b}_1 - \mathbf{B}_2\mathbf{v},
\end{aligned} \tag{23}$$

Furthermore, defining the sliding surface $\mathbf{s} = \varepsilon_2$, with the relative degree from Equation (23) equals to 1, and applying the NSTC Equation (9) in (23) results

$$\begin{aligned}
\dot{\varepsilon}_1 =& -k_1\varepsilon_1 + \mathbf{a}_1\varepsilon_2,\\
\dot{\varepsilon}_2 =& \mathbf{a}_1^{\text{pinv}}[\ddot{\omega}_{rref} + a_1\dot{\omega}_{rref} - b_1 w_1 \psi_1(\chi_1)]\\
& - \mathbf{A}_2\left(\mathbf{a}_1^{\text{pinv}}[\ddot{\omega}_{rref} + a_1\dot{\omega}_{rref} - b_1 w_1 \psi_1(\chi_1)]\right)\\
& + 0.5\begin{bmatrix} a_1k_1 - k_1^2 + a_1a_2k_1 - a_2k_1^2 \\ a_1k_1 - k_1^2 + a_1a_3k_1 - a_3k_1^2 \end{bmatrix}\varepsilon_1 + \mathbf{A}_2\varepsilon_2 - \mathbf{A}_3\mathbf{b}_1 - \mathbf{B}_2(\lambda_s|\mathbf{s}|^{\frac{1}{2}}\text{sign}(\mathbf{s}) + \mathbf{u}_{1s}),
\end{aligned} \tag{24}$$

where $\mathbf{u}_{1s}$ represents the control vector. Note that in Equation (23), the control input vector $\mathbf{v} = \begin{bmatrix} v_d & v_q \end{bmatrix}^\top$ is replaced by the STC resulting Equation (24) under the fact that the tracking error variables ε_1 and ε_2 are decoupled. This latter will guarantee that the tracking errors be steered to zero in finite time. From Equation (24), selecting the sliding surface $\mathbf{s} = \mathbf{x_2}$ arises the system given by [29]

$$\begin{aligned}
\dot{\mathbf{s}} =& \mathbf{a}_1^{\text{pinv}}[\ddot{\omega}_{rref} + a_1\dot{\omega}_{rref} - b_1 w_1 \psi_1(\chi_1)]\\
& - \mathbf{A}_2\left(\mathbf{a}_1^{\text{pinv}}[\ddot{\omega}_{rref} + a_1\dot{\omega}_{rref} - b_1 w_1 \psi_1(\chi_1)]\right)\\
& + 0.5\begin{bmatrix} a_1k_1 - k_1^2 + a_1a_2k_1 - a_2k_1^2 \\ a_1k_1 - k_1^2 + a_1a_3k_1 - a_3k_1^2 \end{bmatrix}\varepsilon_1 + \mathbf{A}_2\varepsilon_2 - \mathbf{A}_3\mathbf{b}_1 - \mathbf{B}_2(\lambda_s|\mathbf{s}|^{\frac{1}{2}}\text{sign}(\mathbf{s}) + \mathbf{u}_{1s}),\\
\dot{\mathbf{u}}_{1s} =& \boldsymbol{\alpha}_s\,\text{sign}(\mathbf{s}).
\end{aligned} \tag{25}$$

where Equation (25) is the NSTC applied to the PMSM through the RWFONN structure.

3.3. Nonlinear Identification and NSTC Applied to DC-Motor

In order to convert the kinetic energy stored by the flywheel into electrical energy, the PMSM must be operated in generator mode. The discharge of the kinetic energy is controlled by the angular velocity of the DC-motor, this can be achieved by decelerating the angular velocity taking into account the reference velocity generated by Equation (2).

3.3.1. DC-Motor Neural Identification

The mathematical model of the DC-motor is taken from [16], which consists of a three state nonlinear system with structure (3). Note that due to the linearity between the effective field current and its constant magnetizing flux, the mathematical model of the DC-motor becomes a linear system of two differential equations. Therefore, the RWFONN here proposed, following the structure (5), to identify the dynamics of the DC-motor mathematical model is defined as

$$\begin{aligned} \dot{x}_{1dc} &= -a_{1dc}x_{1dc} + b_{1dc}w_{1dc}\psi_{1dc}(\chi_{1dc}) + x_{2dc}, \\ \dot{x}_{2dc} &= -a_{2dc}x_{2dc} + b_{2dc}w_{2dc}\psi_{2dc}(\chi_{2dc}) + u_{adc}, \end{aligned} \tag{26}$$

where the RWFONN's states x_{1dc} and x_{2dc} identify the states $\omega_m = \chi_{1dc}$ (angular velocity) and $i_a = \chi_{2dc}$ (armature current) from the DC-motor, respectively. $a_{1dc}, a_{2dc}, b_{1dc}$, and b_{2dc} are positive constant parameters; w_{1dc} and w_{2dc} are the synaptic weights; $\psi_{1dc}(\chi_{1dc}) = e^{(-\chi_{1dc}^2/\beta_{1dc})}\cos(\lambda_{1dc}\chi_{1dc})$ and $\psi_{2dc}(\chi_{2dc}) = e^{(-\chi_{2dc}^2/\beta_{2dc})}\cos(\lambda_{2dc}\chi_{2dc})$ are Morlet wavelet activation functions. The RWFONN's input u_{adc} is the same as u_a (armature voltage) to the DC-motor. The filtered error algorithm used for training the RWFONN (Equation (26)) is performed through $w_{1dc} = \gamma_{1dc}\psi_{1dc}(\chi_{1dc})\xi_{1dc}$ and $w_{2dc} = \gamma_{2dc}\psi_{2dc}(\chi_{2dc})\xi_{2dc}$ with γ_{1dc}, γ_{2dc} adjustable parameters, and $\xi_{1dc} = x_{1dc} - \chi_{1dc}$, $\xi_{2dc} = x_{2dc} - \chi_{2dc}$ are the respective identification errors. From Equation (5), $i = j = k = 1, 2$.

It must be noticed that the proposed RWFONN (Equation (26)) has only one neuron for each state, in a similar way as in Equation (14), which represents a low computational burden in contrast with HONN. The parameters of the DC-motor mathematical model are mentioned in [16], whose values are shown in Table 1.

3.3.2. NSTC Application to DC-Motor

By following the methodology as that in Section 3.2.2, the recursive transformation is now applied to the structure (26), where the generation of the tracking error ζ_1 is defined by

$$\zeta_1 = \omega_{mref} - x_{1dc} \tag{27}$$

where ω_{mref} represents the reference angular velocity of the DC-motor and x_{1dc} is the state variable from Equation (26). The dynamics of the tracking error ζ_1 is given by

$$\dot{\zeta}_1 = \dot{\omega}_{mref} - \dot{x}_{1dc}. \tag{28}$$

By substituting Equation (26) into (28) yields

$$\dot{\zeta}_1 = \dot{\omega}_{mref} - (-a_{1dc}x_{1dc} + b_{1dc}w_{1dc}\psi_{1dc}(\chi_{1dc}) + x_{2dc}) = -k_{1dc}\zeta_1, \tag{29}$$

where $k_{1dc}\zeta_1$ constitutes the desired dynamics to vanish the tracking error. In this way, it is possible to synthesize a control law through the state x_{2dc}, so, to such an end consider

$$x_{2dcref} = \dot{\omega}_{mref} + a_{1dc}\omega_{mref} - a_{1dc}\zeta_1 - b_{1dc}w_{1dc}\psi_{1dc}(\chi_{1dc}) + k_{1dc}\zeta_1. \tag{30}$$

Furthermore, the second error is defined as

$$\zeta_2 = x_{2dcref} - x_{2dc}, \tag{31}$$

with error dynamics

$$\dot{\zeta}_2 = \dot{x}_{2dcref} - \dot{x}_{2dc}. \tag{32}$$

By substituting Equation (26) in (32) it yields

$$\dot{\zeta}_2 = \dot{x}_{2dcref} - (-a_{2dc}x_{2dc} + b_{2dc}w_{2dc}\psi_{2dc}(\chi_{2dc}) + u_{adc}). \tag{33}$$

Moreover, using Equations (29) and (33), the block control form for Equation (26) is defined as

$$\begin{aligned}\dot{\zeta}_1 &= -k_{1dc}\zeta_1 + \zeta_2,\\ \dot{\zeta}_2 &= \dot{x}_{2dcref} + a_{2dc}x_{2dc} - b_{2dc}w_{2dc}\psi_{2dc}(\chi_{2dc}) - u_{adc}.\end{aligned} \tag{34}$$

Thus, based on the error dynamics ζ_i, the Equation (34) represents the block control transformation with structure (26). Furthermore, from the system (26) with relative degree equals to 1, the NSTC is given by

$$\begin{aligned}\dot{\zeta}_1 &= -k_{1dc}\zeta_1 + \zeta_2,\\ \dot{\zeta}_2 &= \ddot{\omega}_{mref} + (-a_{1dc} + a_{2dc})\dot{\omega}_{mref} + (a_{1dc}a_{2dc})\omega_{mref}\\ &\quad - (a_{1dc}k_{1dc} + k_1^2 + a_{1dc}a_{2dc} - a_{2dc}k_{1dc})\zeta_1 - a_{2dc}\zeta_2 - a_{2dc}b_{1dc}w_{1dc}\psi_{1dc}(\chi_{1dc}) -\\ &\quad b_{2dc}w_{2dc}\psi_{2dc}(\chi_{2dc}) - (\lambda_{sdc}|s_{dc}|^{\frac{1}{2}}\mathrm{sign}(s_{dc}) + u_{sdc})\end{aligned} \tag{35}$$

It should be noticed from Equation (35), regarding the tracking error variables ζ_1 and ζ_2, that the control input $u_{sdc} = u_{adc}$ guarantees that these errors will be steered to zero in finite time.

4. Simulation Results

This work presents the simulation results of the NSTC for controlling the angular velocity of a PMSM and a DC-motor to emulate an FESS. The angular velocity of the PMSM is controlled to store kinetic energy in the flywheel, meanwhile the angular velocity of the DC-machine is controlled to emulate the deceleration of the flywheel by transferring kinetic energy to the PMSM working in generator mode, where it is converted into electrical energy and delivered to the utility grid. The used parameters of the motors are shown in Table 1. The simulation was performed using Matlab/Simulink ($Matlab^{TM}$) with a Runge–Kutta solver and a step size of 100 μs.

4.1. Neural Identification

This subsection presents the simulation results of the neural identification of the PMSM state variables, as well as the DC-motor state variables using the RWFONN structure.

4.1.1. PMSM States Identification

For the neural identification of the PMSM states, the parameter values are: $a_1 = 6000$, $a_2 = a_3 = 4000$ and $b_1 = 6000$, $b_2 = b_3 = 4000$; the parameters for the Morlet Wavelet activation functions are: $\beta_1 = 75 \times 10^3$, $\beta_2 = 22 \times 10^3$, $\beta_3 = 35 \times 10^3$, $\lambda_1 = \lambda_2 = \lambda_3 = 0.001$; the parameters of the filtered error are: $\gamma_1 = \gamma_2 = \gamma_3 = 855 \times 10^2$. In Figures 3–5, the neuronal identification of the PMSM states is shown, where, in Figure 3 the identification of the angular velocity is the red dashed line that represents the angular velocity of the PMSM (ω_r) and the blue continuous line represents the state of RWFONN (x_1). In order to show the convergence of these results, the initial conditions are given as: $\omega_r = 0$ and $x_1 = -1$. Note that in the detail of the figure, the identification convergence is given in 0.0002 s, approximately. The transient is shown when the PMSM is started in the energy storage, where the velocity response has a settling time of 0.15 s, approximately.

The identification of current i_d is presented in Figure 4, where the red dashed line represents the current of the PMSM (i_d) and the blue continuous line represents the state of RWFONN (x_2); the initial conditions are given as: $i_d = 0$ and $x_2 = 0.5$. The identification convergence is given in 0.0003 s approximately, as is shown in the detail of the Figure 4. In the transient response, the inrush current i_d achieves a maximum value of the 132 A with an oscillation frequency of 0.0014 Hz. As it is shown, the identification process is effectively realized.

In Figure 5 the identification of current q is shown, where the red dashed line represents the current i_q of the PMSM, and the blue continuous line represents the state of the

RWFONN (x_3), and where the initial conditions are given as: $i_q = 0$ and $x_3 = -0.5$. The identification convergence is given in 0.0003 s, approximately. Figure 5 shows the transient when the PMSM is started, and in the process of energy storage, the inrush current i_q achieves a maximum value of the 153 A with an oscillation frequency of 0.0489 Hz. In the transient, the identification procedure is more demanding and is effectively realized.

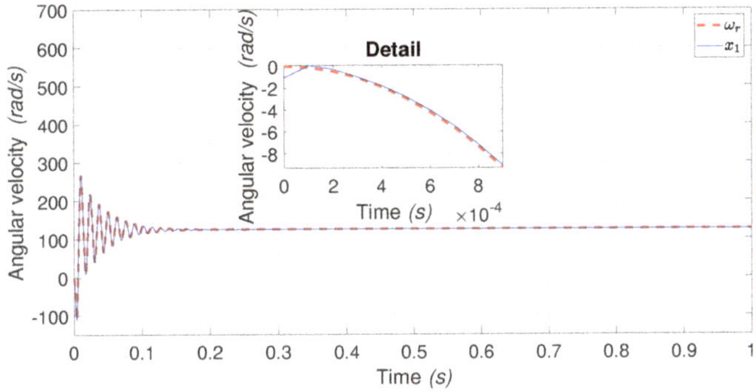

Figure 3. Angular velocity and neural identification behavior: ω_r and x_1.

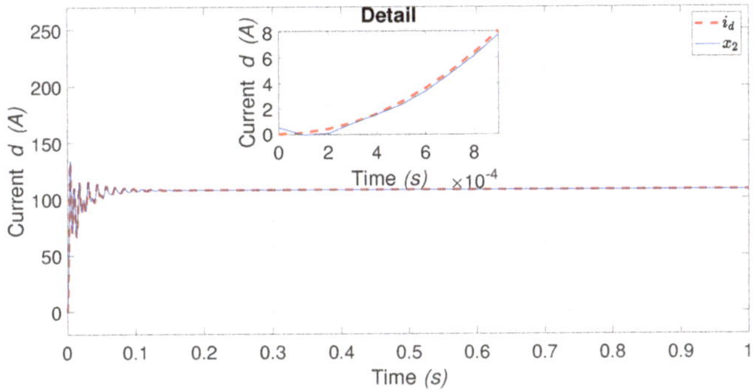

Figure 4. Current and neural identification behavior: i_d and x_2.

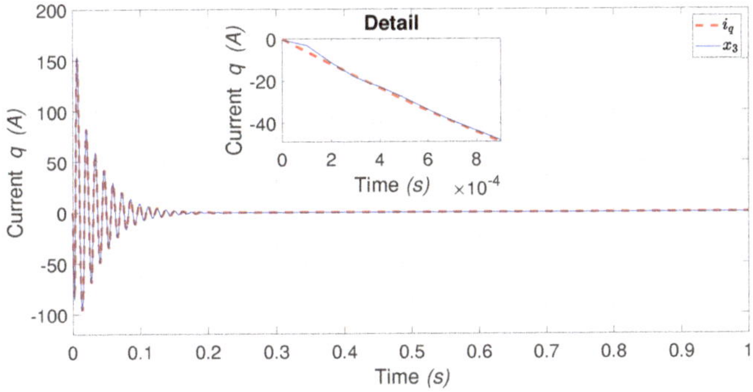

Figure 5. Current and neural identification behavior: i_q and x_3.

4.1.2. DC-motor States Identification

In this section, the neuronal identification of the DC-motor states is carried out, where the values of the parameters used in the RWFONN are defined as follows: $a_{1dc} = a_{2dc} = 6000$ and $b_{1dc} = b_{2dc} = 6000$; the parameters for the Morlet Wavelet activation functions are: $\beta_{1dc} = 85 \times 10^3$, $\beta_{2dc} = 75 \times 10^3$, $\lambda_{1dc} = 0.001$, $\lambda_{2dc} = 0.01$; the parameters of the filtered error are: $\gamma_{1dc} = \gamma_{2dc} = 855 \times 10^2$.

In Figure 6, the neural identification of the angular velocity of the DC-motor is shown. In this figure, ω_m is a state variable of the motor, and x_{1dc} is a state variable of the NN. As it can be seen in the detail of the figure, the NN identifies the state of the motor with a convergence time of 0.0003 s. The initial conditions are: $\omega_m = 0$ and $x_{1dc} = 1$. The figure shows the transient when the DC-motor is starting. The velocity response in the transient has an overshoot of 1.13% and a settling time of 0.05 s.

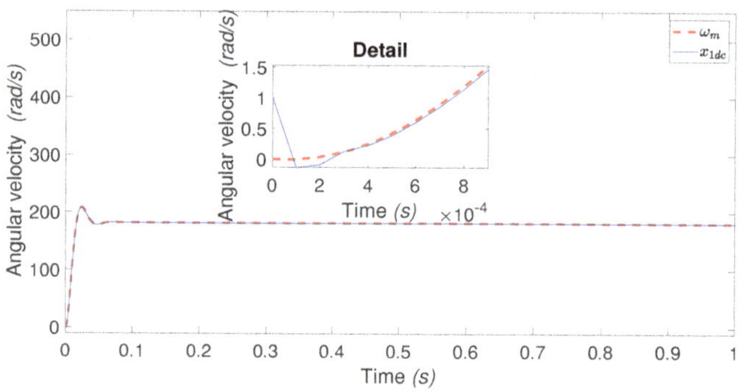

Figure 6. Angular velocity and neural identification behavior: ω_m and x_{1dc}.

Figure 7 shows the neural identification of the armature current of DC-motor, where i_a is the current of the motor and x_{2dc} is the state of the network; the initial conditions are: $i_a = 0$ and $x_{2dc} = 0.5$, and obtaining the neural identification with a time of convergence 0.0004 s, approximately. In this figure, the transient presents the typical inrush current of a DC-motor when it is fed with nominal voltage. The current has a maximum value of 21.7 A and achieves the steady-state value of 0.53 A at 0.08 s.

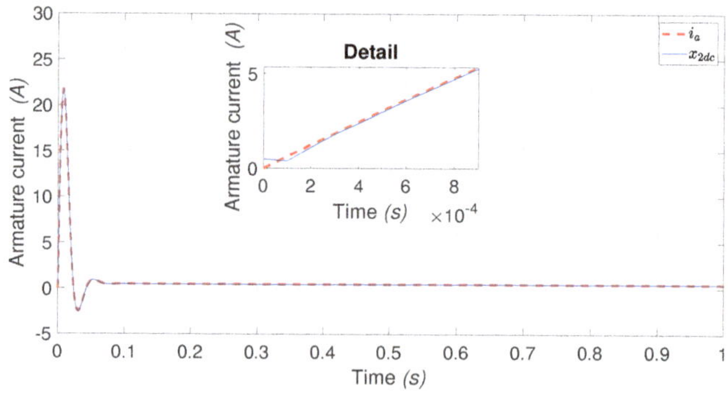

Figure 7. Armature current and neural identification behavior: i_a and x_{2dc}.

In order to make a comparison of the performance of the RWFONN used in this work, numerical simulation results about the identification process from both Recurrent Sigmoid First-Order Neural Network (RSFONN) and RHONN have been obtained. This comparison consists of using the RHONN and RSFONN structures, instead of that for the RWFONN, to the dynamics identification of the plant, training them via a filtered error algorithm. Then, from the data obtained by the identification process, the root mean square (RMS) error is computed, which represents the difference between the actual state and the estimated state when using approaches with different NNs. The comparison is made by estimating the states from both PMSM and DC-motor, where Tables 2 and 3 show the RMS for each state from both motors, respectively. It can be seen in Table 2 that the lowest RMS values (bold values) are for ω_r and i_d from the RSFONN and for i_q from the RHONN. In Table 3, the results show that the lowest RMS values (bold values) are for ω_m from the RSFONN and for i_a from the RWFONN.

Table 2. RMS error for PMSM.

Neural Network Structure	ω_r	i_d	i_q
RWFONN	0.1496	0.0311	0.0263
RSFONN	**0.0105**	**0.0015**	0.0286
RHONN	0.0386	0.0032	**0.0128**

Table 3. RMS error for DC-motor.

Neural Network Structure	ω_m	i_a
RWFONN	0.0018	**0.00004**
RSFONN	**0.0004**	0.00005
RHONN	0.0206	0.0009

Even when different NNs obtain the lowest RMS in the identification process, there are different reasons to consider when choosing the NN structure. Such is the case with the RWFONN, having a very single structure, for which it is easy to apply the STA, contrary to the case with the RHONN or even the RSFONN. Another important thing is that the computational burden is lower for simple structures such as the RWFONN but not for the RHONN or when increasing the high-order terms of the latter.

4.2. PMSM and DC-motor Controller-Emulation of Flywheel

In this subsection, the NSTC for PMSM and a DC-motor controller for emulating the complete system of the flywheel are presented. For the case of the PMSM, the controller is designed to control the angular velocity (ω_m) indirectly through the states of the NN, where the parameters and gains for the NSTC of PMSM are defined as: $k_1 = 8000$, $\lambda_{1s} = 0.1$, $\lambda_{2s} = 2.5$, and $\alpha_{1s} = 0.1$, $\alpha_{2s} = 2.5$.

In Figure 8, the behavior of the PMSM in motor and generator modes is shown. In the case of motor mode, the velocity tracking performances from 0 to 10 and from 30 to 52 s are shown, where the blue dashed line represents ω_{rref}, and black continuous line is ω_m. In this time, the tracking is to $\omega_{rref} = 130$ rad/s, and there is kinetic energy storage during this operation mode. A detail of the transient is shown where a fast convergence can be seen. Due to consideration of a grid failure, the case of PMSM as a generator mode is plotted during the period of time from 10 to 30 and from 52 to 70 s. In this time, the deceleration of the flywheel is emulated by controlling the angular velocity of the DC-motor (see Figure 9 on the sign named Discharge of flywheel) and in this figure, the velocity reference ω_{mref} (blue dashed line) is described by Equation (2) and meanwhile, ω_m is the gray continuous line; a detail is shown in the middle of the figure where the very good results of the tracking can be noted. To ensure the NSTC of the DC-motor, the parameters and gains are defined as: $k_{1dc} = 8000$, $\lambda_{sdc} = 3$ and $\alpha_{sdc} = 3$.

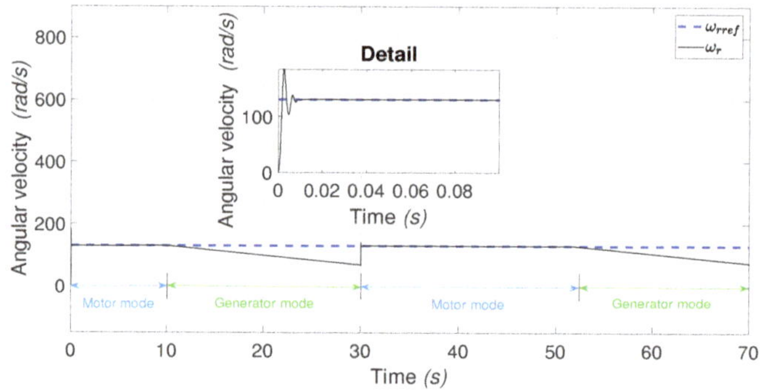

Figure 8. PMSM velocity tracking to emulate the energy storage.

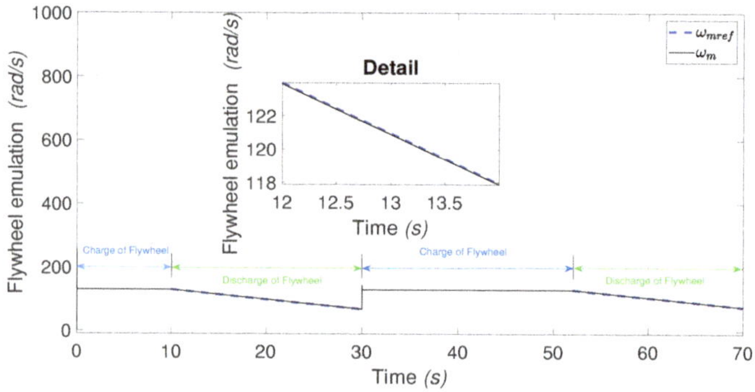

Figure 9. DC-motor velocity tracking to emulate the flywheel deceleration.

Figure 10 shows the PMSM velocity controller performance, where tracking velocity error ε_1 (Figure 10a) is steered to zero effectively by the super-twisting control law v_q (Figure 10a). Note that the transient response at 30 s corresponds to the change from the generator to motor mode. The error diminishes to zero in 0.015 s, as can be seen in the detail of Figure 10a.

Figure 11 shows the DC-motor velocity controller performance. Figure 11a displays the tracking error signal in the intervals (10–30 s) and (52–70 s), which is steered to zero in finite time by the action of the super-twisting control law shown in Figure 11b. When the tracking error is zero, then the velocity control of the DC-motor is achieved, and the control goal is fulfilled, which consists of emulating the deceleration of the flywheel.

4.3. Power Delivery

In Figure 12, the kinetic energy stored in the flywheel is shown from 0 to 10 s and from 30 to 52 s due to the fact that PMSM is in motor mode, generating 760.5 J in steady-state. From 10 to 30 s and from 52 to 70 s, the PMSM is in generator mode, and it shows the discharge of the energy when considering a fault in the utility grid or when the flywheel is required to compensate for the utility grid with active power to solve a problem of the electrical office.

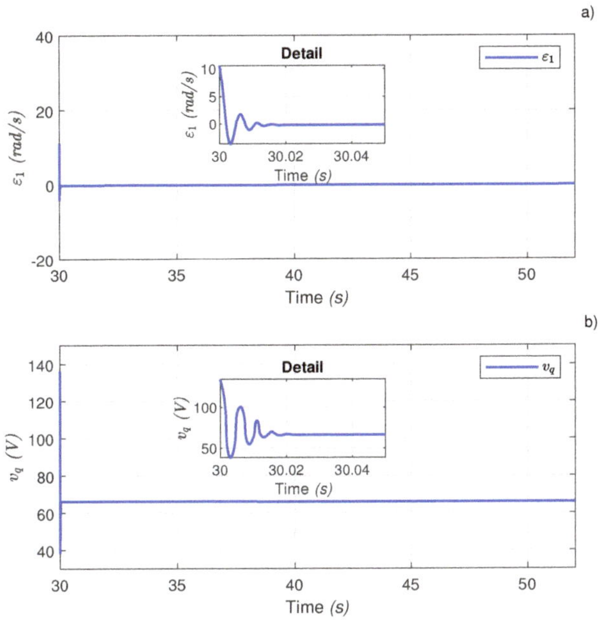

Figure 10. PMSM velocity controller performance: (**a**) velocity error ε_1, (**b**) control signal v_q.

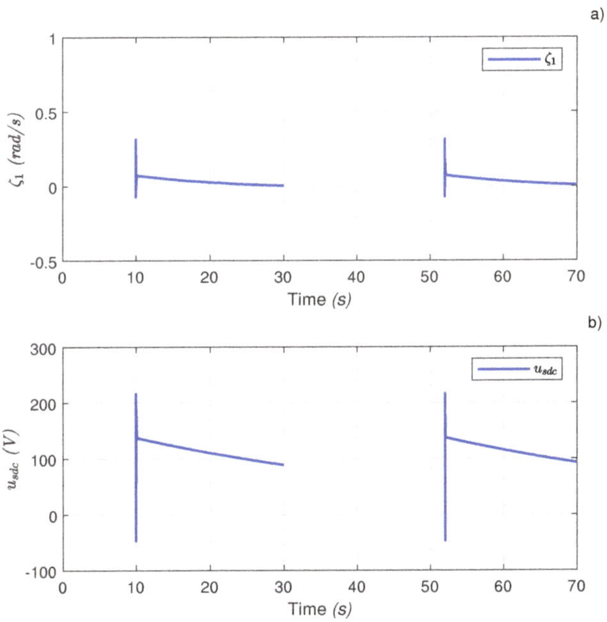

Figure 11. DC-motor velocity tracking performance: (**a**) velocity error ζ_1. (**b**) control signal u_{sdc} (10–30 and 52–70 s).

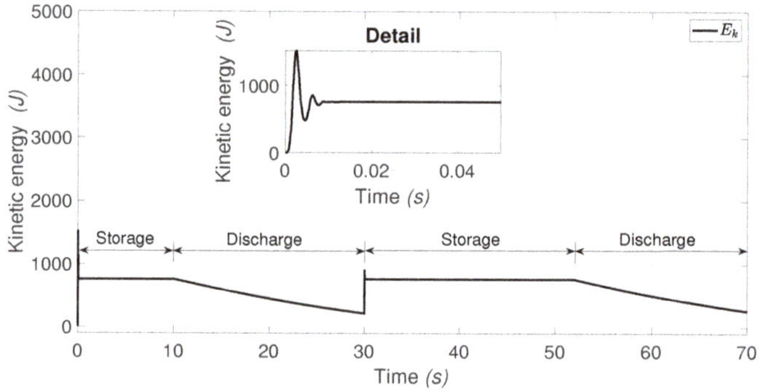

Figure 12. Storage and discharge kinetic energy.

5. Discussion

There exists a variety of three-phase electrical machines that can be implemented in an FESS, such as wound squirrel cage [34] and doubly-fed asynchronous machines [35]. In this work, we used the PMSM to emulate a flywheel storage system because, based on the literature, the small and medium capacity applications of the flywheels make it more convenient to use the PMSM since it is more efficient and less difficult to control. The doubly-fed induction generator is used more in large capacities because the control is carried out by the rotor circuit, where the power flow is proportional to the slip of the machine. Consequently, the electronic power drive is of lower capacity, near 30% of the flywheel capacity, which allows a considerable economic saving in the converter.

This proposal can be extended to other applications in energy storage systems such as power compensators in wind generators as well as wind diesel power systems, electric vehicles, direct electrical grid applications, i.e., applications where it is possible to involve charge/discharge cycles.

Regarding the advantages, this work can be used as a general methodology where it is possible to work as a motor/generator, such as wound squirrel cage, doubly fed asynchronous machines, and of course the PMSM. It is clear that this is possible by making the corresponding changes for the mathematical model of the machine to be used, which all depends on the range of speeds that it is going to operate at, and taking advantage of the benefits offered by the material with which the machine is built (low-speed FESS, or high-speed FESS [35]), and the particular features (electrical, mechanical) of the three-phase machine, etc. It is possible to control nonlinear and linear systems using the NSTC based on RWFONNs trained via a filtered error algorithm. These plants to be controlled are considered unknown regarding the parameters, and only the state-space variables are measured. This work can be extended to other three-phase machines, such as doubly fed asynchronous machines or wound squirrel cage motors.

The main limitation is that this work is an emulated system in simulation, and even when different real scenarios are considered, a real system will always show all the benefits of the proposal. Furthermore, this work is limited by the structure of the RWFONN, as even when the obtained numerical results are good, the computational process can be degraded.

The following interpretation of the results are made: the identification (Figures 3–7) results offer a good approximation of the plants by the RWFONN, and the filtered error algorithm that causes the tracking results in Figures 8 and 9 is good, including the tracking error maintained at zero for ε_1 and ζ_1, Figures 10 and 11, respectively. This demonstrates that the proposed methodology is viable for use in the applications already mentioned. It is clear that they are considered ideal switching processes when the grid failure is presented and there is the change from motor to generator in the PMSM also in the DC-motor, such as in the case in Figures 8 and 9, from which a good tracking is noted when there is a change in the storage–discharge modes.

The perspective of this work is obtaining results when considering more real scenarios in order to have a better approximation of a real-system. This will contribute to making real-time experiments with this proposal considering different structures of NN with the same or even other control algorithms.

6. Conclusions

Simulation results from this work show the emulation of the FESS through identification and control of the states from the models for two different electrical machines, namely, the PMSM and the DC-motor. This is achieved through the use of a RWFONN, the block control linearization technique, and the NSTC, by controlling the velocity of the PMSM in the energy storage stage and controlling the velocity of the DC-motor in the energy feedback stage. The stability analysis demonstrates the convergence of the whole system in a closed-loop using the boundedness of the identification error. It is worth mentioning that the authors intend to continue with the study of this work in real-time until laboratories can be accessed and tests can be performed.

Author Contributions: D.A.M. contributed with the design process of Neural Controller, methodology, validation and editing; C.E.C. contributed with conceptualization, review, and editing; F.J. contributed with conceptualization, review, and editing; O.A.M. validation of simulation results, review and editing. All authors have read and agreed to the published version of the manuscript.

Funding: This research received no external funding.

Data Availability Statement: Not applicable.

Acknowledgments: The first author acknowledges support from Consejo Nacional de Ciencia y Tecnologia (CONACYT) and Universidad de Guadalajara Centro Universitario de los Lagos.

Conflicts of Interest: The authors declare no conflict of interest.

Abbreviations

The following abbreviations are used in this manuscript:

AC	Alternating-Current
DC	Direct-Current
dq	Direct-Quadrature
FESS	Flywheel Energy Storage System
HONN	High Order Neural Network
NN	Neural Network
NSTC	Neural Super-Twisting Control
PMSM	Permanent Magnet Synchronous Machine
RMS	Root Mean Square
RHONN	Recurrent High Order Neural Network
RSFONN	Recurrent Sigmoid First-Order Neural Network
RWFONN	Recurrent Wavelet First-Order Neural Network
STA	Super-Twisting Algorithm
STC	Super-Twisting Control
SVPWM	Space Vector Pulse Width Modulation
UUB	Uniformly Ultimately Bounded

Appendix A. Identification Error Boundedness

Theorem A1. *Suppose that the system (3) and the model (5) are initially at the same state* $\mathbf{x}(0) = \chi(0)$*; then, for any* $\epsilon > 0$ *and any finite* $T > 0$ *there exist an integer L and a matrix* $\mathbf{w}* \in \mathbb{R}^{L\times n}$ *such that the state* $\mathbf{x}(t)$ *of the RWFONN model (5) and weight values* $\mathbf{w} = \mathbf{w}*$ *satisfies*

$$\sup_{0 \le t \le T} |\mathbf{x}(t) - \chi(t)| \le \epsilon.$$

Then, using the Bellman–Gronwall Lemma [36], it is obtained that the identification error $\xi = \mathbf{x} - \chi$ is bounded by

$$\|\xi\| \le \frac{\epsilon}{2}. \tag{A1}$$

Proof. See Reference [37]. □

Appendix B. Closed-Loop Stability Analysis

In order to analyze the stability of Equations (25) and (35), a representation in scalar form is made as follows [29]:

$$\begin{aligned} \dot{s}_h &= -\, k_h |s_h|^{\frac{1}{2}} \mathrm{sign}(s_h) + u_h + f_h \\ \dot{u}_h &= -\, k_h \mathrm{sign}(s_h),\ h = 1, 2, .., d \end{aligned} \tag{A2}$$

Considering $\mu_h^\top = \left[|x_1|^{\frac{1}{2}} \mathrm{sign}(x_1), x_2 \right]$ as the state vector of the super-twisting algorithm, μ_h can be rewritten in terms of s_h, then $\mu_h^\top = \left[|s_h|^{\frac{1}{2}} \mathrm{sign}(s_h) \quad u_h \right]$ and considering that the matrix $\mathbf{P}$ is positive definite, the derivative of μ_h is defined as [38]:

$$\dot{\mu}_h = \begin{bmatrix} \frac{1}{2} \frac{1}{|s_h|^{\frac{1}{2}}} ((-k_{1h}) |s_h|^{\frac{1}{2}} \mathrm{sign}(s_h) + u_h) \\ -k_{2h} \mathrm{sign}(s_h) \end{bmatrix} = \frac{1}{|s_h|} \mathbf{A}_h \mu_h \tag{A3}$$

Equation (A3) can be expressed as the following linear system:

$$\dot{\mu}_h = \mathbf{A}_h \mu_h + \rho_h \tag{A4}$$

From Equation (A2), f_h is bounded with the following restrictions:

$$|f_h| \le \delta_h |s_h|^{\frac{1}{2}},\ |f_h| \le \delta_h |\mu_h|,\ \delta_h \ge 0 \tag{A5}$$

Assuming that f_h in (A2) is an external disturbance bounded by:

$$f_h = \delta_h |s_h|^{\frac{1}{2}} \mathrm{sign}(s_h) = \delta_h \mu_h \tag{A6}$$

To construct a complete family of strong Lyapunov functions for Equation (A2) of the form (A8) the matrix $\mathbf{P} = \mathbf{P}^\top > 0$, the Lyapunov function can be reduced to the solution of an algebraic Lyapunov equation (ALE) [38] as:

$$\mathbf{A}^\top \mathbf{P} + \mathbf{P}\mathbf{A} = -\mathbf{Q} \tag{A7}$$

Lyapunov's candidate function can be written as follows [29,38]:

$$V_h(\mu) = \mu_h^\top \mathbf{P}_h \mu_h \tag{A8}$$

Considering that the matrix $\mathbf{A} = \frac{1}{|\mu_h|} \begin{bmatrix} -\frac{1}{2} k_h & \frac{1}{2} \\ -k_h & 0 \end{bmatrix}$ and $\rho_h = \begin{bmatrix} \frac{1}{2|\mu_h|} f_h \\ 0 \end{bmatrix}$ where $k_h > 0$, then $\mathbf{A}$ is Hurwitz. For every symmetric and positive definite matrix $\mathbf{Q} = \mathbf{Q}^\top > 0$, and $\mathbf{P} = \mathbf{P}^\top > 0$ is the unique symmetric and positve definite solution of the ALE

(Equation (A7)). Moreover, the derivative $\dot{V}$ of the Lyapunov function satisfies the differential inequality almost every where:

$$\dot{V} \leq -\gamma(\mathbf{Q})V^{\frac{1}{2}}(s) \tag{A9}$$

where

$$\gamma(\mathbf{Q}) = \frac{\lambda_{min}(\mathbf{Q})\lambda^{\frac{1}{2}}_{min}(\mathbf{P})}{\lambda_{max}(\mathbf{P})} \tag{A10}$$

is a scalar depending on the selection of the $\mathbf{Q}$ matrix [38].

Proof. The Lyapunov function (A8) is continuous, positive definite and radially unbounded in $\mathbb{R}^2$, it is differentiable everywhere, except on the line $s_1 = 0$. As the trajectories of Equation (A2) cannot stay on this set, before reaching the origin, the derivative of $V(\mu)$ can be written from Equation (A3) as follows:

$$\dot{V}_h(\mu) = \frac{1}{|\mu_h|}\mu_h^\top(\boldsymbol{A}_h^\top \boldsymbol{P}_h + \boldsymbol{P}_h\boldsymbol{A}_h)\mu_h + 2\mu_h^\top \boldsymbol{P}_h\rho_h \tag{A11}$$

Involving Equation (A6) in the term ρ of Equation (A11), it yields:

$$\dot{V}_h(\mu) = -\frac{1}{|\mu_h|}\mu_h^\top Q_h\mu_h \tag{A12}$$

Now, defining $\boldsymbol{P} = \begin{bmatrix} \frac{1}{2}k_h^2 + 2k_h & -\frac{1}{2}k_h \\ -\frac{1}{2}k_h & 1 \end{bmatrix}$ and using the standard inequality for quadratic forms

$$\lambda_{min}\{\boldsymbol{P}\}\|\mu\|_2^2 \leq V_h(\mu) \leq \lambda_{max}\{\boldsymbol{P}\}\|\mu\|_2^2 \tag{A13}$$

where $\|\mu\|_2^2 = \mu_1^2 + \mu_2^2 = |s_1| + s_2^2$ is the Euclidean norm of μ. Equation (A11) along the solutions of the system (A2) can be rewritten as follows:

$$\dot{V}_h(\mu) = -\frac{1}{|s_1|^{\frac{1}{2}}}\mu^\top Q\mu \leq -\frac{1}{|s_1|^{\frac{1}{2}}}\lambda_{min}\{Q\}\|\mu\| \tag{A14}$$

$\dot{V}_h(\mu)$ is negative definite if $Q > 0$, which is the case when $k_h > 0$.

Where $Q = \frac{k_h}{2}\begin{bmatrix} k_h^2 + 2k_h - \delta_h(k_h + 4\frac{k_h}{k_h}) & -k_h \\ -(k_h - \delta_h) & 1 \end{bmatrix}$, therefore the matrix Q is positive definite if k_h satisfies the following condition: $k_h > 2\delta_h$ also $k_h > \frac{1}{2}\frac{k_h^2(\delta_h - k_h)}{k_h - 2\delta_h}$. Using Equations (A13) and (A14) and the fact that $|s_1|^{\frac{1}{2}} \leq \|\mu\|_2 \leq \frac{V^{\frac{1}{2}}(s)}{\lambda^{\frac{1}{2}}_{min}\{P\}}$, the inequality (A9) is satisfied.

Now, the convergence of ξ_j^i should be considered, involving in this way the identification error boundedness (Equation (A1)), then, the Lyapunov function (A8) can be rewritten as follows:

$$V(\mu,\xi) = \mu^\top \boldsymbol{P}\mu + \frac{1}{2}(\xi_1^i)^2 \tag{A15}$$

where the time derivative of Equation (A15) is:

$$\dot{V}(\mu,\xi) = -\frac{1}{|\mu_h|}\mu_h^\top Q_h\mu_h + \xi_1^i\dot{\xi}_1^i \tag{A16}$$

substituting the time derivative of $\xi_1^i = x_1^i - \chi_1^i$, which is obtained through the filtered error algorithm (6) and is applied in Equation (A16)

$$\dot{V}(\mu,\xi) = -\frac{1}{|\mu_h|}\mu_h^\top \mathbf{Q}_h \mu_h + \xi_1^i \left[(-a_1^i x_1^i + b_1^i w_1^i \psi_1^i + x_2^i + x_3^i) - \dot{\chi}_1^i\right] \tag{A17}$$

In order to guarantee that Equation (A17) is negative definite, the desired dynamics for x_{2d} is

$$x_{2d}^i = -c_1\xi_1^i + a_1^i x_1^i - b_1^i w_1^i \psi_1^i - x_3^i + \dot{\chi}_1^i \tag{A18}$$

thus,

$$\begin{aligned}\dot{V}(\mu,\xi) =& -\frac{1}{|\mu_h|}\mu_h^\top \mathbf{Q}_h \mu_h + \\ & \xi_1^i(-a_1^i x_1^i + b_1^i w_1^i \psi_1^i + (-c_1\xi_1^i + a_1^i x_1^i - b_1^i w_1^i \psi_1^i - x_3^i + x_3^i + \dot{\chi}_1^i) - \dot{\chi}_1^i) \\ =& -\frac{1}{|\mu_h|}\mu_h^\top \mathbf{Q}_h \mu_h - c_1\xi_1^2\end{aligned} \tag{A19}$$

with $c_1 > 0$ real value. In this way, by Theorem (1) and Equation (A1), the uniformly ultimately bounded stability of the complete system is ensured. □

References

1. Bolund, B.; Bernhoff, H.; Leijon, M. Flywheel energy and power storage systems. *Renew. Sustain. Energy Rev.* **2005**, *11*, 235–258. [CrossRef]
2. Soomro, A.; Amiryar, M.E.; Pullen, K.R.; Nankoo, D. Comparison of Performance and Controlling Schemes of Synchronous and Induction Machines Used in Flywheel Energy Storage Systems. *Energy Procedia* **2018**, *151*, 100–110. [CrossRef]
3. Ghanaatian, M.; Lotfifard, S. Control of Flywheel Energy Storage Systems in the Presence of Uncertainties. *IEEE Trans. Sustain. Energy* **2019**, *10*, 36–45. [CrossRef]
4. Sebastián, R.; Peña-Alzola, R. Control and simulation of a flywheel energy storage for a wind diesel power system. *Electr. Power Energy Syst.* **2014**, *64*, 1049–1056. [CrossRef]
5. Vafakhah, B.; Masiala, M.; Salmon, J.; Knight, A. Emulation of Flywheel Energy Storage Systems With a PMDC Machine. In Proceedings of the 2008 International Conference on Electrical Machines, Vilamoura, Portugal, 6–9 September 2009.
6. Tria, F.; Srairi, K.; Benchouia, M.; Benbouzid, M.E.H. An integral sliding mode controller with super-twisting algorithm for direct power control of wind generator based on a doubly fed induction generator. *Int. J. Syst. Assur. Eng. Manag.* **2017**, *8*, 762–769. [CrossRef]
7. Wang, J.; Yang, L.; Blalock, C.; Tolbert, L.M. *Flywheel Energy Storage Emulation Using Reconfigurable Hardware Test-Bed of Power Converters*; Energy Storage Applicatations and Technologies: San Diego, CA, USA, 2013.
8. Karrari, S.; Baghaee, H.R.; De Carne, G.; Noe, M.; Geisbuesch, J. Adaptive inertia emulation control for high-speed flywheel energy storage systems. *IET Gener. Transm. Distrib.* **2020**, *14*, 5047–5059. [CrossRef]
9. Hedlund, M.; Oliveira, J.G.; Bernhoff, H. Sliding mode 4-quadrant DCDC converter for a flywheel application. *Control Eng. Pract.* **2013**, *21*, 473–482. [CrossRef]
10. Ahsan, H.; Mufti, M.-U.-D. Dynamic performance improvement of a hybrid multimachine system using a flywheel energy storage system. *Wind Eng.* **2019**, *44*, 239–252. [CrossRef]
11. Bowen, C.; Jihua, Z.; Zhang, R. Modeling and Simulation of Permanent Magnet Synchronous Motor Drives. In Proceedings of the Fifth International Conference on Electrical Machines and Systems (ICEMS'2001), Bali, Indonesia, 4–7 August 2001.
12. He, R.; Han, Q. Dynamics and Stability of Permanent-Magnet Synchronous Motor. *Math. Probl. Eng.* **2017**, *2017*, 4923987. [CrossRef]
13. Gao, S.; Dong, H.; Ning, B.; Tang, T.; Li, Y. Nonlinear mapping-based feedback technique of dynamic surface control for the chaotic PMSM using neural approximation and parameter identification. *IET Control Theory Appl.* **2018**, *12*, 819–827. [CrossRef]
14. Ewert, P.; Orlowska-Kowalska, T.; Jankowska, K. Effectiveness Analysis of PMSM Motor Rolling Bearing Fault Detectors Based on Vibration Analysis and Shallow Neural Networks. *Energies* **2021**, *14*, 712. [CrossRef]
15. Chapman, S.J. *Electric Machines*; Mc Graw Hill: New York, NY, USA, 2000.
16. Castaneda, C.E.; Loukianov, A.G.; Sanchez, E.N.; Castillo-Toledo, B. Discrete-Time Neural Sliding-Mode Block Control for a DC Motor with Controlled Flux. *IEEE Trans. Ind. Electron.* **2012**, *59*, 1194–1207. [CrossRef]
17. Valenzuela, F.A.; Ramírez, R.; Martínez, R.; Morfín, O.A.; Castañeda, C.E. Super-Twisting Algorithm Applied to Velocity Control of DC Motor without Mechanical Sensors Dependence. *Energies* **2020**, *13*, 6041. [CrossRef]

18. Lee, S.-D.; Phuc, B.D.H.; Xu, X.; You, S.S. Roll suppression of marine vessels using adaptive super-twisting sliding mode control synthesis. *Ocean Eng.* **2019**, *195*, 106724. [CrossRef]
19. Amiryar, M.E.; Pullen, K.R. A Review of Flywheel Energy Storage System Technologies and Their Applications. *Appl. Sci.* **2017**, *7*, 286. [CrossRef]
20. Morfín, O.A.; Castañeda, C.E.; Ruiz-Cruz, R.; Valenzuela, F.A.; Murillo, M.A.; Quezada, A.E.; Padilla, N. The Squirrel-Cage Induction Motor Model and Its Parameter Identification Via Steady and Dynamic Tests. *Electr. Power Compon. Syst.* **2018**, *46*, 302–315. [CrossRef]
21. Vázqquez, L.S.; Jurado, F. Continuous-Time Decentralized Wavelet Neural Control for a 2 DOF Robot Manimulator. In Proceedings of the 2014 11th International Conference on Electrical Engineering, Computing Science and Automatic Control (CCE), Ciudad del Carmen, Mexico, 29 September–3 October 2014.
22. Jurado, F.; Lopez, S. A wavelet neural control scheme for a quadrotor unmanned aerial vehicle. *Philos. Trans. R. Soc. A Math. Phys. Eng. Sci.* **2018**, *376*, 20170248. [CrossRef] [PubMed]
23. Magallon, D.A.; Castaneda, C.E.; Jurado, F.; Morfin, O.A. Design of a Morlet wavelet control algorithm using super—Twisting sliding modes applied to an induction machine. In Proceedings of the International Joint Conference on Neural Networks (IJCNN), Glasgow, UK, 19–24 July 2020.
24. Vázquez, L.A.; Jurado, F.; Alanis, A.Y. Decentralized Identification and Control in Real-Time of a Robot Manipulator via Recurrent Wavelet First-Order Neural Network. *Math. Probl. Eng.* **2015**, *2015*, 451049. [CrossRef]
25. Kosmatopoulos, E.B.; Polycarpou, M.M.; Christodoulou, M.A.; Ioannou, P.A. High-order neural network structures for identication of dynamical systems. *IEEE Trans. Neural Netw.* **1995**, *6*, 422–431. [CrossRef] [PubMed]
26. Loukianov, A.G. Robust Block Decomposition Sliding Mode Control Design. *Math. Probl. Eng.* **2002**, *8*, 346–365. [CrossRef]
27. Utkin, V.; Guldner, J.; Shi, J. *Slidin Modes Control in Electromechanical Systems*; Taylor & Francis: Abingdon, UK, 1999.
28. Chalanga, A.; Kamal, S.; Fridman, L.; Bandyopadhyay, B.; Moreno, J.A. Implementation of Super-Twisting Control: Super-Twisting and Higher Order Sliding-Mode Observer-Based Approaches. *IEEE Trans. Ind. Electron.* **2016**, *63*, 3677–3685. [CrossRef]
29. Morfin, O.A.; Valenzuela, F.A.; Betancour, R.R.; Castaneda, C.E.; Ruiz-Cruz, R.; Valderrabano-Gonzalez, A. Real-Time SOSM Super-Twisting Combined with Block Control for Regulating Induction Motor Velocity. *IEEE Access* **2018**, *6*, 25898–25907. [CrossRef]
30. Han, Q.; Ham, C.; Phillips, R. PMSM nonlinear robust control for temperature compensation. In Proceedings of the Thirty-Sixth Southeastern Symposium on System Theory, Atlanta, GA, USA, 16 March 2014.
31. Elbouchikhi, E.; Amirat, Y.; Feld, G.; Benbouzid, M.; Zhou, Z. A Lab-scale Flywheel Energy Storage System: Control Strategy and Domestic Applications. *Energies* **2020**, *13*, 653. [CrossRef]
32. Pillay, P.; Krishnan, R. Modeling, Simulation, and Analysis of Permanent-Magnet Motor Drives, Part I: The Permanent-Magnet Synchronous Motor Drive. *IEEE Trans. Ind. Appl.* **1989**, *25*, 265–273. [CrossRef]
33. Morfin, O.; Ruiz-Cruz, R.; Hernández, J.; Castañeda, C.; Ramírez-Betancour, R.; Valenzuela-Murillo, F. Real-Time Sensorless Robust Velocity Controller Applied to a DC-motor for Emulating a Wind Turbine. *Energies* **2021**, *14*, 868. [CrossRef]
34. Kim, Y.H.; Lee, K.H.; Cho, Y.H.; Hong, Y.K. Comparison of harmonic compensation based on wound/squirrel-cage rotor type induction motors with flywheel. In Proceedings of the Third International Power Electronics and Motion Control Conference (IPEMC 2000), Beijing, China, 15–18 August 2000; Volume 2, pp. 531–536. [CrossRef]
35. Sebastian, R.; Alzola, R.P. Flywheel energy storage systems: Review and simulation for an isolated wind power system. *Renew. Sustain. Energy Rev.* **2012**, *16*, 6803–6813. [CrossRef]
36. Hale J.K. *Ordinary Differential Equations*; Wiley InterScience: Hoboken, NJ, USA, 1969.
37. Rovithakis, G.A.; Christodoulou, M.A. *Adaptive Control with Recurrent High-Order Neural Networks, Theory and Industrial Applications*; Springer: London, UK, 2000.
38. Dávila, A.; Moreno, J.A.; Fridman, L. Optimal Lyapunov function selection for reaching time estimation of Super Twisting Algorithm. In Proceedings of the 48h IEEE Conference on Decision and Control (CDC) held jointly with 2009 28th Chinese Control Conference, Shanghai, China, 15–18 December 2009.

MDPI
St. Alban-Anlage 66
4052 Basel
Switzerland
Tel. +41 61 683 77 34
Fax +41 61 302 89 18
www.mdpi.com

MDPI Books Editorial Office
E-mail: books@mdpi.com
www.mdpi.com/books

www.ingramcontent.com/pod-product-compliance
Lightning Source LLC
Chambersburg PA
CBHW061546220926
51850CB00005B/260
9783036555850